Springer Series on Polymer and Composite Materials

Series editor

Susheel Kalia, Dehradun, India

More information about this series at http://www.springer.com/series/13173

Mohammad Mansoob Khan
Debabrata Pradhan · Youngku Sohn
Editors

Nanocomposites for Visible Light-induced Photocatalysis

Springer

Editors
Mohammad Mansoob Khan
Faculty of Science, Chemical Sciences
Universiti Brunei Darussalam
Gadong
Brunei Darussalam

Debabrata Pradhan
Materials Science Centre
Indian Institute of Technology
Kharagpur, West Bengal
India

Youngku Sohn
Department of Chemistry
Chungnam National University
Yusung
Daejeon
Korea (Republic of)

ISSN 2364-1878 ISSN 2364-1886 (electronic)
Springer Series on Polymer and Composite Materials
ISBN 978-3-319-62445-7 ISBN 978-3-319-62446-4 (eBook)
DOI 10.1007/978-3-319-62446-4

Library of Congress Control Number: 2017946040

Printed on acid-free paper

This Springer imprint is published by Springer Nature
The registered company is Springer International Publishing AG
The registered company address is: Gewerbestrasse 11, 6330 Cham, Switzerland

Preface

The frequent release of hazardous and toxic chemicals into water bodies as well as repeated anthropogenic and industrial activities is of great alarm because these pollutants contaminate rivers, lakes, and underground aquifers. The traces of contaminants ranging from dyes, pharmaceutical drugs, hormones, and sunscreen to pesticides are being spreading in different types of water bodies. Furthermore, most of these contaminants are recalcitrant compounds and cannot be decomposed by the conventional wastewater treatment methods. Therefore, many treated effluents that are considered "safe" for disposal still contain toxic and hazardous pollutants. Generally, these compounds are untraceable when ingested or absorbed by living organisms and are subsequently accumulated, causing adverse health effects. Thus, considerable efforts have been put for the development of suitable, safe, clean, and environment-friendly purification process that can decompose and degrade the recalcitrant organic contaminants from wastewater to reduce negative effect on plants and animals.

Advanced oxidation processes (AOPs) have been considered as an alternate method for the degradation, detoxification, and removal of several toxic organic pollutants in wastewater. The principle of AOPs is to produce superoxide and hydroxyl radicals in water, which are very powerful oxidants capable of oxidizing wide range of organic pollutants without selectivity. Among these AOPs, heterogeneous photocatalysis in the presence of semiconductor or semiconductor-based nanomaterials has shown efficiency in degrading a wide range of indistinct refractory organics into readily biodegradable compounds and eventually mineralizing them to innocuous carbon dioxide and water. Heterogeneous photocatalysis has been actively investigated as a promising self-cleaning, deodorization system, as well as antibacterial agents, and their applications in photocatalytic process are desirable for the purification of water through removing various types of pollutants and bacteria. However, the key part of the heterogeneous photocatalysis is the semiconductor material used as a catalyst. A photocatalyst is defined as a substance that is able to make chemical transformations of the contaminants repeatedly coming to its contact into greener products in the presence of light while regenerating its chemical composition after each cycle of such interactions. The physicochemical properties of the material are crucial for high conversion efficiency stability in the electrolyte which are usually established as per, e.g., composition, size, shape, and morphology. This book comprises following 11 chapters that deal with several types of photocatalyst materials, and their role in several chemical photocatalytic transformation and mechanism:

Chapter 1, "Introduction of Nanomaterials for Photocatalysis", deals with the present research scenario of visible light-induced photocatalysis and its importance. In particular, why nanocomposites are needed to be developed for the visible light-induced photocatalysis and their prime roles in enhancing the performance. In addition, how and where such photocatalysts would find practical and industrial applications is briefly mentioned in this chapter.

Chapter 2, "Basic Principles, Mechanism, and Challenges of Photocatalysis", provides the basic principles and mechanisms that has already been known and developed. It also discusses the role of nanotechnology in the photocatalysis, especially the visible light-induced photocatalysis and present challenges in photocatalysis research.

Chapter 3, "Nanocomposites and Its Importance in Photocatalysis", focuses on the importance of different types of nanocomposites for visible light-induced photocatalysis for possible applications. Nanocomposites include inorganic/organic, inorganic/polymer, and mixed oxides, and photocatalytic performance of those with their merits and demerits.

In Chap. 4, the role of metal nanoparticles and its surface plasmon activity on nanocomposites for visible light-induced catalysis is outlined. The fabrication of different types of nanocomposites involving different metal nanoparticles which are responsible for the enhanced visible light-induced catalysis is thoroughly discussed along with mechanism.

Chapter 5 deals with mixed metal-oxides nanocomposites for visible light-induced photocatalysis. The strategies used for the synthesis of mixed metal oxide nanocomposites and their performance for visible light-induced photocatalysis are delineated in this chapter.

In Chap. 6, synthesis and photocatalytic application of various nanoporous nanocomposite materials are included.

Chapter 7 deals with various polymeric nanocomposites for visible light-induced photocatalysis covering their synthesis and characterizations. Polymer-based nanocomposites include artificial and natural polymer nanocomposites.

In Chap. 8, role of several carbon-based nanocomposites including metal–graphene and metal–CNT nanocomposites in visible light-induced photocatalysis is discussed.

Chapter 9, "g-C_3N_4/Carboneous Polymer-Based Nanocomposites Towards Visible Light-induced Photocatalysis", deals with the nanocomposites of g-C_3N_4 with carbonaceous π-conjugated/polymeric materials for visible light-induced photocatalysis such as NO removal, CO_2 reduction and oxygen reduction reactions, water splitting to liberate H_2 fuel, and degradation of pollutants.

Chapter 10, “Titanium-Based Ternary Mixed Metal Oxide Nanocomposites for Visible Light-induced Photocatalysis”, focuses on the mixed metal oxide nanocomposites for visible light-induced photocatalysis.

Chapter 11 discusses novel applications and future perspectives of nanocomposites. It will also include self-cleaning of glasses (window panes) using photoactive materials, novel paints, tiles, etc.

Gadong, Brunei Darussalam — Mohammad Mansoob Khan
Kharagpur, India — Debabrata Pradhan
Daejeon, Korea (Republic of) — Youngku Sohn

Contents

1 **Introduction of Nanomaterials for Photocatalysis** 1
Diana Vanda Wellia, Yuly Kusumawati, Lina Jaya Diguna and Muhamad Ikhlasul Amal

2 **Basic Principles, Mechanism, and Challenges of Photocatalysis** 19
R. Saravanan, Francisco Gracia and A. Stephen

3 **Nanocomposites and Its Importance in Photocatalysis** 41
Hossam Eldin Abdel Fattah Ahmed Hamed El Nazer and Samir Tawfik Gaballah

4 **Role of Metal Nanoparticles and Its Surface Plasmon Activity on Nanocomposites for Visible Light-Induced Catalysis** 69
Anup Kumar Sasmal and Tarasankar Pal

5 **Mixed Metal Oxides Nanocomposites for Visible Light Induced Photocatalysis** 107
R. Ajay Rakkesh, D. Durgalakshmi and S. Balakumar

6 **Nanoporous Nanocomposite Materials for Photocatalysis** 129
Zahra Hosseini, Samad Sabbaghi and Naghmeh Sadat Mirbagheri

7 **Polymeric Nanocomposites for Visible-Light-Induced Photocatalysis** 175
Chin Wei Lai, Kian Mun Lee and Joon Ching Juan

8 **Carbon-Based Nanocomposites for Visible Light-Induced Photocatalysis** 203
Elaheh Kowsari

9 **Nanocomposites of g-C_3N_4 with Carbonaceous π-conjugated/Polymeric Materials Towards Visible Light-Induced Photocatalysts** 251
Sulagna Patnaik, Dipti Prava Sahoo and Kulamani Parida

10 Titanium-Based Mixed Metal Oxide Nanocomposites for Visible Light-Induced Photocatalysis 295
Soumyashree Pany, Amtul Nashim and Kulamani Parida

11 Novel Applications and Future Perspectives of Nanocomposites 333
Zsolt Kása, Tamás Gyulavári, Gábor Veréb, Gábor Kovács, Lucian Baia, Zsolt Pap and Klára Hernádi

Index 399

Chapter 1
Introduction of Nanomaterials for Photocatalysis

Diana Vanda Wellia, Yuly Kusumawati, Lina Jaya Diguna and Muhamad Ikhlasul Amal

Abstract This introductory chapter discusses the rapid development of nanotechnology for the application of visible light-induced photocatalysis, which is driven by the unique material properties arising from the nanoscale dimensions. It includes the description of the carbon-based nanomaterials developed first in the early development such as fullerene, carbon nanotube, and graphene. Conductive polymers were then described as photocatalysts with different dimensional nanostructures. Moreover, semiconductors were presented as potential materials for photocatalysis. For the practical visible light applications, photocatalysts need to be modified either by narrowing the band gap or by inhibiting the recombination of charge carriers via the formation of heterojunction nanocomposites. As the focus of this book, nanocomposites have been reported as a promising strategy for high-activity visible light-driven photocatalysis. This chapter is also complemented with some examples of industrial applications of photocatalysis for practical use.

Keywords Visible light-induced photocatalyst · Photocatalysis · Nanocomposite · Nanoparticle · Nanomaterial

---

D.V. Wellia
Chemistry Department, Andalas University, Padang, Indonesia

Y. Kusumawati
Chemistry Department, Sepuluh Nopember Institute of Technology, Surabaya, Indonesia

L.J. Diguna
Department of Renewable Energy Engineering, Prasetiya Mulya University, Tangerang 15339, Indonesia

M.I. Amal (✉)
Development and Application Unit for Biocompatible Implant Material in Orthopedics, Indonesian Institute of Sciences, Jl. Sangkuriang, Bandung 40135, Indonesia
e-mail: muha137@lipi.go.id

M.M. Khan et al. (eds.), *Nanocomposites for Visible Light-induced Photocatalysis*, Springer Series on Polymer and Composite Materials,
DOI 10.1007/978-3-319-62446-4_1

Introduction

The interest in nanosciences, nanotechnology, or nanomaterial has been grown in the beginning of 20s era and increased rapidly after 2006 as shown in Fig. 1.1. One of the factors that cause the development in nanosciences is the invention of scanning probe microscopy, for instance, Scanning Tunneling Microscopy (STM) in 1981 and Atomic Force Microscopy (AFM) in 1982 (Binnig and Rohrer 1987). Both STM and AFM have become the instrument that makes researchers able to observe the materials at an atomic level.

The nanosciences also have attracted much researcher attention due to the fact that the materials in nano-size show the unique properties that do not appear in the micro-size. The main reasons that make the nanoscale properties different from the bulk properties are due to the surface effect and quantum size effect (Roduner 2006). Many thermodynamic laws at the macroscopic scale are broken down at the nanoscale size. It is known that thermodynamic phenomena, for instance, a phase transition, are a collective phenomenon. At the nano-size scale, the surface-to-volume ratio is high, and thus it will influence the thermodynamic behavior of the nanomaterials (Yang and Mai 2014). Various experiments have been focused on the study of nano-size effect on the thermodynamic properties of the nanomaterials (Yang and Jiang 2005; Vanithakumari and Nanda 2008; Qi et al. 2005; Guisbiers and Buchaillot 2008; Attarian 2008; Sun et al. 2006).

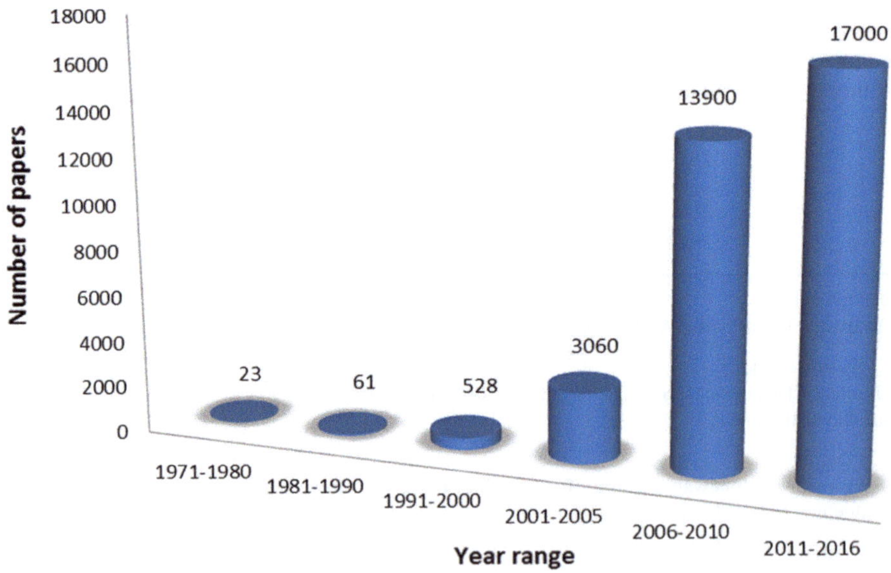

Fig. 1.1 Number of papers on nanosciences published since their discovery. Data is taken from Google Scholar with search keywords: Nanoparticle, nanostructure, nanomaterial, and nanocomposite

The other unique nanoscale property that is surface dominated is the magnetism. Jiang and his coworkers have shown that the magnetic properties of multilayer ferromagnetic films are influenced by the surface effect at nanoscale size (Jiang et al. 2014). Moreover, the atomic interaction at the surface of nanomaterials will be different from macroscopic size (Yang and Mai 2014). It will bring the effect of the difference in the mechanical properties of nanoscale materials compared to the macroscopic ones (Namazu et al. 2000; Li et al. 2004; Chang and Fang 2003; Chang 2003). The nanoscale effects not only influence the physical properties but also the chemical ones. The important impacts are in the catalysis and biochemical reactions. One of the reasons that influence the difference in the catalytic activity of nanoscale materials compared to the macroscopic ones is the existence of polarization changes in the bond to the adsorbate (Roduner 2006), while in a biochemical reaction, nanomaterials become important because their size fits with the size of many biomolecules. The last important reason why nanoscale materials are unique compared to the macroscopic ones is the appearance of quantum size effect. To explain the quantum size effect, bear in mind that in the metal or semiconductor the electrons are delocalized which is illustrated like electrons in a box model. The electron states then combine to extended band structure. In nanoscale size, the box size is close to the atomic size. The particle then becomes pseudo-atom and the electron states are quantized as does in the atoms or molecules. All the properties due to the electron activities, for instance, the electromagnetic wave interaction, ionization potentials, or electron affinities of nanomaterials, show changes from the microscopic ones (Roduner 2006; Liqiang et al. 2003).

Fullerene can be mentioned as one of the first nanomaterials that have been developed in the beginning of nanoscience eras. It has been discovered in 1985 by Sir Harold W. Kroto, Richard E. Smalley and Robert F. Curl, Jr. (Wang 2005). Their discovery has delivered them obtaining the Nobel Prize in 1966. The fullerene structure is relatively stable, but it is chemically reactive due to a very electron-deficient of C=C bonds of C60 (Briggs and Miller 2006; Geckeler and Samal 1999). The invention of fullerene was then followed by the invention in the other nanostructure of carbon allotropes such as one-dimensional carbon nanotubes (CNTs) and two-dimensional graphene. Both CNTs and graphene show a high electron conductivity caused by their one and two-dimensional nanostructure (Wang 2005; Heer et al. 1995; Castro-Neto et al. 2009; Geim and Novoselov 2007).

Recently, it has been observed that the small carbon nanoparticles show interesting optical properties that have not been observed yet at the larger or microscopic particles. The small carbon (Carbon Quantum Dots (CQDs)) which is prepared by surface passivation exhibits fluorescence emissions (Fernando et al. 2015; Lim et al. 2014). The origin of their optical emission ability is due to n-plasmon absorption in the core carbon nanoparticles. They have a broad range of absorption spectrum, covering UV/Vis spectral range (from 300 nm) to the near-IR (800 nm) (Cao et al. 2011; Fernando et al. 2015). Owing to this property, carbon is promising for UV/visible light-induced application, including photocatalysis.

Nanomaterials are not limited to the carbon allotropes, but also the other type of materials such as conductive polymer. This material also has unique properties when the size is reduced to a nanometer scale, for example, an enhanced electrical conductivity. Conductive polymer has been attracted many interests since the discoveries on poly(sulfur nitride), an electronically conducting polymer at low temperature by Greene, Street, and Suter in 1975 (Greene et al. 1975) and on the doping of poly(acetylene), an electronically conducting polymer at room temperature, by Heeger, MacDiarmid, and Shirakawa in 1977 (Chiang et al. 1977). For the contribution to the discovery and development of conductive polymers, the Nobel Prize in Chemistry 2000 was awarded jointly to Heeger, MacDiarmid, and Shirakawa. The ability of conductive polymers to act as a photocatalyst was demonstrated by Yanagida et al. in 1985, whereas the poly(p-phenylene) could catalyze the reduction of protons to hydrogen in the presence of a sacrificial electron donor upon UV irradiation (Yanagida et al. 1985). As a polymer absorbs the light of energy larger than its optical gap, the electrons from the highest occupied molecular orbital (HOMO) get excited to the lowest unoccupied molecular orbital (LUMO), leading to a spatial charge separation and subsequently driving redox reactions. Many researches have been further conducted in developing different polymer nanostructures with enhanced photocatalytic properties, particularly in the visible region. The polymer of graphitic carbon nitride (g-C_3-N_4) with porous structure has recently been demonstrated as a photocatalyst for both reduction of protons in the presence of a sacrificial electron donor and the oxidation of water in the presence of a sacrificial electron acceptor under visible light irradiation (Wang et al. 2009a, b). These sacrificial donors are not restored in the subsequent redox process but destroyed by chemical conversion. After this pioneering work, further progress has been achieved in the nanostructure designs of g-C_3-N_4 in the forms of hollow spheres (Sun et al. 2012) and one-dimensional nanostructure such as nanorods (Li et al. 2011). Moreover, the band gap engineering of g-C_3-N_4 to control its light absorption and redox potential was also conducted at atomic level through the elemental doping with nonmetal (Yan et al. 2010; Liu et al. 2010; Wang et al. 2010) and metal (Ding et al. 2011), and at molecular level such as copolymerization (Zhang et al. 2010). Interestingly, without using the sacrificial donors and in the absence of noble metal cocatalyst, one-dimensional poly(diphenylbutadiyne) nanostructures have shown high and stable photocatalytic activity under visible light (Ghosh et al. 2015). Furthermore, covalent organic frameworks forming highly regular two- or three-dimensional network structures also present an interesting new class of polymeric photocatalysts (Vyas et al. 2015). These network polymers possess the tenability by means of small structural modifications, originated from the electronic and steric variations in the polymer precursors, thus rationally controlling the photocatalytic process. A hybrid of polymers with graphene (Xu et al. 2014) or semiconductor nanostructures (Xiao et al. 2013) is also gaining significant attentions.

The other nanomaterials that have attracted extensive research interest are metal oxide. Most of the metal oxides are semiconductor. Some of them show unique properties capable of absorbing UV light. It is due to the fact that their band gap is

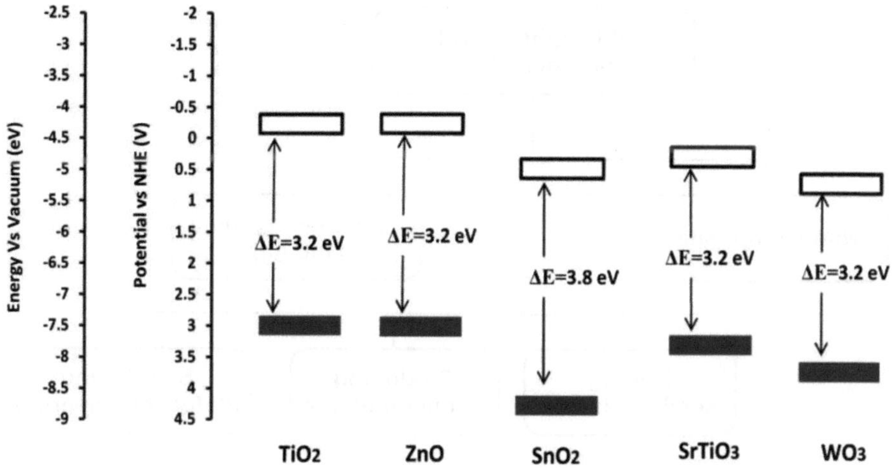

Fig. 1.2 Energy diagram of several semiconductors

located in the UV range. Their properties then can be used to create a UV-induced system. The electronic process happening after the UV light absorption then can be used for many chemical reactions. This makes semiconductor nanomaterials more widely applied as a photocatalyst for many chemical reactions in many fields of area, for instance, environmental, energy, or medicine (Khan and Al-Mayouf 2015). Figure 1.2 shows an energy diagram of alternative semiconductors (Gong et al. 2012). ZnO is the most commonly studied semiconductor after TiO_2. The interest in ZnO is driven by its relatively easy crystallization and anisotropic growth. Recently, various shapes of nano-crystalline ZnO have been prepared, for instance, one-dimensional nanowires of ZnO (Zhang et al. 2009), branched nanowires of ZnO which have higher surface area (Omar and Abdullah 2014), or hexagonal rod-like particles of ZnO that have 35 nm in diameter (Hosni et al. 2014).

Why Materials Such as Nanoparticles, Nanocomposites, Etc. Need to Be Developed for the Visible Light-Induced Photocatalysis?

As described above, due to the wide band gap of the semiconductor, it can only be activated by the UV light. The data shows that UV light is only small part of solar energy (approximately 3–4%) (Zhou et al. 2007). Numerous efforts have been made to produce nanomaterials (nanoparticles, nanocomposites, etc.) that are capable of effectively utilizing the visible light which constitutes the main part of the solar spectrum.

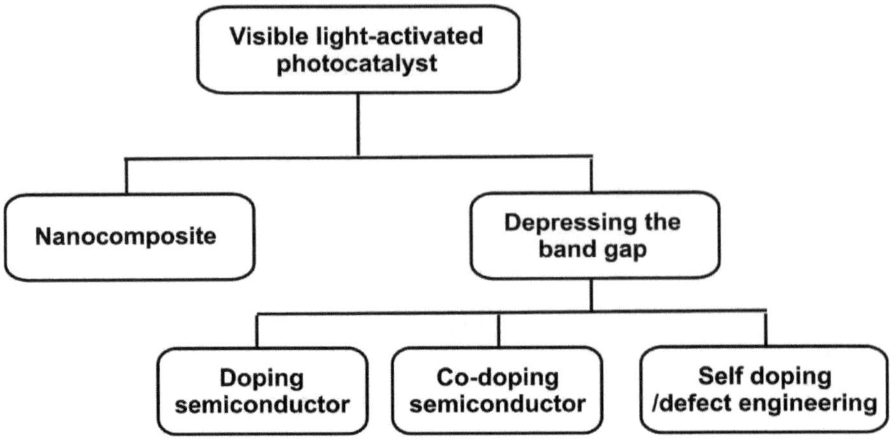

Fig. 1.3 Methods to develop visible light-activated photocatalysts

So far, the effort to produce visible light-induced photocatalyst has been made either by depressing the band gap of photocatalyst material or developing nanocomposite. There are three ways to narrow the band gap: the first is by doping semiconductor photocatalyst with single atom; the second is by doping it with two kinds of atom, called codoping; and the third is by self-doping or defect engineering (see Fig. 1.3).

Cation-Doped Semiconductor Photocatalyst

The addition of dopant to the semiconductors will shift their band gap, either by creating a new state below conduction band or above the valence band. As a consequence, lower energy is required to excite electrons from the valence band to the conduction band. In other words, doping can extend the spectral response of the semiconductor to the visible light region (Yu et al. 2009; Asiltürk et al. 2009). Besides that, metal ion dopant may act as electron or hole traps, so it can reduce recombination rate of photogenerated electron-hole pairs (Yuan et al. 2007; Asiltürk et al. 2009). As an example for TiO_2, they can be doped by cations (metals) to enable photocatalytic activity under visible light. In this case, the principle of doping with cation is the substitution of Ti^{4+} by other cations that have about the same size such as Cr, Fe, Ni, V, or Mn ions. The mechanisms in lowering the band gap are reported by Agrios et al. and Thimsen et al. through different ways. Agrios et al. and Zhang reported that the mixture of the conduction band of Ti (*d*) orbital of TiO_2 and the metal (*d*) orbital of the cation dopant was supposed to be the origin of the decrease in the band gap (Zhang and Lei 2008; Agrios and Pichat 2005), while Thimsen et al. reported that the general consensus of dopant is that the doping introduces additional energy levels into TiO_2 band gap and thus lower the energy required to excite electrons from the valence band to the conduction band (Thimsen et al. 2009). Metal doping also gives another route of the phase transformation which produces crystal defect and surface modifications, and hence changes the activation energy of transformation (Lee et al. 2009).

The photocatalytic activity of cation-doped semiconductor depends on many factors, such as the dopant concentrations, the location of energy levels of dopant in the lattice, *d*-electronic configurations, and distribution of dopants (Kernazhitsky et al. 2013).

Anion-Doped Semiconductor Photocatalyst

Doping of anion, in this case being referred to the nonmetals such as C, N, F, S, and B into the semiconductors, has been developed and considered as a potentially effective method to extend their band gap absorption to the visible light region. For oxide semiconductors, the anion substitutes oxygen in their lattice (Wu et al. 2009).

Sulfur, despite having a larger ionic radius compared to N and C atoms, can be used to synthesize S-doped TiO_2 (Rodríguez and García 2007). For example, Ho et al. prepared S-doped TiO_2 by the one-step low-temperature hydrothermal method. They found that the oxygen atoms in TiO_2 lattice were replaced by sulfur atoms as shown by a peak at about 160–161 eV in XPS, corresponding to the Ti–S bond formation (Ho et al. 2006). Another anion doping is fluorine-doped TiO_2 that has been reported to promote photocatalytic activity (Yu et al. 2002). For example, Yu et al. prepared F-doped TiO_2 by hydrolysis of titanium isopropoxide (TTIP) in the NH_4F–H_2O mixed solution. The prepared sample showed stronger absorption in the UV–visible range and red shift in the band gap transition. The photocatalytic activity was also found to be improved, three times higher than that of Degussa P25 (Yu et al. 2002). Besides S and F, iodine-doped TiO_2 has been reported as a visible light-activated photocatalyst as well. Iodine incorporation causes an absorption in the visible light range with a red shift in the band gap transition (Hong et al. 2005). Moreover, boron atoms can substitute oxygen atoms in the TiO_2 lattice as well to form B-doped TiO_2. The *p* orbital of B is mixed with O *2p* orbital, which results in band gap narrowing (Zhao et al. 2004). Among the anion dopants, C or N atoms have been found to attract more attention due to the superior photocatalytic activity under visible light irradiation (Zhang and Song 2009). For example, Matos et al. prepared C-doped TiO_2 by a solvothermal method and evaluated the photocatalytic activity on methylene blue (MB) degradation. They found that C-doped TiO_2 exhibited first-order rate constant for degradation of MB which showed higher photocatalytic activity than un-doped one. This result was caused by direct optical charge-transfer transition involving both the TiO_2 and carbon phase, keeping the high reactivity of the photogenerated electron and hole (Matos et al. 2010).

Codoping Semiconductor

At present, doping semiconductor with two kinds of atom, called codoping, has gained considerable attention due to the higher photocatalytic activity compared to single-element doping (Zhang et al. 2009). For instance, codoping TiO_2 with N and other anions was reported to show favorable photocatalytic properties including synergetic effect, high surface area, well-crystallized anatase phase, red shift in absorption edge, strong absorbance of light with a longer wavelength, etc. (Wu and Hung 2009).

The mechanism of codoping semiconductor system for visible light activation is different and depends on the element used. For example, in codoping TiO_2, i.e. N-F-codoped TiO_2, Valentin et al. reported that the situation was more complex. The process involved a superposition of two single-doped materials with the simultaneous presence of both shallow and deep localized state into the band gap. Smaller oxygen defects were expected to be present in N-F-codoped TiO_2 bulk samples which probably became a reason for the larger photostability and photocatalytic activity for that codoped sample (Valentin et al. 2008). In the case of N–P-codoped TiO_2, Long et al. mentioned that N and P could act as substitutional dopants and adsorptive dopants (Long and English 2010). When both N and P acted as substitutional dopants, the bandgap narrowed slightly. However, upon N and P absorption on the surface, the bandgap narrowing can be significantly induced even at low dopant concentrations. In another study, Jia et al. reported on the N–Fe-codoped system. They showed that codoping with N and Fe leads to lattice distortion which changes the dipole moments and makes the easier separation of photoexcited electron–hole pairs. Then, a significant red shift occurs resulting in efficient enhancement of photocatalytic activity (Jia et al. 2011).

Self-Doping/Defect Engineering

Another way to decrease the band gap of semiconductor is using electron beam (EB) to create a defect in material. EB-assisted nanofabrication does not involve the use of hazardous chemicals and occurs under ambient condition (Khan et al. 2014). In this method, the interaction of highly reactive electrons from EB with air molecules is expected to give rise to highly reactive and strong oxidizing agents, such as ozone, OH groups, or other radicals, which can react further with TiO_2 nanoparticles (Kim et al. 2009, 2010). The interaction of TiO_2 with the high energy electrons generated by EB and oxidizing agents/ozone can result in the reduction of Ti^{4+} to Ti^{3+}, or the formation of oxygen-deficient/rich species (Kim et al. 2009). This method also gives the enhancement of optical properties, thereby enhancing their photocatalytic activities in visible light (Khan et al. 2014).

Composite Material

Coupling photocatalyst material or catalyst support material with other metal, semiconductor with narrower band gap or polymer forming nanocomposite material has been reported to improve photodegradation efficiency in visible light irradiation. Those nanocomposites as visible light-induced photocatalyst may include mixed metal oxide nanocomposites, nanoporous nanocomposite materials, polymeric nanocomposites, and carbon-based nanocomposites that will be explained briefly below.

1.1 Mixed Metal Oxide Nanocomposites for Visible Light-Induced Photocatalysis

As mentioned above, the coupling of semiconductor with other semiconductors or molecules with narrower band gap has been reported as a promising method to improve visible light photocatalytic activity. The principle for improving the photocatalytic activity in the visible light region is due to increasing the efficiency of charge separation and extending the energy range of photoexcitation for the system (Linsebigler et al. 1995). Biswas et al. prepared annealed CdS–TiO_2 thin film on a glass slide and indium tin oxide (ITO) substrate by chemical bath deposition technique (Biswas et al. 2008). They found that the photocatalytic activity of CdS–TiO_2 thin film for methanol degradation was enhanced due to the improvement of crystallinity in CdS and TiO_2 layers and the increase of roughness of CdS surface after high-vacuum annealing (Biswas et al. 2008). They also reported that the higher photocatalytic activity might be attributed to the fact that the photogenerated electrons and holes can be well separated under UV–visible irradiation due to their suitable valence band and conduction band potentials (Biswas et al. 2008). In another study, Resta et al. found CdS–TiO_2 thin film having higher absorption efficiency in visible light region with respect to TiO_2. They prepared the thin film by a novel in situ approach based on an unimolecular precursor for CdS, [Cd($SBz)_2$]1-methylimidazole (Resta et al. 2010). They also reported that CdS was used as a sensitizer to TiO_2 film leading to improved photocatalytic activity. For WO_3–TiO_2 system, Somasundaram et al. prepared a WO_3–TiO_2 thin film by pulse electrodeposition method and found an optimal condition for minimizing electron–hole recombination (Somasundaram et al. 2006). An excellent photocatalytic activity in the photodecomposition of 2-propanol on WO_3–TiO_2 film was reported by Pan et al., which was prepared via an evaporation-induced self-assembly (EISA) process (Pan and Lee 2006). Many Aurivillius-based compounds also can be coupled with TiO_2 to make nanocomposite as photocatalyst with excellent photocatalytic activity under visible light irradiation (Zhou et al. 2010; Xu et al. 2009). Bi_2WO_6 is the simplest member of the family and the most studied example so far (López et al. 2011). Bi-based oxide couples with TiO_2 will form heterojunction interfaces (Zhou et al. 2010). As a result, they will be bonded tightly to form efficient heterostructure (Shang et al. 2009) and this structure can extend the lifetime of the photon-induced electron–hole pairs (Zhou et al. 2010; Shang et al. 2009).

The other examples of semiconductor composite are the composite of ZnO with Ag, CdO, or both of them. The combining of ZnO with Ag shows an increase in photodegradation activity of textile dye due to the increase of the surface area (Saravanan et al. 2013), while combining ZnO with CdO creates more number of charge carriers due to the retardation of back reaction between the photogenerated charge carriers. This makes the combination of ZnO with CdO promising to shift their optical band gap towards becoming more responsive to the UV–visible light (Saravanan et al. 2011). Combining both of Ag and CdO with ZnO forms a ternary composite that also shows an enhancement in photocatalytic activity. This enhancement is dominantly caused by the increased surface area. Moreover, the existence of Ag in the ternary composite system provides an electron trap which facilitates the electron–hole separation (Saravanan et al. 2015a). The enhancement of photocatalytic activity has also been observed in the ternary nanocomposite $ZnO/Ag/Mn_2O_3$ (Saravanan et al. 2015b).

1.2 Nanoporous Nanocomposite Materials for Photocatalysis

The term of nanoporous generally applies for solids with pore diameters of 2–100 nm. Nanoporous solids are attracted researchers' interests because they owe large surface, three-dimensionally interconnected porous networks, large pore volume, tunable pore size, and nano-sized crystalline walls. These unique properties are potential for many applications, including heterogeneous catalysis and photocatalysis. Numerous systems of metal catalysts, metal oxide catalysts, porous material (as Zeolite), and metals loaded on hierarchically porous inert supports have been fabricated to study for photocatalytic application. These materials are thought to give promising properties improvements, such as high accessibility of bulky molecules, high diffusion rate of reactant and product, and the high catalytic activity due to large active sites (usually heteroatoms of porous material and/or supported nanometal particles) (Su et al. 2016; Luc and Jiao 2016).

There are several reports that showed how the complexity of nanoporous nanocomposite materials structure can improve the photocatalytic activity. Li et al. fabricated porous AgI/Ag nanocomposites with a facile two-step route, involving reactions between dealloyed nanoporous silver and mixed H_2O_2/HCl solution, and reported a dynamic structure coarsening process along the original network structure of nanoparticles (Li and Ding 2010). Deng et al. prepared Ag nanoparticle decorated nanoporous ZnO microrods by solvothermal-assisted heat treatment method. The diameter of rods was 90–150 nm with the length of 0.5–3 μm and composed of ZnO nanoparticles with an average dimension of $\sim$20 nm. Compared to native nanoporous ZnO micrometer rods, the as-prepared Ag nanoparticles/nanoporous ZnO microrods were able to degrade methylene blue twice and 5.6 times faster under the UV and solar light irradiation, respectively. It also showed enhanced photocatalytic activity and improved photostability due to exceptional nano/microconfigured structure, the superior crystallinity of the ZnO rods, and the decorated Ag nanoparticles. The charge separation was promoted by Ag nanoparticles deposited on the ZnO surface which acts as the electron wells.

The Ag nanoparticles also gave plasmatic effect that improves photocatalytic activities in visible region (Deng et al. 2012). Chen et al. synthesized biological hierarchical porous structure of Zn-doped $TiO_2/C@SiO_2$ nanoporous composites by sol–gel method. Rice husk was used as biotemplate and porous catalytic carrier. This hybrid nanoporous material showed high efficiency in pollutants degradation and good absorption properties. The application of rice husk showed improvement with promising in recyclability and durability (Chen et al. 2015).

Even though the breakthrough research has been made especially in structural engineering for light harvesting, separation and transport of photogenerated electron and hole, the application of hierarchically porous structures is still limited. The photocatalytic improvement especially in visible regions should be the main focus of future studies. It can be done by developing advanced hierarchically structured porous photocatalysts, applying dopants, novel template, etc. (Su et al. 2016).

1.3 Polymeric Nanocomposites for Visible Light-Induced Photocatalysis

Polymers have been applied to extend the absorption range of the semiconductor, thereby enhancing the photocatalytic performance under UV or visible light irradiation. If both conduction and valence bands of the polymer are higher or lower than those of semiconductor, then an efficient spatial charge separation could be achieved and subsequently, could enhance the photocatalytic performance such g-C_3N_4-based CdS (Cao et al. 2013a) and In_2O_3 (Cao et al. 2013b) nanocomposites. This spatial charge separation was also found in polymer composites such as g-C_3N_4/poly(3-hexylthiophene) (P3HT) (Yan and Huang 2011) and g-C_9N_{10}/g-$C_{12}N_7H_3$ (Li et al. 2016), in which the charge separation in the former composite is beneficial for photocatalytic H_2 evolution (Yan and Huang 2011). Furthermore, the nanocomposite of polymers and metals, i.e., g-C_3N_4/Au/poly(3-hexylthiophene) (P_3HT)/Pt, has been also reported to have an efficient hydrogen production (Zhang et al. 2015).

1.4 Carbon-Based Nanocomposites for Visible Light-Induced Photocatalysis

Combining of CQDs with TiO_2 or metals to produce carbon-based composite material also shows interesting results. CQDs can enhance the photocatalytic activity of TiO_2. In this composite, carbon has a role as photosensitizing agent. Moreover, the presence of carbon also constructs mid-gap energy levels in TiO_2 (Park et al. 2009; Sun et al. 2013). Besides, combining of CQDs with metals, such as Au, Cu, Ag, or Pt, also gives incremental photocatalytic activity of TiO_2 (Cao et al. 2011; Liu et al. 2014). Liu et al. observed that combining CQDs with metals Au, Cu, and Ag shifts the light absorption into the purple, green, and red light, respectively, with the best performance occurred at Cu/CQDs (Liu et al. 2014).

Aside of CQDs, the composite of graphene or CNTs with the other material also shows interesting properties. Incorporation of graphene or CNTs into semiconductor mostly induces the conductivity enhancement of the semiconductor (Wang et al. 2012; Fan et al. 2012; Kusumawati et al. 2014). The composite of graphene with Ag and WO_3 shows interesting photocatalytic activity toward organic molecules degradation (Khan and Al-Mayouf 2015; Khan et al. 2016a, b). Zouzelka and his coworkers also have observed the satisfying enhancement of TiO_2/MWCNT (multiwalled carbon nanotube) photocatalytic activity toward an eco-persistence pollutant, i.e. the 4-chlorophenol (Zouzelka et al. 2016).

Photocatalyst development has given various colors to the chemical conversion process. This technology can be applied in many fields of industry including medicine, environment, or energy. The most applications of photocatalyst are in environmental fields, for instance, water purification (Saravanan et al. 2014; Zhang et al. 2013), degradation of pollutant molecules (Zouzelka et al. 2016; Saravanan et al. 2015a, b), or self-cleaning technology (Kamegawa et al. 2012; Banerjee et al. 2015). Some photocatalysts also show antibacterial activity (Tobaldi et al. 2016; Lin et al. 2015). Moreover, some photocatalysts have been applied to support the fuel cell devices (Drew et al. 2005; Xia et al. 2016).

Summary

The key to the outstanding performance of nanomaterials is based on their nanoscale. Fullerene has been mentioned as the material introduced in the beginning of nanomaterial development and has inspired researchers to develop further in order to meet requirements of their application. The nanomaterials that have been developed can be categorized by allotrope carbon-based material, conductive polymer, metal, semiconductor, and their composites. Among them, metal oxide semiconductor shows unique properties capable of absorbing UV light due to the fact that its band gap is located in the UV range. Their properties then can be used to create a UV-induced system. However, since UV light is only a small proportion of solar energy (approximately 3–4%), numerous efforts have been made to develop new photocatalyst systems that are capable of effectively utilizing the visible light which constitutes the main part of the solar spectrum. In general, nanocomposites have been reported to excellently improve the photodegradation efficiency under visible light irradiation. This makes photocatalyst nanomaterials more widely applied for many chemical reactions in many fields of the area such as environment, energy, or medicine.

References

Agrios AG, Pichat P (2005) State of the art and perspectives on materials and applications of photocatalysis over TiO_2. J Appl Electrochem 35:655–663

Asiltürk M, Sayilkan F, Arpac E (2009) Effect of Fe^{3+} ion doping to TiO_2 on the photocatalytic degradation of Malachite Green dye under UV and vis-irradiation. J Photochem Photobiol A Chem 203:64–71

Attarian SM (2008) Effective coordination number model for the size dependency of physical properties of nanocrystals. J Phys Condens Matter 20:325237

Banerjee S, Dionysios DD, Pilai SC (2015) Self-Cleaning Applications of TiO_2 by photo-induced hydrophilicity and photocatalysis. Appl Catal B Environ 176–177:396–428
Binnig G, Rohrer H (1987) Scanning tunneling microscopy—from birth to adolescence. Rev Mod Phys 59:615–625
Biswas S, Hossain MF, Takahashi T, Kubota Y, Fujishima A (2008) Photocatalytic activity of high-vacuum annealed CdS/TiO_2 thin film. Thin Solid Films 516:7313–7317
Briggs JB, Miller GP (2006) Fullerene–acene chemistry: a review. Comptes Rendus Chim 9:916–927
Cao L, Sahu S, Anilkumar P, Bunker CE, Xu J, Fernando KAS, Wang P, Guliants EA, Tackett KN, Sun Y-P (2011) Carbon nanoparticles as visible-light photocatalysts for efficient CO_2 conversion and beyond. J Am Chem Soc 13:4754–4757
Cao S-W, Liu X-F, Yuan Y-P, Zhang Z-Y, Liao Y-S, Fang J, Loo SCJ, Sum TC, Xue C (2013a) Solar-to-fuels conversion Over In_2O_3/g-C_3N_4 Hybrid photocatalysts. Appl Catal B 147:940–946
Cao S-W, Yuan Y-P, Fang J, Shahjamali MM, Boey FYC, Barber J, Loo SCJ, Xue C (2013b) In-situ growth of CdS quantum dots on g-C_3N_4 nanosheets for highly efficient photocatalytic hydrogen generation under visible light irradiation. Int J Hydrogen Energy 38:1258–1266
Castro-Neto AH, Guinea F, Peres NMR, Novoselov KS, Geim AK (2009) The electronic properties of graphene. Rev Mod Phys 81:109
Celik E, Yildiz AY, Azem NFA, Tanoglu M, Toparli M, Emrullahoglu OF, Ozdemir I (2006) Preparation and characterization of Fe_2O_3–TiO_2 thin films on glass substrate for photocatalytic applications. Mater Sci Eng B 129:193–199
Chang WJ (2003) Molecular-dynamics study of mechanical properties of nanoscale copper with vacancies under static and cyclic loading. Microelectron Eng 65:239–246
Chang WJ, Fang TH (2003) Influence of temperature on tensile and fatigue behavior of nanoscale copper using molecular dynamics simulation. J Phys Chem Solids 64:1279–1283
Chen H, Zhao L, Xiang Y, He Y, Song G, Wang X, Liang F (2015) A novel Zn-TiO/C@SiO nanoporous material on rice husk for photocatalytic applications under visible light, Desalin Water Treat 57:1–11. (http://dx.doi.org/10.1080/19443994.2015.1035339)
Chiang CK, Fincher CR Jr, Park YW, Heeger AJ, Shirakawa H, Louis EJ, Gau SC, MacDiarmid AG (1977) Electrical conductivity in doped polyacetylene. Phy Rev Lett 39:1098–1101
Deng Q, Duan X, Ng DHL, Tang H, Yang Y, Kong M, Wu Z, Cai W, Wang G (2012) Ag nanoparticle decorated nanoporous ZnO microrods and their enhanced photocatalytic activities. ACS Appl Mater Interfaces 4(11):6030–6037
Ding Z, Chen X, Antonietti M, Wang X (2011) Synthesis of transition metal-modified carbon nitride polymers for selective hydrocarbon oxidation. Chemsuschem 4:274–281
Drew KG, Girishkumar G, Vinodgopal K, Kamat PV (2005) Boosting fuel cell performance with a semiconductor photocatalyst: TiO_2/Pt-Ru hybrid catalyst for methanol oxidation. J Phys Chem B 109:11851–11857
Fan J, Liu S, Yu J (2012) Enhanced photovoltaic performance of dye-sensitized solar cells based on TiO_2 nanosheets/graphene composite films. J Mater Chem 22(33):17027
Fernando KAS, Sahu S, Liu Y, Lewis WK, Guliants EA, Jafariyan A, Wang P, Bunker CE, Sun Y-P (2015) Carbon quantum dots and applications in photocatalytic energy conversion. ACS Appl Mater Interfaces 7:8363–8376
Geckeler KE, Samal S (1999) Syntheses and properties of macromolecular fullerenes, a review. Polym Int 48:743–757
Geim AK, Novoselov KS (2007) The rise of graphene. Nat Mater 6(3):183
Ghosh S, Kouamé NA, Ramos L, Remita S, Dazzi A, Deniset-Besseau A, Beaunier P, Goubard F, Aubert PH, Remita H (2015) Conducting polymer nanostructures for photocatalysis under visible light. Nat Mater 14:505–511
Gong J, Liang J, Sumathy K (2012) Review on dye-sensitized solar cells (DSSCs): fundamental concepts and novel materials. Renew Sustain Energy Rev 16:5848–5860

Greene R, Street G, Suter L (1975) Superconductivity in polysulfur nitride $(SN)_x$. Phys Rev Lett 34:8–10

Guisbiers G, Buchaillot L (2008) Size and shape effects on creep and diffusion at the nanoscale. Nanotechnology 19:435701

Heer WA, Châtelain A, Ugarte D (1995) A carbon nanotube field-emission electron source. Science 270:1179–1180

Ho W, Yu JC, Lee S (2006) Low-temperature hydrothermal synthesis of S-doped TiO_2 with visible light photocatalytic activity. J Solid State Chem 179:1171–1176

Hong X, Wang Z, Cai W, Lu F, Zhang J, Yang Y, Ma N, Liu Y (2005) Visible-light-activated nanoparticle photocatalyst of iodine-doped titanium dioxide. Chem Mater 17:1548–1552

Hosni M, Kusumawati Y, Farhat S, Jouini N, Pauporté T (2014) Effects of oxide nanoparticle size and shape on electronic structure, charge transport, and recombination in dye-sensitized solar cell photoelectrodes. J Phys Chem C 118:16791–16798

Jia L, Wu C, Han S, Yao N, Li Y, Li Z, Chi B, Pu J, Jian L (2011) Theoretical study on the electronic and optical properties of (N, Fe)-codoped anatase TiO_2 photocatalyst. J Alloy Compd 509:6067–6071

Jiang W, Chen J, Ma B, Wang Z (2014) Surface effects on magnetic and thermodynamic properties in nanoscale multilayer ferrimagnetic films. Phys E Low-Dimens Syst Nanostruct 61:101–106

Kamegawa T, Shimizu Y, Yamashita H (2012) Superhydrophobic surfaces with photocatalytic self-cleaning properties by nanocomposites coating of TiO_2 and polytetrafluoroethylene. Adv Mater 24(27):2697–3700

Kernazhitsky L, Shymanovska V, Gavrilko T, Naumov V, Kshnyakin V, Khalyavka T (2013) A comparative study of optical absorption and photocatalytic properties of nanocrystalline single-phase anatase and rutile TiO_2 doped with transition metal cations. J Solid State Chem 198:511–519

Khan MM, Ansari SA, Pradhan D, Ansari MO, Han DH, Lee J, Cho MH (2014) Band gap engineered TiO nanoparticles for visible light induced photoelectrochemical and photocatalytic studies. J Mater Chem A 2(3):637–644

Khan MM, Al-Mayouf SFAA (2015) Metal oxides as photocatalysts. J Saudi Chem Soc 12:462–464

Khan ME, Khan MM, Cho MH (2016a) Biogenic synthesis of a Ag–graphene nanocomposite with efficient photocatalytic degradation, electrical conductivity and photoelectrochemical performance. New J Chem 39:8121–8129

Khan ME, Khan MM, Cho MH (2016b) Fabrication of WO_3 nanorods on graphene nanosheets for improved visible light-induced photocapacitive and photocatalytic performance. RSC Adv 6:20824–20833

Kim KD, Tai WS, Kim YD, Cho SJ, Bae IS, Boo JH, Lee B-C, Yang K-H, pack O-K (2009) Change in water contact angle of carbon contaminated TiO surfaces by high-energy electron beam. B Kor Chem Soc 30(5):1067–1070

Kim MJ, Kim K-D, Tai WS, Seo HO, Luo Y, Kim YD, Lee BC, Park OK (2010) Enhancement of photocatalytic activity of TiO_2 by high-energy electron-beam treatment under atmospheric pressure. Catal Lett 135(1–2):57–61

Kusumawati Y, Martoprawiro MA, Pauporté TH (2014) Effects of graphene in graphene/TiO composite flms applied to solar cell photoelectrode. J Phys Chemi C 118(19):9974–9981

Lee HY, Lan WY, Tseng TY, Hsu D, Chang YM, Lin JG (2009) Optical control of phase transformation in Fe-doped TiO_2 nanoparticles. Nanotechnology 20:315702–315706

Li Y, Ding Y (2010) Porous AgCl/Ag nanocomposites with enhanced visible light photocatalytic properties. J Phys Chem C 114(7):3175–3179

Li X, Chang WC, Chao YJ, Wang R, Chang M (2004) Nanoscale structural and mechanical characterization of a natural nanocomposite material: the shell of red abalone. Nano Lett 4:613–617

Li XH, Zhang J, Chen X, Fischer A, Thomas A, Antonietti M, Wang X (2011) Condensed graphitic carbon nitride nanorods by nanoconfinement: promotion of crystallinity on photocatalytic conversion. Chem Mater 23:4344–4348

Li H, Hu H, Bao C, Guo F, Zhang X, Liu X, Hua J, Tan J, Wang A, Zhou H, Yang B, Qu Y, Liu X (2016) Forming heterojunction: an effective strategy to enhance the photocatalytic efficiency of a new metal-free organic photocatalyst for water splitting. Scientific Reports 6:29327

Lim SY, Shen W, Gao Z (2014) Carbon quantum dots and their applications. Chem Soc Rev 44:362–381

Lin H, Deng W, Zhou T, Ning S, Long J, Wang X (2015) Iodine modified nanocrsytalline titania for photo-catalytic antibacterial application under visible light illumination. Appl Catal B Environ 176–177:36–43

Linsebigler AL, Lu G, Yates JT Jr (1995) Photocatalysis on TiO_2 surfaces: principles, mechanisms, and selected results. Chem Rev 95:735–758

Liqiang J, Xiaojun S, Jing S, Weimin C, Zili X, Yaoguo D, Honggang F (2003) Review of surface photovoltage spectra of nano-sized semiconductor and its applications in heterogeneous photocatalysis. Sol Energy Mater Sol Cells 79:133–151

Liu G, Niu P, Sun C, Smith SD, Chen Z, Lu GQ, Cheng HM (2010) Unique electronic structure induced high photoreactivity of sulfur-doped graphitic C_3N_4. J Am Chem Soc 132:11642–11648

Liu R, Huang H, Li H, Liu Y, Zhong J, Li Y, Zhang S, Kang Z (2014) Metal nanoparticle/carbon quantum dot composite as a photocatalyst for high-efficiency cyclohexane oxidation. ACS Catal. 4:328–336

Long R, English NJ (2010) First-principles calculation of synergistic (N, P)-codoping effects on the visible-light photocatalytic activity of anatase TiO_2. J Phys Chem C 114:11984–11990

López SM, Hidalgo MC, Navío JA, Colón G (2011) J Hazard Mater 185:1425–1434

Luc W, Jiao F (2016) Synthesis of nanoporous metals, oxides, carbides, and sulfides: beyond nanocasting. Acc Chem Res 49(7):1351–1358

Matos J, Garcia A, Zhao L, Titirici MM (2010) Solvothermal carbon-doped TiO_2 photocatalyst for the enhanced methylene blue degradation under visible light. J Appl Electrochem 35:655–663

Namazu T, Isono Y, Tanaka T (2000) Evaluation of size effect on mechanical properties of single crystal silicon by nanoscale bending test using AFM. J Microelectromech Syst 9:450–459

Omar A, Abdullah H (2014) Electron transport analysis in zinc oxide-based dye-sensitized solar cells: a review. Renew Sustain Energy Rev 31:149–157

Pan JH, Lee WI (2006) Preparation of highly ordered cubic mesoporous WO_3/TiO_2 films and their photocatalytic properties. Chem Mater 18:847–853

Park Y, Kim W, Park H, Tachikawa T, Majima T, Choi W (2009) Carbon-doped TiO_2 photocatalyst synthesized without using an external carbon precursor and the visible light activity. Appl Catal B Environ 95:355–361

Qi WH, Wang MP, Zhou M, Hu WY (2005) Surface-area-difference model for thermodynamic properties of metallic nanocrystals. J Phys D: Appl Phys 38 (9):1429–1436

Resta V, Laera AM, Piscopiello E, Capodieci L, Ferrara MC, Tapfer L (2010) Synthesis of CdS/TiO_2 nanocomposites by using cadmium thiolate derivatives as unimolecular precursors. Phys Status Solidi A 207:1631–1635

Rodríguez JA, García MF (2007) Synthesis, properties and applications of oxide nanomaterials. Wiley, USA

Rodner E (2006) Size matters why nanomaterials are different. Chem Soc Rev 35(7):583–592

Saravanan R, Shankar H, Prakash T, Narayanan V, Stephen A (2011) ZnO/CdO composite nanorods for photocatalytic degradation of methylene blue under visible light. Mater Chem Phys 125(1–2):277–280

Saravanan R, Karthikeyen N, Gupta VK, Thirumai E, Thangdurai P, Narayanang V, Stephen A (2013) ZnO/Ag nanocomposite: an efficient catalyst for degradation studies of textile effluents under visible light. Mater Sci Eng C 33(4):2235–2244

Saravanan R, Gupta VK, Narayanan V, Stephen A (2014) Visible light degradation of textile effluent using novel catalyst ZnO/I^3-Mn_2O_3. J Taiwan Inst Chem Eng 45(4):1910–1917

Saravanan R, Khan MM, Gupta VK, Mosquera E, Gracia F, Narayanang V, Stephen A (2015a) ZnO/Ag/CdO nanocomposite for visible light-induced photocatalytic degradation of industrial textile effluents. J Colloid Interface Sci 452:126–133

Saravanan R, Khan MM, Gupta VK, Mosquera E, Gracia F, Narayanang V, Stephen A (2015b) $ZnO/Ag/Mn_2O_3$ nanocomposite for visible light-induced industrial textile effluent degradation, uric acid and ascorbic acid sensing and antimicrobial activity. RSC Adv 5:34645–34651

Shang M, Wang W, Zhang L, Sun S, Wang L, Zhou L (2009) 3D Bi_2WO_6/TiO_2 hierarchical heterostructure: controllable synthesis and enhanced visible photocatalytic degradation performances. J Phys Chem C 113:14727–14731

Somasundaram S, Chenthamarakshan CR, de Tacconi NR, Basit NA, Rajeshwar K (2006) Composite WO_3/TiO_2 films: Pulsed electrodeposition from a mixed bath versus sequential deposition from twin baths. Electrochem Commun 8:539–543

Sun CQ, Shi Y, Li CM, Li S, AuYeung TC (2006) Size-induced undercooling and overheating in phase transitions in bare and embedded clusters. Phys Rev B 73:75408

Sun J, Zhang J, Zhang M, Antonietti M, Fu X, Wang X (2012) Bioinspired hollow semiconductor nanospheres as photosynthetic nanoparticles. Nat Commun 3:1139

Sun M, Ma X, Chen X, Sun Y, Cui X, Lin Y (2013) A nanocomposite of carbon quantum dots and TiO_2 nanotube arrays: enhancing photoelectrochemical and photocatalytic properties. RSC Adv 4:1120–1127

Su C, Tandiana R, Tian B, Sengupta A, Tang W, Su J, Loh KP (2016) Visible-light photocatalysis of aerobic oxidation reactions using carbazolic conjugated microporous polymers. ACS Catal 6 (6):3594–3599

Thimsen E, Biswas S, Lo C, Biswas P (2009) Predicting the band structure of mixed transition metal oxides: theory and experiment. J Phys Chem C 113:2014–2021

Tobaldi DM, Piccirillo C, Rozman N, Pullar RC, Seabra MP, Sever Škapin A, Castro PML, Labrincha JA (2016) Effects of Cu, Zn and Cu-Zn addition on the microstructure and antibacterial and photocatalytic functional properties of Cu-Zn modified TiO_2 nano-heterostructures. J Photochem Photobiol A: Chem 330:44–54

Valentin CD, Finazzi E, Pacchioni G (2008) Density functional theory and electron paramagnetic resonance study on the effect of N− F codoping of TiO_2. Chem Mater 20:3706–3714

Vanithakumari SC, Nanda KK (2008) A universal relation for the cohesive energy of nanoparticles. Phys Lett A 372:6930–6934

Vyas VS, Haase F, Stegbauer L, Savasci G, Podjaski F, Ochsenfeld C, Lotsch BV (2015) A tunable azine covalent organic framework platform for visible light-induces hydrogen generation. Nat Commun 6:8508

Wang J (2005) Carbon-nanotube based electrochemical biosensors: a review. Electroanalysis 17(1):7–14

Wang X, Maeda K, Chen X, Takanabe K, Domen K, Hou Y, Fu X, Antonietti M (2009a) Polymer semiconductors for artificial photosynthesis: hydrogen evolution by mesoporous graphitic carbon nitride with visible light. J Am Chem Soc 131:1680–1681

Wang X, Maeda K, Thomas A, Takanabe K, Xin G, Carlsson JM, Domen K, Antonietti M (2009b) A metal-free polymeric photocatalyst for hydrogen production from water under visible light. Nat Mater 8:76–80

Wang Y, Di Y, Antonietti M, Li H, Chen X, Wang X (2010) Excellent visible-light photocatalysis of fluorinated polymeric carbon nitride solids. Chem Mater 22:5119–5121

Wang H, Leonard SL, Hu YH (2012) Promoting effect of graphene on dye-sensitized solar cells. Ind Eng Chem Res 51(32):10613–10620

Wu KR, Hung CH (2009) Characterization of N, C-codoped TiO_2 films prepared by reactive DC magnetron sputtering. Appl Surf Sci 256:1595–1603

Xia L, Bai J, Li J, Zeng Q, Li X, Zou BX (2016) A highly efficient $BiVO_4/WO_3/W$ heterojunction photoanode for visible light responsive dual photoelectrode photocatalytic fuel cell. Appl Catal B Environ 183:224–230

Xiao J, Chen W, Wang F, Du J (2013) Polymer/TiO hybrid nanoparticles with highly effective UV-screening but eliminated photocatalytic activity. Macromolecules 46(2):375–383

Xu C, Wei X, Ren Z, Wang Y, Xu G, Shen G, Han G (2009) Solvothermal preparation of $Bi_2 WO_6$ nanocrystals with improved visible light photocatalytic activity. Mater Lett 26:2194–2197

Xu J, Luo L, Xiao G, Zhang Z, Lin H, Wang X, Long J (2014) Layered C N S polymer/graphene hybrids as metal-free catalysts for selective photocatalytic oxidation of benzylic alcohols under visible light. ACS Catal 4(9):3302–3306

Yan H, Huang Y (2011) Polymer composites of carbon nitride and poly(3-hexylthiophene) to achieve enhanced hydrogen production from water under visible light. Chem Commun 47:4168–4170

Yan SC, Li ZS, Zou ZG (2010) Photodegradation of rhodamine B and methyl orange over boron-doped g-C_3N_4 under visible light irradiation. Langmuir 26:3894–3901

Yanagida S, Kabumoto A, Mizumoto K, Pac C, Yoshino K (1985) Poly(p-phenylene)-catalysed photoreduction of water to hydrogen. J Chem Soc Chem Commun 8:474–475

Yang CC, Jiang Q (2005) Size and interface effects on critical temperatures of ferromagnetic, ferroelectric and superconductive nanocrystals. Acta Mater 53:3305–3311

Yang CC, Mai YW (2014) Thermodynamics at the nanoscale: a new approach to the investigation of unique physicochemical properties of nanomaterials. Mater Sci Eng R Rep 79:1–40

Yu JC, Yu JG, Ho WK, Jiang ZT, Zhang LZ (2002) Effects of F-doping on the photocatalytic activity and microstructures of nanocrystalline TiO_2 powders. Chem Mater 14:3808–3816

Yu J, Xiang Q, Zhou M (2009) Preparation, characterization and visible-light-driven photocatalytic activity of Fe-doped titania nanorods and first-principles study for electronic structures. Appl Catal B Environ 90:595–602

Yuan Z, Zhang J, Lin B, Li J (2007) Effect of metal ion dopants on photochemical properties of anatase TiO_2 films synthesized by a modified sol-gel method. Thin Solid Films 515:7091–7095

Zhang X, Lei L (2008) One step preparation of visible-light responsive Fe–TiO_2 coating photocatalysts by MOCVD. Mater Lett 62:895–897

Zhang S, Song L (2009) Preparation of visible-light-active carbon and nitrogen codoped titanium dioxide photocatalysts with the assistance of aniline. Catal Commun 10:1725–1729

Zhang Q, Dandeneau CS, Zhou X, Cao G (2009) ZnO nanostructures for dye-sensitized solar cells. Adv Mater 21:4087–4108

Zhang J, Chen X, Takanabe K, Maeda K, Domen K, Epping JD, Fu X, Antonietti M, Wang X (2010) Synthesis of a carbon nitride structure for visible-light catalysis by copolymerization. Angew Chem Int Ed 49:441–444

Zhang N, Yang MQ, Tang ZR, Xu YJ (2013) CdS-Graphene nanocomposites as visible light photocatalysis for redox reaction in water: a green route for selective transformation and environmental remediation. J Catal 303:60–69. (https://doi.org/10.1016/j.jcat.2013.02.026)

Zhang Y, Mao F, Yan H, Liu K, Cao H, Wu J, Xiao D (2015) A polymer-metal-polymer-metal heterostructure for enhanced photocatalytic hydrogen production. J Mater Chem A 3:109–115

Zhao W, Ma WH, Chen CC, Zhao JC, Shuai ZG (2004) Efficient degradation of toxic organic pollutants with $Ni_2O_3/TiO_{2-x}B_x$ under visible irradiation. J Am Chem Soc 16:4782–4783

Zhou L, Tan X, Zhao L, Sun M (2007) Photocatalytic degradation of NO_x over platinum and nitrogen codoped titanium dioxide under visible light irradiation. Collect Czech Chem Commun 72:379–391

Zhou Y, Krumeich F, Heel A, Patzke GR (2010) One-step hydrothermal coating approach to photocatalytically active oxide composites. Dalton Trans 39:6043–6048

Zouzelka R, Kusumawati Y, Remzova M, Rathousky J, Pauporté T (2016) Photocatalytic activity of porous multiwalled carbon nanotube-TiO_2 composite layers for pollutant degradation. J Hazard Mater 317:52–59

Chapter 2
Basic Principles, Mechanism, and Challenges of Photocatalysis

R. Saravanan, Francisco Gracia and A. Stephen

Abstract Photocatalyst is a gifted method which can be used for various purposes like degradation of various organic pollutants in wastewater, production of hydrogen, purification of air, and antibacterial activity. When compared with other methods, photocatalysis is rapidly growing and gaining more attention from the researchers due to its several advantages such as low cost and attractive efficiency. Photocatalysis is a unique process for rectifying energy and environmental issues. In this connection, this chapter deals with basic principles, classification, mechanism, limitations, and operating parameters of photocatalytic processes. Furthermore, the most efficient photocatalytic materials, its mechanism, its challenges, and their solution of rectification were discussed in detail.

Keywords Photocatalyst · Mechanism · Semiconductors · Oxidation · Reduction

Introduction

Removal of pollution is of worldwide importance as such it goes beyond the national borders of each and every nation. In recent times, the essentiality to maintain a good eco-friendly nature has come up with the destruction of various pollutions in the atmosphere due to environmental contamination which occurs in any form of untreated hazards disposal or discharge of material into water, land, or air that causes or may cause acute (short-term) or chronic (long-term) detriment to the Earth's ecological balance or that lowers the quality of life (Chong et al. 2010; Pelaez et al. 2012; Schwarzenbach et al. 2010). The spectrum of pollutants present

R. Saravanan (✉) · F. Gracia
Department of Chemical Engineering and Biotechnology, University of Chile, Beauchef 850, Santiago, Chile
e-mail: saravanan3.raj@gmail.com

A. Stephen
Department of Nuclear Physics, University of Madras, Guindy Campus, Chennai 600 025, India

M.M. Khan et al. (eds.), *Nanocomposites for Visible Light-induced Photocatalysis*, Springer Series on Polymer and Composite Materials,
DOI 10.1007/978-3-319-62446-4_2

in the environment happens as a result of excessive human needs due to over-population, and so the skills related to the technologies have been developed for the researchers being necessary to carry out the pollution rectification processes (Pelaez et al. 2012; Schwarzenbach et al. 2010). A wide range of chemical contaminants flowing freely from industries and agricultural activities have become an important issue that is affecting the ecological safety (Jegannathan and Nielsen 2013; Freeman et al. 1992; Chong et al. 2010).

Water is one of the essential enablers of life on earth. Worldwide, 3.2 million people die per year because of unsafe water, poor sanitation, and inadequate hygiene (Liu et al. 2012). Thus, the quality of water directly affects the life standard of human beings and animals. The main water contamination sources are from industrial discharge of chemicals, agricultural movements, and other environmental changes. In spite of fresh water being a renewable source of energy, the requirement of good water quality is essential and the same is needed for healthy life. Water pollutants may exist in different hazardous wastes like pharmaceutical wastes, pesticides, herbicides, textile dyes, resins, and phenolic compounds (Chong et al. 2010; Konstantinou and Albanis 2004; Zhang and Fang 2010).

In the modern era, water pollution turns out to be the finest topic to discuss because of the depletion of underground water resources and the lack of managing wastewater, which ensures an unsustainable life with safe water. Even, very lesser amounts of water effluents create adverse health effects in humans and other ecosystems. Hence, industrial discharge of wastewater requires proper sewage treatment plants for the essential wastewater management (Akpan and Hameed 2009; Gupta et al. 2012; Konstantinou and Albanis 2004). Wastewater treatment technologies have come up with various efficient methods, but cheaper and less time-consuming method is a major tool to access safe water (Gupta et al. 2012). Initially, this chapter deals with the main source of water pollution and their rectifying methods with merits and demerits. When compared with other methods, photocatalysis is one of the distinctive methods for remedying energy and environmental issues due to several advantages. The main core of this chapter presents the description of the more attractive photocatalytic method and their basic principles, mechanism, limitations, and operating parameters in photocatalytic processes and the challenges of photocatalysis in detail.

Main Source of Water Pollution

Water as an environmental resource is regenerative in the sense that it could absorb the pollution loads up to certain levels without affecting its quality. In fact, there could be a problem of water pollution only if the pollution loads exceed the natural regenerative capacity of a water resource (Schwarzenbach et al. 2010). In the last decades, water pollution has moved to the top of the international political agenda due to its undesirable health and environmental effects. The contaminants released from industries and agricultural activities have become the main source, which affects most of the water bodies in the ecological system (Chong et al. 2010; Gupta et al. 2012). It has already been reported that the major organic compounds responsible for increasing environmental threat are the industrial dyes and textile dyes. About 10–20% of total dye products in the world is lost in textile wastes during the manufacturing process and released as effluents into the green environment (Konstantinou and Albanis 2004; Akpan and Hameed 2009). Organic dyes are one of the leading groups of pollutants released into wastewaters from textile and other industrial processes (Gupta et al. 2012; Chong et al. 2010). On a global scale, over 0.7 million tons of organic synthetic dyes are manufactured every year mainly for use in the textile, leather goods, industrial painting, food, plastics, cosmetics, and consumer electronic sectors (Rajeshwar et al. 2008). There are at least million colored chemical substances that were produced during the last century, out of these about 10,000 were industrially produced (Akpan and Hameed 2009; Konstantinou and Albanis 2004). Major pollutants in textile wastewaters are highly suspended solids, chemical oxygen demand, heat, color, acidity, and other soluble substances (Gupta et al. 2012; Freeman et al. 1992). Out of which, textile industries discharge a large quantity of dyes into water bodies which pose serious ecological problems (Schwarzenbach et al. 2010). Many industries use dyes extensively for various operations such as textile, paper, plastic, leather, tanning, etc. These industries discharge mixing of pollutants in a different process (Gupta et al. 2012). The main pollution in textile wastewater came from dyeing and finishing processes. The textile industry uses approximately 21–377 m^3 of water per ton of textile produced and thus generates large quantities of wastewater from different steps of dyeing and finishing process (Gupta et al. 2012).

Colorants or additive substances causing variation in color or visible light absorption can be divided into two categories: dyes and pigments (Rajeshwar et al. 2008; Pagga and Brown 1986). The distinct delineation between dyes and pigments is as follows: Dyes are soluble or partly soluble organic (carbon-based plant and animal extracted) colored compounds suspended in a medium that represents one type of colorant (Pagga and Brown 1986). The process of dyeing can be loosely defined as imparting color to the textile fiber or leather. On the other hand, typically pigments are complete insoluble substances that have no chemical affinity for the substrate to be colored (Pagga and Brown 1986).

Industrial wastewater is becoming more and more contaminated with the increasing number of industrial chemical products. The wastewater generated by the textile industry is rated as the most polluting source among all industrial sectors. The textile industry utilizes about 10,000 different dyes and pigments and more than 50% of which are azo dyes. Dyes can be classified on the basis of structure, function, or both. Dyes can also be classified as acid, basic, direct, disperse, reactive, anionic, cationic, etc., and indeed this notation is often simultaneously used with the dye chemical structure type (for example Basic Blue 41 and Acid Yellow 23 are both mono azo dyes) of the synthetic dyes manufactured today; azo compounds are considered dominant ($\sim$50–70%) with anthraquinone dyes being a distant second (Akpan and Hameed 2009; Konstantinou and Albanis 2004; Pagga and Brown 1986).

On the other hand, pigments are categorized into two main groups—the organic and other one is inorganic pigments (Rajeshwar et al. 2008; Pagga and Brown 1986). The organic pigments are established by carbon chains and carbon rings. Which are classified into following classes such as azo pigments, polycyclic pigments, anthraquinone pigments, dioxazine pigments, triaryl carbonium pigments, and quinophthalone pigments

Examples:

Artificial—Prussian blue, verdigris.
Animal—Indian yellow, carmine, sepia.
Vegetable—gamboge, sap green, indigo.

For inorganic pigments basically, chemical compounds not based on carbon, are usually metallic salts precipitated from solutions. Inorganic pigments are classified into following: white pigments, black pigments, special pigments, and color pigments

Examples:

Artificial Pigments: aureolin, viridian, cobalt blue
Natural Pigments: ochre, terre verte, ultramarine.

The textile wastewater treatment is a very serious problem due to several reasons which are listed as follows:

- High total dissolved solids (TDS) content of the wastewater,
- Presence of toxic heavy metals such as Cr, As, Cu, Cd, etc.
- Nonbiodegradable nature of organic dyestuffs present in the effluent, and
- The presence of free chlorine and dissolved silica. Most dyes used in the textile industries are stable to light and not biologically degradable. Because dyes usually have a synthetic origin and complex aromatic molecular structures which make them more stable and more difficult to biodegrade (Jegannathan and Nielsen 2013; Gupta et al. 2012). Pagga et al. reported that out of 87% dyestuffs, only 47% are biodegradable. It was estimated that about 12–15% of these dyes are released as effluents during the making processes which cause unfavorable environmental pollution (Pagga and Brown 1986).

Wastewater Treatment Methods

Wastewater treatment and recycling is an essential component and the researchers are coming forward to carry out with convenient economical technologies. The wastewater has been treated by different physical, chemical, and biological processes (Gupta et al. 2012). The major environmental aspect is the removal of color from textile and dyestuff manufacturing wastewater. A large number of conservative treatment processes have been employed in various industrial wastewaters such as chemical, biological, food, pharmaceutical, pulp and paper, dye processing, and textile wastes (Esplugas et al. 2007; Gupta et al. 2012; Moo-Young 2007; Gogate and Pandit 2004). Conventional biological treatment processes are not successful because of the recalcitrant nature of synthetic dyes and the high salinity of wastewater containing dyes (Johnson and Mehrvar 2008; Rajeshwar et al. 2008; Gupta et al. 2012). Chlorination and ozonation are also quite incapable owing to their high operating costs (Rajeshwar et al. 2008; Coleman et al. 2000). The traditional physical techniques such as adsorption on activated carbon, ultrafiltration, reverse osmosis, coagulation by chemical agents, ion exchange on synthetic adsorbent resins, etc., have been used for the removal of dye pollutants (Coleman et al. 2000; Esplugas et al. 2007; Gogate and Pandit 2004; Gupta et al. 2012; Johnson and Mehrvar 2008; Moo-Young 2007). These methods are successful only in transferring organic compounds from water to another phase, thus creating secondary pollution which requires further treatment of solid wastes and regeneration of the adsorbent which makes the process costlier.

The following essential factors must be considered in the wastewater treatment technologies (Gogate and Pandit 2004; Oller et al. 2011; Serpone et al. 2010):

- Treatment flexibility.
- Mineralization of parent and intermediate contaminants.
- The final efficiency of wastewater treatment.
- Recycling capacity and potential use of treated water.
- Cost-effectiveness and eco-friendliness.

Therefore, substantial attention has been focused on complete oxidation of organic compounds to harmless products such as CO_2 and H_2O by the advanced oxidation process (AOP) and appears as one of the most confidential technologies.

AOPs can be classified into two groups:

(1) Non-photochemical AOPs; Non-photochemical AOPs include cavitations, Fenton and Fenton-like processes, ozonation, ozone/hydrogen peroxide, wet air oxidation (Fujishima et al. 2000; Malato et al. 2009; Nakata and Fujishima 2012), etc.,
(2) Photochemical AOPs and photochemical oxidation processes include homogeneous (UV/hydrogen peroxide, UV/ozone, UV/ozonc/hydrogen peroxide, photo-Fenton, homo and heterogeneous (photocatalysis) processes) (Parida and Parija 2006; Rehman et al. 2009; Gupta et al. 2012). The intention of any AOPs design is to produce and use hydroxyl free radical ($OH^{\bullet}$) as a strong oxidant to destroy compounds that cannot be oxidized by the conventional oxidant. Hydroxyl radicals are nonselective in nature and they can react without any other additives with a series of contaminants whose rate constants are usually in the order of 10^6–10^9 mol L^{-1} S^{-1}(Rehman et al. 2009).

The main and short mechanism of AOPs can be explained as follows:

- Initially, the light energy fall on the surface of a catalyst, the valence band electrons are agitated and move to the conduction band. Holes would be left in the valance band of the catalyst. These holes in the valence band can oxidize donor molecules and react with water molecules to generate hydroxyl radicals (The hydroxyl radicals have strong oxidizing power responsible for the degradation of pollutants).
- The oxidative reaction of these radicals with organic compounds in the water producing biodegradable intermediates.
- The reaction of biodegradable intermediates with oxidants is referred to as mineralization (i.e., production of water, carbon dioxide, and inorganic ions).

Discovery of Photocatalysis

In 1972, ***Fujishima*** and ***Honda*** discovered the phenomenon of a photocatalytic splitting of water on a TiO_2 electrode under ultraviolet (UV) light (Fujishima and Honda 1972). This event becomes noticeable as the beginning of a new era in heterogeneous photocatalysis. However, this discovery was not accepted at first by electrochemists because at that time, the idea that light could also be used as an energy source had not yet taken hold among electrochemists who maintained that oxygen could not be generated at such a low voltage, because water electrolysis takes place at 1.5–2 V or even higher. Since then, for understanding the fundamental processes and enhancing the photocatalytic efficiency of TiO_2, extensive research was performed by chemists, physicists, and chemical engineers. Such studies are frequently related to energy renewal and energy storage (Fujishima and Honda 1972; Fujishima et al. 2000). TiO_2-based photocatalysts are considered as an attractive approach growing rapidly for the total destruction of organic compounds in polluted air and wastewater (Fujishima et al. 2000; Fujishima and Honda 1972; Rajeshwar et al. 2008).

Prof. Fujishima stresses that benefits of science and technology should be shared by everyone. He says "***The primary objective of science and technology is to create a society where people can have healthy, comfortable and long lives. The crucial thing in science and technology is to develop a new concept that can be applied to actual products and services, and these new products and services will eventually make people happy***". By making use of his discovery of photocatalyst, he came forward to create such a society (Fujishima et al. 2000).

Definition of Photocatalysis

Photocatalysis is the amalgamation of photochemistry and catalysis. The word ***"photocatalysis"*** is derived from the Greek language and composed of two parts:

- The prefix photo means ***light***
- ***Catalysis*** is the process where a substance involves in altering the rate of a chemical transformation of the reactants without being altered in the end. The substance which is known to be a catalyst increases the rate of the reaction by reducing the activation energy.

Hence, photocatalysis is a process where light and catalysts are concurrently used to support or speed up a chemical reaction. So, photocatalysis can be defined ***as "catalysis driven acceleration of a light-induced reaction."***

Photocatalyst is classified into two categories: homo- and heterogeneous process (Rajeshwar et al. 2008; Rehman et al. 2009; Fujishima et al. 2000).

Homogeneous photocatalytic processes mostly are used with metal complexes as catalysts (transition metals complexes like iron, copper, chromium, etc.). In this process, under the photon and thermal condition, the higher oxidation state of metal ion complexes generated hydroxyl radicals. Then, these hydroxyl radicals react with organic matter which leads destruction of toxic matters. While the comparing homogeneous and heterogeneous; heterogeneous photocatalysis is a technically gifted method which can be used for the degradation of various organic pollutants in wastewater (Rajeshwar et al. 2008; Rehman et al. 2009; Fujishima et al. 2000). This process has several advantages over the competing processes (Fujishima and Honda 1972; Fujishima et al. 2000; Rajeshwar et al. 2008). They are (i) complete mineralization, (ii) no waste disposal problem, (iii) low cost, and (iv) necessity of mild temperature and pressure conditions only.

For example, semiconducting materials (TiO_2, ZnO, SnO_2, and CeO_2) mainly act as heterogeneous photocatalysts, because of its favorable combination of electronic structures which is characterized by a filled valence band and an empty conduction band, light absorption properties, charge transport characteristics and excited states lifetime (Khan et al. 2015b; Rehman et al. 2009; Konstantinou and Albanis 2004; Fujishima et al. 2000; Nakata and Fujishima 2012). An excellent semiconductor photocatalyst should be (i) photoactive, (ii) able to utilize visible and/or near-UV light, (iii) biologically and chemically inert, (iv) photostable (i.e., stability toward photo corrosion), (v) inexpensive, and (vi) nontoxic. Semiconductor photocatalysis emerges to be a promising technology that has a number of applications in environmental systems (Khan et al. 2015b; Rehman et al. 2009; Konstantinou and Albanis 2004; Fujishima et al. 2000; Nakata and Fujishima 2012).

The photocatalyst is an extraordinary method which can be used for various purposes like degradation of various organic pollutants in wastewater, production of hydrogen, purification of air, and antibacterial activity. Recently, the photocatalytic process is attaining more concentration in the field of wastewater treatment to obtain complete mineralization of the pollutant achieved under mild conditions of temperature and pressure. The noteworthy features of these processes include undisposed of waste and cost-effectiveness when sunlight or near-UV light can be used as a source of irradiation. Photocatalyst is a term which means photon assisted generation of catalytically active species. In general, Photocatalysis can be defined as "*a change in the rate of chemical reactions or their generation under the action of light in the presence of substances called photocatalyst that absorbs light quanta and is involved in the chemical transformations of the reactants*" (Hagen 2006).

Photocatalytic Mechanism

Photocatalytic reaction primarily depends on wavelength or light (photon) energy and the catalyst. In general, semiconducting materials are used as a catalyst which performs as sensitizers for the irradiation of light stimulated redox process due to their electronic structure, which is characterized by a filled valence band and a vacant conduction band (Hagen 2006; Khan et al. 2015b). Figure 2.1 shows the schematic representation of semiconductor photocatalytic mechanism.

The fundamental steps in the process of semiconductor photocatalysis are as follows (Khan et al. 2015b; Rajeshwar et al. 2008; Rehman et al. 2009; Fujishima et al. 2000; Hagen 2006):

- When the light energy in terms of photons fall on the surface of a semiconductor and if the energy of incident ray is equivalent or more than the bandgap energy

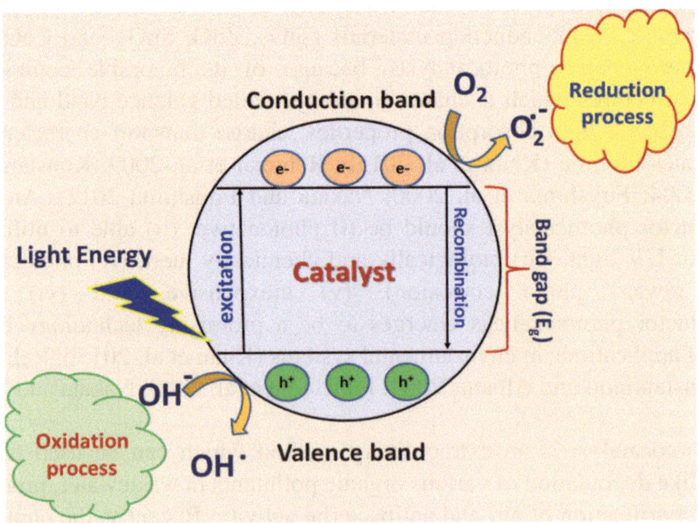

Fig. 2.1 Schematic representation of semiconductor photocatalytic mechanism

of the semiconductor, the valence band electrons are agitated and move to the conduction band of the semiconductor.

- Holes would be left in the valence band of the semiconductor. These holes in the valence band can oxidize donor molecules and react with water molecules to generate hydroxyl radicals (The hydroxyl radicals have strong oxidizing power responsible for the degradation of pollutants).
- The conduction band electrons react with dissolved oxygen species to form superoxide ions. These electrons induce the redox reactions.

These holes and electrons could undergo successive oxidation and reduction reactions with any species, which might be adsorbed on the surface of the semiconductor to give the necessary products.

Description of Oxidation Mechanism

The photocatalyst surface contains water, which is mentioned as "absorbed water." This water is oxidized by positive holes created in the valence band due to the electrons shift to the conduction band as a result of light irradiation, thus making way for the formation of hydroxyl (OH•) radicals (agents which have strong oxidative decomposing power). Afterward, these hydroxyl radicals react with organic matter present in the dyes. If oxygen is present when this process happens, the intermediate radicals in the organic compounds along with the oxygen mole-

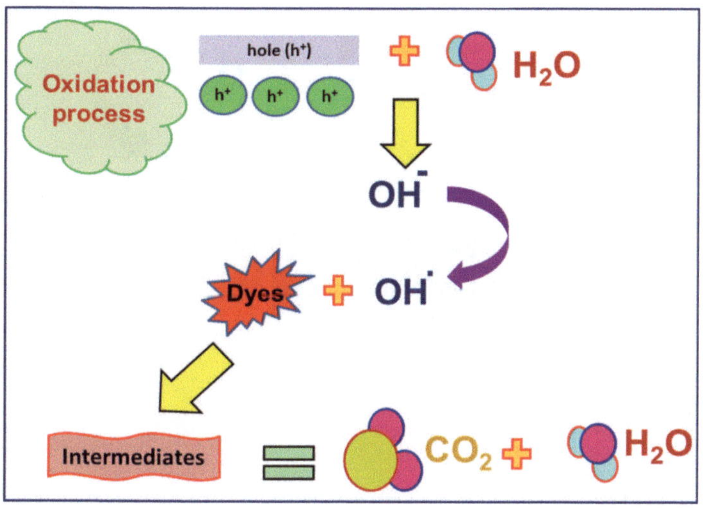

Fig. 2.2 Schematic representation of oxidation mechanism

cules can experience radical chain reactions and consume oxygen in some cases. In such a case, the organic matter finally decomposes ultimately becoming carbon dioxide and water (Khan et al. 2015b; Fujishima and Honda 1972; Rajeshwar et al. 2008). Under such circumstances, organic compounds can react straightly with the positive holes, resulting in oxidative decomposition. The complete oxidation processes were shown in Fig. 2.2.

Description of Reduction Mechanism

Figure 2.3 represents the reduction process, the reduction of oxygen contained in the air occurs as a pairing reaction (Rajeshwar et al. 2008; Rehman et al. 2009). Reduction of oxygen takes place as an alternative to hydrogen generation due to the fact that oxygen is an easily reducible substance. The conduction band electrons react with dissolved oxygen species to form superoxide anions. These superoxide anions attach to the intermediate products in the oxidative reaction, forming peroxide or changing to hydrogen peroxide and then to water. The reduction is likely to occur more easily in organic matter than in water. Therefore, the higher concentration of organic matter tends to increase the number of positive holes. This reduces the carrier recombination and enhances the photocatalytic activity (Nakata and Fujishima 2012; Rajeshwar et al. 2008; Khan et al. 2015b).

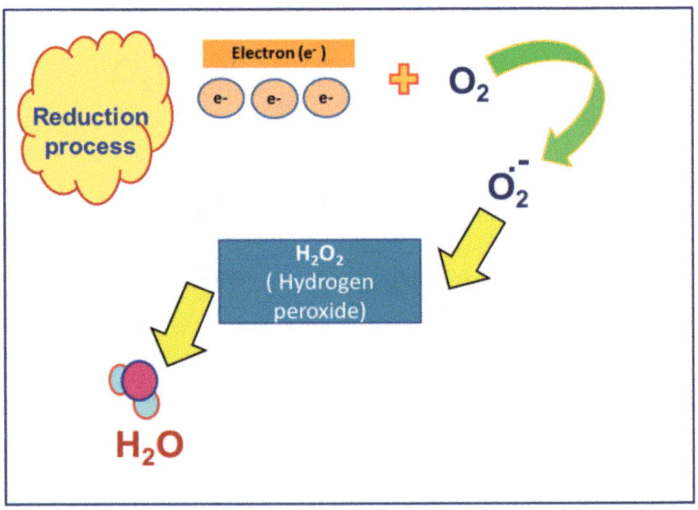

Fig. 2.3 Schematic representation of reduction mechanism

Operating and Affecting Parameters of Photocatalysis

The rate of photo mineralization of an organic compound by photocatalysis method primarily depends on the following parameters: structure, shape, size, and surface area of the catalyst, reaction temperature, pH, light intensity, amount of catalyst, and concentration of wastewater (Saravanan et al. 2011b; Rajeshwar et al. 2008; Rehman et al. 2009; Fujishima et al. 2000; Wang et al. 2007, 2008).

Crystal Structure, Shape, Size, and Surface Area of Catalyst

The structure of catalyst plays a key role in achieving superior photocatalytic activity. For example, TiO_2 material has three phases such as anatase, rutile, and brookite. But the most sensitive and attractive phase is the anatase phase having a notable photocatalytic activity due to its stability, the position of the conduction band, the higher degree of hydroxylation, and adsorption power (Khan et al. 2014b; Gnanasekaran et al. 2015). On the other hand, morphology also acts as a potential factor that influences the final degradation efficiency which was reported earlier (Saravanan et al. 2011b; Wang et al. 2007, 2008). Saravanan et al. reported that spherical-shaped ZnO samples show higher efficiency compared with the spindle-and rod-shaped ZnO samples due to its large surface area (Saravanan et al. 2013a). Nanomaterials having large surface area and smaller size are compared with bulk materials so that it can effectively show higher efficiency in the photocatalytic reaction. When compared with bulk TiO_2, the nanosized TiO_2 material shows more efficient for water purification and recycling ability due to their smaller size (Han et al. 2014; Cernuto et al. 2011). When the size of the catalyst is smaller, a huge number of atoms are accumulated on the surface of a catalyst which leads to increase in surface to volume ratio. This property enhances number of active sites and interfacial charge carrier transfer rates thereby achieving higher catalytic activities (Cernuto et al. 2011). And also it is well known that the photocatalytic redox reaction mainly takes place on the surface of the photocatalysts and so the surface properties significantly influence the efficiency of catalyst (Saravanan et al. 2013a; Khan et al. 2015b).

Effect of Reaction Temperature

A number of researchers have been demonstrated to study the dependence of photocatalytic activity on reaction temperature (Malato et al. 2009; Rajeshwar et al. 2008; Rehman et al. 2009). During the photocatalytic reaction of TiO_2 material, when the temperature is raised above 80 °C, it will enhance the electron–hole recombination and desorption process of adsorbed reactant species, resulting in the decrease of photocatalytic activity. The degradation rate dependency on temperature is reflected by the low activation energy (5–20 kJ mol^{-1}) compared with ordinary thermal reactions. Due to photonic activation, heat is not required for photocatalytic systems and can operate at room temperature. The optimum reaction temperature for photocatalytic activity of TiO_2 material is reported to be in the range of 20–80 °C. When the temperature is at 0 °C, there is an increase in the apparent activation energy (Chatterjee and Dasgupta 2005). This optimum range mainly depends on the activation energy of the material in the photocatalytic reaction (Chatterjee and Dasgupta 2005).

Effect of pH

In the photocatalytic reactions, the pH of the solution is a vital factor, since it makes a clear explanation on the surface charge properties of the photocatalyst. Shourong et al. reported that the change in pH affects the efficiency of degradation of organic pollutants (Castillo-Ledezma et al. 2011; Kazeminezhad and Sadollahkhani 2016; Reza et al. 2015). The photocatalytic degradation of TiO_2 material for Reactive Blue 4 was done under different pH conditions (Neppolian et al. 2002). The results were clearly explained that the lower degradation efficiency in the acidic solutions ($pH < 5$) due to the degradation of the dye is lagging by the high concentration of proton. However, in alkaline medium ($pH > 10$), the existence of hydroxyl ions defuses the acidic end products that are produced by the photodegradation reaction. Furthermore, an unexpected drop of degradation has been detected in the alkaline range (pH 11–13) because of hydroxyl radicals (OH) are rapidly scavenged and does not react with dyes (Neppolian et al. 2002; Reza et al. 2015). The effect of pH on the rate of reaction can be interpreted in terms of electrostatic interactions between charged particles and the contaminants. These influence the adsorption and subsequently the surface properties.

Effect of Light Intensity

The degradation rate of photocatalytic reaction mostly depends upon the light intensity. The quanta of light absorbed by any photocatalyst or reactant are given by the quantum yield which is the ratio of the rate of reaction to the rate of absorption of radiation. The result of photocatalytic reaction responses varied under different wavelengths of the light source (Malato et al. 2009; Reza et al. 2015; Chatterjee and Dasgupta 2005; Akpan and Hameed 2009). The catalyst TiO_2 having a large bandgap (3.2 eV) which absorb mostly in the UV region (Reza et al. 2015). The degradation reaction rate of TiO_2 varies for different intensities of light as follows; the reaction rate increases with increasing light intensity in the range of 0–20 mW/cm^2. Certainly, the reaction rate depends on the square root of the light intensity (half order) above the certain value ($\sim$25 mW/cm^2) of intermediate light intensity. The reaction rate decreases at high-intensity light irradiation due to the favoring of more electron–hole recombination. The excessive light intensity promotes more electron–hole recombination thereby causing decrease in the reaction rate (Malato et al. 2009; Reza et al. 2015; Chatterjee and Dasgupta 2005; Akpan and Hameed 2009).

Effect of Amount of Catalyst

The amount of catalyst (sample) also influences the efficiency of photocatalytic degradation. If there is an increase in the quantity of catalyst, the number of active sites on the semiconductor surface increases moreover, which in turn produces number of $OH^{\bullet}$ and $O_2^{\bullet -}$ radicals (Malato et al. 2009; Rajeshwar et al. 2008). As a result, the photocatalytic degradation rate is increased. Konstantinou et al. elucidated that the degradation rate is directly proportional to the catalyst concentration in any reactor system (Konstantinou and Albanis 2004). However, as the catalyst loading is improved beyond an optimum concentration, the degradation rate is unfavorable because there will be decrease in the light penetration depth into the solution and consequently diminishing of light scattering occurs.

Concentration of Pollutants in Wastewater

Another main factor to find out the degradation rate is the pollutant type and their concentration (Rajeshwar et al. 2008; Rehman et al. 2009; Malato et al. 2009; Chatterjee and Dasgupta 2005). Many researchers have accounted for the photocatalytic activity under similar operating conditions and using similar catalyst, but the variation in the preliminary concentration of water contaminants results with different irradiation time necessary to attain complete mineralization (Chong et al. 2010; Rajeshwar et al. 2008; Rehman et al. 2009; Malato et al. 2009; Chatterjee and Dasgupta 2005). Kiriakidou et al. reported that the TiO_2 material shows different efficiency under similar operating conditions with use of different concentrations of Acid Orange 7 (25–600 mg/L) and the results indicated that complete degradation rate was achieved (25–100 mg/L). After that, the degradation rate diminishes at higher concentration of dye (200–600 mg/L) (Kiriakidou et al. 1999) The above observation has been completely agreed with similar type of several previous reports (Malato et al. 2009; Reza et al. 2015; Chatterjee and Dasgupta 2005; Akpan and Hameed 2009).

Major Advantages of Photocatalysis

The advantages of this photocatalytic technology are as follows (Nakata and Fujishima 2012; Konstantinou and Albanis 2004; Fujishima et al. 2000; Rajeshwar et al. 2008; Rehman et al. 2009):

(i) Photocatalysis offers a good replacement for the energy-intensive conventional treatment methods (adsorption on activated carbon, ultrafiltration, reverse osmosis, coagulation by chemical agents, ion exchange on synthetic adsorbent resins) with the capacity for using renewable and pollution-free solar energy.
(ii) Photocatalysis leads to the formation of harmless products, unlike conventional treatment measures which transfer pollutants from one phase to another.
(iii) The photocatalytic process can be used in the destruction of a variety of hazardous compounds in different wastewater streams.
(iv) The reaction conditions for photocatalysis are mild, the reaction time is modest and a lesser chemical input is required.
(v) Minimal of secondary waste generation and
(vi) It can be applied to hydrogen generation, gaseous phase, and aqueous treatments as well for solid (soil) phase treatments to some extent.

Limitations of Photocatalysis

The photocatalytic activity depends on the following limitations. (Rehman et al. 2009; Rajeshwar et al. 2008; Fujishima et al. 2000):

- interfacial charge transfer
- improve the charge separation and
- inhibition of charge carrier recombination.

These are essential for enhancing the efficiency of the photocatalytic process.

Semiconductor Photocatalyst and Its Challenges

During the last three decades, the researchers have much focusing attention on the reactions that take place on the illuminated surface of semiconductor metal oxides, sulfides, and selenides (Fig. 2.4), which have a modest bandgap energy of 1.1–3.8 eV between their valence and conduction bands (Khan et al. 2015b). The most efficient photocatalytic materials found in the literature are metal oxides such as TiO_2, ZnO, and CeO_2 because metal sulfides and metal selenides are not stable,

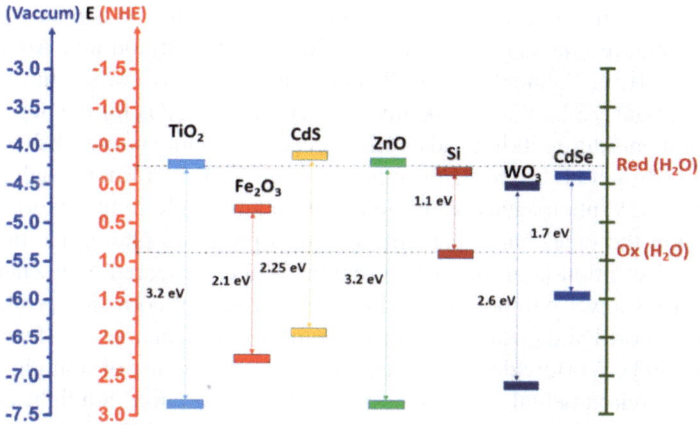

Fig. 2.4 Bandgaps and redox potentials, using the normal hydrogen electrode (NHE) as a reference for several semiconductors

photo anodic corrosive and also toxic (Zhou et al. 2012; Saravanan et al. 2013a; Khan et al. 2014b, c, 2015b).

In general, the photocatalytic reaction mechanism of semiconductors is explained by the following equation based on the earlier reports (Konstantinou and Albanis 2004; Saravanan et al. 2013a).

$$\text{Semiconductor} + \text{Light Energy} \rightsquigarrow \text{Semiconductor}\left(e_{cb}^{-} + h_{vb}^{+}\right) \quad (2.1)$$

$$\text{Dye} + \text{Semiconductor}\left(h_{vb}^{+}\right) \rightarrow \text{Oxidation process} \quad (2.2)$$

$$\text{Semiconductor}\left(h_{vb}^{+}\right) + H_2O \rightarrow \text{Semiconductor} + H^{+} + OH^{\cdot} \quad (2.3)$$

$$\text{Semiconductor}\left(h_{vb}^{+}\right) + OH^{-} \rightarrow \text{Semiconductor} + OH^{\cdot} \quad (2.4)$$

$$\text{Dye} + \text{Semiconductor}\left(e_{cb}^{-}\right) \rightarrow \text{Reduction process} \quad (2.5)$$

$$\text{Semiconductor}\left(e_{cb}^{-}\right) + O_2 \rightarrow \text{Semiconductor} + O_2^{\cdot -} \quad (2.6)$$

$$O_2^{\cdot -} + H^{+} \rightarrow HO_2^{\cdot} \quad (2.7)$$

$$HO_2^{\cdot} + HO_2^{\cdot} \rightarrow H_2O_2 + O_2 \quad (2.8)$$

$$H_2O_2 + O_2^{\cdot -} \rightarrow OH^{\cdot} + OH^{-} + O_2 \quad (2.9)$$

$$\text{Dye} + OH^{\cdot} \rightarrow \text{Degradation products} \quad (2.10)$$

When compared with other metal oxides, titanium dioxide (3.32 eV) is considered better and also a hopeful candidate for the photocatalytic devastation of organic pollutants due to its high quantum efficiency, high stability in aqueous media and nontoxic in nature (Nakata and Fujishima 2012; Konstantinou and Albanis 2004; Khan et al. 2015b; Schneider et al. 2014). ZnO and CeO_2 which have a similar bandgap of about 3.32 eV are sometimes preferred over TiO_2 for the degradation of organic pollutants due to its high adsorption properties (Saravanan et al. 2013c; Zhang et al. 2009; Choi et al. 2016). Unfortunately, both TiO_2 and ZnO which are highly evaluated for UV photocatalysis are inactive under visible light due to their wide bandgaps. On the other hand, hematite is also a preferred photocatalytic material, because its absorption is in the visible region. While compared with ZnO or TiO_2, hematite shows lower photocatalytic efficiency because of corrosion property or the formation of short-lived metal-to-ligand or ligand-to-metal charge transfer states (Fox and Dulay 1993). Worldwide efforts are in progress to make use of sunlight for energy production, environmental protection, and water purification. Sunlight contributes about 5–7% ultraviolet light, 46% visible light, and 47% infrared radiation (Saravanan et al. 2011a; Rehman et al. 2009; Khan et al. 2015b). Hence, inherent are the TiO_2 and ZnO semiconductors which have the inability to make use of the vast potential solar photocatalysis (Rehman et al. 2009). Various technical methods have been employed to make them absorb photons of lower energy as well (Rehman et al. 2009; Nakata and Fujishima 2012). Researchers have been focusing on achieving higher degradation efficiency with these materials, particularly under visible light. In the next section, a brief of explanation about the methods of improving photocatalytic activity is presented.

Methods of Improving Photocatalytic Activity

In order to improve the photocatalytic activity, the way of modifying the surface of the particles is an essential step to prevent the electron–hole recombination via metal and nonmetal doping, coupling with various metal and metal oxides and surface sensitization by the polymer.

Composite System

A mixture of different oxides can diminish the bandgap, expanding the absorbance range to visible light region accordingly to achieve a higher photocatalytic activity (Saravanan et al. 2011a, 2013a, b; Khan et al. 2014d; Khoa et al. 2015). The composite system is based on the principle of dye sensitization due to their smaller (narrow) bandgap that can be used as sensitizer rather than organic dyes. The coupled materials have two different energy level systems which play an important role in accomplishing charge separation (Rehman et al. 2009). Some of the coupled systems such as semiconductor/metal, semiconductor/semiconductor, or semiconductor/polymer were successfully synthesized (Khan et al. 2013, 2015a;

Ansari et al. 2015). Many researchers have prepared the coupled systems that contain Au/TiO_2 (Khan et al. 2014d), ZnO/CdO (Saravanan et al. 2015) and CeO_2/Au (Khan et al. 2014a) PANI/ZnO (Saravanan et al. 2016). The conjugated polymer acts as a sensitizer and this conjugated polymer [poly-fluorine-co-thiophene (PFT)] shows more stability in water when compared with organic dyes, and the reductive potential of PFT is weaker than that of semiconductors (TiO_2 and ZnO) and support injection of its excited electrons to the conduction band of these semiconductors (Qiu et al. 2008; Song et al. 2007). Cun et al. explained that the ZnO/SnO_2 system shows greater photocatalytic activity, because the conduction band of SnO_2 is lower than that of ZnO so that the former can act as a sink for the photogenerated electrons. Since the holes move in the opposite direction from the electrons, photogenerated holes might be caught within the ZnO particle, making charge separation resulting in more degradation rates (Cun et al. 2002).

Metal Ion Dopants

The doping of an appropriate material into a catalyst can enhance the photocatalytic performance. Kanade et al. explained that the synthetic strategy for doping of transition metal ions in semiconductor nanostructures would be useful for the improvement of visible light photocatalysts and photovoltaic devices (Kanade et al. 2007). The transition metal ions (Cu, Co, Mn, and Fe) substitute for Zn ions with tetrahedral O coordination in ZnO lattice and the result gives the narrow bandgap (visible region) due to the spin exchange interactions which leads the enhancement of photocatalytic activity under visible light (Milenova et al. 2014). The co-doped ZnO sample having maximum surface oxygen defects proves degrading more methylene blue under visible light (Xiao et al. 2007). When transition metal incorporated in the particles, the d-electronic configuration of the dopant and its energy level within the TiO_2 lattice also seem to significantly influence the photoactivity (Ekambaram et al. 2007). When vanadium ion is implanted on TiO_2 surface, the absorption edge of the sample is in the visible region and consequently, the enhanced photocatalytic activity is examined at a lower concentration. At higher concentration, the excess vanadium ion wraps up the TiO_2 surface and acts as recombination centers leading to lower visible light activity (Yamashita et al. 2002).

Nonmetal Doping

Doping with nonmetals [B, C, N, and S] in TiO_2 promote the photocatalytic activity in visible light due to the synergetic effect (In et al. 2007; Sakthivel and Kisch 2003). Shifu et al. explained that the photocatalytic activity of N doped ZnO sample is higher than that of pure ZnO because conductivity conversion of the samples from zinc oxide to nitrogen leads red shift and subsequently improves the photocatalytic activity under visible light (Shifu et al. 2009). The degradation of methyl orange using N–S co-doped TiO_2 catalyst shows improved visible light activity due to excess of oxygen vacancies (Wei et al. 2008). The B and N co-doping with TiO_2 samples shows visible light absorption since the synergetic effect modifying the electronic structure of TiO_2 was reported (Ling et al. 2008).

Dye Sensitization

Different dyes such as acid red 44, eosin-Y, merbromine, rhodamine B, and rhodamine 6G, 8-hydroxyquinoline have been used to sensitize TiO_2 particles under visible light (Moon et al. 2003; Abe et al. 2000; Rehman et al. 2009). When there is visible light illumination, the dye molecules are excited and assist in promoting electrons, these excited electrons shift into the conduction band of the semiconductor. These reactions make number of holes and electrons which are capable of the efficient decomposition of organic pollutants through oxidation and reduction reactions (Rehman et al. 2009).

Summary

In this chapter, we concluded that the basic theory of photocatalysis, mechanism, and its advantages clearly indicated that photocatalysis is the simply powerful emerging and promising technology that holds a number of applications in environmental systems which are effectively utilized for the industrial applications including wastewater treatment, hydrogen generation, air purification, antibacterial activity, and so on. Photocatalysts have several advantages, but however an ideal photocatalyst should be inexpensive, nontoxic, long-term stability, easily reproducible on separation, and also has a highly effective photocatalytic activity.

References

Abe R, Hara K, Sayama K, Domen K, Arakawa H (2000) Steady hydrogen evolution from water on Eosin Y-fixed TiO_2 photocatalyst using a silane-coupling reagent under visible light irradiation. J Photochem Photobiol A 137(1):63–69

Akpan UG, Hameed BH (2009) Parameters affecting the photocatalytic degradation of dyes using TiO_2-based photocatalysts: a review. J Hazard Mater 170(2–3):520–529

Ansari MO, Khan MM, Ansari SA, Cho MH (2015) Electrically conductive polyaniline sensitized defective-TiO_2 for improved visible light photocatalytic and photoelectrochemical performance: a synergistic effect. New J Chem 39(11):8381–8388

Castillo-Ledezma JH, Sánchez Salas JL, López-Malo A, Bandala ER (2011) Effect of pH, solar irradiation, and semiconductor concentration on the photocatalytic disinfection of Escherichia coli in water using nitrogen-doped TiO_2. Eur Food Res Technol 233(5):825–834

Cernuto G, Masciocchi N, Cervellino A, Colonna GM, Guagliardi A (2011) Size and shape dependence of the photocatalytic activity of TiO_2 nanocrystals: a total scattering Debye function study. J Am Chem Soc 133(9):3114–3119

Chatterjee D, Dasgupta S (2005) Visible light induced photocatalytic degradation of organic pollutants. J Photochem Photobiol C 6(2–3):186–205

Choi YI, Lee S, Kim SK, Kim Y, Cho DW, Khan MM, Sohn Y (2016) Fabrication of ZnO, ZnS, Ag-ZnS, and Au-ZnS microspheres for photocatalytic activities, CO oxidation and 2-hydroxyterephthalic acid synthesis. J Alloy Compd 675:46–56

Chong MN, Jin B, Chow CWK, Saint C (2010) Recent developments in photocatalytic water treatment technology: a review. Water Res 44(10):2997–3027

Coleman HM, Eggins BR, Byrne J, Palmer FL, King E (2000) Photocatalytic degradation of 17-β-oestradiol on immobilised TiO_2. Appl Catal B 24(1):L1–L5

Cun W, Jincai Z, Xinming W, Bixian M, Guoying S, Ping'an P, Jiamo F (2002) Preparation, characterization and photocatalytic activity of nano-sized ZnO/SnO_2 coupled photocatalysts. Appl Catal B 39(3):269–279

Ekambaram S, Iikubo Y, Kudo A (2007) Combustion synthesis and photocatalytic properties of transition metal-incorporated ZnO. J Alloy Compd 433(1–2):237–240

Esplugas S, Bila DM, Krause LGT, Dezotti M (2007) Ozonation and advanced oxidation technologies to remove endocrine disrupting chemicals (EDCs) and pharmaceuticals and personal care products (PPCPs) in water effluents. J Hazard Mater 149(3):631–642

Fox MA, Dulay MT (1993) Heterogeneous photocatalysis. Chem Rev 93(1):341–357

Freeman H, Harten T, Springer J, Randall P, Curran MA, Stone K (1992) Industrial pollution prevention!: a critical review. J Air Waste Manage Assoc 42(5):618–656

Fujishima A, Honda K (1972) Electrochemical photolysis of water at a semiconductor electrode. Nature 238(5358):37–38

Fujishima A, Rao TN, Tryk DA (2000) Titanium dioxide photocatalysis. J Photochem Photobiol C 1(1):1–21

Gnanasekaran L, Hemamalini R, Ravichandran K (2015) Synthesis and characterization of TiO_2 quantum dots for photocatalytic application. J Saudi Chem Soc 19(5):589–594

Gogate PR, Pandit AB (2004) A review of imperative technologies for wastewater treatment II: hybrid methods. Adv Environ Res 8(3–4):553–597

Gupta VK, Ali I, Saleh TA, Nayak A, Agarwal S (2012) Chemical treatment technologies for waste-water recycling—an overview. RSC Adv 2(16):6380

Hagen J (2006) Industrial catalysis: a practical approach/Jens Hagen, 2nd edn. Wiley, Weinheim

Han G, Wang L, Pei C, Shi R, Liu B, Zhao H, Yang H, Liu S (2014) Size-dependent optical properties and enhanced visible light photocatalytic activity of wurtzite CdSe hexagonal nanoflakes with dominant 001 facets. J Alloy Compd 610:62–68

In S, Orlov A, Berg R, Garcia F, Pedrosa-Jimenez S, Tikhov MS, Wright DS, Lambert RM (2007) Effective visible light-activated B-doped and B, N-codoped TiO_2 photocatalysts. J Am Chem Soc 129(45):13790–13791

Jegannathan KR, Nielsen PH (2013) Environmental assessment of enzyme use in industrial production—a literature review. J Clean Prod 42:228–240

Johnson MB, Mehrvar M (2008) Aqueous metronidazole degradation by UV/H_2O_2 process in single-and multi-lamp tubular photoreactors: kinetics and reactor design. Ind Eng Chem Res 47 (17):6525–6537

Kanade KG, Kale BB, Baeg JO, Lee SM, Lee CW, Moon SJ, Chang H (2007) Self-assembled aligned Cu doped ZnO nanoparticles for photocatalytic hydrogen production under visible light irradiation. Mater Chem Phys 102:98–104

Kazeminezhad I, Sadollahkhani A (2016) Influence of pH on the photocatalytic activity of ZnO nanoparticles. J Mater Sci Mater Electron 27(5):4206–4215

Khan ME, Khan MM, Cho MH (2015a) Biogenic synthesis of a Ag–graphene nanocomposite with efficient photocatalytic degradation, electrical conductivity and photoelectrochemical performance. New J Chem 39(10):8121–8129
Khan MM, Adil SF, Al-Mayouf A (2015b) Metal oxides as photocatalysts. J Saudi Chem Soc 19 (5):462–464
Khan MM, Ansari SA, Amal MI, Lee J, Cho MH (2013) Highly visible light active Ag@TiO_2 nanocomposites synthesized using an electrochemically active biofilm: a novel biogenic approach. Nanoscale 5(10):4427–4435
Khan MM, Ansari SA, Ansari MO, Min BK, Lee J, Cho MH (2014a) Biogenic fabrication of Au@CeO_2 nanocomposite with enhanced visible light activity. J Phys Chem C 118(18):9477–9484
Khan MM, Ansari SA, Pradhan D, Ansari MO, Han DH, Lee J, Cho MH (2014b) Band gap engineered TiO_2 nanoparticles for visible light induced photoelectrochemical and photocatalytic studies. J Mater Chem A 2(3):637–644
Khan MM, Ansari SA, Pradhan D, Han DH, Lee J, Cho MH (2014c) Defect-induced band gap narrowed CeO_2 nanostructures for visible light activities. Ind Eng Chem Res 53(23):9754–9763
Khan MM, Lee J, Cho MH (2014d) Au@TiO_2 nanocomposites for the catalytic degradation of methyl orange and methylene blue: an electron relay effect. J Ind Eng Chem 20(4):1584–1590
Khoa NT, Kim SW, Yoo D, Cho S, Kim EJ, Hahn SH (2015) Fabrication of Au/graphene-wrapped ZnO-nanoparticle-assembled hollow spheres with effective photoinduced charge transfer for photocatalysis. ACS Appl Mater Interfaces 7(6):3524–3531
Kiriakidou F, Kondarides DI, Verykios XE (1999) The effect of operational parameters and TiO_2-doping on the photocatalytic degradation of azo-dyes. Catal Today 54:119–130
Konstantinou IK, Albanis TA (2004) TiO_2-assisted photocatalytic degradation of azo dyes in aqueous solution: kinetic and mechanistic investigations. Appl Catal B 49(1):1–14
Ling Q, Sun J, Zhou Q (2008) Preparation and characterization of visible-light-driven titania photocatalyst co-doped with boron and nitrogen. Appl Surf Sci 254(10):3236–3241
Liu L, Johnson HL, Cousens S, Perin J, Scott S, Lawn JE, Rudan I, Campbell H, Cibulskis R, Li M, Mathers C, Black RE (2012) Global, regional, and national causes of child mortality: an updated systematic analysis for 2010 with time trends since 2000. The Lancet 379 (9832):2151–2161
Malato S, Fernández-Ibáñez P, Maldonado MI, Blanco J, Gernjak W (2009) Decontamination and disinfection of water by solar photocatalysis: recent overview and trends. Catal Today 147(1):1–59
Milenova K, Avramova I, Eliyas A, Blaskov V, Stambolova I, Kassabova N (2014) Application of activated M/ZnO (M = Mn Co, Ni, Cu, Ag) in photocatalytic degradation of diazo textile coloring dye. Environ Sci Pollut Res Int 21(21):12249–12256
Moon J, Yun CY, Chung K, Kang M, Yi J (2003) Photocatalytic activation of TiO_2 under visible light using Acid Red 44. Catal Today 87(1–4):77–86
Moo-Young HK (2007) Pulp and paper effluent management. Water Environ Res 79(10):1733–1741
Nakata K, Fujishima A (2012) TiO_2 photocatalysis: design and applications. J Photochem Photobiol C 13(3):169–189
Neppolian B, Choi HS, Sakthivel S, Arabindoo B, Murugesan V (2002) Solar light induced and TiO_2 assisted degradation of textile dye reactive blue 4. Chemosphere 46:1173–1181
Oller I, Malato S, Sanchez-Perez JA (2011) Combination of advanced oxidation processes and biological treatments for wastewater decontamination—a review. Sci Total Environ 409 (20):4141–4166
Pagga U, Brown D (1986) The degradation of dyestuffs: part II behaviour of dyestuffs in aerobic biodegradation tests. Chemosphere 15(4):479–491
Parida KM, Parija S (2006) Photocatalytic degradation of phenol under solar radiation using microwave irradiated zinc oxide. Sol Energy 80(8):1048–1054

Pelaez M, Nolan NT, Pillai SC, Seery MK, Falaras P, Kontos AG, Dunlop PS, Hamilton JW, Byrne J, O'Shea K, Entezari MH, Dionysiou DD (2012) A review on the visible light active titanium dioxide photocatalysts for environmental applications. Appl Catal B 125:331–349
Qiu R, Zhang D, Mo Y, Song L, Brewer E, Huang X, Xiong Y (2008) Photocatalytic activity of polymer-modified ZnO under visible light irradiation. J Hazard Mater 156(1–3):80–85
Rajeshwar K, Osugi ME, Chanmanee W, Chenthamarakshan CR, Zanoni M, Kajitvichyanukul P, Krishnan-Ayer R (2008) Heterogeneous photocatalytic treatment of organic dyes in air and aqueous media. J Photochem Photobiol C 9(4):171–192
Rehman S, Ullah R, Butt AM, Gohar ND (2009) Strategies of making TiO_2 and ZnO visible light active. J Hazard Mater 170(2–3):560–569
Reza KM, Kurny AS, Gulshan F (2015) Parameters affecting the photocatalytic degradation of dyes using TiO_2: a review. Appl Water Sci. doi:10.1007/s13201-015-0367-y
Sakthivel S, Kisch H (2003) Daylight photocatalysis by carbon-modified titanium dioxide. Angew Chem Int Ed Engl 42(40):4908–4911
Saravanan R, Gracia F, Khan MM, Poornima V, Gupta VK, Narayanan V, Stephen A (2015) ZnO/CdO nanocomposites for textile effluent degradation and electrochemical detection. J Mol Liq 209:374–380
Saravanan R, Gupta VK, Narayanan V, Stephen A (2013a) Comparative study on photocatalytic activity of ZnO prepared by different methods. J Mol Liq 181:133–141
Saravanan R, Karthikeyan S, Gupta VK, Sekaran G, Narayanan V, Stephen A (2013b) Enhanced photocatalytic activity of ZnO/CuO nanocomposite for the degradation of textile dye on visible light illumination. Mater Sci Eng C Mater Biol Appl 33(1):91–98
Saravanan R, Sacari E, Gracia F, Khan MM, Mosquera E, Gupta VK (2016) Conducting PANI stimulated ZnO system for visible light photocatalytic degradation of coloured dyes. J Mol Liq 221:1029–1033
Saravanan R, SHANKAR H, Prakash T, Narayanan V, Stephen A (2011a) ZnO/CdO composite nanorods for photocatalytic degradation of methylene blue under visible light. Mater Chem Phys 125(1–2):277–280
Saravanan R, Shankar H, Rajasudha G, Stephen A, Narayanan V (2011b) Photocatalytic degradation of organic dye using nano ZnO. Int J Nanosci 10(01–02):253–257
Saravanan R, Thirumal E, Gupta VK, Narayanan V, Stephen A (2013c) The photocatalytic activity of ZnO prepared by simple thermal decomposition method at various temperatures. J Mol Liq 177:394–401
Schneider J, Matsuoka M, Takeuchi M, Zhang J, Horiuchi Y, Anpo M, Bahnemann DW (2014) Understanding TiO_2 photocatalysis: mechanisms and materials. Chem Rev 114(19):9919–9986
Schwarzenbach RP, Egli T, Hofstetter TB, von Gunten U, Wehrli B (2010) Global water pollution and human health. Annu Rev Environ Resour 35(1):109–136
Serpone N, Horikoshi S, Emeline AV (2010) Microwaves in advanced oxidation processes for environmental applications. A brief review. J Photochem Photobiol C 11(2–3):114–131
Shifu C, Wei Z, Sujuan Z, Wei L (2009) Preparation, characterization and photocatalytic activity of N-containing ZnO powder. Chem Eng J 148(2–3):263–269
Song L, Qiu R, Mo Y, Zhang D, Wei H, Xiong Y (2007) Photodegradation of phenol in a polymer-modified TiO_2 semiconductor particulate system under the irradiation of visible light. Catal Commun 8(3):429–433
Wang H, Xie C, Zhang W, Cai S, Yang Z, Gui Y (2007) Comparison of dye degradation efficiency using ZnO powders with various size scales. J Hazard Mater 141(3):645–652
Wang Y, Li X, Lu G, Chen G, Chen Y (2008) Synthesis and photo-catalytic degradation property of nanostructured-ZnO with different morphology. Mater Lett 62(15):2359–2362
Wei F, Ni L, Cui P (2008) Preparation and characterization of N-S-codoped TiO_2 photocatalyst and its photocatalytic activity. J Hazard Mater 156(1–3):135–140
Xiao Q, Zhang J, Xiao C, Tan X (2007) Photocatalytic decolorization of methylene blue over $Zn_{1-x}Co_xO$ under visible light irradiation. Mater Sci Eng B 142(2–3):121–125

Yamashita H, Harada M, Misaka J, Takeuchi M, Ikeue K, Anpo M (2002) Degradation of propanol diluted in water under visible light irradiation using metal ion-implanted titanium dioxide photocatalysts. J Photochem Photobiol A 148(1–3):257–261

Zhang L, Fang M (2010) Nanomaterials in pollution trace detection and environmental improvement. Nano Today 5(2):128–142

Zhang L, Yang H, Xie X, Zhang F, Li L (2009) Preparation and photocatalytic activity of hollow ZnSe microspheres via Ostwald ripening. J Alloy Compd 473(1–2):65–70

Zhou H, Qu Y, Zeid T, Duan X (2012) Towards highly efficient photocatalysts using semiconductor nanoarchitectures. Energy Environ Sci 5(5):6732

Chapter 3
Nanocomposites and Its Importance in Photocatalysis

Hossam Eldin Abdel Fattah Ahmed Hamed El Nazer and Samir Tawfik Gaballah

Abstract Photocatalysis is a promising technique for solving the worldwide energy and environmental crisis. The key challenge in this technique is to develop efficient photocatalysts that have to satisfy several criteria such as high chemical and photochemical stability as well as effective charge separation and strong light absorption. Synthesis of semiconducting nanocomposites is considered to be a promising way to achieve efficient photocatalysts. This improved photocatalytic activity of the nanocomposite photocatalysts is attributed to the enhancement of the charge separation, irradiation absorption, and photo and chemical stability. This chapter summarizes many research studies on semiconducting nanocomposites for different photocatalytic applications. Different consistencies for photocatalytic organic transformations have been discussed herein.

Keywords Photocatalysis · Nanocomposites · Semiconducting · Selective organic transformation

Introduction

The photocatalysis researches have progressed rapidly in the recent years. This is because of its applications in several fields such as medicine, cosmetics, agricultural, electronics, coatings, plastics, etc. The semiconductor nanoparticles are considered to be important materials for many applications, i.e., optoelectronic devices, photonic transducers, and photoluminescent tags in biological studies.

The photocatalytic mechanism follows the principle that the electrons at the valence band of semiconducting photocatalyst transfer to the conduction band by absorbing photons when the irradiation energy is equal to or greater than the band gap

H.E.A.F.A.H. El Nazer (✉) · S.T. Gaballah
Chemical Industries Research Division, Photochemistry Department, National Research Centre, 33 El Bohouth St. (Former El-Tahrir st.), Dokki, 12622 Giza, Egypt
e-mail: dr_hosamnazer@yahoo.com; chestg@yahoo.com

M.M. Khan et al. (eds.), *Nanocomposites for Visible Light-induced Photocatalysis*, Springer Series on Polymer and Composite Materials,
DOI 10.1007/978-3-319-62446-4_3

of the semiconductor and hence leave holes in the valence band. Photocatalysts exhibit different band gaps and band positions, so the photogenerated electron–hole pairs have different oxidation and reduction potentials. The recombination of electron–hole pairs would occur when the charge carrier (electron and hole) migrates to the surface of the photocatalyst that may decrease the photocatalytic efficiency. Several methods have been suggested for enhancing the separation of electron–hole pairs as well as increasing the photocatalytic reaction rate. These methods include fabricating nanocomposites of photocatalyst with metals (Kraeutler and Bard 1978; Lee and Choi 2005) and/or other electroconductive/photoactive materials (Xiang et al. 2012; Yang et al. 2014).

Since the dawn of history, humans exploited the available natural resources to build civilizations. They also tried to mimic the nature and other creatures in many ways. For instance, they mimicked birds via inventing the airplanes to overcome the gravity of the Earth and flew across the globe. They also mimicked fish by inventing watercrafts such as ships, submarines, boats, etc. Similarly, they tried to mimic the plants after the scientists solved the mechanism of photosynthesis in which the plant converts water and carbon dioxide (reactants) into carbohydrates (products) in the presence of chlorophyll (photocatalyst) using the energy of sunlight.

Organic chemists have been always interested in synthesizing novel organic compounds using thermal energy. Recently, they incorporated organic chemical reactions that proceed in the presence of light (photochemical reactions). During the 50s and 60s of the twentieth century, photochemistry has become an appreciated tool for organic synthesis

Heterogeneous photocatalysis regime has been mainly focusing on the utilization of the semiconductor for the photodegradation of water and air pollutants. Recently, extensive researches on the preparation of new photoactive semiconductors have produced a considerable number of new nanocomposite systems which have a wide range of applications in many areas such as organic transformation, hydrogen production via water splitting, and reduction of carbon dioxide to produce methane. The work on enhancing the photocatalysis potency has shown that the deposition of a noble metal on the semiconductor particle positively affects the electron transfer processes at the semiconductor interface and hence it greatly enhances the photocatalytic efficiency (Greeley et al. 2006; Kudo and Miseki 2009).

Semiconducting Nanocomposites as Efficient Photocatalysts

Photocatalyst performance mainly depends on how efficiently it separates photogenerated electrons and holes. Photocatalytic activity is traditionally improved by depositing noble metals (e.g., Pt, Ag, Pd, Au) or nonmetal anions, or by doping with metal cations (Su et al. 2014; Xing et al. 2013; Yu et al. 2005, 2010, 2011, 2013a). In a noble metal composite system, photogenerated electrons accumulate on the metal, and holes remain on the photocatalyst surface, suppressing the recombination of charge carriers. Doping with nonmetal anions (e.g., C, N, S, F) (Barolo et al. 2012; Yu et al. 2012a; Yu and Yu (2009); Yu et al. 2010) could also slow the recombination of photogenerated electrons and holes or extend the light absorption of titania into the visible region.

The formation of a well-defined heterojunction between two semiconductors with matching electronic band structures may also suppress the recombination of photogenerated electrons and holes (Heremans et al. 2009; Yu et al. 2014; Zhang et al. 2008a).

Figure 3.1 illustrates how p-n heterojunctions enhance the separation efficiency of electron-hole pairs. The p-n heterojunction is a junction between two semiconductors, one doped with a donor (n-type) and one with an acceptor (p-type). A strong local electric field exists near the junction, pointing from n toward p, because of the juxtaposition of high concentrations of negatively and positively charged ions. The difference of the electric potential in the electric field can enhance the separation of photogenerated electrons and holes, increasing the quantum efficiency of the photocatalytic reactions. Other types of heterojunctions could be used for photocatalysis. For example, heterojunctions between two semiconductors with matching electronic band structures could similarly enhance the separation of photogenerated electron-hole pairs. Thus, engineering the junction between semiconductors is essential for improving photocatalytic activity.

Titanium dioxide (TiO_2) has dominated the field of photocatalysis in terms of research, characterization, and applications. The reason for TiO_2's widespread use comes from its moderate band gap, non-toxicity, high surface area, low cost, recyclability, high photoactivity, wide range of processing procedures, and its excellent chemical and photochemical stability. It should also be noted that TiO_2 is ranked as one of the top 50 most available materials. With a band gap of 3.2 eV, a photon would need a wavelength equal to or shorter than 385 nm to electronically excite this semiconductor, meaning that it needs UV energy. TiO_2's band gap, although favorable for UV photocatalysis, subjects TiO_2 to low efficiency yields in solar applications (its largest potential market) since less than 5% of the sun's

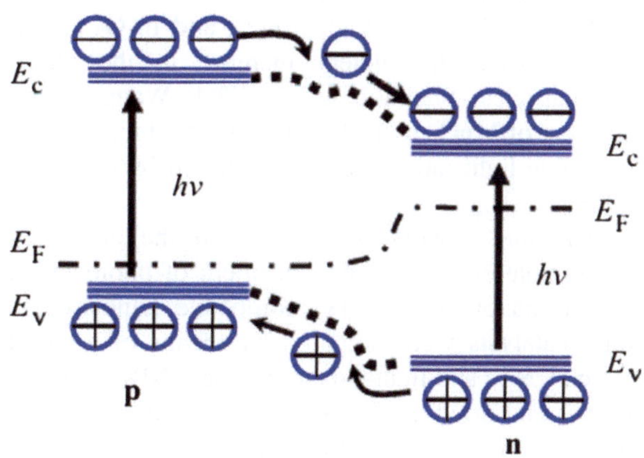

Fig. 3.1 The schematic diagram represents the role of the p-n heterojunction in enhancing the separation of electron-hole pairs

energy is emitted at wavelengths below 385 nm. Therefore, though the anatase form of titanium dioxide (TiO_2) is considered an ideal photocatalyst for UV applications, in its unmodified form it is rendered highly inefficient for visible light applications. Silver (Ag) has been deposited onto TiO_2, primarily not only because it is more cost-effective than gold (Au) and platinum (Pt), but also because it has an intrinsic ability to prevent bacterial growth, as well as an effective photocatalytic ability at the nanoscale (Cozzoli et al. 2004). Previously, Ag has been added to TiO_2 nanoparticles, TiO_2 nanorods, and TiO_2 nanofilms. In fact, studies conducted by Li and colleagues have shown that Ag-deposited TiO_2 anatase nanoparticles have an improved photoresponse compared to that of anatase TiO_2 nanoparticles, Degussa P25 TiO_2 nanoparticles, and mixed anatase—rutile TiO_2 nanoparticles (Li and Gray 2007). The use of Ag–TiO_2 nanocomposite films has also been shown to have an increased photocatalytic reactivity compared to the unmodified material. For example, UV-illuminated Ag–TiO_2 nanocomposite films are 6.3-fold more effective than UV-illuminated pure TiO_2 films for photodegrading methyl orange (Yu et al. 2005).

The deposition of Au and Pt onto TiO_2 nanoparticles has also demonstrated an increase in the photocatalytic reactivity of TiO_2. Yu and coworkers have reported an improved photocatalytic reactivity of Au–TiO_2 nanocomposite microspheres compared to TiO_2 microspheres and Degussa P25 TiO_2 nanoparticles (Yu et al. 2009). In addition, UV-illuminated TiO_2 nanofilms embedded with Au nanostructures have a better photonic efficiency than UV-illuminated pure TiO_2 films (Bannat et al. 2009).

The coupling of TiO_2 to a narrow-gap semiconductor material can result in an increase in photocatalytic reactivity, as well as an increase in photoresponse. When a narrow-gap semiconductor coupled to a TiO_2 nanoparticle is exposed to visible light, it produces reactive electrons that can travel through the semiconductor to the nonactivated TiO_2 nanoparticle. This process extends the photoresponse of the TiO_2 to visible light wavelengths. Coupling TiO_2 to a semiconductor also decreases charge recombination, because the heterojunction space between the two semiconductors allows for a more efficient separation of reactive electrons and electropositive holes (Zhu et al. 2008).

The metal oxides ZnO, MnO_2, and In_2O_3 coupled to TiO_2 nanoparticles have also showed an efficient increase in the photocatalytic capability of TiO_2 nanoparticles (Shchukin et al. 2004; Wang et al. 2009; Xue et al. 2008). The addition of MnO_2 to TiO_2 nanoparticles broadened the excitation spectrum of TiO_2 to visible light ranges, as demonstrated by methylene blue degradation (Xue et al. 2008).

Yao and coworkers investigated the efficiency of the reduction of charge recombination and the enhancement of photocatalytic activity by anatase TiO_2—carbon nanotubes (CNTs) composite nanostructures (Yao et al. 2008). These photocatalysts were prepared by means of a simple low-temperature process in which CNTs and titania nanoparticles (NPs) were dispersed in water, dehydrated at 80 °C, and dried at 104 °C. Charge recombination was investigated by measuring photoluminescence spectra of selected composite.

Following mechanism for the enhanced photocatalysis of single-wall CNT (SWCNT)/TiO_2 composite was proposed: each anatase NP is in intimate contact with SWCNTs. Under UV-visible irradiation the electrons are excited from valence band to conduction band of anatase, creating holes in the valence band. In the absence of SWCNTs most of these charges quickly recombine. When SWCNTs are attached to the surface of anatase, the relative position of SWCNT conduction band edge permits the transfer of electrons from anatase surface into SWCNT, allowing charge separation, stabilization, and hindering recombination. The longer life of holes in the valence band of anatase accounts for the higher photocatalytic activity. Although the multi-wall CNT (MWCNT)/TiO_2 composites behave similarly, they do not enhance the photocatalytic activity to the same extent as the SWCNT/TiO_2 composites do, because there are much less individual contacts between MWCNT and anatase surface.

The mechanism of the increase of photocatalytic activity when graphene was used to form a nanocomposite together with an oxide semiconductor such as TiO_2 or the titania/silica composite: graphene transferred or/and trapped electrons photogenerated in the oxide semiconductor, leaving the holes to form the reactive species (Fig. 3.2). Therefore the charge recombination was suppressed, leading to the improvement of the photocatalytic performance. Nanocomposites comprising titania and reduced graphene oxide (rGO) were prepared by (Sher Shah et al. 2012) by a simple one-step hydrothermal reactions using titania precursor, $TiCl_4$, and graphene oxide (GO) without reducing agents. Hydrolysis of Ti–Cl_4 and mild reduction of GO were simultaneously carried out under hydrothermal conditions.

Graphene oxide was prepared from graphite powder using a modified Hummer's method. Composites of TiO_2-reduced graphene oxide (rGO) were synthesized by simultaneously carrying out the reduction of GO, hydrolysis of $TiCl_4$, and crystallization of produced TiO_2 in a single-step hydrothermal reaction. The photocatalytic activity of the as-prepared composite catalyst shows good activity towards the photodegradation of a rhodamine B (RhB) solution under the irradiation by visible light at ambient temperature (Stankovich et al. 2007).

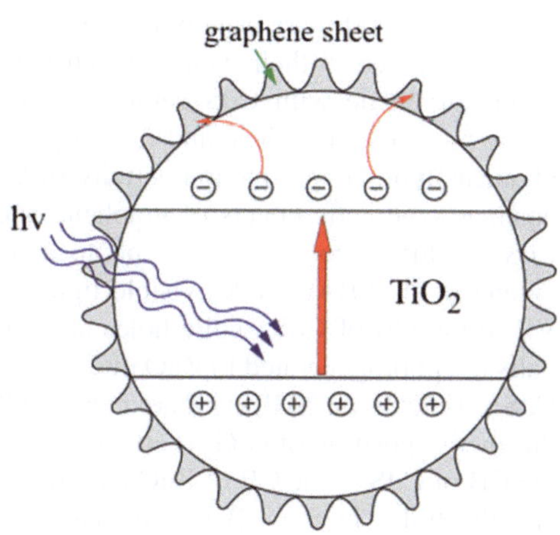

Fig. 3.2 Schematic diagram for the charge transfer and separation in the TiO_2–graphene composites under UV-visible light irradiation

Copper oxide nanoparticles based on the silica matrix have been synthesized. The exact composition of structures depends on the copper source, the molar ratio of the components, and the type of the precursor. In a set of synthesized samples, Cu incorporated into gel skeleton, probably because polyethylene glycol (PEG) may interact with both silica and Cu^{2+} cations which inhibit the movement of Cu^{2+} cations from the gel skeleton to the outer solution. The result shows that Cu^{2+} cations in complex form could interact with silica gel network in the absence of PEG. Thus, copper is entrapped as small particles in the gel skeleton. Although the interaction between copper nanocomplex and silica gel network has not been made clear, the existence of PEG may affect the interaction between copper complex and silica gel network. The photocatalytic studies on copper nanocomplex synthesized in the absence of PEG using copper oxide nanoparticles showed the best results (Kakhki et al. 2016).

Co-precipitation method was used to synthesize ZnO, ZnO–GO, Nd–ZnO, and Nd–ZnO–GO nanocomposites (with variable concentrations of Nd). XRD and TEM results showed that the nanocomposites were crystalline and consisted of mainly wurtzite hexagonal phase. The UV-vis spectra showed that the dopants (GO and Nd) in the composite caused an increase in the absorption edge of ZnO to shift to lower energy region due to a significant reduction in the band gap. The nanocomposite [Nd–ZnO–GO (0.3% Nd)] showed higher photocatalytic efficiency compared to the bare ZnO. Results from dopant levels in the composite showed that prevention of electron–hole pair's recombination depends on its suitable amount, as higher amounts become recombination centers and inhibit photocatalytic activity. The combined effect of Nd and GO led to the separation of electron–hole pairs leading to high degradation of indigo carmine (IC) under the simulated solar light irradiation. The photocatalytic degradation of organic pollutants shows a considerable reduction in the presence of radical scavengers confirming the effect of hydroxyl and superoxide radicals as highly reactive species. A high degree of complete mineralization of organics that decreases the formation of possible toxic degradation products was obtained by TOC analysis (Oppong et al. 2016).

Specific semiconductors such as TiO_2 (Qi et al. 2011; Yu et al. 2013c) and ZnO (Wang et al. 2009) are usually utilized to combine with CdS to form heterojunction composites due to their well-matched band edge positions. Simultaneously, combining graphene with CdS–semiconductor hybrids can enlarge the surface area, create more reactive sites, and further promote the electron transfer. Therefore, the photocatalytic H_2 production activity of CdS can be greatly enhanced through the positive synergetic effects of graphene and heterojunctions. Wang et al. prepared CdS–ZnO/rGO composites for photocatalytic H_2 production from water splitting (Wang et al. 2014). Under visible light irradiation, electrons are excited from the VB to the CB of CdS, while holes are generated in the VB, and no electron–hole pairs are photogenerated in ZnO due to its wide band gap. Since the CB potential of CdS (−0.52 V vs. NHE) is more negative than that of ZnO (−0.22 V vs. NHE) and the energy position of G/G—(−0.08 V vs. NHE), the electrons tend to transfer from the CB of CdS to the CB of ZnO and then to the rGO surface for the H_2 production, and the holes in the VB of CdS are consumed by the sacrificial reagent in the reaction solution. Therefore, the electron–hole pairs are completely isolated in space, and transfer efficiently to participate in the photocatalytic redox reactions, resulting in a 34-times higher photocatalytic H_2 production rate of the CdS–ZnO/GO sample than that of the CdS NPs. Similar to the case of CdS–ZnO/GO, CdS–TaON/rGO (Hou et al. 2012) and CdS–$ZnIn_2S_4$/rGO (Hou et al. 2013) heterostructure systems were also established for efficient H_2 production from water splitting. Besides inorganic semiconductors, the organic graphic carbon nitride (g-C_3N_4) semiconductor is also a promising candidate for the development of CdS-/graphene-based heterogeneous photocatalysts due to its unique 2D structure, narrow band gap (ca. 2.7 eV), and high thermal and chemical stability (Cao et al. 2013; Xiang et al. 2011; Yu et al. 2013b, 2012b). For example, Xiang et al. found that the g-C_3N_4/rGO composite could act as an ideal visible light-responsive photocatalyst for H_2 production (Xiang et al. 2011). In the preparation procedure of the g-C_3N_4/rGO composite, g-C_3N_4 was sandwiched between the rGO nanosheets. Therefore, intimate contact was formed between the facet-to-facet layers, which was beneficial for the charge separation and transfer. In addition, Cao et al. found that CdS/g-C_3N_4 composite photocatalyst exhibited remarkable efficiency for photocatalytic H_2 production under visible light (Cao et al. 2013). The heterojunction formed between g-C_3N_4 and CdS favors the efficient charge separation. These results indicate that designing inorganic–organic heterogeneous system such as CdS/g-C_3N_4/graphene composite photocatalyst is a promising research direction in this area.

High-performance photocatalyst of ZnO/graphene oxide (ZnO/GO) nanocomposite was synthesized via a facile chemical deposition route and used for the photodegradation of organic dye under visible light (Li et al. 2012). The ZnO/GO composite shows absorption in the whole visible region using UV–visible diffuse reflectance spectroscopy analysis (Fig. 3.3). ZnO/GO nanocomposite exhibits much higher photocatalytic efficiency than GO sheets and ZnO particles separately.

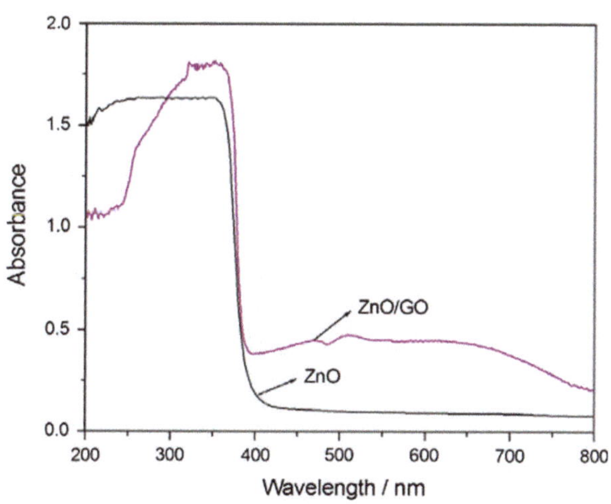

Fig. 3.3 UV-vis diffuse reflection spectra of pure ZnO and ZnO/GO nanocomposite

The synthesis of a Ag/MoS_2 nanocomposite photocatalyst for visible light-driven hydrogen gas evolution has been studied (Cheah et al. 2015). It is noticed that all the Ag/MoS_2 nanocomposites show an absorption peak centered at 606 nm, which is slightly blueshifted as compared to that of the pure MoS_2. Moreover, the nanocomposites also exhibit two small humps located at 400 nm and 438 nm, which are the characteristic peaks of the localized surface plasmon resonance from the Ag nanoparticles. Both of these peaks are due to the collective oscillation of the conduction band electrons after the Ag nanoparticles interact with incident visible light. The photocatalytic evaluation indicates that hydrogen gas production activities have Ag-loading dependence, where 20 wt% Ag/MoS_2 shows optimum performance with a 95% enhancement in hydrogen gas evolution rate if compared to that of commercial MoS_2 nanopowder.

$BiOIO_3$/reduced graphene oxide (rGO) nanocomposites were prepared by a simple one-pot hydrothermal method, during which $BiOIO_3$ nanoplates were formed in situ on rGO sheets resulting from partial reduction of rGO (Xiong et al. 2015) combination of rGO with $BiOIO_3$ is considered to be effective for the visible light response. In contrast to the pure $BiOIO_3$, there is an enhanced absorption in the visible light region with increasing GO content resulting from the broad background absorption of rGO for $BiOIO_3$/rGO nanocomposites. Increased visible light absorption generally leads to a high visible light photocatalytic activity.

A ternary ZnO/Ag/CdO nanocomposite was synthesized using thermal decomposition method (Saravanan et al. 2015). The optical band gaps of the synthesized catalysts were examined using UV–vis absorption spectroscopy. Figure 3.4 shows the absorption edges of the pure ZnO and binary (ZnO/Ag and ZnO/CdO) nanocomposites which lie between 350 and 400 nm. The absorption bands of the ternary (ZnO/Ag/CdO) catalysts are wider (400–600 nm) and it clearly indicates that ZnO/Ag/CdO nanocomposite showing absorption which is red shifted

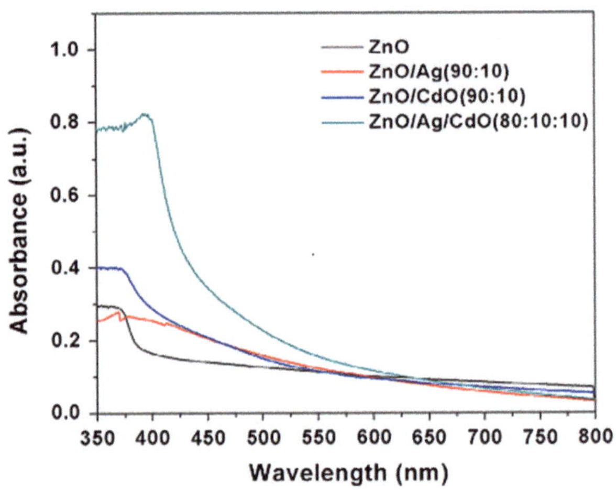

Fig. 3.4 UV-vis absorption spectra of the prepared catalysts

compared with pure ZnO which lies in the blue region of the spectrum. The small size, high surface area, and synergistic effect in the ZnO/Ag/CdO nanocomposite are responsible for high photocatalytic activity.

Nanocomposites Photocatalysts for Organic Transformation

(i) Selective photocatalytic oxidation reactions

Oxidation of alcohols

Organic chemists are always interested in the synthesis of carbonyl compounds due to their importance as intermediates in organic synthesis. Several avenues were established in the literature to obtain carbonyl compounds. The most common method is to oxidize alcohols with an oxidizing agent such as potassium permanganate, manganese dioxide, potassium dichromate, or chromium dioxide. The use of TiO_2 in photocatalytic reactions under UV irradiation is a clean catalytic system widely applied for degradation of organic pollutants by dioxygen. However, its selectivity for organic synthesis is unsatisfactory (Clark and Miller 1977; Palmisano et al. 2007a; Pichat et al. 1987; Yurdakal et al. 2008). In photocatalytic oxidation of alcohols, the high oxidative potential ($E^o = + 2.7$ V vs. NHE at pH 7) (Bard et al. 1985) of the holes generated in the VB of TiO_2 is sufficient enough to force the oxidation of the alcohols (reactants) and the aldehydes or ketones (products) resulting in a low selectivity for the product aldehydes or ketones (Clark and Miller 1977; Hoffmann et al. 1995; Kamat, 1993; Palmisano et al. 2007a; Pichat et al. 1987; Yurdakal et al. 2008). To improve the selectivity for alcohols oxidation into aldehydes using TiO_2 photocatalyst, the formation of strongly oxidative holes in the VB of TiO_2 must be avoided. This can be achieved by modifications of TiO_2 (Anpo and Takeuchi 2003) with a metal ion doping such as Fe^{3+}, V^{4+}, Cr^{3+}, Sn^{4+}, and Sm^{3+}, (Lin and Lin 2011; Ma et al. 2010; Tiana et al. 2009; Zhao et al. 2011; Zhu et al. 2004) nonmetal ion doping such as C, N, S, and F, (Asahi et al. 2001; Nagaveni et al. 2004; Ohno et al. 2004; Shifu et al. 2011) or dye-sensitization (Zhang et al. 2008b). These modifications have become essential to prolong the photoresponse of TiO_2 to the visible region.

Photocatalytic oxidation of alcohols to the corresponding aldehydes and ketones have been achieved in both liquid (Mohamed et al. 2002; Palmisano et al. 2007b) and gas phases (Pillai and Sahle-Demessie 2002) with high selectivity. The selective oxidation reaction of benzylic alcohols to the corresponding benzaldehydes catalyzed by the co-catalytic system of dye-sensitized TiO_2 and (2,2,6,6-tetramethylpiperidin-1- yl)oxyl (TEMPO) was reported by Zhao et al. (2013) (Zhang et al. 2008b) (Scheme 3.1).

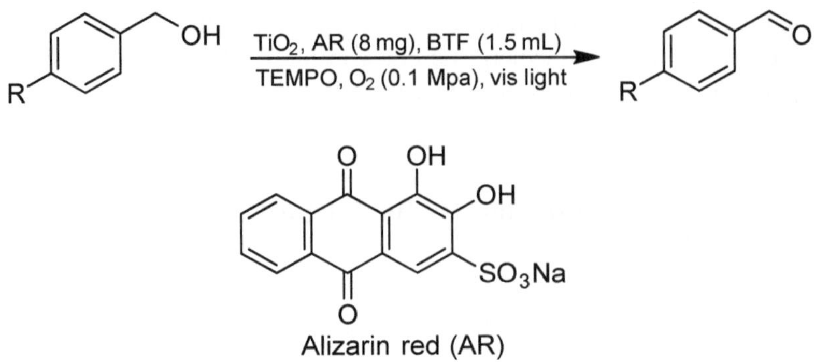

Scheme 3.1 Selective photocatalytic oxidation of substituted benzyl alcohols over the TiO_2/Ar/TEMPO under the irradiation of visible light in the presence of oxygen

The proposed mechanism is illustrated in Scheme 3.2. Accordingly, under visible light irradiation and in the presence of O_2, the excited alizarin red transferred electrons to the conduction band of TiO_2 and TEMPO subsequently reduced the dye radical cation to complete the dye photocatalytic cycle. The process was then followed by the direct oxidation of alcohols to aldehydes. Another system was employed by Robinson et al. (Jeena et al. 2012a, b) using AR/ZnO/TEMPO system with $AgNO_3$ as an internal oxidant and successfully oxidized benzylic alcohols to aromatic aldehydes.

Proposed mechanism for selective photocatalytic oxidation of substituted alcohols to aldehydes or ketones over the TiO_2/Ar/TEMPO system under the irradiation of visible light in presence of oxygen (Jeena et al. 2012a, b; Zhang et al. 2008b).

Metal–organic frameworks (MOFs) are class of compounds consisting of metal ions coordinated to organic ligands via self-assembly to form coordination polymers. MOFs have recently attracted much attention because of their high porosity, specific surface area, and other outstanding properties (Fu et al. 2012; Lang et al. 2014; Li et al. 1999). Wu and coworkers have successfully prepared UiO–66-NH_2–CdS MOFs

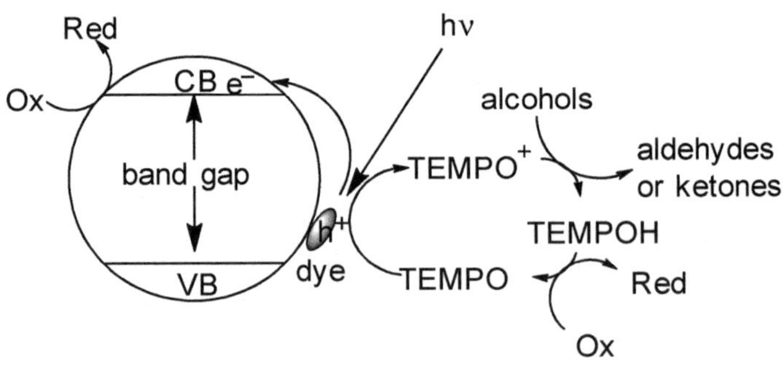

Scheme 3.2 The proposed mechanism for the selective photocatalytic transformation of alcohols to aldehydes or ketones in the presence of alizarin red dye/TiO_2/TEMPO system under the irradiation of visible light in presence of oxygen (Jeena et al. 2012a, b; Zhang et al. 2008b)

nanocomposite via photodeposition technique at room temperature (Shen et al. 2013). The photocatalytic properties of the prepared UiO–66–NH_2–CdS nanocomposite were evaluated and the results showed that selective oxidation of benzyl alcohol derivatives solution in benzotrifluoride (BTF) using molecular oxygen occurred, which may be attributed to the large specific surface area and the charge injection from CdS into UiO–66–NH_2 which leads to considerable long-lived charged ion pairs by reducing the recombination of electron–hole pairs (Scheme 3.3).

Plasmonic photocatalysts are a class of photocatalysts in which noble metals nanoparticles are deposited on the semiconductor surface. They are considered as a promising form of visible light photocatalysts. In this nanocomposite, the noble metal is responsible for light harvesting. In addition to the most widely used metal oxide support TiO_2, other reducible metal oxides such as CeO_2 or inert supports like Al_2O_3, SiO_2, or ZrO_2, have been affirmed to assist good performance for plasmonic NPs. The CB of TiO_2 can facilitate the electron transfer to an electron acceptor such as O_2, which is valuable for effective selective oxidation. The aerobic photooxidation of alcohols in toluene under the sunlight irradiation was effected in the presence of Au nanoparticles (<5 nm diameter) loaded on anatase/rutile TiO_2 (Degussa P25) interface which is utilized as plasmonic photocatalysts (Scheme 3.6) (Tsukamoto et al. 2012). However, different TiO_2 samples were tried as a support for Au NPs, and the results showed that Degussa P25 TiO_2 provides the best result. The photocatalytic activity unfavorably depends on the structural design of the catalyst. Thus, Au NPs, the active sites in this photocatalytic system, functionalize via plasmon activation by visible light followed by successive electron transfer in the Au/rutile/anatase contact site. The activated Au NPs injects their conduction electrons to rutile and then to adjacent anatase TiO_2. This catalyzes the oxidation of the alcohol substrates by the positively charged Au NPs accompanied by the reduction of O_2 by the conduction band electrons on the surface of anatase TiO_2 (Scheme 3.4).

Au/TiO_2, prepared by the deposition–precipitation method from H_4AuCl_4 and TiO_2, can act as an efficient photocatalyst on exposure to visible light (>450 nm) irradiation from a 2000 W Xe lamp. When exposed to sunlight, Au/TiO_2 can promote the conversion of very dilute alcohols (2 mM) into aldehydes or ketones in toluene. Au/TiO_2 plasmon can also act as an efficient photocatalyst for the chemoselective oxidation of alcohols with O_2 in water under the visible light irradiation using a 300 W Xe lamp (Scheme 3.5) (Naya et al. 2010). When the heterocatalyst composed of Au/TiO_2 NPs and the cationic surfactant trimethylsterylammonium chloride ($C_{18}TAC$) was used as the photocatalytic system, the reaction rate was dramatically enhanced by 3.3–5.7-fold. In a special case, a 29.6-fold increase of the reaction rate could be achieved relative to the

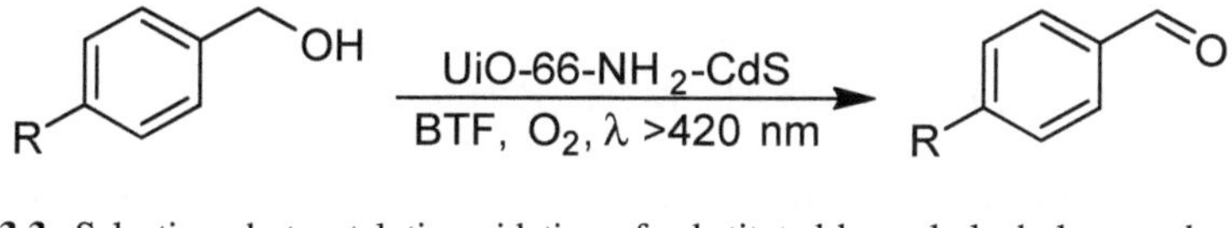

Scheme 3.3 Selective photocatalytic oxidation of substituted benzyl alcohols over the UiO–66–NH_2–CdS nanocomposite under the irradiation of visible light in presence of oxygen

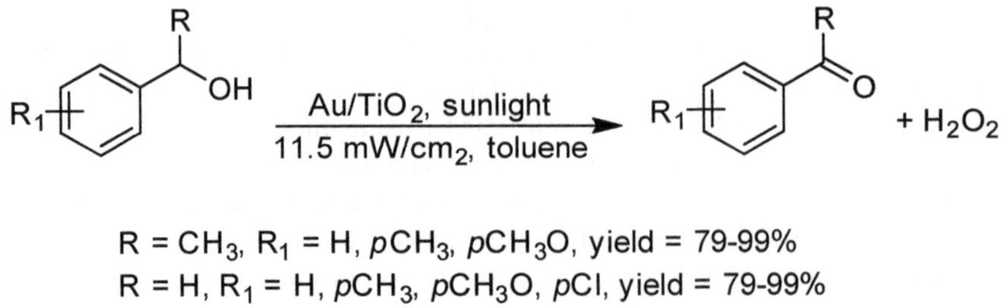

Scheme 3.4 Selective photocatalytic oxidation of primary and secondary benzyl alcohols over Au/TiO_2 nanocomposite under sunlight irradiation

surfactant-free system. The kinetic studies for the conversion of substituted aromatic alcohols to aldehydes showed a linear Hammett plot. On the contrary, Au/ZrO_2 NPs, which is another photocatalytic system based on an inert support, did not show response towards visible light due to the high CB edge of ZrO_2, therefore, inhibition of the electron transfer from Au nanoparticles to ZrO_2 occurred.

Au/CeO_2 nanocomposite, prepared by the photodeposition of H_4AuCl_4 on CeO_2 in the presence of a reducing agent such as citric acid, showed strong absorbance at around 550 nm due to surface plasmon resonance. In this photocatalyst, CeO_2 plays the role of TiO_2 as Au NPs support. Au/CeO_2 NPs selectively oxidized aromatic alcohols to the corresponding aldehydes quantitatively in an aqueous suspension under irradiation of green light in the presence of O_2 (Scheme 3.6) (Tanaka et al. 2011). It was reported that the activity of the photocatalyst is governed by the external surface area of Au NPs instead of the amount of Au loaded as indicated by a linear correlation detected between the external surface area of Au NPs loaded on CeO_2 and the rate of photocatalytic benzaldehyde production.

Sahle-Demessie and coworkers (Pillai and Sahle-Demessie 2002) reported that gas phase photocatalytic reaction with TiO_2-coated pads at 463 K in the presence of O_2 oxidizes various kinds of aliphatic and benzylic alcohols (e.g., 1-pentanol, cyclohexanol, benzyl alcohol, and 1-phenylethanol) to the corresponding carbonyl

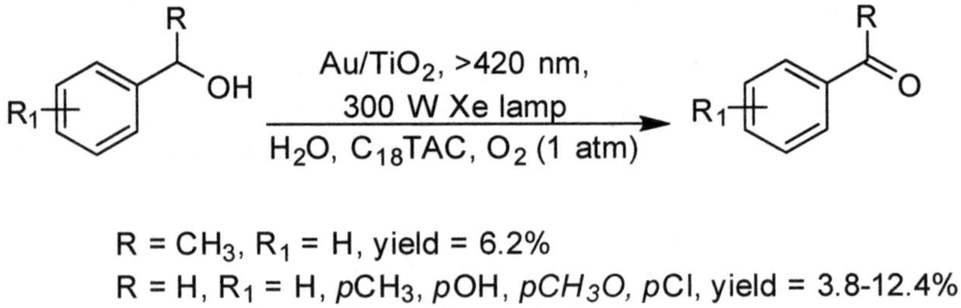

Scheme 3.5 Selective photocatalytic oxidation of alcohols with O_2 in water over Au/TiO_2 nanocomposite in the presence of the cationic surfactant trimethylsterylammonium chloride ($C_{18}TAC$) under the visible light irradiation using a 300 W Xe lamp (>420 nm)

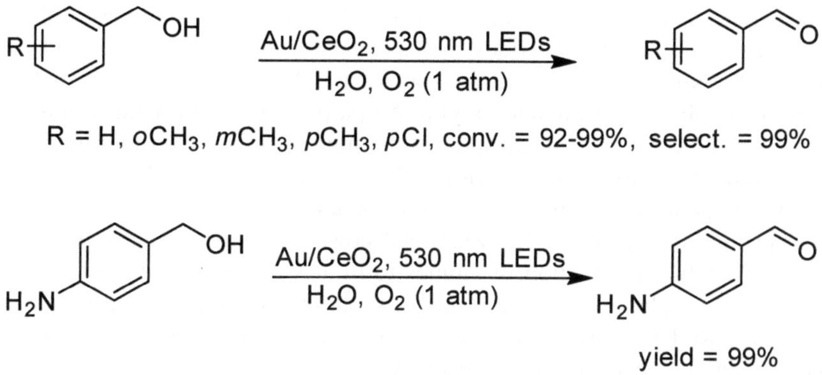

Scheme 3.6 Selective photocatalytic oxidation of alcohols with O_2 in water over Au/CeO_2 under green light irradiation (530 nm)

compounds with moderate yields (~35%) but with very high selectivity (>95%). An interesting feature of this gas phase reaction is that the reaction of 1-phenylethanol gives rise to styrene with very high selectivity (83%) in 97% substrate conversion. In contrast, the photocatalytic reaction in an acetonitrile solution containing 1-phenylethanol gave acetophenone as a major product, where styrene is detected with only a very small amount. Although the detailed mechanism is not clarified, the temperature is probably an important factor for the selective styrene formation.

Oxidation of Primary Benzylamines (C–N Bond Formation)

The amines oxidation is an essential chemical process towards the synthesis of intermediates for fine chemicals suitable in drug and pesticide industry (Murahashi 1995). The classical efficient oxidants used for this transformation are 2-iodylbenzoic acid, (Nicolaou et al. 2003) *tert*-butyl hydroperoxide, (Zhu et al. 2014) and *N-tert*-butylbenzenesulfinimidoyl chloride (Matsuo and Mukaiyama 2001). Lately, heterogeneous photocatalytic selective oxidation of amines using visible light has been widely used in the presence of O_2 which is used as the oxidant to receive electrons under photocatalysis conditions.

The visible light activity of Au NPs (particle size of ~7 nm at 298 K) supported on a metal oxide such as rutile TiO_2 displayed a particularly extraordinary level of heterogeneous photocatalytic selective aerobic oxidation of amines to yield the corresponding imines on a synthetic scale with great selectivity (Scheme 3.7) (Naya et al. 2013). However, the aerobic oxidation of the secondary amines on Au/rutile TiO_2 nanocomposite was also successful to form the corresponding imines under the same conditions (Scheme 3.9, bottom). The study showed that the photocatalytic activity was governed by the Au loading quantity as well as its particle size. Different metal oxides such as anatase and rutile TiO_2, $SrTiO_3$, ZnO, WO_3, In_2O_3, Nb_2O_5 were tested as supports for Au NPs where the Au loading quantity maintained constant and varying Au particle size. Surprisingly, among the various Au/MOs, Au/rutile TiO_2 showed the highest level of visible light activity for the amine oxidation. The experimental results have proposed that the reaction proceeds via the localized surface plasmon resonance-excited electron transfer from the Au nanoparticle to the TiO_2. Zhao and coworkers have recently reported the TiO_2-photocatalyzed oxidation of amines to imines under visible light irradiation via a surface complex mechanism (Lang et al. 2012). Accordingly, adsorbed amine molecules onto TiO_2 form a surface complex that absorbs visible light and subsequently initiate electron transfer reactions.

In another report, the transformation of primary amines to imines by a silicate supported anatase photocatalyst was effected by visible light (Zavahir and Zhu 2015). Thus, a series of primary benzylic amines was oxidative coupled into corresponding imines with dioxygen as the benign oxidant over composite catalysts of TiO_2 (anatase)-silicate under visible light irradiation of $\lambda > 460$ nm. The visible light response of this system is believed to be a result of a high population of defects and contacts between silicate and anatase crystals in the composite and the strong interaction between the benzylic amine and the catalyst. It is found that tuning the intensity and wavelength of the light irradiation and the reaction temperature can remarkably enhance the reaction activity. Water can also act as a green medium for the reaction with an excellent selectivity. This report contributes to the use of readily synthesized, environmentally benign, TiO_2-based composite photocatalyst and solar energy to realize the transformation of primary amines to imine compounds.

Using visible light plasmonic photocatalysis, both aliphatic and aromatic amines are unquestionably very useful starting substrates for the production of pharmaceutically important and structurally complicated molecules. TiO_2 support provides better results than other supports. The presence of Ti^{3+} of TiO_2, arising from nitrogen doping, provides more coordination sites for the alkyne, thereby prompting a better performance than that of pure TiO_2. Imines can also be produced by

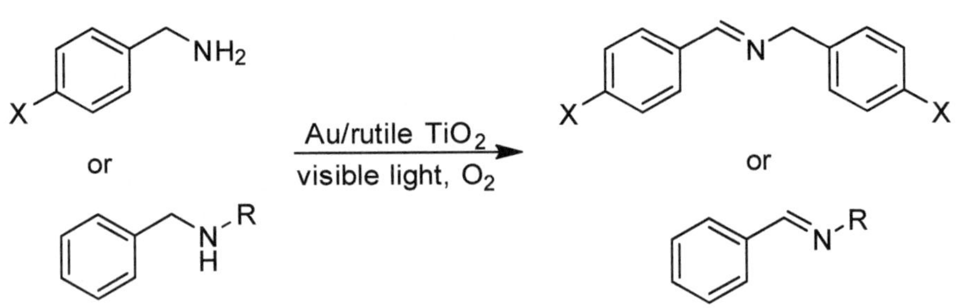

Scheme 3.7 Aerobic photocatalytic oxidation of the secondary amines on Au/rutile TiO_2 under visible light irradiation

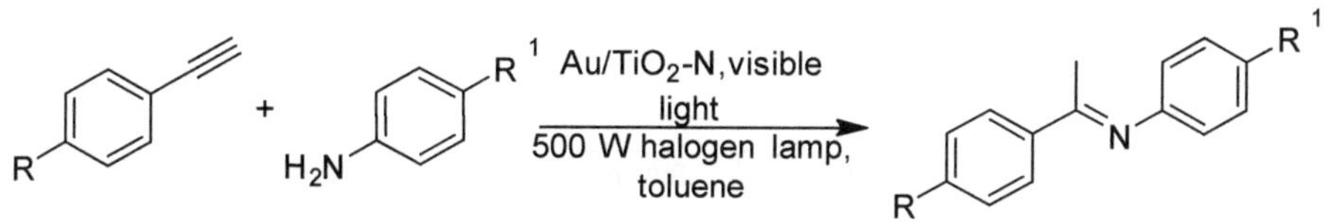

Scheme 3.8 Selective photocatalytic coupling of alkynes with anilines to produce imines on Au/TiO_2-N under visible light irradiation

reacting alkynes with anilines to hydroamination products with Au NPs supported on nitrogen-doped TiO_2 (Au/TiO_2–N), which serves as the visible light plasmonic photocatalyst (Scheme 3.8) (Zhao et al. 2013).

The electronic heterogeneity of the surface of Au–Pd alloy nanoparticles that originates from the difference in the electronegativity of Au and Pd causes the surface charge distribution of the Au–Pd nanoparticles to be un-uniform (Sarina et al. 2013; Zhang et al. 2012). This characteristic is beneficial in improving the interaction between the alloy and the reactant, in which Au NPs serve as the plasmonic metal excited by visible light, and Pd NPs act as the catalytic active sites directly. By utilizing this feature, an Au–Pd alloy loaded on ZrO_2 was applied to selectively oxidize amines into imines in acetonitrile in the presence of oxygen (Scheme 3.9) (Sarina et al. 2013).

Besides ion doping and dye-sensitization, surface modification of TiO_2 is another expedient method for extending the TiO_2 light response to the visible light region. Highly dispersed NiO particles can be used as a surface modification agent. NiO/TiO_2 can catalyze the cyclization reaction of *N,N*-dimethylaniline derivatives with maleimide (Scheme 3.10) under visible light irradiation (Tang et al. 2015). *N, N*-dimethylaniline derivatives first underwent oxidative dehydrogenation by the TiO_2 VB hole to generate α-amino alkyl radical intermediates, which subsequently added to maleimide to produce new radicals, and finally, intramolecular cyclization and dehydrogenation took place to give the target products.

Aromatic oxidation

Phenol, one of the important intermediates in the chemical industry world, is commonly prepared via a multistep process producing acetone as a by-product because direct oxidation of benzene to phenol under mild conditions is difficult. This is understood in terms of the fact that the sp^2 C–H bond of benzene is more

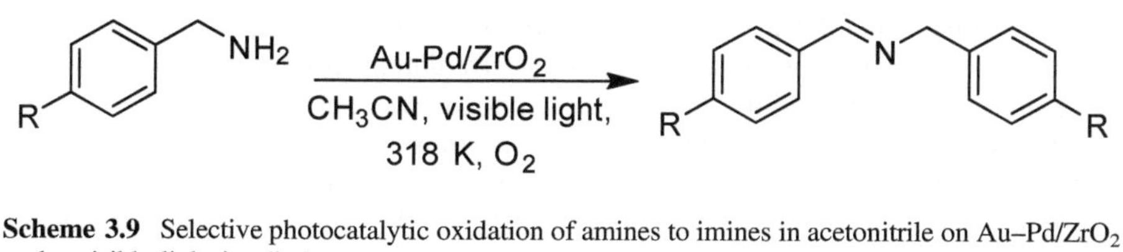

Scheme 3.9 Selective photocatalytic oxidation of amines to imines in acetonitrile on Au–Pd/ZrO_2 under visible light irradiation

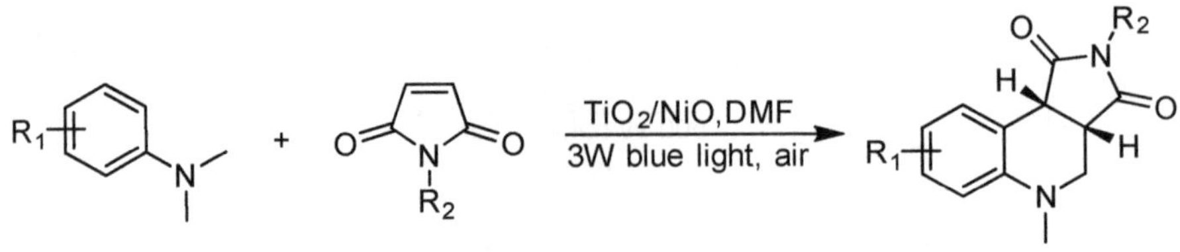

Scheme 3.10 Selective photocatalytic cyclization reaction of N,N-dimethylaniline derivatives with maleimide on NiO/TiO_2 under blue light (3W) irradiation

stable than sp^3 C–H bonds, as a result the hydroxylation is more difficult. Consequently, a straightforward oxidation of benzene to phenol under mild conditions is desirable. Photocatalytic oxidation is one of the direct methods that could fulfill the required circumstances. One possibility of preparing phenol with high yield and selectivity can be achieved by photocatalytic reactions of various metal/metal ions impregnated TiO_2 to this type of reactions under UV or visible light irradiation. The modification of TiO_2 particles by selective metal ion doping extends the absorption spectrum to the visible range through breaking down the recombination rate of the electron–hole pairs and hence enhancing the interfacial charge transfer efficacy (Litter and Navio 1996; Yamashita et al. 1999). Many factors are strongly controlling the efficacy of metal ion doping such as the concentration of the dopant, distribution of the dopants, electronic configuration of the doping ions, and ionic radius of the metals. Noble metals, for instance, Cu, Pt, Au, Ag, Pd, and Rh-doped on TiO_2 surface act as a good co-catalyst in improving the photocatalytic activity of TiO_2, (Yamashita et al. 1999) whereas V, Fe, and Cr-doped TiO_2 have been revealed to increase the quantum efficiency in several cases (Scheme 3.13) (Palmisano et al. 1988). The photocatalytic oxidation of benzene and ethylbenzene was studied using Ag and Cu loaded on TiO_2 (Einaga 2006; Habibi et al. 2004) and oxidation of toluene to benzyl alcohol were also achieved using Fe-doped TiO_2 (Muoz et al. 2007). It has been found that deposition of Rh and Ag on TiO_2 surface improves the photocatalytic oxidation of benzene (Einaga et al. 2004). A comparative study on the role of oxidation states of Fe, Ag, Au, and Cu deposited on TiO_2 to form nanocomposites for benzene selective photooxidation to phenol was performed (Gupta et al. 2015). It was reported that the activity of metal ions co-catalyst deposited onto TiO_2 enhanced benzene oxidation in the order of $Fe^{+3} > Ag^{+} > Au^{+3} > Cu^{+2}$ at small reactive metal loading.

The photocatalytic selective oxidation of benzene using Au/TiO_2 nanoparticles was performed in an aqueous medium (Ide et al. 2011). Au/TiO_2 (3%) was prepared in anhydrous ethanol by the reductive deposition of $HAuCl_4$ using $NaBH_4$ on Degussa P25 TiO_2. Under simulated solar irradiation containing UV light, the aqueous selective oxidation of benzene to phenol was performed. Under the oxidizing CO_2 atmosphere (230 kPa), a selectivity of 89% and benzene conversion of 13% were achieved on Au/TiO_2 (Ide et al. 2011). Visible light-induced oxidation of benzene was also attainable with nanostructured Au prepared under somewhat different environments. Au nanodisk at the interlayer between layers of titanate could serve as a visible light photocatalyst, and was prepared by the modification of layered titanate with (3-mercaptopropyl)trimethoxysilane. The thiol-modified layered titanate was first mixed with $HAuCl_4$, followed by reduction with $NaBH_4$. Under visible light irradiation, a remarkable selectivity of 96% for phenol was reported (Ide et al. 2010). It should be mentioned that the addition of phenol in the photocatalytic system can enhance both the reaction rate and the product selectivity. Besides the use of $NaBH_4$, the reduction of $HAuCl_4$ can also be conducted with ethanol under photochemical conditions. Under UV irradiation in N_2 atmosphere and ethanol solution, the TiO_2 will be reduced by ethanol, and upon adding $HAuCl_4$ to the reaction system, Au NPs will be loaded in situ on TiO_2. Pt and Ag NPs may also be loaded on TiO_2 via a similar method. Such synthesized materials, especially Au/TiO_2, are effective visible light photocatalysts for the selective oxidation of benzene to phenol in the presence of O_2 in water (Zheng et al. 2011) (Scheme 3.11).

(ii) Selective Photocatalytic reduction reactions

Preparation of aniline via reduction of nitrobenzene

Aniline is a key compound in chemical and pharmaceutical industries because it can be used as an intermediate and precursor for preparation of several products and drugs. The simplest way to obtain aniline is the reduction of nitrobenzene which can be accomplished via catalytic hydrogenation. An environmentally friendly method to obtain aniline is to use photocatalysts in the nanoscale to activate the reduction process. Some of the common working photocatalysts are TiO_2, ZnO, and CdS nanoparticles which are characterized by their high surface energies and areas. However, semiconductor NPs only gives good reactivity in the UV region by generating electron–hole pairs. Surface modification of the semiconductor NPs with metal or metal oxide has been applied to improve productivity. Kominami and coworkers successfully prepared Au/TiO_2 with Ag NPs as a co-catalyst and studied its photocatalytic reactivity towards reduction of nitrobenzene with 2-propanol and they obtained a quantitative yield of aniline with acetone under visible light irradiation (Scheme 3.12) (Tanaka et al. 2013).

Another visible light photocatalytic system is achieved by forming a surface complex via adsorbing 2,3-dihydroxynaphthalene (**2,3-DN**) on the surface of anatase TiO_2 via a robust bonding of the *o*-dihydroxyl groups. The surface complex in combination with a reduction co-catalyst Pt NPs is used to photocatalyze the reduction of nitrobenzene to aniline with triethanolamine (TEOA) on exposure to visible light irradiation (>420 nm) from a 500 W Xe lamp (Scheme 3.15) (Kamegawa et al. 2012). The combination of a metal complex organic dye with a

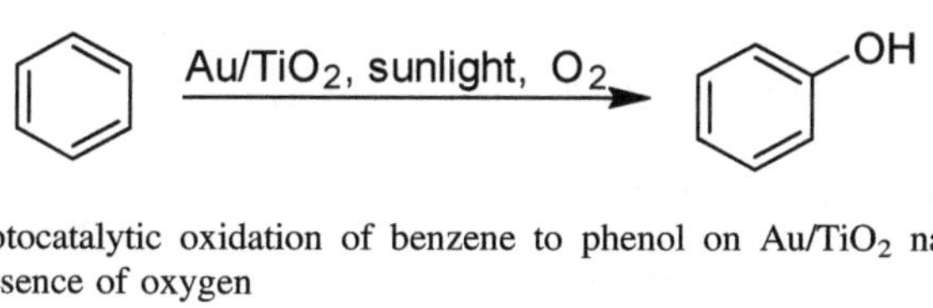

Scheme 3.11 Photocatalytic oxidation of benzene to phenol on Au/TiO_2 nanoparticles under sunlight in the presence of oxygen

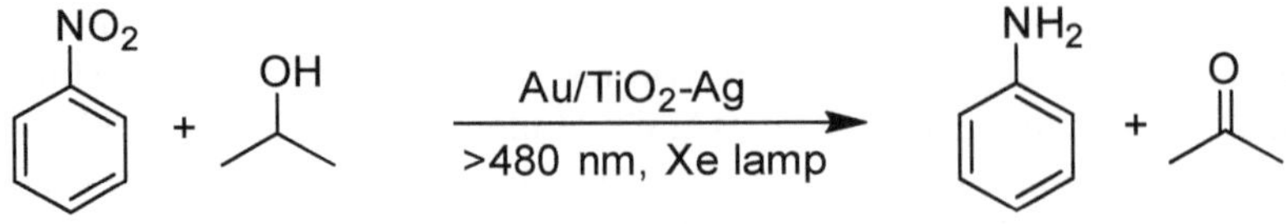

Scheme 3.12 Selective photocatalytic reduction of nitrobenzene to aniline on Au/TiO_2 with Ag NPs as a co-catalyst under visible light irradiation

TiO_2 photocatalyst could efficiently reduce nitrobenzenes to aniline with TEOA under visible light irradiation. An organic dye **N3** (Scheme 3.13) in combination with TiO_2 (Degussa P25) along with the assistance of Pt NPs as a reducing co-catalyst, produced in situ from K_2PtCl_6, can effectively reduce nitrobenzene into aniline with TEOA when exposed to the visible light irradiation from 530 nm LEDs (Fuldner et al. 2010). On the contrary, in the absence of K_2PtCl_6, the reaction rate for the reduction of nitrobenzene to aniline could be improved by adding trace amount of urea derivatives as the co-catalyst, which is attributed to the providing of additional proton shuttling channels by urea derivatives (Fuldner et al. 2011).

Chang and coworkers (Roy et al. 2013) succeeded in preparing highly efficient photocatalytic graphene-ZnO-Au nanocomposites (G-ZnO–Au NCs) by a modest hydrothermal method. First, graphene-ZnO nanospheres with an average diameter of (45.3 ± 3.7) nm via reducing zinc acetate and graphene oxide by catechin in the presence of ethylenediamine as a stabilizing agent and gold nanorods at 300 °C. Second, Au nanorods are deposited onto G-ZnO nanospheres to form G-ZnO–Au NCs. Upon UV irradiation in methanol, G-ZnO–Au NCs produces electron-hole pairs. The holes were trapped by methanol enabling released photogenerated electrons to catalyze the reduction of nitrobenzene (NB) to aniline in 97.8% yield. The efficiency of G-ZnO–Au NCs is, respectively, 3.5- and 4.5-fold greater than those obtained by commercial TiO_2 and ZnO NSs. The mechanism of the

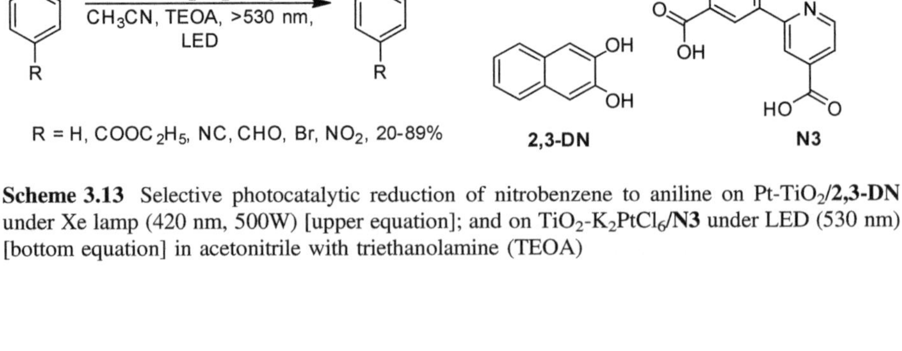

Scheme 3.13 Selective photocatalytic reduction of nitrobenzene to aniline on Pt-TiO_2/**2,3-DN** under Xe lamp (420 nm, 500W) [upper equation]; and on TiO_2-K_2PtCl_6/**N3** under LED (530 nm) [bottom equation] in acetonitrile with triethanolamine (TEOA)

photoreduction process was studied by surface-assisted laser desorption/ionization mass spectrometry to detect the major product (aniline), intermediates (nitrosobenzene and phenylhydroxylamine), and of nitrobenzene through photoelectrocatalytic or photocatalytic reactions. The result reveals that the reduction of nitrobenzene to aniline is through nitrosobenzene to phenylhydroxylamine in the photoelectrocatalytic reaction, while via nitrosobenzene directly in the photocatalytic reaction.

(iii) Coupling photocatalytic reactions

C–C bond formation reactions

A wider interest in the utilization of the solar energy including C–C bond formation as Suzuki reaction has emerged recently due to environmental concern. Wang et al. investigated the application of visible-to-near-infrared harvested light in Suzuki coupling reactions by the use of plasmonic Au–Pd and Au–TiO_x–Pd nanostructures. The incorporation of plasmonic Au nanorods with catalytic Pd nanoparticles through seeded growth empowered effective light harvesting for catalytic reactions on the nanostructures for Suzuki coupling of bromobenzenes and aromatic boronic acids to biphenyls under the irradiation of 809 nm laser (Scheme 3.14) (Wang et al. 2013). Upon plasmon excitation, Au nanorods absorb visible light while the Pd shell performs as the direct catalyst resulting in inducing and accelerating the coupling reaction through both plasmonic photocatalysis and photothermal conversion. The reaction efficiency was dependent on the size of the Au–Pd nanorods, with the smaller nanostructured Au–Pd nanorods delivering the best performance. The Suzuki coupling reaction yield using Au–Pd nanorods was ~2 times that obtained when the reaction was thermally heated to the same temperature. Moreover, the yield was also ~2 times that obtained from Au–TiO_x–Pd nanostructures under the same laser illumination, where a TiO_x shell (25-nm) was introduced to avoid the photocatalysis process. These findings represent a comparison between the effect of mutual plasmonic photocatalysis and photothermal conversion with that of sole photothermal conversion. The plasmonic photocatalysis contribution was higher when the laser illumination was at the plasmon resonance wavelength. It was also concluded that the conduction electron of the SPR Au nanocrystals produces energetic electrons at the surface Pd sites, accordingly enhancing the intrinsic catalytic activity of Pd in stimulating the coupling.

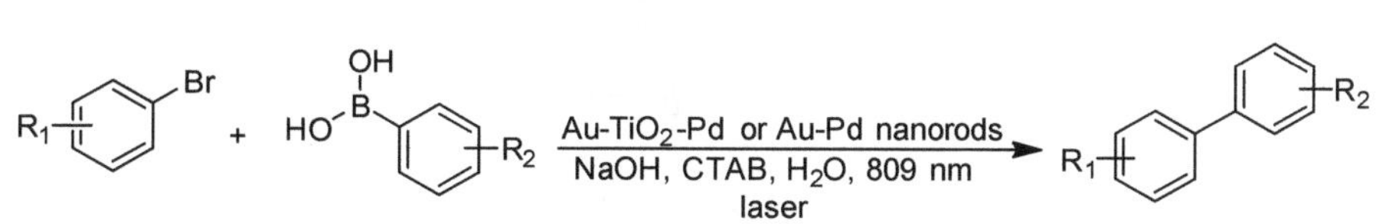

Scheme 3.14 Photocatalytic Suzuki coupling reaction on Au-TiO_2-Pd or Au–Pd nanorods in aqueous basic medium in the presence cetrimonium bromide (CTAB) under laser light (809 nm) irradiation

C–N Bond Formation Reactions: Photocatalytic Synthesis of Benzimidazole

Derivatives of benzimidazole are fundamental pharmaceutical materials which play an important role in the manufacturing of imperative drugs such as omeprazole and bendamustine which is the generic name for the chemotherapy drug Treanda®. The Philips method for the synthesis of benzimidazoles is generally affected by refluxing *o*-phenylenediamines along with organic acid in acidic medium to affect the cyclization process. However, benzimidazoles were synthesized using environmentally friendly and benign method from *o*-phenylenediamines or *N*-

Scheme 3.15 Mechanistic pathway of the formation of 1,2-disubstituted benzimidazole via the photocatalytic effect of the Pt–TiO_2 nanocomposite

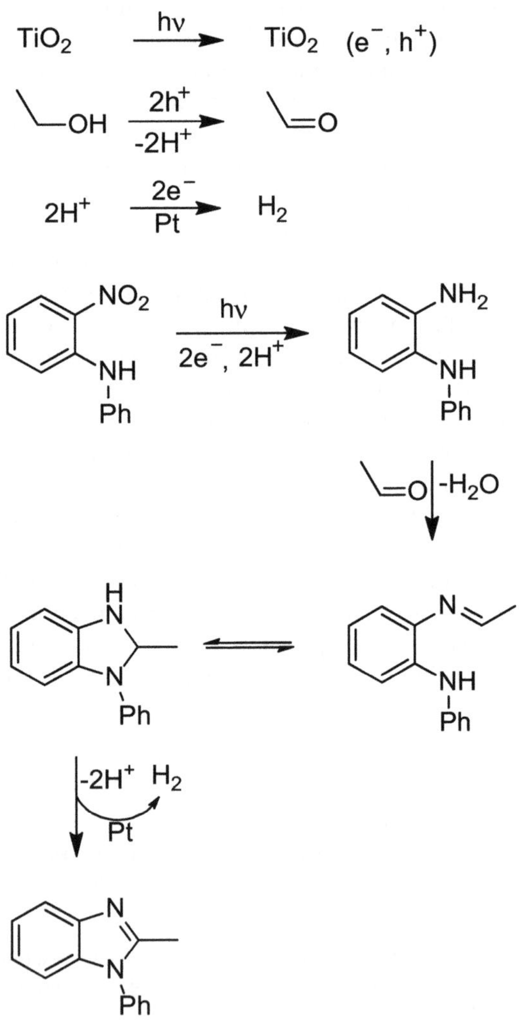

substituted 2-nitroanilines with the nanosized photocatalyst. Selvam and Swaminathan were able to synthesize benzimidazole derivatives by photocatalytic cyclization of *o*-phenylenediamine and various alcohols with irradiated TiO_2/acidic clay composite catalyst using UV-A and solar light (Selvam and Swaminathan 2007). They also concluded that doping Ag on TiO_2 enhanced the product yield and the selectivity of the photocatalytic system. The higher efficiency of Ag–TiO_2 nanoparticles in solar light laid a new method for the synthesis of benzimidazoles using a green chemical process. A few years later, Pt–TiO_2 nanocomposite having different Pt loadings was used to prepare 2-substituted benzimidazoles using visible and UV light (Shiraishi et al. 2010). In this study, it was found that the highest conversion was achieved with Pt(0.2)–TiO_2, whereas the catalysts with larger Pt loading showed lower conversion. This was attributed to the prevention of the incident light absorption by TiO_2 due to excessive amounts of Pt (Zhao and Miyauchi 2008). *N*-Aryl-2-alkylbenzimidazoles were also obtained by photocatalytic reactions of 2-nitrodiphenylamines or *o*-phenylenediamine with alcohols using 3–12 nm-sized Pt–TiO_2 particles using solar and UV-A light (Selvam and Swaminathan 2011). On the other hand, irradiation of ethanolic solution of *o*-phenylenediamine in the presence of TiO_2 did not afford 2-methylbenzimidazole due to the lack of oxidizing center (Wang et al. 1997). Mechanistically, (Selvam and Swaminathan 2011; Shiraishi et al. 2010) the photocatalyst is providing a heterogeneous center for oxidizing the alcohol into the corresponding aldehyde with the liberation of gaseous hydrogen which in turn reduces the nitro to amine group to form *N*-aryl-2-alkylphenylenediamines. Condensation of the latter and the aldehyde would be followed by catalytic dehydrogenation on Pt–TiO_2. Whereas in the diamines case, this reaction is advanced by photocatalytic oxidation of the alcohol by Pt–TiO_2 and a catalytic dehydrogenation of the intermediate on the surface of Pt nanoparticles. In all cases, benzimidazole was formed via photocatalytic reactions on the surface of Pt–TiO_2 (Scheme 3.15).

Summary

Photocatalysis is considered to be one of the clean technologies that has been used in many medical and environmental applications. The photocatalytic reactions involve in using efficient semiconducting photocatalysts that could absorb light energy and convert it to chemical reactions. Synthesis of nanocomposites photocatalysts is considered to be a promising way to achieve efficient photocatalysts. This is attributed to the enhancement of the charge separation, irradiation absorption, and photo and chemical stability of the nanocomposite photocatalysts. Photocatalysts exhibit different band gap widths and band positions, so the photogenerated electron and hole pairs have different oxidation and reduction potentials. The recombination of electron–hole pairs occurs when the charge carrier (electron and hole) migrates to the surface of the photocatalyst that decreases the photocatalytic efficiency. Several methods have been suggested for enhancing the separation of electron–hole pairs as well as increasing the photocatalytic reaction rate. These methods include synthesizing photocatalytic nanocomposites containing metals. Photocatalytic activity is traditionally improved by depositing noble metals (e.g., Pt, Ag, Pd, Au) or nonmetal anions, or by doping with metal cations. In a noble metal composite system, photogenerated electrons accumulate on the metal, and holes remain on the photocatalyst surface, suppressing the recombination of charge carriers. Doping with nonmetal anions (e.g., C, N, S, and F) could also slow the recombination of photogenerated electrons and holes or extend the light absorption of titania into the visible region. The photocatalytic activity of anatase TiO_2 was enhanced by synthesis of TiO_2/carbon nanotubes (CNTs) composites. Carbon nanotubes increase charge transfer between TiO_2 and species in solution. When graphene is involved in a photocatalytic nanocomposite system, it increases the activity of the catalyst as it reduces the band gap energy of the main photocatalyst. Ternary nanocomposites from different oxide have displayed good photocatalytic activity in the presence of visible light due to the enhancement of charge separation and decrement of band gap. Recently, there is a great interest in synthesizing novel organic compounds by light. During the 50s and 60s of the twentieth century, photochemistry has become an appreciated tool for recent organic synthesis. Photocatalytic transformation of some organic compounds has been discussed. Nanocomposites of some semiconducting photocatalysts, i.e., TiO_2, ZnO, etc., have shown good photocatalytic activity and selectivity toward some organic reactions. These reactions include oxidation, reduction, and coupling reactions. The selectivity of these organic reactions depends on photocatalyst composition, pH, additives as well as the light source type.

References

Anpo M, Takeuchi M (2003) The design and development of highly reactive titanium oxide photocatalysts operating under visible light irradiation. J Catal 216:505–516

Asahi R, Morikawa T, Ohwaki T, Aoki K, Taga Y (2001) Visible-light photocatalysis in nitrogen-doped titanium oxides. Science 293:269–271

Bannat I, Wessels K, Oekermann T, Rathousky J, Bahnemann D, Wark M (2009) Improving the photocatalytic performance of mesoporous titania films by modification with gold nanostructures. Chem Mater 21:1645–1653

Bard AJ, Parsons R, Jordan J (1985) Standard potentials in aqueous solutions. Marcel Dekker, New York

Barolo G, Livraghi S, Chiesa M, Paganini MC, Giamello E (2012) Mechanism of the photoactivity under visible light of N-doped titanium dioxide. charge carriers migration in irradiated N–TiO2 investigated by electron paramagnetic resonance. J Phys Chem. C 116:20887–20894

Cao S-W, Yuan Y-P, Fang J, Shahjamali MM, Boey FYC, Barber J, Joachim Loo SC, Xue C (2013) In-situ growth of CdS quantum dots on g-C_3N_4 nanosheets for highly efficient photocatalytic hydrogen generation under visible light irradiation. Int J Hydrogen Energy 38:1258–1266

Cheah AJ, Chiu WS, Khiew PS, Nakajima H, Saisopa T, Songsiriritthigul P, Radiman S, Hamid MAA (2015) Facile synthesis of a Ag/MoS2 nanocomposite photocatalyst for enhanced visible-light driven hydrogen gas evolution. Catal Sci Technol 5:4133–4143

Clark JH, Miller JM (1977) Hydrogen bonding in organic synthesis V: potassium fluoride in carboxylic acids as an alternative to crown ether with acid salts in the preparation of phenacyl esters. Tetrahedron Lett 18:599–602

Cozzoli PD, Comparelli R, Fanizza E, Curri ML, Agostiano A, Laub D (2004) Photocatalytic synthesis of silver nanoparticles stabilized by TiO_2 Nanorods: a semiconductor/metal nanocomposite in homogeneous nonpolar solution. J Am Chem Soc 126:3868–3879

Einaga H (2006) Effect of silver deposition on TiO_2 for photocatalytic oxidation of benzene in the gas phase. React Kinet Catal Lett 88:357–362

Einaga H, Ibusuki T, Futamura S (2004) Improvement of catalyst durability by deposition of Rh on TiO_2 in photooxidation of aromatic compound. Environ Sci and Technol 38:285–289

Fu Y, Sun D, Chen Y, Huang R, Ding Z, Fu X, Li Z (2012) An amine-functionalized titanium metal—organic framework photocatalyst with visible-light-induced activity for CO_2 reduction. Angew Chem 124:3420–3423

Fuldner S, Mild R, Siegmund HI, Schroeder JA, Gruber M, Konig B (2010) Green-light photocatalytic reduction using dye-sensitized TiO_2 and transition metal nanoparticles. Green Chem 12:400–406

Fuldner S, Mitkina T, Trottmann T, Frimberger A, Gruber M, Konig B (2011) Urea derivatives enhance the photocatalytic activity of dye-modified titanium dioxide. Photochem Photobiol Sci 10:623–625

Greeley J, Jaramillo TF, Bonde J, Chorkendorff I, Norskov JK (2006) Computational high-throughput screening of electrocatalytic materials for hydrogen evolution. Nat Mater 5:909–913

Gupta N, Bansal P, Pal B (2015) Metal ion-TiO_2 nanocomposites for the selective photooxidation of benzene to phenol and cycloalkanol to cycloalkanone. J Exp Nanosci 10:148–160

Habibi MH, Isfahani AZ, Mohammadkhani A, Montazerozohori M (2004) Photooxidation of ethylbenzene with TiO_2 and metal coated TiO_2 and its kinetics. Monatsh Chem 135:1121–1127

Heremans P, Cheyns D, Rand BP (2009) Strategies for increasing the efficiency of heterojunction organic solar cells: material selection and device architecture. Acc Chem Res 42:1740–1747

Hoffmann MR, Martin ST, Choi W, Bahnemann DW (1995) Environmental applications of semiconductor photocatalysis. Chem Rev 95:69–96

Hou J, Wang Z, Kan W, Jiao S, Zhu H, Kumar RV (2012) Efficient visible-light-driven photocatalytic hydrogen production using CdS@TaON core-shell composites coupled with graphene oxide nanosheets. J Mater Chem 22:7291–7299

Hou J, Yang C, Cheng H, Wang Z, Jiao S, Zhu H (2013) Ternary 3D architectures of CdS QDs/graphene/$ZnIn_2S_4$ heterostructures for efficient photocatalytic H_2 production. Phys Chem Chem Phys 15:15660–15668

Ide Y, Matsuoka M, Ogawa M (2010) Efficient visible-light-induced photocatalytic activity on gold-nanoparticle-supported layered titanate. J Am Chem Soc 132:16762–16764

Ide Y, Nakamura N, Hattori H, Ogino R, Ogawa M, Sadakane M, Sano T (2011) Sunlight-induced efficient and selective photocatalytic benzene oxidation on TiO_2-supported gold nanoparticles under CO_2 atmosphere. Chem Commun 47:11531–11533

Jeena V, Robinson RS (2012a) Convenient photooxidation of alcohols using dye sensitised semiconductors in combination with silver nitrate and TEMPO—an electron paramagnetic resonance study. Dalton Trans 41:3134–3137

Jeena V, Robinson RS (2012b) Convenient photooxidation of alcohols using dye sensitised zinc oxide in combination with silver nitrate and TEMPO. Chem Commun 48:299–301

Kakhki RM, Ahsani F, Mir N (2016) Enhanced photocatalytic activity of CuO-SiO_2 nanocomposite based on a new Cu nanocomplex. J Mater Sci: Mater in Electron:1–9

Kamat PV (1993) Photochemistry on nonreactive and reactive (semiconductor) surfaces. Chem Rev 93:267–300

Kamegawa T, Seto H, Matsuura S, Yamashita H (2012) Preparation of hydroxynaphthalene-modified TiO_2 via formation of surface complexes and their applications in the photocatalytic reduction of nitrobenzene under visible-light irradiation. ACS Appl Mater Interfaces 4:6635–6639

Kraeutler B, Bard AJ (1978) Heterogeneous photocatalytic preparation of supported catalysts. Photodeposition of platinum on titanium dioxide powder and other substrates. J Am Chem Soc 100:4317–4318

Kudo A, Miseki Y (2009) Heterogeneous photocatalyst materials for water splitting. Chem Soc Rev 38:253–278

Lang X, Chen X, Zhao J (2014) Heterogeneous visible light photocatalysis for selective organic transformations. Chem Soc Rev 43:473–486

Lang X, Ma W, Zhao Y, Chen C, Ji H, Zhao J (2012) Visible-light-induced selective photocatalytic aerobic oxidation of amines into imines on TiO_2. Chem—A Eur J 18:2624–2631

Lee J, Choi W (2005) Photocatalytic reactivity of surface platinized TiO_2: substrate specificity and the effect of Pt oxidation state. J Phys Chem B 109:7399–7406

Li B, Liu T, Wang Y, Wang Z (2012) ZnO/graphene-oxide nanocomposite with remarkably enhanced visible-light-driven photocatalytic performance. J Colloid Interface Sci 377:114–121

Li GH, Gray KA (2007) The solid-solid interface: explaining the high and unique photocatalytic reactivity of TiO_2-based nanocomposite materials. Chem Phys 339:173–187

Li H, Eddaoudi M, O'Keeffe M, Yaghi OM (1999) Design and synthesis of an exceptionally stable and highly porous metal-organic framework. Nature 402:276–279

Lin W-C, Lin Y-J (2011) Effect of vanadium(IV)-doping on the visible light-induced catalytic activity of titanium dioxide catalysts for methylene blue degradation. Environ Eng Sci 29:447–452

Litter MI, Navio JA (1996) Photocatalytic properties of iron-doped titania semiconductors. J Photochem Photobiol A: Chem 98:171–181

Ma Y, Zhang J, Tian B, Chen F, Wang L (2010) Synthesis and characterization of thermally stable Sm, N co-doped TiO_2 with highly visible light activity. J Hazard Mater 182:386–393

J-i Matsuo, Mukaiyama T (2001) *N-tert*-Butylbenzenesulfinimidoyl chloride. Encyclopedia of reagents for organic synthesis. Wiley, Hoboken

Mohamed OS, Gaber AE-AM, Abdel-Wahab AA (2002) Photocatalytic oxidation of selected aryl alcohols in acetonitrile. J Photochem Photobiol, A 148:205–210

Muoz MJL, Aguado J, Ruprez B (2007) The influence of dissolved transition metals on the photocatalytic degradation of phenol with TiO_2. Res Chem Intermediat 33:377–392

Murahashi S-I (1995) Synthetic aspects of metal-catalyzed oxidations of amines and related reactions. Angew Chem, Int Ed Engl 34:2443–2465

Nagaveni K, Hegde MS, Ravishankar N, Subbanna GN, Madras G (2004) Synthesis and structure of nanocrystalline TiO_2 with lower band gap showing high photocatalytic activity. Langmuir 20:2900–2907

S-i Naya, Inoue A, Tada H (2010) Self-assembled heterosupramolecular visible light photocatalyst consisting of gold nanoparticle-loaded titanium(IV) dioxide and surfactant. J Am Chem Soc 132:6292–6293

S-i Naya, Kimura K, Tada H (2013) One-step Selective aerobic oxidation of amines to imines by gold nanoparticle-loaded rutile titanium(IV) oxide plasmon photocatalyst. ACS Catal 3:10–13

Nicolaou KC, Mathison CJN, Montagnon T (2003) New reactions of IBX: oxidation of nitrogen- and sulfur-containing substrates to afford useful synthetic intermediates. Angew Chem Int Ed 42:4077–4082

Ohno T, Akiyoshi M, Umebayashi T, Asai K, Mitsui T, Matsumura M (2004) Preparation of S-doped TiO_2 photocatalysts and their photocatalytic activities under visible light. Appl Catal A 265:115–121

Oppong SOB, Anku WW, Shukla SK, Govender PP (2016) Synthesis and characterisation of neodymium doped-zinc oxide—graphene oxide nanocomposite as a highly efficient photocatalyst for enhanced degradation of indigo carmine in water under simulated solar light. Res Chem Intermed: 1–21

Palmisano G, Augugliaro V, Pagliaro M, Palmisano L (2007a) Photocatalysis: a promising route for twenty-first century organic chemistry. Chem Commun: 3425–3437

Palmisano G, Yurdakal S, Augugliaro V, Loddo V, Palmisano L (2007b) Photocatalytic selective oxidation of 4-methoxybenzyl alcohol to aldehyde in aqueous suspension of home-prepared titanium dioxide catalyst. Adv Synth Catal 349:964–970

Palmisano L, Augugliaro V, Sclafani A, Schiavello M (1988) Activity of chromium-ion-doped titania for the dinitrogen photoreduction to ammonia and for the phenol photodegradation. J Phy Chem 92:6710–6713

Pichat P, Mozzanega M-N, Courbon H (1987) Investigation of the mechanism of photocatalytic alcohol dehydrogenation over Pt/TiO_2 using poisons and labelled ethanol. J Chem Soc, Faraday Trans 1: Phys Chem Condens Phases 83:697–704

Pillai UR, Sahle-Demessie E (2002) Selective oxidation of alcohols in gas phase using light-activated titanium dioxide. J Catal 211:434–444

Qi L, Yu J, Jaroniec M (2011) Preparation and enhanced visible-light photocatalytic H_2-production activity of CdS-sensitized Pt/TiO_2 nanosheets with exposed (001) facets. Phys Chem Chem Phys 13:8915–8923

Roy P, Periasamy AP, Liang C-T, Chang H-T (2013) Synthesis of graphene-ZnO-Au nanocomposites for efficient photocatalytic reduction of nitrobenzene. Environ Sci Technol 47:6688–6695

Saravanan R, Mansoob Khan M, Gupta VK, Mosquera E, Gracia F, Narayanan V, Stephen A (2015) ZnO/Ag/CdO nanocomposite for visible light-induced photocatalytic degradation of industrial textile effluents. J Colloid Interface Sci 452:126–133

Sarina S, Zhu H, Jaatinen E, Xiao Q, Liu H, Jia J, Chen C, Zhao J (2013) Enhancing catalytic performance of palladium in gold and palladium alloy nanoparticles for organic synthesis reactions through visible light irradiation at ambient temperatures. J Am Chem Soc 135:5793–5801

Selvam K, Swaminathan M (2007) A green chemical synthesis of 2-alkylbenzimidazoles from 1,2-phenylenediamine and propylene glycol, or alcohols mediated by Ag–TiO_2/clay composite photocatalyst. Chem Lett 36:1060–1061

Selvam K, Swaminathan M (2011) An easy one-step photocatalytic synthesis of 1-aryl-2-alkylbenzimidazoles by platinum loaded TiO_2 nanoparticles under UV and solar light. Tetrahedron Lett 52:3386–3392

Shchukin D, Poznyak S, Kulak A, Pichat P (2004) TiO_2-In_2O_3 photocatalysts: preparation, characterisations and activity for 2-chlorophenol degradation in water. J Photochem Photobiol, A 162:423–430

Shen L, Liang S, Wu W, Lianga R, Wu L (2013) CdS-decorated UiO–66(NH_2) nanocomposites fabricated by a facile photodeposition process: an efficient and stable visible-light-driven photocatalyst for selective oxidation of alcohols. J Mater Chem A 1:11473–11482

Sher Shah MSA, Park AR, Zhang K, Park JH, Yoo PJ (2012) Green synthesis of biphasic TiO_2—reduced graphene oxide nanocomposites with highly enhanced photocatalytic activity. ACS Appl Mater Interfaces 4:3893–3901

Shifu C, Yunguang Y, Wei L (2011) Preparation, characterization and activity evaluation of TiN/F–TiO_2 photocatalyst. J Hazard Mater 186:1560–1567

Shiraishi Y, Sugano Y, Tanaka S, Hirai T (2010) One-pot synthesis of benzimidazoles by simultaneous photocatalytic and catalytic reactions on Pt@TiO_2 nanoparticles. Angew Chem Int Ed 49:1656–1660

Stankovich S, Dikin DA, Piner RD, Kohlhaas KA, Kleinhammes A, Jia Y, Wu Y, Nguyen ST, Ruoff RS (2007) Synthesis of graphene-based nanosheets via chemical reduction of exfoliated graphite oxide. Carbon 45:1558–1565

Su R, Tiruvalam R, Logsdail AJ, He Q, Downing CA, Jensen MT, Dimitratos N, Kesavan L, Wells PP, Bechstein R et al. (2014) Designer titania-supported Au-Pd nanoparticles for efficient photocatalytic hydrogen production. ACS Nano 8:3490–3497

Tanaka A, Hashimoto K, Kominami H (2011) Selective photocatalytic oxidation of aromatic alcohols to aldehydes in an aqueous suspension of gold nanoparticles supported on cerium(IV) oxide under irradiation of green light. Chem Commun 47:10446–10448

Tanaka A, Nishino Y, Sakaguchi S, Yoshikawa T, Imamura K, Hashimoto K, Kominami H (2013) Functionalization of plasmonic Au/TiO_2 photocatalyst with an Ag co-catalyst for quantitative reduction of nitrobenzene to aniline in 2-propanol suspensions under irradiation of visible light. Chem Commun (Cambridge, United Kingdom) 49:2551–2553

Tang J, Grampp G, Liu Y, Wang B-X, Tao F-F, Wang L-J, Liang X-Z, Xiao H-Q, Shen Y-M (2015) Visible light mediated cyclization of tertiary anilines with maleimides using nickel(II) oxide surface-modified titanium dioxide catalyst. J Org Chem 80:2724–2732

Tiana B, Li C, Gua F, Jianga H, Hua Y, Zhang J (2009) Flame sprayed V-doped TiO_2 nanoparticles with enhanced photocatalytic activity under visible light irradiation. Chem Eng J 151:220–227

Tsukamoto D, Shiraishi Y, Sugano Y, Ichikawa S, Tanaka S, Hirai T (2012) Gold nanoparticles located at the interface of anatase/rutile TiO_2 particles as active plasmonic photocatalysts for aerobic oxidation. J Am Chem Soc 134:6309–6315

Wang F, Li C, Chen H, Jiang R, Sun L-D, Li Q, Wang J, Yu JC, Yan C-H (2013) Plasmonic harvesting of light energy for Suzuki coupling reactions. J Am Chem Soc 135:5588–5601

Wang H, Partch RE, Li Y (1997) Synthesis of 2-alkylbenzimidazoles via TiO_2-mediated photocatalysis. J Org Chem 62:5222–5225

Wang X, Liu G, Chen Z-G, Li F, Wang L, Lu GQ, Cheng H-M (2009) Enhanced photocatalytic hydrogen evolution by prolonging the lifetime of carriers in ZnO/CdS heterostructures. Chem Commun:3452–3454

Wang X, Yin L, Liu G (2014) Light irradiation-assisted synthesis of ZnO-CdS/reduced graphene oxide heterostructured sheets for efficient photocatalytic H_2 evolution. Chem Commun 50:3460–3463

Xiang Q, Yu J, Jaroniec M (2011) Preparation and enhanced visible-light photocatalytic H_2-production activity of graphene/C_3N_4 composites. J Phys Chem C 115:7355–7363

Xiang Q, Yu J, Jaroniec M (2012) Graphene-based semiconductor photocatalysts. Chem Soc Rev 41:782–796

Xing M-Y, Yang B-X, Yu H, Tian B-Z, Bagwasi S, Zhang J-L, Gong X-Q (2013) Enhanced photocatalysis by Au nanoparticle loading on TiO_2 single-crystal (001) and (110) facets. J Phys Chem Lett 4:3910–3917

Xiong T, Dong F, Zhou Y, Fu M, Ho W-K (2015) New insights into how RGO influences the photocatalytic performance of BiOIO3/RGO nanocomposites under visible and UV irradiation. J Colloid Interface Sci 447:16–24

Xue M, Huang L, Wang J-Q, Wang Y, Gao L, J-h Zhu, Zou Z-G (2008) The direct synthesis of mesoporous structured MnO_2/TiO_2 nanocomposite: a novel visible-light active photocatalyst with large pore size. Nanotechnology 19:185604

Yamashita H, Ichihashi Y, Takeuchi M, Kishiguchi S, Anpo M (1999) Characterization of metal ionimplanted titanium oxide photocatalysts operating under visible light irradiation. Radiat J Synchroton 6:451–452

Yang M-Q, Zhang N, Pagliaro M, Xu Y-J (2014) Artificial photosynthesis over graphene-semiconductor composites. Are we getting better? Chem Soc Rev 43:8240–8254

Yao Y, Li G, Ciston S, Lueptow RM, Gray KA (2008) Photoreactive TiO_2/carbon nanotube composites: synthesis and reactivity. Environ Sci Technol 42:4952–4957

Yu C, Fan C, Meng X, Yang K, Cao F, Li X (2011) A novel Ag/BiOBr nanoplate catalyst with high photocatalytic activity in the decomposition of dyes. React Kinet, Mech Catal 103:141–151

Yu C, Fan Q, Xie Y, Chen J, shu Q, Yu JC (2012a) Sonochemical fabrication of novel square-shaped F doped TiO_2 nanocrystals with enhanced performance in photocatalytic degradation of phenol. J Hazard Mater 237–238:38–45

Yu C, Li G, Kumar S, Yang K, Jin R (2014) Phase transformation synthesis of novel Ag_2O/Ag_2CO_3 heterostructures with high visible light efficiency in photocatalytic degradation of pollutants. Adv Mater 26:892–898

Yu C, Yang K, Zhou WQ, Fan QZ, Wei LF, Yu JC (2013a) Preparation, characterization and photocatalytic performance of noble metals (Ag, Pd, Pt, Rh) deposited on sponge-like ZnO microcuboids. J Phys Chem Solids 74:1714–1720

Yu C, Yu JC (2009) A simple way to prepare C-N-codoped TiO_2 photocatalyst with visible-light activity. Catal Lett 129:462–470

Yu C, Yu JC, Fan C, Wen H, Hu S (2010) Synthesis and characterization of Pt/BiOI nanoplate catalyst with enhanced activity under visible light irradiation. Mater Sci Eng, B 166:213–219

Yu J, Wang S, Low J, Xiao W (2013b) Enhanced photocatalytic performance of direct Z-scheme g-C_3N4–TiO_2 photocatalysts for the decomposition of formaldehyde in air. Phys Chem Chem Phys 15:16883–16890

Yu J, Xiong J, Cheng B, Liu S (2005) Fabrication and characterization of Ag–TiO_2 multiphase nanocomposite thin films with enhanced photocatalytic activity. Appl Catal B 60:211–221

Yu J, Yang B, Cheng B (2012b) Noble-metal-free carbon nanotube-$Cd_{0.1}Zn_{0.9}S$ composites for high visible-light photocatalytic H_2-production performance. Nanoscale 4:2670–2677

Yu J, Yue L, Liu S, Huang B, Zhang X (2009) Hydrothermal preparation and photocatalytic activity of mesoporous Au–TiO_2 nanocomposite microspheres. J Colloid Interface Sci 334:58–64

Yu Y, Ren J, Meng M (2013c) Photocatalytic hydrogen evolution on graphene quantum dots anchored TiO_2 nanotubes-array. Int J Hydrogen Energy 38:12266–12272

Yurdakal S, Palmisano G, Loddo V, Augugliaro V, Palmisano L (2008) Nanostructured rutile TiO_2 for selective photocatalytic oxidation of aromatic alcohols to aldehydes in water. J Am Chem Soc 130:1568–1569

Zavahir S, Zhu H (2015) Visible light induced green transformation of primary amines to imines using a silicate supported anatase photocatalyst. Molecules 20:1941–1954

Zhang H, Watanabe T, Okumura M, Haruta M, Toshima N (2012) Catalytically highly active top gold atom on palladium nanocluster. Nat Mater 11:49–52

Zhang J, Xu Q, Feng Z, Li M, Li C (2008a) Importance of the relationship between surface phases and photocatalytic activity of TiO_2. Angew Chem Int Ed 47:1766–1769

Zhang M, Chen C, Ma W, Zhao J (2008b) Visible-light-induced aerobic oxidation of alcohols in a coupled photocatalytic system of dye-sensitized TiO_2 and TEMPO. Angew Chem 120:9876–9879

Zhao J, Zheng Z, Bottle S, Chou A, Sarina S, Zhu H (2013) Highly efficient and selective photocatalytic hydroamination of alkynes by supported gold nanoparticles using visible light at ambient temperature. Chem Commun 49:2676–2678

Zhao Y, Liu J, Shi L, Yuan S, Fang J, Wang Z, Zhang M (2011) Solvothermal preparation of Sn^{4+} doped anatase TiO_2 nanocrystals from peroxo-metal-complex and their photocatalytic activity. Appl Catal B 103:436–443

Zhao Z-G, Miyauchi M (2008) Nanoporous-walled tungsten oxide nanotubes as highly active visible-light-driven photocatalysts. Angew Chem, Int Ed Engl 47:7051–7055

Zheng Z, Huang B, Qin X, Zhang X, Dai Y, Whangbo M-H (2011) Facile in situ synthesis of visible-light plasmonic photocatalysts M@TiO_2 (M = Au, Pt, Ag) and evaluation of their photocatalytic oxidation of benzene to phenol. J Mater Chem 21:9079–9087

Zhu C, Xia J-B, Chen C (2014) Vanadium-catalyzed oxidative Strecker reaction: α-C–H cyanation of para-methoxyphenyl (PMP)-protected primary amines. Tetrahedron Lett 55:232–234

Zhu J, Yang D, Geng J, Chen D, Jiang Z (2008) Synthesis and characterization of bamboo-like CdS/TiO_2 nanotubes composites with enhanced visible-light photocatalytic activity. J Nanopart Res 10:729–736

Zhu J, Zheng W, He B, Zhang J, Anpo M (2004) Characterization of Fe–TiO_2 photocatalysts synthesized by hydrothermal method and their photocatalytic reactivity for photodegradation of XRG dye diluted in water. J Mol Catal A: Chem 216:35–43

Chapter 4
Role of Metal Nanoparticles and Its Surface Plasmon Activity on Nanocomposites for Visible Light-Induced Catalysis

Anup Kumar Sasmal and Tarasankar Pal

Abstract Heterogeneous photocatalysis has become an encouraging reaction technique to combat energy crisis and global environmental issues. Visible light ($\sim$400 nm–750 nm)-driven photocatalysis is the most imperative heterogeneous photocatalysis because of its selective product delivery, easy operation, and utilization of abundant available clean energy resource. In this context, utilization of clean, and available sunlight (having 44% visible light) could be a pleasant platform for solving energy and environmental problems. Thus visible light-driven photocatalysis is highly demanding, and so designing of such photocatalysts and their exploitation in catalysis under visible light has become a central research theme in catalysis. Surface plasmon resonance (SPR) active nanomaterials or composites are very effective to carry out catalytic redox reactions in presence of visible light due to the electron–hole formation, and termed as visible light plasmonic photocatalyst. Processes can be demonstrated through oxidation by "hole" and reduction by "hot electron". Herein, we discussed on fabrication or synthesis of visible light plasmonic photocatalysts, and their application on catalytic reaction under visible light illumination. Visible light-induced SPR with detailed understanding of the fate of generated electron and hole on the redox reactions has been discussed. We have depicted various types of catalytic reactions such as photodegradation of large organic dyes (organic transformation), oxidation reaction, reduction reaction, hydroxylation, imine synthesis, water splitting reaction, biaryl synthesis, and CO_2 reduction.

Keywords Plasmonic photocatalyst · Visible light · Surface plasmon resonance · Electron · Hole · Catalysis

---

A.K. Sasmal · T. Pal (✉)
Department of Chemistry, Indian Institute of Technology,
Kharagpur 721302, India
e-mail: tpal@chem.iitkgp.ernet.in

M.M. Khan et al. (eds.), *Nanocomposites for Visible Light-induced Photocatalysis*,
Springer Series on Polymer and Composite Materials,
DOI 10.1007/978-3-319-62446-4_4

Introduction

Heterogeneous catalysis has vital applications in chemical and energy industries. However, such reactions need higher temperature for the effective catalysis to beat the activation energy, i.e. energy consumption is devoted for the heterogeneous catalysis. It is important to mention that heterogeneous photocatalysis has become a promising and efficient technology towards environmental remediation and energy solution (Zhou et al. 2012). Further, solar-driven heterogeneous photocatalysis has attracted considerable attention for its application in global environmental pollution (Hoffmann et al. 1995; Zou et al. 2001; Asahi et al. 2001; Khan et al. 2002; Maeda et al. 2006). Photocatalytic water splitting reaction, an eco-friendly chemical reaction, generates hydrogen gas which is considered as an excellent fuel (Maeda et al. 2006; Warren and Thimsen 2012). Photocatalysts convert organic pollutants into H_2O and CO_2 mineralization while reduce CO_2 into alkanes derivatives to diminish the CO_2 from the atmosphere (Fujishima and Honda 1972; Tang et al. 2004; Liu et al. 2010). So it is highly demanding to develop sunlight (abundance of light)-active efficient photocatalysts (Wang et al. 2012). The traditional photocatalyst TiO_2 or ZnO requires UV light ($\lambda \leq 400$ nm) for its higher band gap value (≥ 3 eV) (Bavykin et al. 2006; Cheng et al. 2015). So these catalysts can utilize only about 4% of sunlight. In this regard, the development of visible light active photocatalysts (with lower band gap) is significant while approximate 44% of the whole solar spectrum is visible light (λ of 400–800 nm) (Zou et al. 2001). Additionally, photocatalysis under low wavelength UV light can only be carried out in special glass apparatus. Another drawback of the UV active phtocatalysis is the non-selectivity because of the huge generation of active oxygen species such as the superoxide anion radical ($O_2^{-\cdot}$), hydroxyl radical (·OH), peroxide radical (·OOH) which decomposes organic molecules into small inorganic species (CO_2 and H_2O) leading to uncontrolled oxidation. And UV-induced hole (h_{vb}^{+}, vacancy after electron leaving) is also highly oxidative which generates free radical oxygen species responsible for the undesired products (Lang et al. 2014). It is noteworthy to mention that these drawbacks of UV light-driven photocatalysis are resolved significantly in case of visible light-driven photocatalysis which is subjected to the better selectivity, easy manipulation in common glass apparatus, and utilization of available energy. Indeed, scientists paid their endeavour in the recent years towards the visible light active photocatalyst fabrication for dye degradation, generation of H_2 fuel by water splitting, organic transformations, etc. Consequently, various types of visible light active photocatalysts have been discovered such as heterojunction photocatalysts, semiconductor photocatalysts, dye-sensitized photocatalysts, etc. It is believed that utilization of clean, abundant and renewable sunlight (having 44% visible light) could be a gigantic platform for solving energy and environmental problems. Because of the above mentioned beneficial effect of visible light driven photocatalysis, it is highly demanding for designing and application of efficient visible light active heterogeneous photocatalysts.

For the development of efficient visible light-driven photocatalysts, it is necessary to consider the absorption energy of ≤ 3 eV and also significant separation of photogenerated electrons and holes to be utilized for reduction and/or oxidation reactions before their recombination. In this regard, nanoparticles (NPs) of noble metals (i.e. Ag, Au, Pt) are efficient to absorb visible light strongly (Sarina et al. 2013a, b; Watanabe et al. 2006; Chulkov et al. 2006; Stewart et al. 2008; Ghosh and Pal 2007; Wang et al. 2007) due to their surface plasmon resonance (SPR). SPR can be tuned by adjusting their size, shape and also surrounding (Zhou et al. 2012; Wang et al. 2012; Cheng et al. 2015; Lang et al. 2014; Murray and Barnes 2007; Lerme et al. 2010; Skrabalak et al. 2008; Daniel and Astruc 2004; Zhao et al. 2008). Additionally, NPs of noble metals can trap electron and works as active reaction sites (Cao et al. 2006). Generally, noble metal NPs deposition on semiconductor (i.e. metal–semiconductor nanocomposite) could be a novel notion for the development of visible light-active catalyst which can absorb visible light strongly, and metal–semiconductor interface are suitable for effective separation of photogenerated electrons and holes. Such composites, comprised of semiconductor and noble metal, are SPR active because of the noble metal under visible light. The nanomaterials or composites, SPR active due to the metal and capable of catalytic redox reaction in presence of visible light, are known as visible light plasmonic photocatalyst. However, noble metal NPs are visible light absorber as well as thermal redox active centres. Semiconductors are also active sites for the redox reactions. It is important to mention that solely noble metal or noble metal deposited on support is also efficient for the redox reactions under visible light because of the SPR effect.

SPR is the collective oscillations of electrons of conduction band in a metal particle generated by the electromagnetic field of incident light (Sarina et al. 2013a, b; Watanabe et al. 2006; Chulkov et al. 2006; Stewart et al. 2008; Ghosh and Pal 2007; Wang et al. 2007; Link and El-Sayed 1999). The SPR effect of noble metal in their nanostructures or nanocomposites (i.e. plasmonic photocatalysts) is accountable for the various applications. Plasmonic nanostructures attested wide range of application such as biotechnology, nearfield optics, catalytic sensors, solar cells, single-molecule spectroscopy, surface-enhanced Raman spectroscopy (Zheng et al. 2011; Renger et al. 2010; Maier et al. 2001; Larsson et al. 2009; Yu et al. 2010; Atwater and Polman 2010; Nakayama et al. 2008; Brus 2008; Kneipp et al. 1997; Nie and Emory 1997). In this chapter, we have focussed mainly on the following aspects: (i) the fundamentals of SPR, (ii) fabrication or synthesis of plasmonic photocatalyst, (iii) application of plasmonic photocatalysts on catalytic reaction by the visible light-induced SPR with detail understanding how the SPR works in plasmonic photocatalysts. It is worthwhile to mention that coinage metals (Cu, Ag and Au) are very effective for their plasmonic activity. However, copper in this regard needs a special mention for its cost effectiveness and it needs expert manipulation to protect Cu from oxidation. We have discussed various types of catalytic reactions in last section (application of plasmonic photocatalysts) such as photodegradation of large organic dyes (organic transformation), oxidation reaction, reduction reaction, hydroxylation, imine synthesis, cross-coupling reaction, water splitting reaction, and CO_2 reduction.

Fundamentals of Surface Plasmon Resonance (SPR)

Gustav Mie solved Maxwell's equations on optical absorption and scattering, and established Mie theory in the year 1908 rendering the idea of "plasmonics", mainly valid for spherical particles (Mie 1908). "Plasmonics" deals on the interaction between electromagnetic field and free electrons of metal. It is described that SPR is the result of interaction between light and metal (Kreibig and Vollmer 1995; Bohren and Huffman 1998; Kelly et al. 2003; Schuller et al. 2010; Hartland 2011; Halas et al. 2011; Roy and Pal 2015;). Simply, SPR is the collective oscillations of electrons of conduction band in a metal particle powered by electromagnetic field of incident light (Sarina et al. 2013a, b; Watanabe et al. 2006; Chulkov et al. 2006; Stewart et al. 2008; Ghosh and Pal 2007; Wang et al. 2007; Link and El-Sayed 1999). SPR is attributed upon illumination of light on NPs when the particle sizes are comparable or smaller than the wavelength of illuminated light. Upon light illumination there occurs polarization of the free electron cloud of the particle and this causes the accumulation of small negatively charged centres against the positive nuclei. And hence (by the accumulation of charge), electron density is increased on one side of NP (negative charge away from the line of propagation of light; shown as green colour) and decreased on the other side of NP (positive charge towards the line of propagation of light; shown as violet colour) as shown in Fig. 4.1a. So there is generated a new electric dipole in the particle upon light illumination (Fig. 4.1a). This is also considered as displacement of electron density in the NP. However, this newly distributed charge density generates an electric field inside and outside of the NP with a direction opposite to the direction of the electric field of the light. Concurrently, during this circumstance, the coulombic restoring force (Henglein 1999) of NP is established which acts on the electron to pull back on its earlier position (before redistribution of charge density). So two phenomena simultaneously occur on the NP under light illumination: one is new generation of electric dipole (or displacement of electron density), and another is the pulling back of electron by the coulombic restoring force. The simultaneous occurrence of these two phenomena leads to the oscillation of electrons on the NP which is known as surface plasmon resonance [or localized surface plasmon resonance (LSPR)] (Zhou et al. 2012; Cheng et al. 2015; Sarina et al. 2013a, b; Watanabe et al. 2006; Chulkov et al. 2006; Stewart et al. 2008; Ghosh and Pal 2007; Wang et al. 2007; Link and El-Sayed 1999; Kreibig and Vollmer 1995; Bohren and Huffman 1998; Kelly et al. 2003; Schuller et al. 2010; Hartland 2011; Halas et al. 2011; Roy and Pal 2015). This oscillation of electron of SPR is subjected to the electrons of conduction band.

Many metal NPs could endow LSPR under UV to visible light region because of the presence of huge free electrons in the conduction band. However, Au, Ag and Cu show distinct plasmon absorption in the visible region as mentioned earlier. LSPR leads the NP with very large extinction cross-section (sum of absorption and scattering cross-sections) and enhancement of local electromagnetic field near the surface of plasmonic NP (Zhou et al. 2012). The local field strength can be

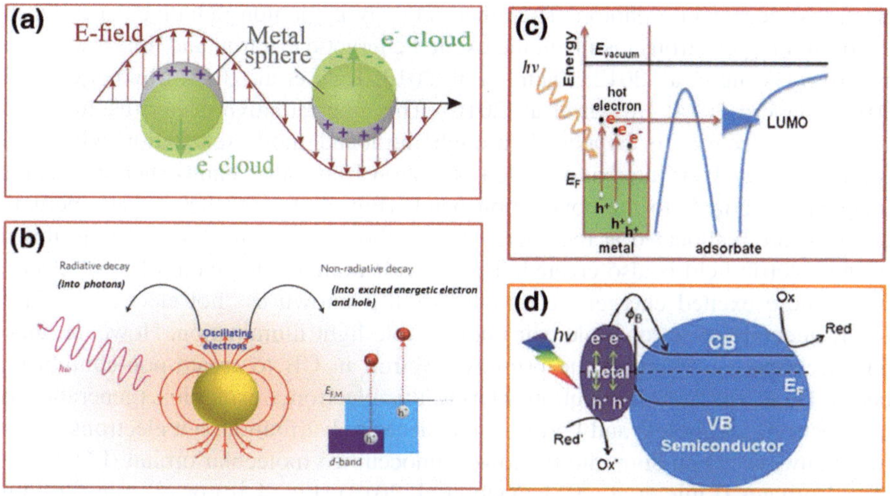

Fig. 4.1 **a** Schematic representation of SPR in spherical metallic NP under light illumination. From reference Kelly et al. (2003), Copyright 2003 American Chemical Society; **b** Surface plasmon decay processes: *Left* localized surface plasmons can decay radiatively via re-emitted photons, or *Right* non-radiatively via excitation of hot electrons (i.e. generation of electron and hole). From reference Clavero (2014), Copyright 2014 Nature Publishing Group; **c** Passing of hot electrons to LUMO of the adsorbates. From reference Xiao et al. (2013), Copyright 2013 Royal Society of Chemistry; **d** Transfer of hot electrons to the connected semiconductor (n-type). From references Cheng et al. (2015), Copyright 2014 Royal Society of Chemistry and Cheng et al. (2014), Copyright 2013 Royal Society of Chemistry

increased up to 500 times than the applied field during LSPR on NP of different shapes (cubes, nanowires, triangular plates) and junctions having sharp tips, edges and concave curvatures (Henglein 1999).

LSPR is strongly dependent on many factors such as the values of the dielectric constants of NP and the surrounding material, the particle size and shape of NP, and the surrounding environment (solvent and surfactants) (Wang et al. 2012). For example, LSPR absorption is not observed by Au NPs having diameter <5 nm, while sharp absorption band in the range of 520–530 nm is observed by the Au NPs with diameter of 5–50 nm (Chen et al. 1998; Yamada et al. 2007). All these factors influence the oscillation frequency of the conduction band electrons and enhancement of the local electromagnetic fields near the rough surfaces of NPs through LSPR (Kreibig and Vollmer 1995; Bohren and Huffman 1998; Kelly et al. 2003; Schuller et al. 2010; Hartland 2011; Halas et al. 2011; Roy and Pal 2015; Eustis and El-Sayed 2006).

Most important aspect of LSPR is the decay process of oscillating electron. The coherent oscillation of electrons (plasmon quantum) decays through two combative pathways (Fig. 4.1b) (Wang and Astruc 2014; Zhang et al. 2013; Kale et al. 2014; Mukherjee et al. 2013; Clavero 2014; Brown et al. 2016). One is the radiative pathway resulting scattering. In this process plasmon quantum (oscillating electrons) decays into photon. The other pathway is the non-radiative decay where the oscillating electrons results in the excited energetic electron and hole (lack of an electron) (Wang et al. 2012; Zhang et al. 2013; Kale et al. 2014; Mukherjee et al. 2013; Clavero 2014; Brown et al. 2016). This non-radiative decay (excitation of oscillating electron) is happened through intraband mode (excitation within the conduction band) or interband mode (excitation from other bands such as d bands (magenta coloured) to the conduction band) (Fig. 4.1b). So the excited energetic electrons are reallocated at the energy levels above the Fermi level. Due to this, a strong electric field is also created. Thus, LSPR renders increment of local electric field. These excited energetic electrons are also known as "hot electrons". Thus, LSPR generates electron–hole pair upon visible light illumination. However, these hot electrons can energize the primary electron in CB to create non-equilibrium Fermi–Dirac electron distribution. During this electronic excitation (generation of hot electrons and holes) and Fermi–Dirac electron distribution, hot electrons follow two pathways (1) passing into the lowest unoccupied molecular orbital (LUMO) of the adsorbates (Linic et al. 2011; Xiao et al. 2013) (Fig. 4.1c) or (2) transferred to the connected semiconductors beating the interfacial Schottky barrier (electrostatic depletion layer) (Fig. 4.1d) (Zhou et al. 2012; Wang et al. 2012; Warren and Thimsen 2012; Xiao et al. 2013; Jiang et al. 2014; Lou et al. 2014; Cheng et al. 2014; Hou and Cronin 2013). Simultaneously, there are generated holes when electrons are transferred to the attached semiconductor or adsorbed molecules to keep the charge balance. Then these transferred energetic electrons and/or hole play the role for catalytic redox reactions i.e. photocatalysis.

During the visible light-driven photocatalysis, there occur not only the charge carriers (energetic electron and hole) transfer but also plasmonic heating (generation of heat by SPR) because of the electron–electron and electron–phonon collisions (Clavero 2014; Inagaki et al. 1981). Thus the charge carrier transfer (hot electron and hole transfer) and thermal heating processes because of the LSPR are effective for the catalytic chemical reactions (redox reaction) of the adsorbates. Thus this is known as plasmon-enhanced photocatalysis.

Fabrication Processes of Plasmonic Photocatalysts

Numerous synthetic methods for the fabrication or preparation of plasmonic photocatalysts have been developed. General method of the fabrication is to anchoring the noble metal NPs or their films on various substrates such as metal oxides, zeolite, halides, graphene oxides, reduced graphene oxides, carbon nanotubes, etc. The reported methods include deposition–precipitation, photoreduction, ion-exchange, impregnation, chemical reduction, physical vapour deposition, hydrothermal, encapsulation, etc. Herein, it has been discussed the most important fabrication methods of plasmonic photocatalysts.

(i) Deposition–precipitation

One of the most commonly used easy and facile methods for the preparation of plasmonic photocatalyst is deposition–precipitation method (Zhou et al. 2012). Herein, an example has been illustrated for the preparation of Au/TiO_2 plasmonic photocatalyst (Primo et al. 2011). The procedure for the preparation of TiO_2-supported gold NPs is illustrated in Fig. 4.2. In this synthetic method, TiO_2 was added to the aqueous solution of hydrogen tetrachloroaurate(III) ($HAuCl_4$) of pH = 5–8 and stirred for 24 h (step-II) leading to a solid product. In this step (step-II), Au $(OH)_x$ is deposited on TiO_2 and thus known as deposition step. Subsequently, the as-obtained solid product was thoroughly washed with water to remove chloride ions completely. Finally, reduction was carried out (step-III) by thermal treatment under hydrogen gas (or any another appropriate reducing atmosphere), or through reduction in alcohols or reducing liquid. In this step, Au^{3+} is reduced to metallic Au^0 and precipitated on TiO_2 to provide Au/TiO_2 plasmonic photocatalyst. The main parameter of the deposition–precipitation fabrication method is the pH value of the medium at the deposition step. The deposition–precipitation procedure is a very simple and convenient approach.

(ii) Photoreduction or photodeposition

This is the fabrication methodology where electrons and holes are produced first upon light irradiation on semiconductor, and thereby the electrons are used for the reduction of metal precursor (such as salt of the metal) to metal. Then the reduced metal particles are deposited on the surface of the used semiconductor (Kochuveedu et al. 2013). This method is known as photodeposition. It is also called as photoreduction. Usually TiO_2 or ZnO nanomaterials are used as semiconductor for these processes. The metal precursors (salts or ions) are generally allowed for the adsorption on the surface of the support before the photoreduction. After that the complex is irradiated with light for the reduction of metal precursor into its metal NPs and their simultaneous deposition on the semiconductor to make metal/semiconductor composite photocatalysts. For example, Ag/TiO_2 was prepared by this method (Hu et al. 2006). P-25 TiO_2, $AgNO_3$, CH_3OH and NaCl (22.2 mg) were added to an aqueous solution for the adsorption of $AgNO_3$ on TiO_2 (Fig. 4.3). Then under nitrogen atmosphere the solution was irradiated with UV light (4-W black light lamp, wavelength ~365 nm) for 3 h for Ag^+ reduction. After

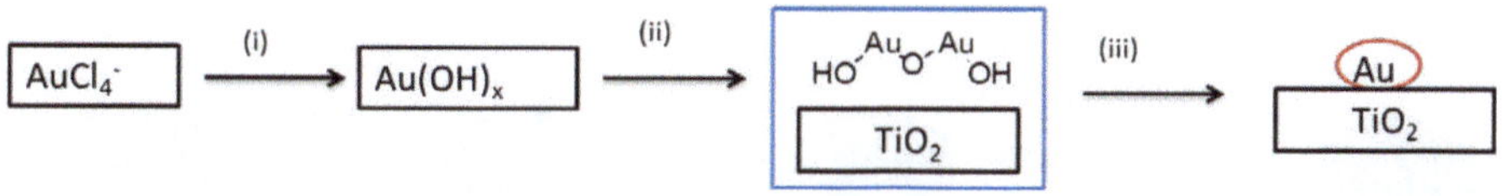

Fig. 4.2 Steps of the deposition–precipitation method. **i** pH = 5-8, **ii** $Au(OH)_x$ deposition on TiO_2 (24 at rt); **iii** reduction and precipitation: Thermal treatment with H_2 gas/reducing atmosphere or reduction with alcohol/reducing liquid media

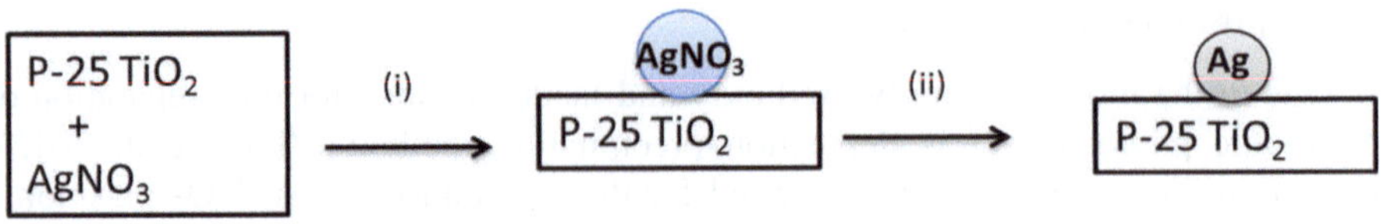

Fig. 4.3 Photodeposition method for Ag/TiO_2 preparation. **i** Methanol/brine solution (adsorption), **ii** UV illumination at wavelength ~365 nm (photoreduction)

illumination, the material was separated by filtration and washed with deionized water, and dried to furnish Ag/TiO_2 composite.

(iii) Ion-exchange method

Ion-exchange methods have also been exploited to fabricate the photocatalysts. Commercial ion exchange method has been developed by our group for the synthesis mono-, bi-, and tri-metallic nanoparticle (Praharaj et al. 2004, 2006; Dutta et al. 2016) which was further exploited for metal oxide preparation (Sinha et al. 2009). Through this method the precursor salts are generally allowed for the chemical exchange reaction in the solution. After the chemical exchange reaction a partial reduction of a metal salt leads to the metal particle which gets deposited onto the unreacted salt to lead the metal/metal salt composite photocatalyst. However, the solubility of products should be lower than reactants. The facile in situ ion-exchange route was adopted for the preparation of ternary hierarchical architecture photocatalysts Ag/AgX(X = Cl, Br, I)/$AgIO_3$ (Zeng et al. 2016). Figure 4.4 illustrates the fabrication method of the Ag/AgX (X = Cl, Br, I)/$AgIO_3$ ternary hierarchical composites. In this method, the KX (X = Cl, Br, I) solution was added into $AgIO_3$ suspension in water and AgX (X = Cl, Br, I) nanosheets was generated according to the ion-exchange reaction (Eq. 4.1). After the reaction AgX was deposited on unreacted $AgIO_3$. In this case KX and KIO_3 are soluble in solution. However, in the interim, as-generated AgX sheets would decompose partially into Ag particles by the atmosphere light. As Ag particles were produced due to the outer light and these were deposited on the surface of AgX. Presence of light for longer time leaded more generation of Ag particles from AgX and deposited on unreacted AgX. Consequently, the Ag/AgX(X = Cl, Br, I)/$AgIO_3$ ternary hierarchical architectures are fabricated (Fig. 4.4).

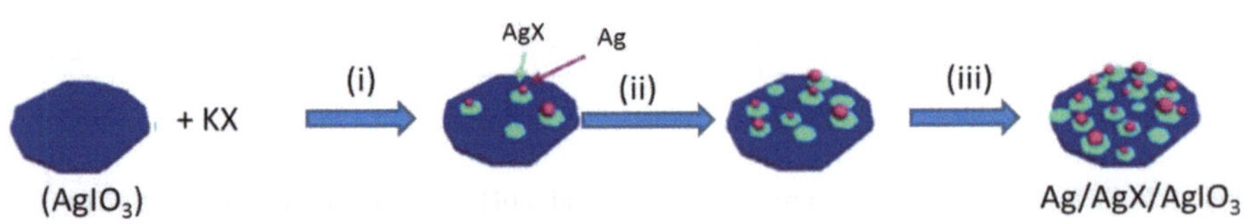

Fig. 4.4 Fabrication of Ag/AgX(X = I, Br, Cl)/$AgIO_3$. **i** Magnetic stirring in water, **ii** Light illumination, **iii** More light illumination. From reference Zeng et al. (2016), Copyright 2016 American Chemical Society

$$KX + AgIO_3\,(\text{excess}) = AgX + KIO_3 + AgIO_3\,(\text{unreacted}) \tag{4.1}$$

$$Ag^+\,(\text{of AgX}) \rightarrow Ag\,(\text{Through reduction}) \tag{4.2}$$

$$Ag\ +\ AgX\,(\text{unreacted}) +\ AgIO_3(\text{unreacted}) \rightarrow Ag/AgX/AgIO_3 \tag{4.3}$$

(iv) Chemical method

Chemical method for the plasmonic composite photocatalyst synthesis is the simplest methodology. This method is consisting of two steps like earlier methods such as ion-exchange method and deposition–precipitation method. The steps are adsorption of metal precursors on the surface of a semiconductor and followed by chemical reduction of metal precursor which furnishes the metal NPs–semiconductor composite photocatalyst. Otherway, metal NPs are prepared and followed by these preformed metal NPs is used for the preparation of metal–semiconductor photocatalysts just through mixing. Linic and his co-workers prepared Ag–TiO_2 and Au–TiO_2 adopting the later method (Fig. 4.5) (Ingram and Linic 2011). They prepared Ag (or Au NPs) through the addition of $AgNO_3$ (or $HAuCl_4$ for Au NPs synthesis) and polyvinylpyrrolidone (PVP) (as stabilizer) to the solution of ethylene glycol (as solvent and reductant) and dilute HCl. The reaction produced Ag NPs (or Au NPs) with a layer of PVP. Herein, PVP stabilizes the surface of Ag NPs (or Au NPs) by preventing agglomeration. Then the pre-synthesized Ag NPs (or Au NPs) were used for the preparation of Ag/N–TiO_2 (or Au/N–TiO_2) by mixing the Ag (or Au NPs) solution with the N–TiO_2 in ethanol mixture and followed by sonicating for one hour furnishing Ag/N–TiO_2 (or Au/N–TiO_2).

(v) Sputtering method

This method is a highly energetic process to prepare metal–semiconductor photocatalyst composite. Highly energetic particles are pitched onto the target material. As a result metal atoms are ejected due to the bombardment from the solid target material. These sputtered atoms, ejected into gas phase, are not in thermodynamic

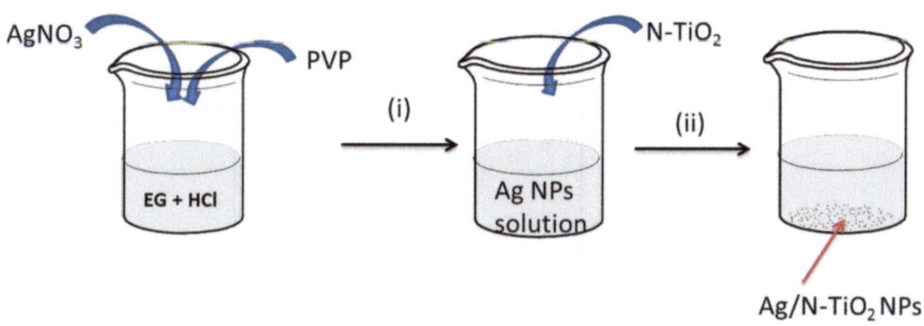

Fig. 4.5 Synthesis of Ag–TiO_2 plasmonic photocatalyst. **i** ethylene glycol/HCl (preparation of Ag NPs), **ii** mixing and sonication for 1 h (preparation of Ag–TiO_2 nanocomposite)

equilibrium state, and easily deposited onto the substrate (such as semiconductor) as described in Fig. 4.6. Generally, these processes are carried out in a vacuum using argon plasma. An intriguing characteristic of this method is that a thin metal layer could easily be deposited on substrate to cover it. Au-decorated ZnO has been prepared by this method. In this preparation method, first ZnO was deposited on Si (100) and then it was used as the substrate for sputtering. As the result, ultrathin Au layer (of thickness ~2 nm) was fabricated on ZnO/Si material to provide the Au/ZnO/Si composite (Dhara and Giri 2011).

(vi) Encapsulation

This method describes the covering of plasmonic metal nanoparticles by semiconductor. So the encapsulation of metal nanoparticles with semiconductor leads to the core–shell nanostructure of metal–semiconductor plasmonic photocatalysts, where metal NP positioned in core and semiconductor positioned in the outer layer (Fig. 4.7). Encapsulation has a very useful significance because it prevents the agglomeration and also surface oxidation of plasmonic metal NPs. Hence, the composite gains long-term stability. For plasmonic metals which are prone towards oxidation, the encapsulation is very much useful.

Cushing and his co-workers (Cushing et al. 2012) demonstrated the synthesis of Au nanospheres (average diameter ~20 nm). Then they prepared Au@SiO_2 and Au@Cu_2O core–shell structures successfully by coating the synthesized Au NPs with SiO_2 layer (~5 nm thick) and Cu_2O layer (~25 nm thick), separately. The synthetic diagram is given in (Fig. 4.7).

(vii) Thermal method

This is a wet chemical method. There are two types of thermal method, hydrothermal and solvothermal. These methods are simple and cost-effective methods. In the hydrothermal method, metal (or metal precursor) and semiconductor (or semiconductor precursor) are taken in water in a test tube and is closed tightly. Then the test tube was placed in an autoclave or hydrothermal chamber for

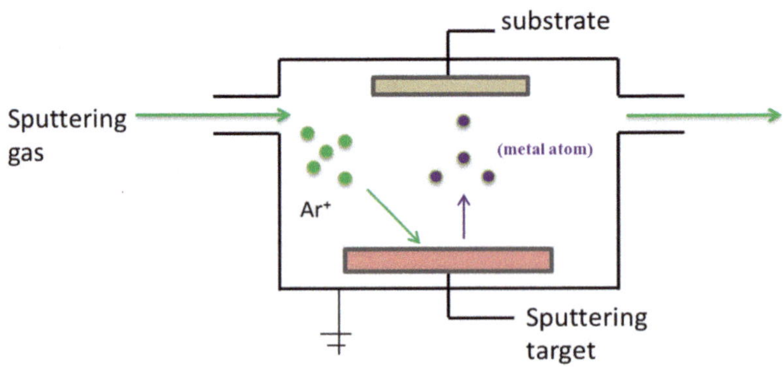

Fig. 4.6 Sputtering method for metal deposition on substrate

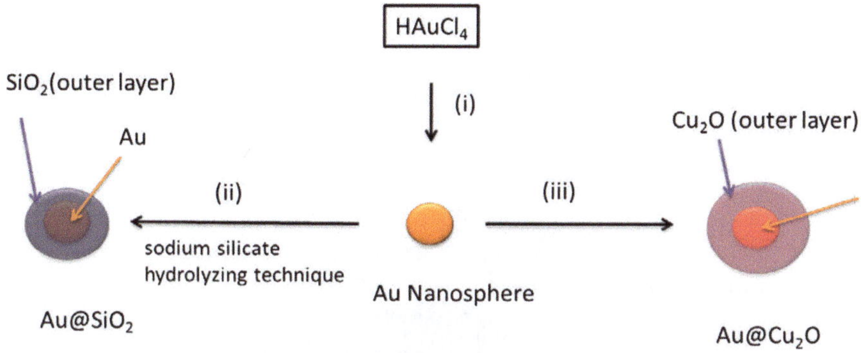

Fig. 4.7 Synthesis of Au@SiO_2 and Au@Cu_2O core–shell nanoparticles where SiO_2 and Cu_2O located at outer layer. **i** Sodium citrate, **ii** aminopropyltrimethoxysilane/sodium silicate/ethanol, 2d kept, **iii** $CuCl_2$, SDS, NaOH, $NH_4OH{\cdot}HCl$, stirring

thermal treatment. In the autoclave or hydrothermal chamber (developed indigenously in our laboratory) (Sinha et al. 2010), temperature and pressure of the reaction system are dramatically increased which could be suitable for the reaction. During the process metal or metal precursors or metal NPs (converted from metal precursor) adsorbed on the surface of semiconductors or semiconductor precursor. Through this adsorption (or through adsorption and reduction/decomposition in case of precursors), the composite is transformed into the metal–semiconductor photocatalyst composite on thermal treatment. In the solvothermal process, any other solvent(s) is used instead of water. However, there occurs some aggregation of metal NPs. We reported the synthesis Au-ZnO plasmonic photocatalyst using modified hydrothermal (MHT) method (Mondal et al. 2014). In this process, triethylamine (as reducing and hydrolysing agent), $HAuCl_4$ (Au metal precursor) and $ZnSO_4$ (semiconductor precursor) in aqueous solution was placed in a screw capped test tube. The thermal treatment for 24 h at 100 °C furnished the Au–ZnO photocatalyst (Fig. 4.8). The hydrothermal chamber is a wooden box fitted with thermometer, 100 W bulb, and test tube stand (Fig. 4.8) (Sinha et al. 2010).

(viii) Template method

In this method metal–semiconductor composite is prepared using another material which plays important role to combine the metal and semiconductor for the composite formation. So the method leads to the ternary-type composite (metal–semiconductor–template or metal–semiconductor–functionaliser) material. So it is required third-type material having functionalizing property known as templating material. This method is known as template method. A sandwich-like structure composed of SiO_2/Au/TiO_2 was prepared applying this method (Zhang et al. 2011). In this template method (Fig. 4.9), SiO_2 particles were first synthesized and functionalized with (3-Aminopropyl)triethoxysilane (APTES). Then this APTES functionalized SiO_2 were attached with Au NPs to generate SiO_2-APTES/Au. Again,

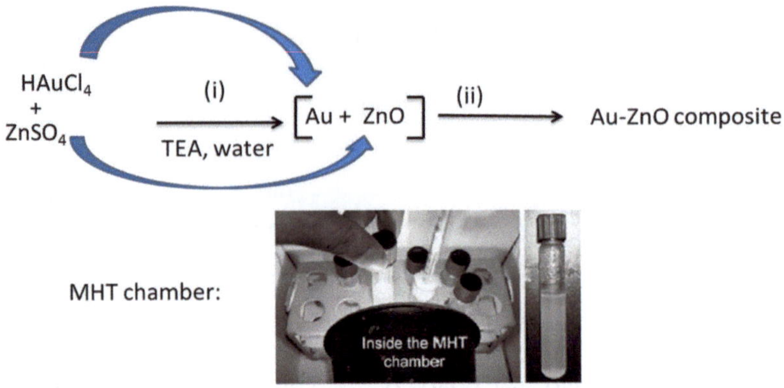

Fig. 4.8 Preparation Au-ZnO composite through modified hydrothermal (MHT) method at 100 °C. **(i)** Thermal treatment under MHT (100 °C), **(ii)** Growth of ZnO petals on the Au particles under MHT condition [(100 °C) to furnish Au–ZnO. Image of MHT chamber taken from reference Sinha et al. (2010), Copyright 2010 John Wiley & Sons]

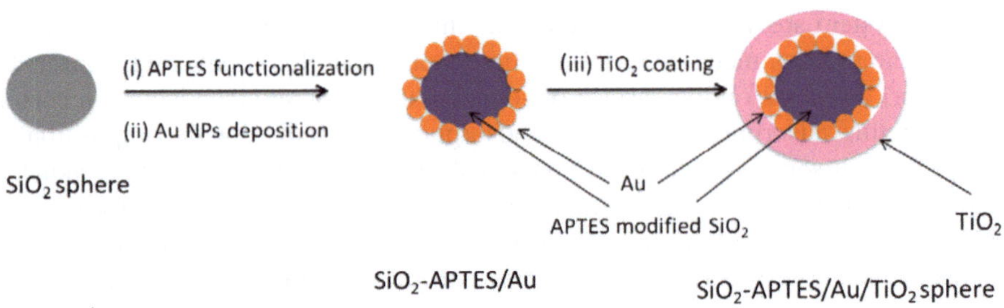

Fig. 4.9 Template method for the fabrication of SiO_2-APTES/Au/TiO_2 sphere composite

SiO_2-APTES/Au NPs were coated with amorphous TiO_2 to provide SiO_2-APTES/Au/TiO_2 hybrid particles. TiO_2 is located as shell of the particle. Here in APTES acts as template.

Catalytic Applications of Plasmonic Catalysts in Presence of Visible Light

Cu, Ag, Au, Pd, Pt NPs-based plasmonic photocatalysts have been chiefly used to catalyze various chemical redox reactions using the SPR effect of these metals under visible light irradiation. Herein, we have discussed major applications of SPR on the catalytic organic transformations (such as degradation of organic pollutants, organic synthetic methods, e.g. oxidation reactions, reduction reactions, imine formation, coupling reaction, hydroxylation, etc.) and clean energy conversion (such as water splitting, CO_2 reduction). Detail discussion of these reactions has been interpreted here.

(A) **Organic Transformations**

In this section, it has been discussed transformations of small as well as large organic molecules by the plasmonic photocatalysts under visible light illumination. Some important reactions have been depicted such as degradation of organic dyes, oxidation reactions, reduction reactions, imine formation, hydroxylation, etc. However, oxidation reactions, reduction reactions, hydroamination, hydroxylation, and coupling reactions have been classified as organic synthesis.

(i) *Degradation of Organic Pollutants.*

Organic dyes with large amount are used in textile, printing, photographic and other industries which leads to environmental pollution. A substantial fraction of dyes are lost in the dying process and thus these have been released into water to make it pollute. These are some notable examples. In China, it has been found 1.6×10^9 m^3 of dye-containing wastewater. Generally, dye pollutants, the large organic molecules, are hard to be biodegraded or oxidized under environmental condition or with chemicals. Henceforth, there is an urgent need to apply appropriate photocatalysts to degrade these dyes under aqueous medium. Furthermore, possible degraded products of the dyes along with dyes are carcinogenic, mutagenic, toxic and teratogenic for living organisms of ecosystem. For these reason, scientists have paid their attention and effort for the development of nanomaterials to degrade organic pollutants under visible light. Visible light is more cost effective than UV light or other rays. Furthermore, sunlight contains 44% of visible light. So development of visible light active photocatalyst is highly demanding. Towards the purpose of fabrication of visible light-active photocatalysis, plasmonic photocatalyst has attracted enormous attention and considerable efforts are devoted for the development of nanomaterials to degrade organic pollutants under visible light (Zhou et al. 2012; Brown et al. 2011).

Zhu et al. prepared Au/TiO_2 nanosheets with exposed {001} facets and applied the plasmonic photocatalyst for novel visible light-driven degradation of rhodamine B (RhB) in aqueous solution (Zhu et al. 2012). AuNPs (average diameter ~ 5 nm) were deposited on the {001} facet of TiO_2 nanosheets leading to Au/TiO_2 plasmonic photocatalyst which displayed very efficient photocatalytic activity in visible light region due to the SPR effect of Au NPs. Au/TiO_2-001 (Au/TiO_2 nanosheets with exposed {001} facets) attested superior photocatalytic activities than bare TiO_2-001. Such efficient photocatalytic activity occurred because of higher electron mobility and better adsorption of pollutant molecule by {001} facets. Significantly, the Au NPs play an essential role to generate photoelectrons and enhance the visible light absorption intensity for the enhanced photocatalysis.

Mechanistically, the photogenerated excited electrons from Au NP because of the SPR under visible light illumination would transfer (higher negative potential than CB level of TiO_2-001) to CB of TiO_2 particle. Simultaneously, holes are also generated in the Au NPs. Consequently, the holes (i.e. $^{\cdot}Au^{+}$) attack adsorbed RhB to degrade it. On the other hand, the photogenerated hot electrons (already passed to the CB of TiO_2) may be trapped by oxygen to generate $^{\cdot}O_2^{-}$ radicals which react with H^+ (of H_2O) to produce H_2O_2 (through ·OOH) in aqueous solution as depicted in Fig. 4.10. Again, H_2O_2 would capture electron (of CB of TiO_2) and transforms into ·OH radicals. These reactive species ·OH, $^{\cdot}O_2^{-}$, H_2O_2 are responsible for the degradation of RhB. However, photoexcited RhB* radical may transfers electron to the CB of TiO_2 to form $RhB^{\cdot+}$ which is easily degraded. The mechanism with

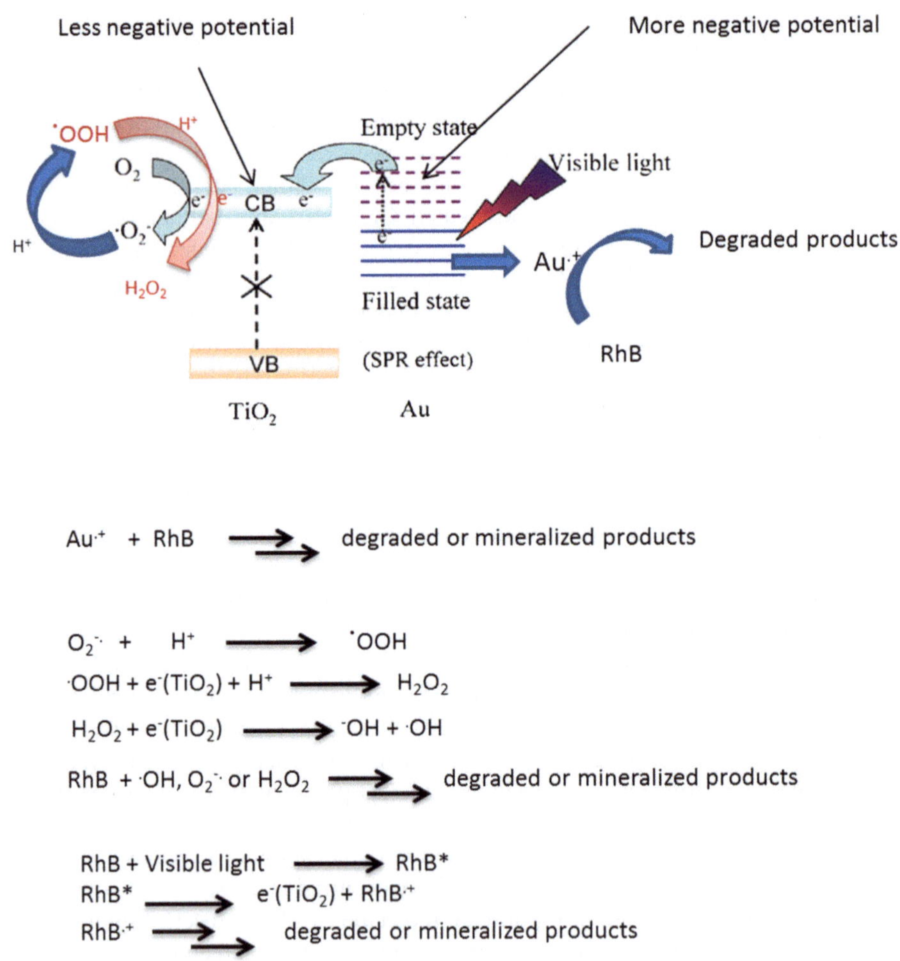

Fig. 4.10 Mechanism of photocatalytic degradation of RhB by plasmonic Au–TiO_2 under visible light. From reference Zhu et al. (2012), Copyright 2012 Elsevier

equations depicted in Fig. 4.10. TiO_2 is not effective to absorb visible light to generate hot electrons and hole.

Ternary composite Ag/AgCl/$AgIO_3$ has been found to be very efficient plasmonic photocatalyst for the photodegradation of methyl orange (MO) under visible light ($\lambda > 420$ nm) irradiation (Zeng et al. 2016). The SPR effect of Ag species in the ternary composite plays major role for such efficiency of photocatalysis. Mechanistic diagram has been presented in Fig. 4.11. AgCl has large band gap and thus it has no ability to absorb visible light ($\lambda > 420$ nm). Ag particle has the ability to absorb visible light and thus it induces the photogeneration of electrons and holes due to the dipolar character and SPR of metallic Ag. Then the electrons go to the CB of $AgIO_3$ (through the CB of AgCl). As the CB of $AgIO_3$ at positive state, it reduces the reducing property of electrons. The redox potential of CB in $AgIO_3$ is more positive than the redox potential of $O_2/^{\cdot}O_2^-$ to produce $^{\cdot}O_2^-$. So the Fermi level of $AgIO_3$ would aligned such that the reduction reaction of O_2 to $^{\cdot}O_2^-$ can take place smoothly. Then $^{\cdot}O_2^-$ plays the role for the degradation of MO according to the reaction discussed earlier. Since reducing capacity of electrons to reduce O_2 has declined, herein holes play a more important role than electrons (or $^{\cdot}O_2^-$) for photodegradation of MO. Holes convert MO to $MO^{+\cdot}$ which degraded or mineralized into CO_2 and water.

We have recently reported the SPR of Au induced enhanced photocatalytic efficiency of Au-ZnO for RhB degradation in the sun light (Mondal et al. 2014). The probable mechanism for the dyes degradation is illustrated in Fig. 4.12. Upon sunlight exposure, electrons (e^-) and holes (h^+) are generated in the CB and VB on the surfaces of ZnO, respectively (Fig. 4.12 and Equations). Then the photogenerated hot electrons (transferred from CB of ZnO to Au) react with the adsorbed O_2 to produce $^{\cdot}O_2^-$ (Fig. 4.12). This superoxide radicals react with water to generate H_2O_2, ^-OH and $^{\cdot}OH$ according to equations discussed earlier. These radicals ($\cdot OH$, $^{\cdot}O_2^-$) and H_2O_2 are responsible for the degradation of organic dyes molecule. In our work, we have shown that photocatalytic activity of Au–ZnO is much greater than that of ZnO suggesting presence of the Au NP on ZnO enhances photodegradation

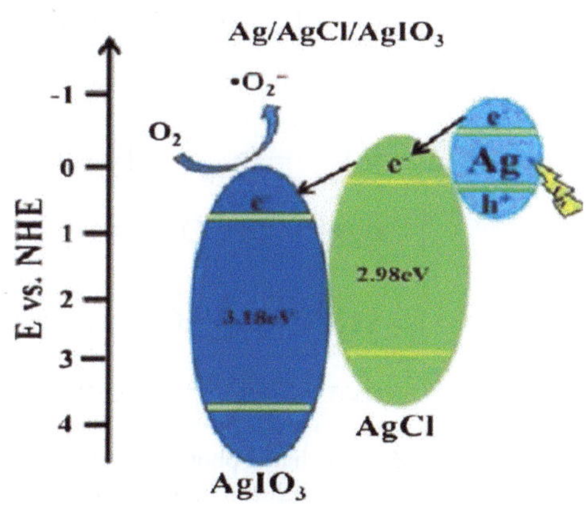

Fig. 4.11 Evidence of SPR effect of Ag for the photocatalytic degradation of MO by Ag/AgCl/$AgIO_3$. From reference Zeng et al. (2016), Copyright 2016 American Chemical Society

efficiency. For such successful plasmonic catalysis, there have been illustrated two main key factors. They are Schottky junction and localized surface plasmon resonance (LSPR). The Schottky junction between Au and ZnO creates an internal electric field which guides the movement of electrons and holes in the opposite directions. Thus, charge separation and charge transfer occurs smoothly. Whereas LSPR is responsible for strong absorption of visible light and excitation of charge carriers (electron and hole) takes place. Herein, the surface plasmon frequency of Au and the LSPR play the contribution for visible light absorption process. Presumably, the SPR effect of Au helps the excitation of more electrons and holes.

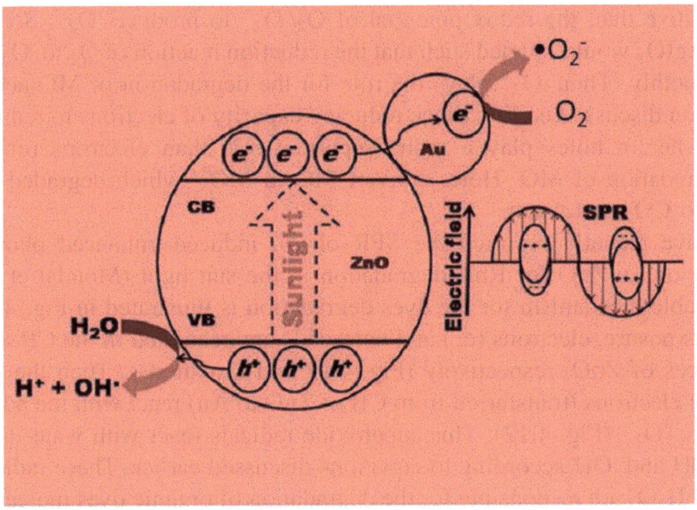

$$ZnO + h\nu \rightarrow ZnO\,[e^- + h^+]$$

$$ZnO\,[e^-] + Au \rightarrow Au\,[e^-] + ZnO$$

$$Au + h\nu\,(SPR) \rightarrow Au\,[e^- + h^+]$$

$$Au[e^-] + O_2 \rightarrow Au + {}^{\bullet}O_2^-$$

$$\bullet O_2^- \rightleftharpoons H_2O_2 + {}^-OH + OH^{\bullet}$$

$$dye + {\cdot}OH,\ O_2^{-\cdot}\ or\ H_2O_2 \rightarrow degraded\ products$$

$$H_2O + h^+\,[ZnO] \rightarrow H^+ + OH^{\bullet}$$

$$dye + OH^{\bullet} \rightarrow degraded\ products$$

Fig. 4.12 Au-ZnO mediated photocatalysis of dyes (RhB, MB, MO, CR and Rose Bengal). From reference Mondal et al. (2014), Copyright 2014 Royal Society of Chemistry

Kominami et al. presented Au/CeO_2 catalyzed mineralization of few organic acids (oxalic acid, formic acid and acetic acid) in aqueous medium in presence of visible light (wavelength ~520 nm) (Kominami et al. 2011). These acids were decomposed to carbon dioxide efficiently. The plasmonic effect of Au plays important role for such efficient photocatalytic degradation. The mechanism is presented in Fig. 4.13. First, incident light was absorbed by Au nanoparticles for the LSPR excitation, and the excited electrons are passed from Au particle to the CB of CeO_2. Consequently, electron-deficient Au NPs (i.e. $Au^{+}_{(n)}$ because of hole) could oxidize the organic acid into carbon dioxide. On the other hand, electron of conduction band in CeO_2 generates ·OH, $^{\cdot}O_2^{-}$ and H_2O_2 (as discussed earlier). Such active radicals (·OH, $^{\cdot}O_2^{-}$) and H_2O_2 are also effective for the oxidation of organic acid into carbon dioxide.

Zhu and his co-workers demonstrated an efficient surface plasmon resonance effect of Au NPs (Zhu et al. 2009). They reported how Au NPs can strongly absorb visible light which results to the degradation of sulforhodamine-B (SRB) (wavelength of used blue light ~400–500 nm). Herein, Au NPs absorb visible light strongly through the SPR effect, and thus the 6sp electrons gain energy which migrate to higher intraband energy levels. So the oscillating electrons decay (through the excitation of intraband mode) results hot electrons and holes (Fig. 4.14), and hot electrons are arrested by oxygen molecules to generate $^{\cdot}O_2^{-}$ radicals. These $^{\cdot}O_2^{-}$ radicals generates further ·OOH, ·OH and H_2O_2 (as discussed earlier) which play the role for the degradation of dye. Additionally, SRB is excited by visible light to SRB* which also donates electron to the gold's 6sp band (positively charged due to the SPR, i.e. hole generation) and $SRB^{\cdot+}$ is produced to be degraded. This facilitates the photocatalytic degradation. Thus there is the combination between SPR of Au and dye sensitization by SRB. When more powerful visible light illuminated, more hot electrons (thus more $^{\cdot}O_2^{-}$ generation) and more holes (positive charge) in the 6sp band of Au are produced. Thereby enhanced $^{\cdot}O_2^{-}$ and $SRB^{\cdot+}$ generation leads to the better degradation.

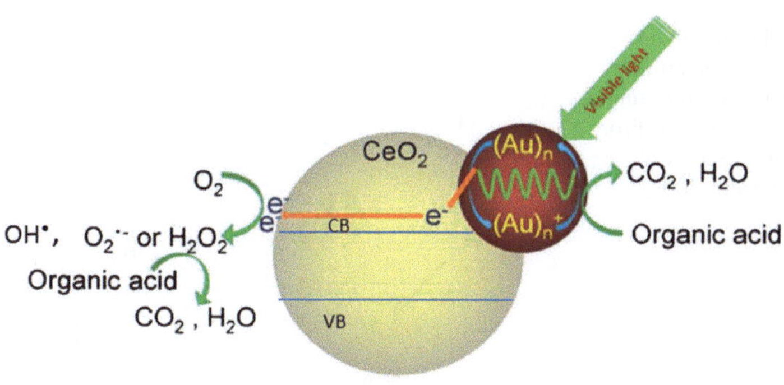

Fig. 4.13 Mechanism for SPR induced mineralization of organic acids in aqueous medium by Au/CeO_2 in presence of visible light. From reference Kominami et al. (2011), Copyright 2011 Elsevier

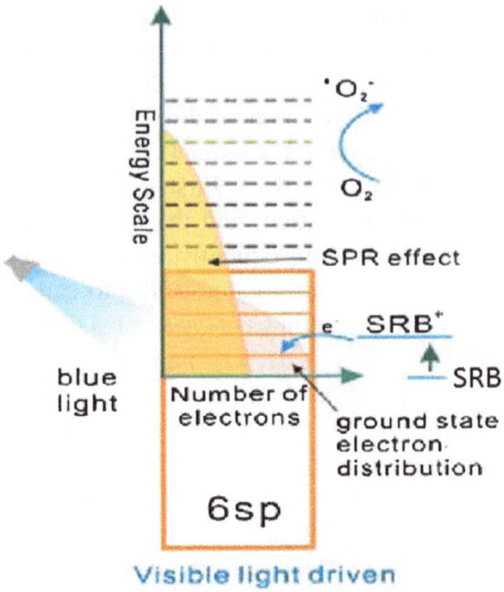

Fig. 4.14 Photocatalytic degradation of SRB (sulforhodamine-B) by Au NPs in presence of visible light. From reference Zhu et al. (2009), Copyright 2009 Royal Society of Chemistry

Photocatalytic degradation of rhodamine B (RhB) in aqueous medium by Ag-deposited TiO_2 nanoparticles under visible light has been described (Sung-Suh et al. 2004). The enhancement of photocatalytic activity can be explained by SPR effect of Ag as well as visible light absorption by RhB on catalyst surface (Fig. 4.15). The SPR of Ag generates electron and hole by visible light. Again RB*, excited by the visible light, transfers electron to the CB of TiO_2 and thus generates $RB^{+\cdot}$ which is easily degraded through oxidation. The electrons (electron on CB on TiO_2) again are transferred to Ag particle. The collective electron generated by SPR of Ag and transferred electron from CB of TiO_2 (from RB*) are absorbed by the O_2 which further produces $^{\cdot}O_2^-$. The superoxide radicals ($^{\cdot}O_2^-$) further generates ·OOH, ·OH, & H_2O_2 which play the role for the degradation of dye.

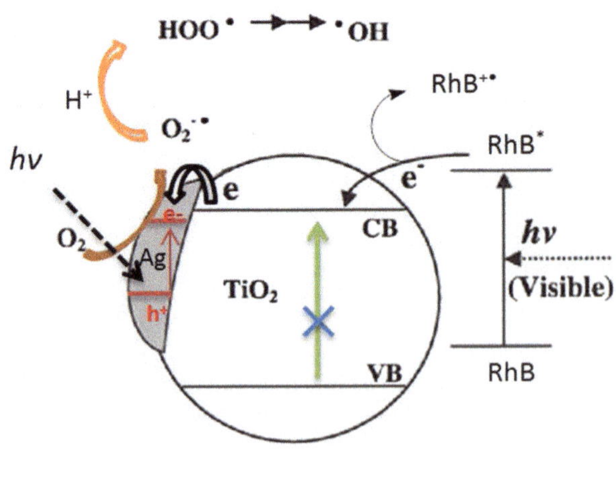

Fig. 4.15 Photocatalytic degradation of rhodamine B (RhB) by Ag–TiO_2 NPs under visible light irradiation. From reference Sung-Suh et al. (2004), Copyright 2004 Elsevier

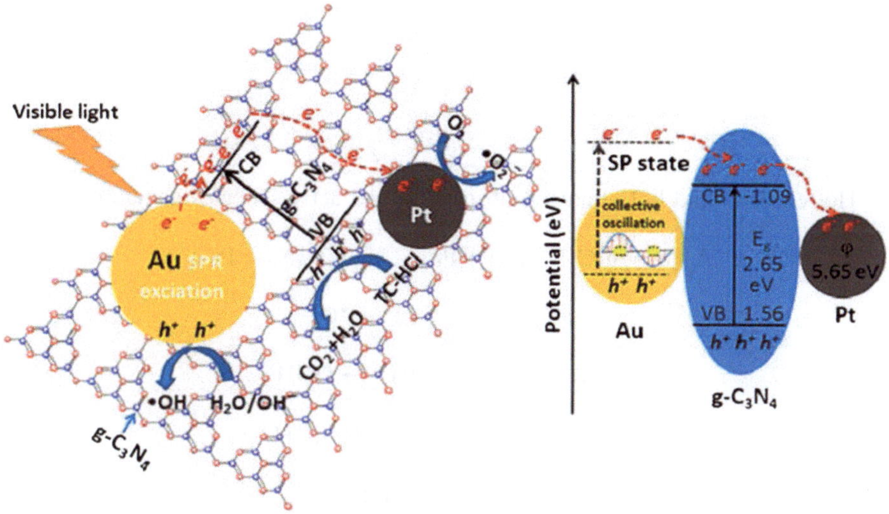

Fig. 4.16 Proposed mechanism for degradation of tetracycline hydrochloride (TC–HCl) by Au/Pt/g-C_3N_4 nanocomposites under visible light irradiation. From reference Xue et al. (2015), Copyright 2015 American Chemical Society

Enhanced photocatalytic activity of Au/Pt/g-C_3N_4 on tetracycline hydrochloride (TC–HCl) caused by the SPR effect of Au as well as electron-sink function of Pt nanoparticles has been investigated (Xue et al. 2015). These effects improve light absorption property and photogenerated charge carriers separation of g-C_3N_4 which work synergistically leading to the efficient photocatalysis progression (Fig. 4.16).

Mechanistically, because of the SPR excitation of Au NPs there are generated hot electron–hole pairs. Consequently, the hot electrons passed to the CB of g-C_3N_4 (CB of Au is above CB of g-C_3N_4) and hot holes are left on Au NP transiently. Then g-C_3N_4, a photoexcited semiconductor, rapidly transfers hot electron to the Pt NPs. However, under visible light illumination electron and hole are also generated in g-C_3N_4. The electrons transfer process from CB of g-C_3N_4 to Pt NP is feasible since the CB of g-C_3N_4 (−1.09 eV) is lower than the work function of Pt (5.65 eV). The collective transferred electrons on Pt NPs react with O_2 to generate superoxide radicals ($^{\cdot}O_2^-$) which further generates ·OOH, ·OH, & H_2O_2. Concurrently, photogenerated holes (h +) left on the Au NP reacted with H_2O or OH^- to produce ·OH. But in the contrary, photogenerated holes (h^+) left on the VB of g-C_3N_4 cannot react with H_2O or OH^- to produce ·OH because the E_{VB} (+1.57 eV vs. SHE) of g-C_3N_4 is lower than the standard redox potentials of ·OH/H_2O (+2.68 V vs. SHE) and ·OH/OH^- (+1.99 V vs. SHE). Henceforth, the holes (h^+) on the VB of g-C_3N_4 (large catalytic site) are spent to decompose TC–HCl according to the equations as follows. Huge generation of ·OH are also highly effective for the decomposition process.

Photoexcitation:

$$Au + h\nu \rightarrow Au\ (e^- + h^+)$$

$$g\text{-}C_3N_4 + h\nu \rightarrow g\text{-}C_3N_4(e^- + h^+)$$

Hot electron and hole transfer, absorption and active species generation:

$$Au(e^-)/g\text{-}C_3N_4(e^-) + Pt \rightarrow Au/g\text{-}C_3N_4 + Pt(e^-)$$

$$Pt(e^-) + O_2 \rightarrow Pt + {}^{\bullet}O_2^-$$

$${}^{\bullet}O_2^- + H_2O \rightarrow \bullet OH + \bullet OOH$$

$$Au\ (h^+) + H_2O \rightarrow \bullet OH + H^+$$

$${}^{\bullet}O_2^- + H^+ \rightarrow \bullet OOH$$

$$2 \bullet OOH \rightarrow H_2O_2 + O_2$$

$$H_2O_2 + {}^{\bullet}O_2^- \rightarrow \bullet OH + OH^- + O_2$$

$$Au\ (h^+) + OH^- \rightarrow Au + \bullet OH$$

Degradation:

$$\bullet OH + TC - HCl \rightarrow \text{degradation products.}$$

$$g\text{-}C_3N_4(h^+) + TC\text{ - }HCl \rightarrow \text{degradation products.}$$

Recently, we have developed copper plasmonic catalysts Cu-ZnO and Cu_2O-Cu-ZnO for efficient photocatalytic degradation of methylene blue (MB) (Pal et al. 2015). The enhanced photodegradation was caused by the SPR effect of copper (Fig. 4.17). ZnO, an n-type semiconductor, has the wide band gap value (3.3 eV) and large excitation energy (60 meV). Hence, the use of ZnO is restricted up to the UV region only but not in the visible region. Thus, the photocatalytic efficiency of ZnO is negligible in the presence of visible light. We observed no noticeable degradation of MB by ZnO in presence of visible light and under the same experimental condition. On electronic contact between Cu and ZnO, electron is transferred from Cu Fermi level to ZnO Fermi level until thermodynamic equilibrium is reached (Fermi level of Cu > Fermi level ZnO). Consequently, CB of ZnO is induced and a Schottky barrier is generated at the interface between Cu metal and ZnO. During photocatalysis under visible light illumination, hot electron is produced in Cu particle due to SPR and the photoexcited SPR band of Cu stays at higher potential than the CB of ZnO. So, visible light-induced hot electrons on the SPR band of Cu are transferred to the CB of ZnO leaving the holes in metal (Cu).

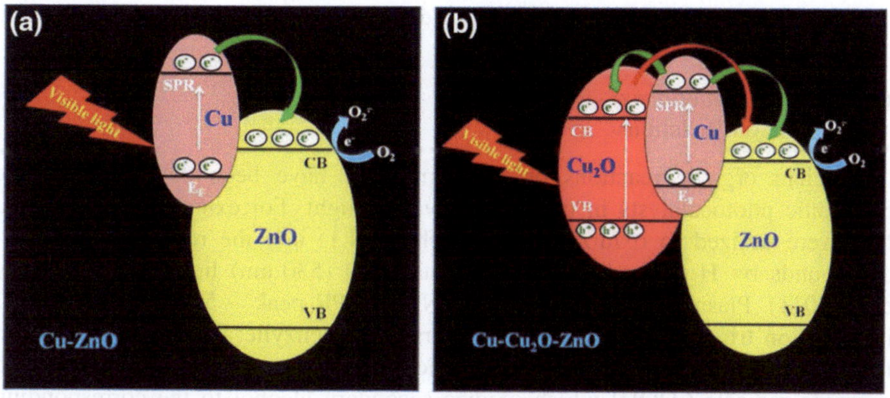

Fig. 4.17 Photocatalytic degradation of methylene blue under visible light by Cu–ZnO and Cu_2O–Cu–ZnO. From reference Pal et al. (2015), Copyright 2015 American Chemical Society

Then the hot electron in CB of ZnO are absorbed by oxygen to produce superoxide radical anions ($^{\cdot}O_2^{-}$) which further generates ·OOH, ·OH and H_2O_2 on reacting with water (as per the earlier discussed). These active species bear the responsibility for photodegradation of MB dye.

For the Cu-Cu_2O-ZnO ternary nanocomposite, all three Fermi levels are equilibrated upon electronic contact by transferring electrons and Schottky barriers are formed at the interfaces between metal and semiconductors. However, here is present an n-p heterojunction between p-type semiconductor Cu_2O and n-type semiconductor ZnO. So upon visible light absorption electrons–hole pair is generated by Cu_2O and the electrons on CB of Cu_2O are transferred to the CB of ZnO leaving holes in the VB of Cu_2O. Simultaneously, upon visible light absorption hot electron is produced in Cu particle due to the SPR of Cu. These hot electrons are passed from SPR band of Cu to the CB of ZnO through two different pathways. One is the direct electron transfer from SPR band of Cu to the CB of ZnO. While another is the electron transfer from SPR band of Cu to the CB of Cu_2O followed by CB of Cu_2O to CB of ZnO. Then the huge excited electrons in CB of ZnO are absorbed by dissolve oxygen to produce ($^{\cdot}O_2^{-}$) which further generates ·OOH, ·OH and H_2O_2 on reacting with water. These play the role for photodegradation of MB dye as discussed earlier. The direct SPR effect of Cu of Cu_2O-Cu-ZnO has been proved by comparing its photocatalytic activity with Cu_2O-ZnO. We observed that Cu_2O–ZnO has lower photocatalytic activity than that of Cu_2O–Cu–ZnO.

There are many reports on the SPR induced photocatalytic degradations of dyes under visible light. For example, Ag/AgCl hybrid nanostructure has been applied in the degradation of methyl orange (Shahzad et al. 2016). The high efficiency of the catalysis was observed because of the strong SPR of Ag. Lim and co-workers developed an Ag–AgBr/TiO_2/RGO nanocomposite for the efficient photocatalytic degradation of penicillin G where RGO (reduced graphene oxide) and AgBr play the major role as the reaction sites (Wang et al. 2013a). Pt/Bi_2O_3 NP has also shown the SPR effect of Pt under visible light irradiation for the decomposition of organic pollutants (Li et al. 2010).

(ii) *Organic Reactions*
(a) Oxidation Reactions

Efficient organic reactions or transformations have been developed by the plasmonic photocatalysts in presence of visible light. For example, colloidal Au NPs were utilized to oxidize alcohol (Scheme 4.1) into the respective carbonyl compounds by H_2O_2 upon irradiation with LED (530 nm) light (Hallett-Tapley et al. 2011). Plasmonic excitation of Au NPs (LSPR peak ~522 nm) under LED illumination triggers the significant conversion of benzylic alcohols to carbonyls. Plausibly, in presence of light, SPR-induced hot electrons convert H_2O_2 to the peroxide radicals (·OOH) which oxidize secondary alcohol to the corresponding carbonyls through radical [ArR(OH)C·] generation followed by H^+ expulsion as depicted in the Scheme 4.1. Oxygen is produced due to the thermal effect and plasmonic excitation effect of Au on H_2O_2.

Au NPs deposited anatase/rutile TiO_2-attested plasmonic photocatalysts for the oxidation of alcohols in organic solvent (toluene) in presence of sunlight as visible light source (Scheme 4.2) (Tsukamoto et al. 2012). Herein, Au NP deposited at the interface of anatase/rutile TiO_2 particles which are really the active sites for the reaction. In presence of visible light, due to the SPR effect, the Au NP produces hot electron which is passed (passing through the Au/rutile/anatase contact site) to CB of rutile followed by CB of rutile to adjacent CB of anatase TiO_2. However, the

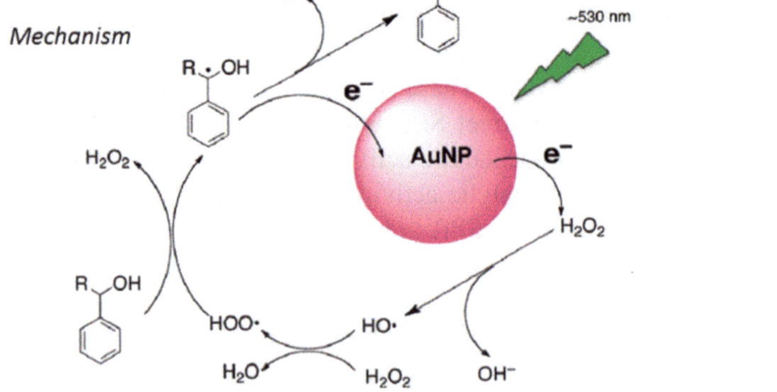

Scheme 4.1 SPR of Au NPs induced alcohol oxidation in presence of H_2O_2 under visible light (LED). From references Hallett-Tapley et al. (2011), Copyright 2011 American Chemical Society

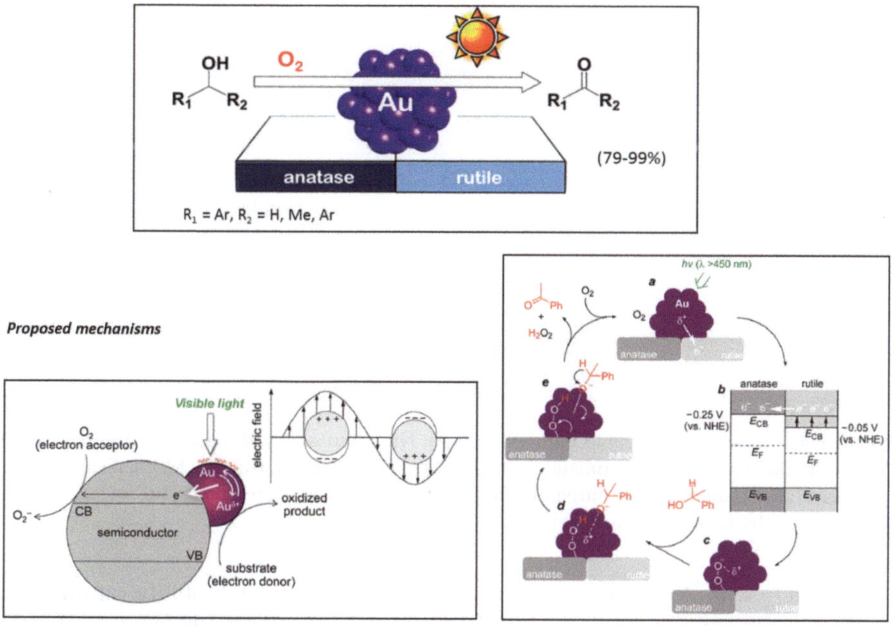

Scheme 4.2 Au NP deposited at the interface of anatase/rutile TiO_2 particles: an efficient plasmonic photocatalyst for aerobic oxidation of alcohol. From reference Tsukamoto et al. (2012), Copyright 2012 American Chemical Society

reduction of O_2 by electrons occurs on the surface of anatase TiO_2 (CB of anatase TiO_2). Thereby $^{\cdot}O_2^-$ is generated. Simultaneously, SPR-induced hot electron transfer leads to the positively charged Au particle. The generated $^{\cdot}O_2^-$ (abstracts H from alcohol) and the positively charged Au particle (because of the presence of hole) catalyses the oxidation of alcohol.

(b) Reduction Reactions

SPR effect by the noble metal NPs stimulates reduction reactions of organic molecules under mild conditions. Zhu et al. reported that Au/ZrO_2 displayed high photocatalytic activity for the reduction of nitroarenes to generate azobenzenes (Zhu et al. 2010). ZrO_2 is used as the support for the nanocomposite. The photocatalytic reduction followed as per the following reaction (Scheme 4.3).

Mechanistically, there occurs H–AuNP species generation from the oxidation of isopropyl alcohol which plays the major role for the reduction. However, KOH might help in the hydrogen abstraction from isopropyl alcohol. Then upon visible light illumination, SPR-induced excited electrons on AuNPs transferred to the nitro group of nitroarene to cleave the N–O bonds by H–AuNP. Then decomposition of HO–AuNP produces H–AuNP species and O_2. As a result, nitrobenzene generates azobenzene through azoxybenzene. The reaction may further proceed. The azobenzene, an extremely unstable compound under thermal condition, rapidly transforms into aniline. However, nitroso compound was also reduced under the similar conditions.

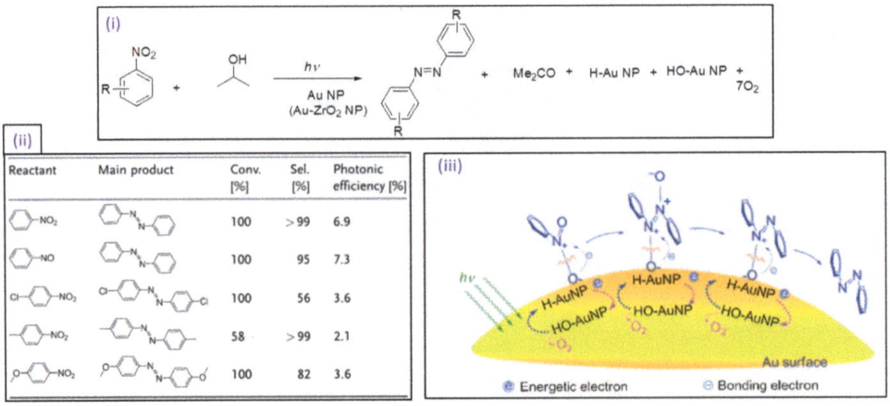

Scheme 4.3 **i** Scheme of visible light-induced Au NP photocatalyzed reduction of nitroarenes. **ii** Table: Reduction of nitroaromatic compounds of the reactions; **iii** Mechanism for the photocatalytic reduction of nitroaromatic compounds. From reference Zhu et al. (2010), Copyright 2010 John Wiley & Sons

Au/TiO_2 with Ag NPs as a co-catalyst displayed reducing ability on nitrobenzene towards aminobenzene under visible light irradiation (Scheme 4.4) (Tanaka et al. 2013). The following processes occur for the reduction. Au NPs through the SPR excitation absorbs the visible light (450–600 nm) and produces hot electrons which are transferred into the conduction band of TiO_2. Consequently, Au NP becomes electron-deficient to oxidize 2-propanol into acetone and thereby electron-deficient Au is reduced into original metallic state (Au^0 NPs). However, electrons in the conduction band of TiO_2 are transferred to Ag NP (co-catalyst as the reduction site). Then the electrons on Ag NP reduce nitrobenzene (NB) into aniline (AN). TiO_2 itself can not activate the electron–hole separation under visible light.

Scheme 4.4 Visible light-induced SPR of Au in Au–TiO_2–Ag for the reduction of nitrobenzene

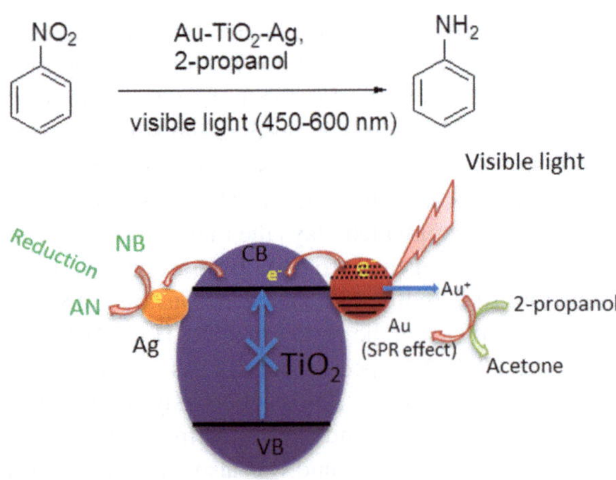

Ag (Ag coated AFM tips), as SPR photocatalyst, utilized for the reduction of nitroarene to azo-compound under visible light irradiation (532 nm laser) monitored by tip-enhanced Raman spectroscopy (Lantman et al. 2012). Klinkova et al. reported the plasmonically enhancement of catalytic performance structurally rationalized palladium NP for the nitroarene reduction (Klinkova et al. 2016).

(c) Hydroxylation

Hydroxylation of aromatic compounds is indispensable because of their industrial and medicinal importance. SPR effect of metal leaded for the efficient hydroxylation of benzene. Ide et al. reported sunlight promoted photocatalytic hydroxylation of benzene (aqueous solution) on TiO_2-supported Au NPs (Scheme 4.5) (Ide et al. 2011; Wang and Astruc 2014). The reaction displayed higher efficiency under CO_2 atmosphere. However, this could also be considered as oxidation reaction. The reaction was occurred because of the strong SPR effect as well as enhanced visible light absorption by Au and thereby possible oxidation of benzene by generated hydroxyl radical (having oxidizing power) on TiO_2 of Au–TiO_2 nanoparticle. Strong SPR generated hot electrons are passed to the CB of TiO_2 which produces hydroxyl radicals as discusses earlier. However, the oxidation process might follow the free radical pathway. Interestingly, higher CO_2 pressure reduces the successive oxidation of phenol (into catechol, hydroxyquinone, trihydroxybenzenes and finally mineralization to CO_2) leading to the better yield of phenol.

(d) Imines Synthesis

Since imine has immense importance as versatile intermediates for pharmaceuticals as well as fine chemicals preparation, their synthesis is highly valuable in the synthetic community. Tada and co-workers demonstrated the high efficiency of Au/rutile–TiO_2 for the aerobic oxidation of amines to their corresponding imines at 25 °C (Scheme 4.6) (Naya et al. 2013). On the basis of experimental result, it has been suggested that the LSPR-excited electron transfer from Au NP to rutile–TiO_2 lowers the Fermi level of Au NPs. Eventually, Au becomes Au^+ and thereby strongly adsorbed secondary benzyl amines on the Au surface are oxidized to imines. Simultaneously, the electrons of CB on rutile–TiO_2 reduce oxygen.

Au-Pd@ZrO_2 catalyzed oxidative coupling of benzylamine (or aromatic-substituted benzylamine) into imines under visible light irradiation proved an elegant method for the imine synthesis (Scheme 4.7) (Sarina et al. 2013b). The reaction attests a positive alloy effect. However, the reaction occurs on Au–Pd alloy surface where ZrO_2 is the support only. Mechanistically, oxidative coupling of

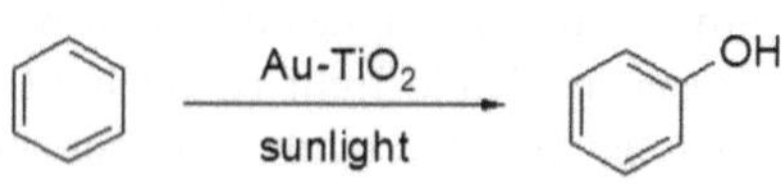

Scheme 4.5 Surface plasmon resonance of Au under sunlight: an selective hydroxylation of aqueous benzene using Au–TiO_2

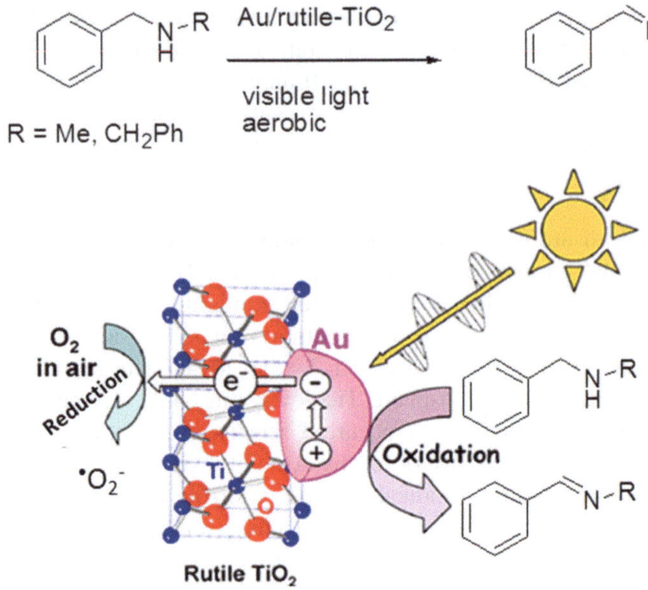

Scheme 4.6 SPR-induced amine to imine conversion by Au/rutile–TiO_2 under visible light. From reference Naya et al. (2013), Copyright 2013 American Chemical Society

benzylamine into imines follows the most probable mechanism as depicted in Scheme 4.7. Herein, LSPR of gold nanocrystals enhances the catalytic efficiency in presence of visible light. Benzylamine is initially oxidized into dehydrogenated benzylamine [PhC·(H)(NH_2)] through the abstraction of α-H from the –CH_2— group and forms also alloy-H species. The LSPR of Au induced electron transferred to Pd to form alloy-H bond formation through palladium possibly. Then the dehydrogenated benzylamine [PhC·(H)(NH_2)] absorbs oxygen to generate [Ph(H)(NH_2)COO·] radical. Immediately, [Ph(H)(NH_2)COO·] radical breaks into nascent benzaldehyde and (NH_2–O·) which forms alloy-(NH_2–O–) species coordinating with nitrogen and oxygen atom. However, alloy–(NH_2–O) and alloy–H render NH_2OH which is desorbed from alloy surface to provide free alloy surface for the next cycle reaction. Alongside, nascent PhCHO combines with unreacted amines to afford the final product benzyl group substituted imine. So the SPR of Au-induced conduction electrons on the Au–Pd alloy play the important role for the enhanced oxidative catalytic transformation towards imine synthesis through oxidative coupling under visible light irradiation.

Surface plasmon excitation of Au NPs on ZnO is amiable for effective tertiary coupling of aldehyde, amine and phenylacetylene towards rapid synthesis of propargylamines in the presence visible light (LED light) (Gonzalez-Bejar et al. 2013). The reactions, known as also A^3-coupling, occur with good yields at room temperature. The proposed mechanism for the plasmon-mediated catalysis by AuNP@ZnO is illustrated in Scheme 4.8. First, the alkynes are adsorbed on the

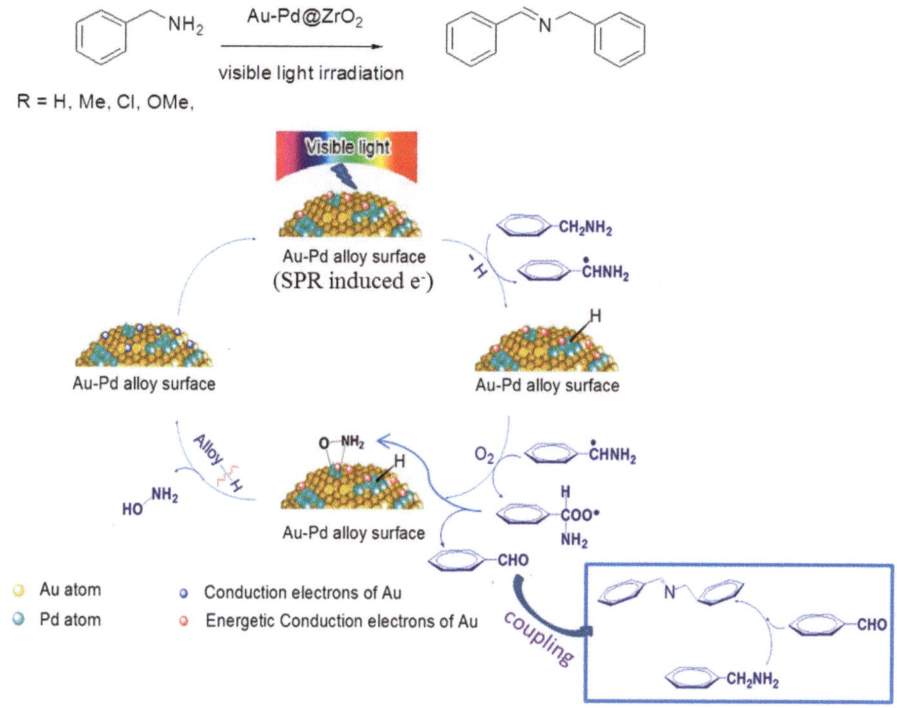

Scheme 4.7 SPR of Au-assisted oxidative coupling of amines into imine by Au–Pd@ZrO_2 under visible light. From reference Sarina et al. (2013b), Copyright 2013 American Chemical Society

surface of Au@ZnO generating alkynyl-[Au@ZnO] complex because of the alkynophilicity of Au. Simultaneously, aldehyde and the amine forms enamine. Then, the alkynyl-[Au@ZnO] complex interacts with enamine to produce the propargylamine through transfer of alkynyl group. However, it is important to mention that SPR-induced AuNP excitation causes thermal effect and/or charge transfer processes which could bear the responsibility for effective catalysis. Thus, the SPR of Au produces the local surface temperature of the AuNP which beats the activation energy for coupling reaction. However, charge transfer processes (electron and hole transfer processes) may also operate for the reaction.

(e) Biaryl Synthesis

The Miyaura–Suzuki coupling reaction (Miyaura et al. 1979) has attracted much attention in the past few decades because of its application for the synthesis of biaryls. The LSPR effect of Pd hexagonal nanoplates under visible to near-infrared light utilized for catalytic cross-coupling reactions (Trinh et al. 2015). Upon plasmon excitation in the presence of visible light, the coupling reaction between iodobenzene and phenylboronic acid occurs by the plasmon induced hot electrons (Scheme 4.9). The hot electrons are responsible for the C-I bond breaking which is

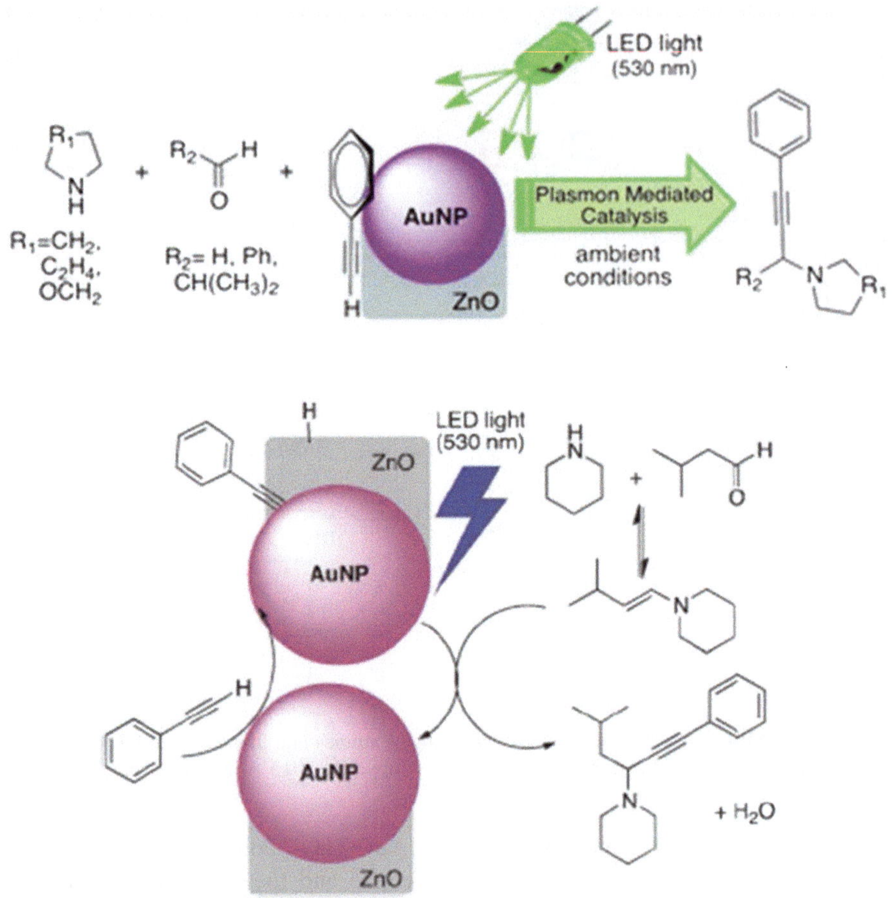

Scheme 4.8 Proposed mechanism for the plasmon mediated A^3-coupling by AuNP@ZnO in presence of visible light. From reference Gonzalez-Bejar et al. (2013), Copyright 2013 Royal Society of Chemistry

the rate-determining step. However, the rate-determining/activation steps of iodobenzene and phenylboronic acid are influenced by electrons and holes, respectively. These phenomena occur by the surface plasmon in the presence of visible light on hex-Pd NPs.

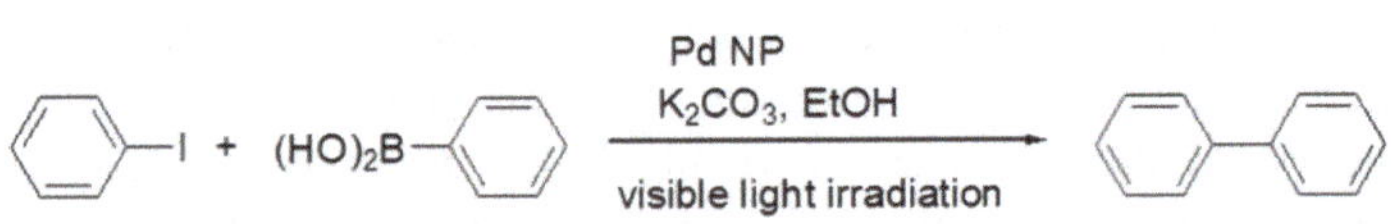

Scheme 4.9 Miyaura–Suzuki coupling reaction catalyzed by plasmonic Pd hexagonal nanoplate under visible light illumination

Wang et al. reported Miyaura–Suzuki coupling reactions under laser and solar irradiation on Au–Pd nanostructures (Scheme 4.10) (Wang et al. 2013b). In the nanostructure, PdNPs were deposited on the edge surface of AuNRs. It has been shown that Au nanocrystal absorbs the light to generate hot electron through SPR while PdNPs promote the coupling reactions. Another, interesting feature is that very high catalytic activity appears because of the plasmonic photocatalysis contribution rather than photothermal heating contribution. This has been confirmed by comparing the activities of Au–Pd and Au–TiO_x–Pd catalysts where the Au–TiO_x–Pd possesses only the photothermal heating effect while the Au–Pd nanorods possesses both the plasmonic photocatalysis and the photothermal heating. However, the hot electrons having energy easily adsorbed by the haloaryl molecules to break the C–X bond to proceed the reaction.

Au–Pd@ZrO_2 (Au-Pd alloy NPs supported on ZrO_2) under visible light illumination was also applied for the Miyaura–Suzuki coupling reaction affording high yields of the biaryls (Sarina et al. 2013b). Bhalla and co-workers have also demonstrated Ag@Cu_2O core–shell NPs catalyzed Suzuki and Suzuki type coupling reactions under visible light illumination at room temperature (Sharma et al. 2015).

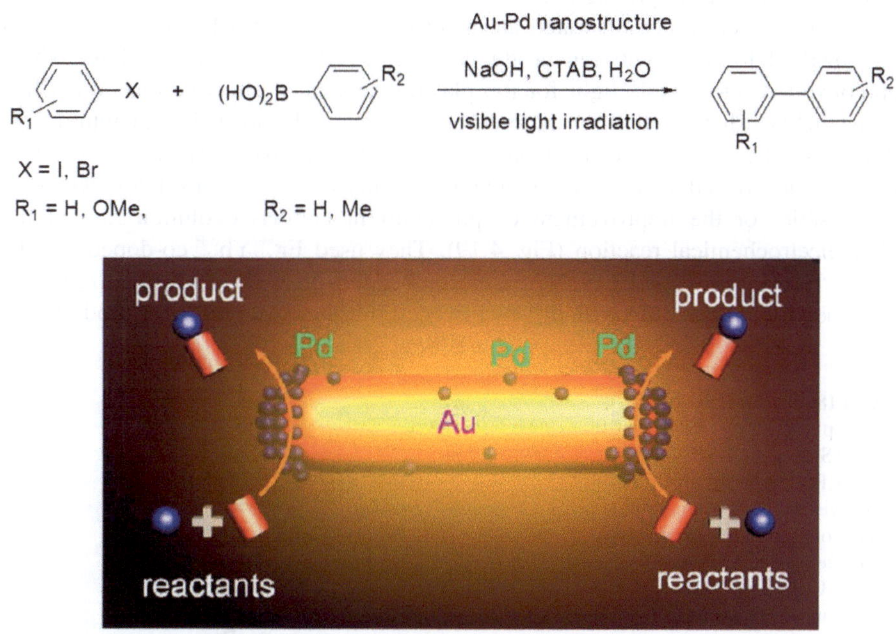

Scheme 4.10 Au–Pd nanostructure catalyzed Miyaura–Suzuki coupling under visible light illumination. From reference Wang et al. (2013b), Copyright 2013 American Chemical Society

(B) **Clean Energy Conversion**

(a) *Water Splitting*

Photocatalytic water splitting conveys the water conversion into hydrogen and oxygen. Hydrogen production is highly demanding in renewable clean and efficient energy research and thus photocatalytic water splitting becomes a most promising field. Fujishima and Honda first reported photocatalytic water splitting for hydrogen generation in 1972 using n-type TiO_2 semiconductor electrode (Fujishima and Honda 1972). Then scientists paid their extensive effort for the solar energy scavenging for hydrogen generation. Torimoto et al. synthesized an efficient photocatalyst for efficient water splitting (Torimoto et al. 2011). They prepared a novel hybrid nanostructure comprised of two core–shell nanostructures ($CdS@SiO_2 \parallel Au@SiO_2$), where only 0.37% rhodium (Rh) is deposited on CdS and SiO_2 layer (insulating layer) prevents direct electron transfer from CdS to Au. However, photoexcitation of the surface plasmon of Au NPs under visible light irradiation harvests the locally enhanced electric field (Fig. 4.18). This locally enhanced electric field influences the nearby CdS core (intrinsic bandgap of 2.40 eV) to generate electron and hole pair within CdS. Consequently, generated electrons on CdS particles are transferred to Rh (co-catalyst deposited on CdS) for the H^+ reduction to H_2. So LSPR induced electric field enhances the photocatalytic water splitting for H_2 production.

Ru-Shi Liu et al. demonstrated an interesting photoelectrochemical water splitting method for the hydrogen production using NIR (near-infrared) light which up-converted into visible light for the plasmon-enhanced effective reaction (Chen et al. 2013). They used ZnO nanorod-array decorated with CdTe quantum dots (QD) and plasmon-enhanced upconversion (UCN) nanoparticles for the reaction. Herein, the Au-induced plasmon quantum enhances the upconversion which is responsible for the improvement of photocurrent and H_2 evolution rate of the photoelectrochemical reaction (Fig. 4.19). They used Er^{3+}/Yb^{3+} co-doped $NaYF_4$ as the UCN NPs. The UCN NPs were embedded on CdTe QD-sensitized ZnO nanorods to construct the NIR-driven PEC cell. Further, Au NPs were modified on

Fig. 4.18 Photocatalytic water splitting by ($CdS@SiO_2 \parallel Au@SiO_2$) (0.37% Rh deposited on CdS) under visible light irradiation in aqueous propanol. From reference Torimoto et al. (2011), Copyright 2011 American Chemical Society

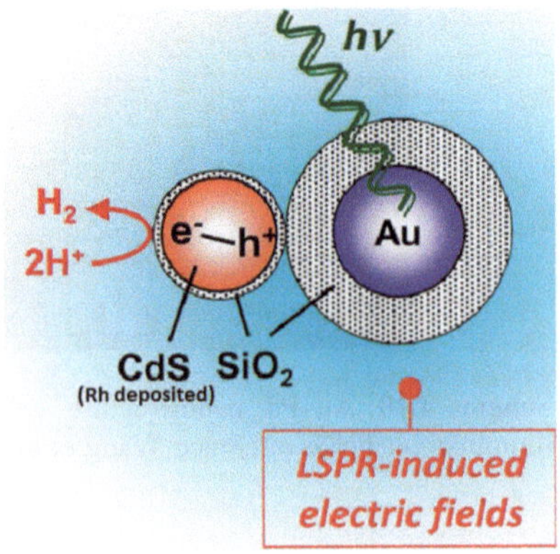

the UCN surface, thus the Au-induced SPR enhances the intensity of upconversion. UCNs convert NIR light into visible light of high-energy photons which excites the CdTe QDs to generate high-energy electron–hole pairs. Then the excited electrons are transferred to the CB of ZnO, and further to the Pt foil (through FTO/external circuit) and reacted with water to generate hydrogen. Hence, the production of the photocurrent occurs. Concurrently, photogenerated holes oxidize water to generate oxygen. This reaction leads to the high photocurrent and gas evolution rate of the photoelectrochemical reaction because of the Au-induced surface plasmon resonance.

Iron oxide (Fe_2O_3) coated on Au nanopillars was found to be an efficient for the water splitting (Gao et al. 2012). There occurs an increment of optical absorption in the nanostructured topography for the efficient water splitting reaction. This increase of optical absorption is activated by the SPR effect of Au as well as photonic-mode light trapping in the middle of Fe_2O_3 layers.

(b) *Photocatalytic* CO_2 *Conversion to Hydrocarbon Fuels.*

Plasmonic photocatalyst have also established for the efficient reduction of CO_2 by H_2O which is an energy-conversion reaction (Wang and Astruc 2014).

$$CO_2 + 2H_2O \rightarrow CH_4 + 2O_2$$

Feng et al. demonstrated an efficient conversion of CO_2 into solar fuel using double-shelled plasmonic hollow Ag–TiO_2 spheres in presence of water vapour under visible light (420 nm cut off filter used to obtain visible light) (Feng et al. 2015). The photoreduction of CO_2 gas and oxidation of H_2O vapour was carried out under gas–solid system over the Ag–TiO_2. The mechanism has been depicted in Fig. 4.20. Under visible light illumination, SPR of Ag produces hot electrons which are transferred to the CB of TiO_2. Then these electrons reduce carbon dioxide into methane. Simultaneously, water is oxidized to oxygen by the photogenerated holes in the Ag NP.

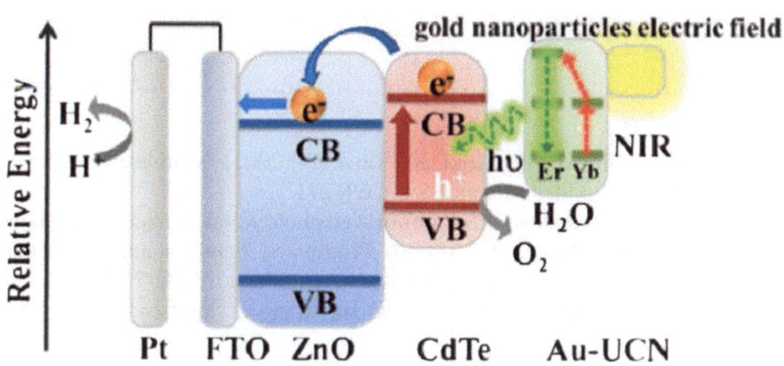

Fig. 4.19 The mechanism of Au-induced plasmon enhancement towards photochemical water splitting though upconversion of NIR to visible light. From reference Chen et al. (2013), Copyright 2013 Royal Society of Chemistry

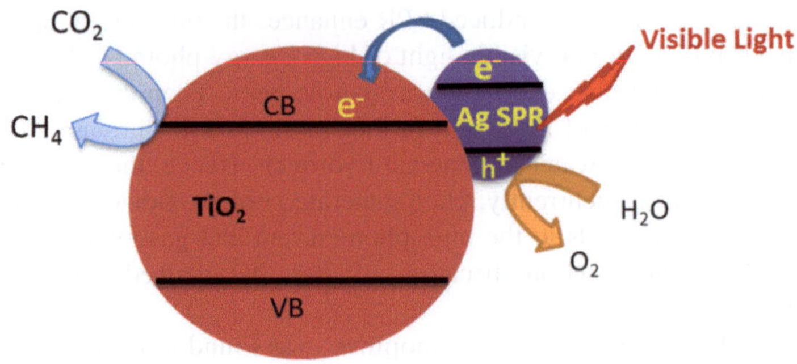

Fig. 4.20 SPR-induced photoreduction of carbon dioxide under visible light by Ag–TiO_2

Hou et al. demonstrated such reaction using Au/TiO_2 as plasmonic photocatalyst for the improved conversion of CO_2 into hydrocarbon fuels under visible light illumination (λ = 532 nm) (Hou et al. 2011).

Summary

In this chapter, it has been discussed the basic principle of SPR and its application in various catalytic reactions. Plasmonic photocatalysts fabrication and their applications in organic reaction with mechanism have been focused primarily. We discussed various types of reaction triggered by SPR effect of various metal (such as Cu, Ag, Au, etc.) using the concept of plasmonic photocatalysis under visible light illumination. Detail discussion have been interpreted about the effect of SPR induced electron–hole generation under visible light illumination by plasmonic photocatalysts on the catalytic reaction such as dye degradation, oxidation, reduction, imine formation, cross-coupling reaction, water splitting, carbon dioxide reduction, etc.

References

Asahi R, Morikawa T, Ohwaki T, Aoki K, Taga Y (2001) Visible-light photocatalysis in nitrogen-doped titanium oxides. Science 293:269–271

Atwater HA, Polman A (2010) Plasmonics for improved photovoltaic devices. Nat Mater 9:205–213

Bavykin DV, Friedrich JM, Walsh FC (2006) Protonated titanates and TiO_2 nanostructured materials: synthesis, properties, and applications. Adv Mater 18:2807–2824

Bohren CF, Huffman DR (1998) Absorption and scattering of light by small particles. Wiley, Weinheim

Brown MD, Suteewong T, Kumar RSS, D'Innocenzo V, Petrozza A, Lee M, Wiesner U, Snaith HJ (2011) Plasmonic dye-sensitized solar cells using core-shell metal-insulator nanoparticles. Nano Lett 11:438–445

Brown AM, Sundararaman R, Narang P, Goddard WA III, Atwater HA (2016) Nonradiative plasmon decay and hot carrier dynamics: effects of phonons, surfaces, and geometry. ACS Nano 10:957–966

Brus L (2008) Noble metal nanocrystals: plasmon electron transfer photochemistry and single-molecule raman spectroscopy. Acc Chem Res 41:1742–1749

Cao XB, Gu L, Zhuge LJ, Gao WJ, Wang WC, Wu SF (2006) Template-free preparation of hollow Sb2S3 microspheres as supports for Ag nanoparticles and photocatalytic properties of the constructed metal-semiconductor nanostructures. Adv Funct Mater 16:896–902

Chen S, Ingram RS, Hostetler MJ, Pietron JJ, Murray RW, Schaaff TG, Khoury JT, Alvarez MM, Whetten R (1998) Gold nanoelectrodes of varied size: transition to molecule-like charging. Science 280:2098–2101

Chen CK, Chen HM, Chen C-J, Liu R-S (2013) Plasmon-enhanced near-infrared-active materials in photoelectrochemical water splitting. Chem Commun 49:7917–7919

Cheng H, Huang B, Dai Y (2014) Engineering BiOX (X = Cl, Br, I) nanostructures for highly efficient photocatalytic applications. Nanoscale 6:2009–2026

Cheng H, Fuku K, Kuwahara Y, Moriab K, Yamashita H (2015) Harnessing single-active plasmonic nanostructures for enhanced photocatalysis under visible light. J Mater Chem A 3:5244–5258

Chulkov EV, Borisov AG, Gauyacq JP, Sanchez-Portal D, Silkin VM, Zhukov VP, Echenique PM (2006) Electronic excitations in metals and at metal surfaces. Chem Rev 106:4160–4206

Clavero C (2014) Plasmon-induced hot-electron generation at nanoparticle/metal-oxide interfaces for photovoltaic and photocatalytic devices. Nat Photonics 8:95–103

Cushing SK, Li JT, Meng F, Senty TR, Suri S, Zhi MJ, Li M, Bristow AD, Wu NQ (2012) Photocatalytic activity enhanced by plasmonic resonant energy transfer from metal to semiconductor. J Am Chem Soc 134:15033–15041

Daniel M-C, Astruc D (2004) Gold nanoparticles: assembly, supramolecular chemistry, quantum-size-related properties, and applications toward biology, catalysis, and nanotechnology. Chem Rev 104:293–346

Dhara S, Giri P (2011) On the origin of enhanced photoconduction and photoluminescence from Au and Ti nanoparticles decorated aligned ZnO nanowire heterostructures. J Appl Phys 110:124317

Dutta S, Ray C, Sasmal AK, Negishi Y, Pal T (2016) Fabrication of dog-bone shaped Au NR_{core}-Pt/Pd_{shell} trimetallic nanoparticle-decorated reduced graphene oxide nanosheets for excellent electrocatalysis. J Mater Chem A 4:3765–3776

Eustis S, El-Sayed MA (2006) Why gold nanoparticles are more precious than pretty gold: Noble metal surface plasmon resonance and its enhancement of the radiative and nonradiative properties of nanocrystals of different shapes. Chem Soc Rev 35:209–217

Feng S, Wang M, Zhou Y, Li P, Tu W, Zou Z (2015) Double-shelled plasmonic Ag–TiO_2 hollow spheres toward visible light-active hotocatalytic conversion of CO_2 into solar fuel. APL Mater 3:104416

Fujishima A, Honda K (1972) Electrochemical photolysis of water at a semiconductor electrode. Nature 238:37–38

Gao H, Liu C, Jeong HE, Yang P (2012) Plasmon-enhanced photocatalytic activity of iron oxide on gold nanopillars. ACS Nano 6:234–240

Ghosh SK, Pal T (2007) Interparticle coupling effect on the surface plasmon resonance of gold nanoparticles: from theory to applications. Chem Rev 107:4797–4862

Gonzalez-Bejar M, Peters K, Hallett-Tapley GL, Grenier M, Scaiano JC (2013) Rapid one-pot propargylamine synthesis by plasmon mediated catalysis with gold nanoparticles on ZnO under ambient conditions. Chem Commun 49:1732–1734

Halas NJ, Lal S, Chang WS, Link S, Nordlander P (2011) Plasmons in strongly coupled metallic nanostructures. Chem Rev 111:3913–3961

Hallett-Tapley GL, Silvero MJ, Gonzalez-Bejar M, Grenier M, Netto-Ferreira JC, Scaiano JC (2011) Plasmon-mediated catalytic oxidation of sec-phenethyl and benzyl alcohols. J Phys Chem C 115:10784–10790

Hartland GV (2011) Optical studies of dynamics in noble metal nanostructures. Chem Rev 111:3858–3887
Henglein A (1999) Radiolytic preparation of ultrafine colloidal gold particles in aqueous solution: optical spectrum, controlled growth, and some chemical reactions. Langmuir 15:6738–6744
Hoffmann MR, Martin ST, Choi W, Bahnemann DW (1995) Environmental applications of semiconductor photocatalysis. Chem Rev 95:69–96
Hou W, Cronin SB (2013) A review of surface plasmon resonance-enhanced photocatalysis. Adv Funct Mater 23:1612–1619
Hou WB, Hung WH, Pavaskar P, Goeppert A, Aykol M, Cronin SB (2011) Photocatalytic conversion of CO_2 to hydrocarbon fuels via plasmon-enhanced absorption and metallic interband transitions. ACS Catal 1:929–936
Hu C, Lan Y, Qu J, Hu X, Wang A (2006) Ag/AgBr/TiO_2 visible light photocatalyst for destruction of azodyes and bacteria. J Phys Chem B 110:4066–4072
Ide Y, Nakamura N, Hattori H, Ogino R, Ogawa M, Sadakane M, Sano T (2011) Sunlight-induced efficient and selective photocatalytic benzene oxidation on TiO_2-supported gold nanoparticles under CO_2 atmosphere. Chem Commun 47:11531–11533
Inagaki T, Kagami K, Arakawa ET (1981) Photoacoustic observation of nonradiative decay of surface plasmons in silver. Phys Rev B 24:3644–3646
Ingram DB, Linic S (2011) Water splitting on composite plasmonic-metal/semiconductor photoelectrodes: evidence for selective plasmon-induced formation of charge carriers near the semiconductor surface. J Am Chem Soc 133:5202–5205
Jiang R, Li B, Fang C, Wang J (2014) Metal/semiconductor hybrid nanostructures for plasmon-enhanced applications. Adv Mater 26:5274–5309
Kale MJ, Avanesian T, Christopher P (2014) Direct photocatalysis by plasmonic nanostructures. ACS Catal 4:116–128
Kelly KL, Coronado E, Zhao LL, Schatz GC (2003) The optical properties of metal nanoparticles: the influence of size, shape, and dielectric environment. J Phys Chem B 107:668–677
Khan SUM, Al-Shahry M, Ingler WB Jr (2002) Efficient photochemical water splitting by a chemically modified n–TiO_2. Science 297:2243–2245
Klinkova A, Ahmed A, Choueiri RM, Guestb JR, Kumacheva E (2016) Toward rational design of palladium nanoparticles with plasmonically enhanced catalytic performance. RSC Adv 6:47907–47911
Kneipp K, Wang Y, Kneipp H, Perelman LT, Itzkan L, Dasari RR, Feld MR (1997) Single molecule detection using surface-enhanced raman scattering (SERS). Phys Rev Lett 78:1667–1670
Kochuveedu ST, Jang YH, Kim DH (2013) A study on the mechanism for the interaction of light with noble metal-metal oxide semiconductor nanostructures for various photophysical applications. Chem Soc Rev 42:8467–8493
Kominami H, Tanaka A, Hashimoto K (2011) Gold nanoparticles supported on cerium(IV) oxide powder for mineralization of organic acids in aqueous suspensions under irradiation of visible light of λ = 530 nm. Appl Catal A 397:121–126
Kreibig U, Vollmer M (1995) Optical properties of metal clusters. Springer, Berlin
Lang X, Chen X, Zhao J (2014) Heterogeneous visible light photocatalysis for selective organic transformations. Chem Soc Rev 43:473–486
Lantman EMV, Deckert-Gaudig T, Mank AJG, Deckert V, Weckhuysen BM (2012) Catalytic processes monitored at the nanoscale with tip-enhanced Raman spectroscopy. Nat Nanotechnol 7:583–586
Larsson EM, Langhammer C, Zori I, Kasemo B (2009) Nanoplasmonic probes of catalytic reactions. Science 326:1091–1094
Lerme J, Baida H, Bonnet C, Broyer M, Cottancin E, Crut A, Maioli P, Fatti ND, Vallee F, Pellarin M (2010) Size dependence of the surface plasmon resonance damping in metal nanospheres. J Phys Chem Lett 1:2922–2928
Li R, Chen W, Kobayashib H, Ma C (2010) Platinum-nanoparticle-loaded bismuth oxide: an efficient plasmonic photocatalyst active under visible light. Green Chem 12:212–215

Linic S, Christopher P, Ingram DB (2011) Plasmonic-metal nanostructures for efficient conversion of solar to chemical energy. Nat Mater 10:911–921

Link S, El-Sayed MA (1999) Spectral properties and relaxation dynamics of surface plasmon electronic oscillations in gold and silver nanodots and nanorods. J Phys Chem B 103:8410–8426

Liu Q, Zhou Y, Kou JH, Chen XY, Tian ZP, Gao J, Yan SC, Zou ZG (2010) High-yield synthesis of ultralong and ultrathin Zn_2GeO_4 nanoribbons toward improved photocatalytic reduction of CO_2 into renewable hydrocarbon fuel. J Am Chem Soc 132:14385–14387

Lou Z, Wang Z, Huang B, Dai Y (2014) Synthesis and activity of plasmonic photocatalysts. ChemCatChem 6:2456–2476

Maeda K, Teramura K, Lu DL, Takata T, Saito N, Inoue Y, Domen K (2006) Photocatalyst releasing hydrogen from water. Nature 440:295

Maier SA, Brongersma ML, Kik PG, Meltzer S, Requicha AAG, Koel BE, Atwater HA (2001) Plasmonics-a route to nanoscale optical devices. Adv Mater 13:1501–1505

Mie G (1908) Beiträge zur Optik trüber Medien, speziell kolloidaler Metallösungen. Ann Phys 330:377–445

Miyaura N, Yamada K, Suzuki A (1979) A new stereospecific cross-coupling by the palladium-catalyzed reaction of 1-alkenylboranes with 1-alkenyl or 1-alkynyl halides. Tet Lett 20:3437–3440

Mondal C, Pal J, Ganguly M, Sinha AK, Jana J, Pal T (2014) A one pot synthesis of Au-ZnO nanocomposites for plasmon-enhanced sunlight driven photocatalytic activity. New J Chem 38:2999–3005

Mukherjee S, Libisch F, Large N, Neumann O, Brown LV, Cheng J, Lassiter JB, Carter EA, Nordlander P, Halas NJ (2013) Hot electrons do the impossible: plasmon-induced dissociation of H_2 on Au. Nano Lett 13:240–247

Murray WA, Barnes WL (2007) Plasmonic materials. Adv Mater 19:3771–3782

Nakayama K, Tanabe K, Atwater HA (2008) Plasmonic nanoparticle enhanced light absorption in GaAs solar cells. Appl Phys Lett 93:121904

Naya S, Kimura K, Tada H (2013) One-step selective aerobic oxidation of amines to imines by gold nanoparticle-loaded rutile titanium(IV) oxide plasmon photocatalyst. ACS Catal 3:10–13

Nie S, Emory SR (1997) Probing single molecules and single nanoparticles by surface-enhanced Raman scattering. Science 275:1102–1106

Pal J, Sasmal AK, Ganguly M, Pal T (2015) Surface plasmon effect of cu and presence of n-p heterojunction in oxide nanocomposites for visible light photocatalysis. J Phys Chem C 119:3780–3790

Praharaj S, Nath S, Ghosh SK, Kundu S, Pal T (2004) Immobilization and recovery of Au nanoparticles from anion exchange resin: resin-bound nanoparticle matrix as a catalyst for the reduction of 4-nitrophenol. Langmuir 20:9889–9892

Praharaj S, Nath S, Panigrahi S, Ghosh SK, Basu S, Pande S, Jana S, Pal T (2006) Layer-by-layer deposition of bimetallic nanoshells on functionalized polystyrene beads. Inorg Chem 45:1439–1441

Primo A, Corma A, Garcıa H (2011) Titania supported gold nanoparticles as photocatalyst. Phys Chem Chem Phys 13:886–910

Renger J, Quidant R, Hulst NV, Novotny L (2010) Surface-enhanced nonlinear four-wave mixing. Phys Rev Lett 104:046803

Roy A, Pal T (2015) Nucleophile-induced shift of surface plasmon resonance and its implication in chemistry. Sur Modif Biopolymers (Thakur VK, Singha AS (eds), Willey)

Sarina S, Waclawik ER, Zhu H (2013a) Photocatalysis on supported gold and silver nanoparticles under ultraviolet and visible light irradiation. Green Chem 15:1814–1833

Sarina S, Zhu HY, Jaatinen E, Xiao Q, Liu HW, Jia JF, Chen C, Zhao J (2013b) Enhancing catalytic performance of palladium in gold and palladium alloy nanoparticles for organic synthesis reactions through visible light irradiation at ambient temperatures. J Am Chem Soc 135:5793–5801

Schuller JA, Barnard ES, Cai W, Jun YC, White JS, Brongersma ML (2010) Plasmonics for extreme light concentration and manipulation. Nat Mater 9:193–204

Shahzad A, Kim W-S, Yu T (2016) A facile synthesis of Ag/AgCl hybrid nanostructures with tunable morphologies and compositions as advanced visible light plasmonic photocatalysts. Dalton Trans 45:9158–9165

Sharma K, Kumar M, Bhalla V (2015) Aggregates of the pentacenequinone derivative as reactors for the preparation of Ag@Cu_2O core—shell NPs: an active photocatalyst for Suzuki and Suzuki type coupling reactions. Chem Commun 51:12529–12532

Sinha AK, Jana S, Pande S, Sarkar S, Pradhan M, Basu M, Saha S, Pal A, Pal T (2009) New hydrothermal process for hierarchical TiO_2 nanostructures. CrystEngComm 11:1210–1212

Sinha AK, Basu M, Pradhan M, Sarkar S, Pal T (2010) Fabrication of large-scale hierarchical ZnO hollow spheroids for hydrophobicity and photocatalysis. Chem Eur J 16:7865–7874

Skrabalak SE, Chen JY, Sun YG, Lu XM, Au L, Cobley CM, Xia YN (2008) Gold nanocages: synthesis, properties, and applications. Acc Chem Res 41:1587–1595

Stewart ME, Anderton CR, Thompson LB, Maria J, Gray SK, Rogers JA, Nuzzo RG (2008) Nanostructured plasmonic sensors. Chem Rev 108:494–521

Sung-Suh HM, Choi JR, Hah HJ, Koo SM, Bae YC (2004) Comparison of Ag deposition effects on the photocatalytic activity of nanoparticulate TiO_2 under visible and UV light irradiation. J Photochem Photobiol, A 163:37–44

Tanaka A, Nishino Y, Sakaguchi S, Yoshikawa T, Imamura K, Hashimoto K, Kominami H (2013) Functionalization of a plasmonic Au/TiO_2 photocatalyst with an Ag co-catalyst for quantitative reduction of nitrobenzene to aniline in 2-propanol suspensions under irradiation of visible light. Chem Commun 49:2551–2553

Tang JW, Zou ZG, Ye JH (2004) Efficient photocatalytic decomposition of organic contaminants over $CaBi_2O_4$ under visible-light irradiation. Angew Chem Int Ed 43: 4463–4466

Torimoto T, Horibe H, Kameyama T, Okazaki K, Ikeda S, Matsumura M, Ishikawa A, Ishihara H (2011) Plasmon-enhanced photocatalytic activity of cadmium sulfide nanoparticle immobilized on silica-coated gold particles. J Phys Chem Lett 2:2057–2062

Trinh TT, Sato R, Sakamoto M, Fujiyoshi Y, Haruta M, Kurata H, Teranishi T (2015) Visible to near-infrared plasmon-enhanced catalytic activity of Pd hexagonal nanoplates for the Suzuki coupling reaction. Nanoscale 7:12435–12444

Tsukamoto D, Shiraishi Y, Sugano Y, Ichikawa S, Tanaka S, Hirai T (2012) Gold nanoparticles located at the interface of anatase/rutile TiO_2 particles as active plasmonic photocatalysts for aerobic oxidation. J Am Chem Soc 134:6309–6315

Wang C, Astruc D (2014) Nanogold plasmonic photocatalysis for organic synthesis and clean energy conversion. Chem Soc Rev 43:7188–7216

Wang H, Brandl DW, Nordlander P, Halas NJ (2007) Plasmonic nanostructures: artificial molecules. Acc Chem Res 40:53–62

Wang P, Huang B, Daia Y, Whangbo M-H (2012) Plasmonic photocatalysts: harvesting visible light with noble metal nanoparticles. Phys Chem Chem Phys 14:9813–9825

Wang P, Tang Y, Dong Z, Chenc Z, Lim T-T (2013a) Ag–AgBr/TiO_2/RGO nanocomposite for visible-light photocatalytic degradation of penicillin G. J Mater Chem A 1:4718–4727

Wang F, Li CH, Chen HJ, Jiang RN, Sun LD, Li Q, Wang JF, Yu JC, Yan CH (2013b) Plasmonic harvesting of light energy for Suzuki coupling reactions. J Am Chem Soc 135:5588–5601

Warren SC, Thimsen E (2012) Plasmonic solar water splitting. Energy Environ Sci 5:5133–5146

Watanabe K, Menzel D, Nilius N, Freund H-J (2006) Photochemistry on metal nanoparticles. Chem Rev 106:4301–4320

Xiao M, Jiang R, Wang F, Fang C, Wang J, Yu JC (2013) Plasmon-enhanced chemical reactions. J Mater Chem A 1:5790–5805

Xue J, Ma S, Zhou Y, Zewu Z, He M (2015) Facile photochemical synthesis of Au/Pt/g-C_3N_4 with plasmon enhanced photocatalytic activity for antibiotic degradation. ACS Appl Mater Interfaces 7:9630–9637

Yamada K, Miyajima K, Mafun F (2007) Thermionic emission of electrons from gold nanoparticles by nanosecond pulse-laser excitation of interband. J Phys Chem C 111:11246–11251

Yu JG, Tao HZ, Cheng B (2010) In situ monitoring of heterogeneous catalytic reactions. ChemPhysChem 11:1617–1618

Zeng C, Hu Y, Guo Y, Zhang T, Dong F, Zhang Y, Huang H (2016) Facile in situ self-sacrifice approach to ternary hierarchical architecture Ag/AgX (X = Cl, Br, I)/$AgIO_3$ distinctively promoting visible-light photocatalysis with composition-dependent mechanism. ACS Sustainable Chem Eng 4:3305–3315

Zhang Q, Lima DQ, Lee I, Zaera F, Chi M, Yin Y (2011) A highly active titanium dioxide based visible-light photocatalyst with nonmetal doping and plasmonic metal decoration. Angew Chem Int Ed 50:7088–7092

Zhang XM, Chen YL, Liu RS, Tsai DP (2013) Plasmonic photocatalysis. Rep Prog Phys 76:046401

Zhao J, Pinchuk AO, McMahon JM, Li SZ, Ausman LK, Atkinson AL, Schatz GC (2008) Methods for describing the electromagnetic properties of silver and gold nanoparticles. Acc Chem Res 41:1710–1720

Zheng XX, Liu Q, Jing C, Li Y, Li D, Luo WJ, Wen YQ, He Y, Huang Q, Long YT, Fan CH (2011) Catalytic gold nanoparticles for nanoplasmonic detection of DNA hybridization. Angew Chem Int Ed 50:11994

Zhou X, Liu G, Yu J, Fan W (2012) Surface plasmon resonance-mediated photocatalysis by noble metal-based composites under visible light. J Mater Chem 22:21337–21354

Zhu H, Chen X, Zheng Z, Ke X, Jaatinen E, Zhao J, Guo C, Xied T, Wang D (2009) Mechanism of supported gold nanoparticles as photocatalysts under ultraviolet and visible light irradiation. Chem Commun 7524–7526

Zhu H, Ke X, Yang X, Sarina S, Liu H (2010) Reduction of nitroaromatic compounds on supported gold nanoparticles by visible and ultraviolet light. Angew Chem Int Ed 49:9657–9661

Zhu SY, Liang SJ, Gu Q, Xie LY, Wang JX, Ding ZX, Liu P (2012) Effect of Au supported TiO_2 with dominant exposed {001} facets on the visible-light photocatalytic activity. Appl Catal B 119:146–155

Zou Z, Ye J, Sayama K, Arakawa H (2001) Direct splitting of water under visible light irradiation with an oxide semiconductor photocatalyst. Nature 414:625–627

Chapter 5
Mixed Metal Oxides Nanocomposites for Visible Light Induced Photocatalysis

R. Ajay Rakkesh, D. Durgalakshmi and S. Balakumar

Abstract Mixed metal oxide nanocomposite assisted photocatalysis has gained enormous interest among the scientists as a potential candidate for degrading environmentally harmful pollutants. This chapter reviews the recent advancement in the field of photocatalysis, focusing on the scientific challenges and opportunities offered by semiconducting mixed metal oxide materials. This review begins with a literature review to explore the suitable material and to optimize their energy band configurations for visible light active photocatalytic applications. This continues with examining the design and fabrication of hybrid nanocomposite materials for efficient photocatalytic performance. Finally, the discussion is meant on the synthesis methods for understanding the key aspects to engineer the nanocomposites for its use as an efficient and sustainable photocatalytic materials. This chapter also emphasizes vital problem that should be noted in upcoming research activities.

Keywords Nanocomposites · Metal oxide nanostructures · Photocatalysis · Water remediation · Charge recombination

Introduction

To date, synthesis and rational design of semiconducting mixed metal oxide nanocomposite materials have gained enormous attention among nanotechnologists due to their excellent physiochemical properties. The interaction between two different metal oxides can significantly increase the performance of the nanocomposites and still develop series of opportunities in the field of visible light active photocatalytic water remediation applications (Zhang et al. 2007; Zhao et al. 2011). Particularly, mixed metal oxide photocatalysts show enhanced photocatalytic performance due to their proper heterojunction between two different materials. They

R.A. Rakkesh · D. Durgalakshmi · S. Balakumar (✉)
National Centre for Nanoscience and Nanotechnology, University of Madras, Guindy Campus, 600 025 Chennai, India
e-mail: balasuga@yahoo.com

M.M. Khan et al. (eds.), *Nanocomposites for Visible Light-induced Photocatalysis*, Springer Series on Polymer and Composite Materials,
DOI 10.1007/978-3-319-62446-4_5

are prone to greatly absorb the wide range of visible light and can generate charge separation between the metal oxides that hinders the recombination rate of the electron–hole pairs (Greene et al. 2007; Wang et al. 2010). The design and fabrication of mixed metal oxide nanocomposite is highly attractive now and also in the near future for visible light active environmental cleaning applications.

Metal oxides have been extensively studied and well demonstrated as an effective photocatalyst for the degradation of harmful pollutants in our environment (Fig. 5.1). Currently available photocatalyst in the market is titanium dioxide (TiO_2), which is highly stable, low cost, biocompatibility and chemically inert to our environment. However, it has some drawbacks for using it in photocatalytic field, which are (i) wider bandgap energy ($\sim$3.2 eV), i.e., it can work only under ultraviolet region (<390 nm), which is only 4% of incoming sunlight, for practical applications, the TiO_2 absorption band should be tuned to visible range and (ii) rapid recombination rate of electron–hole pairs (Xu and Schoonen 2000; Fujishima and Honda 1972; Mills and Hunte 1997).

In order to overcome such essential drawbacks, many techniques have been implemented, such as modulation of band structure, metal and nonmetal ion doping and defect inducing on the host materials. One of the well-known methods is to alter the band structure by doping with metals or nonmetals in TiO_2, which leads to redshift in the band position to visible light active photocatalytic material. If the

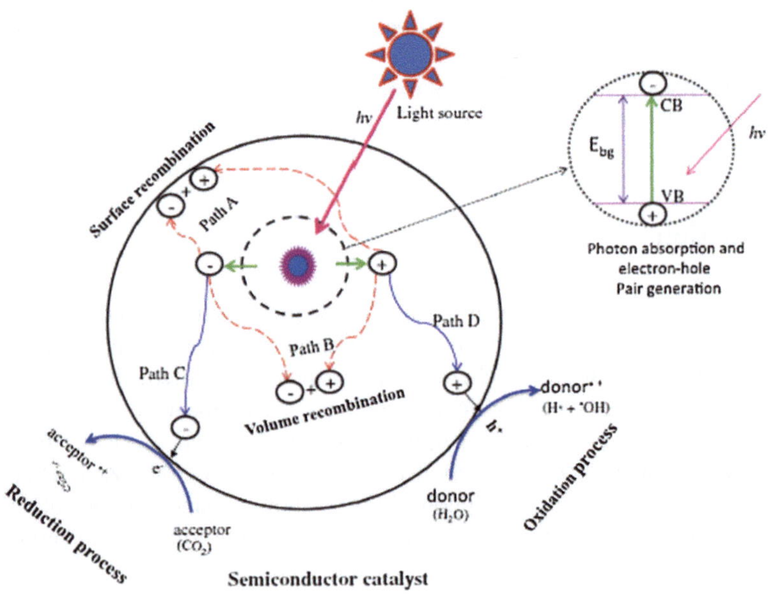

Fig. 5.1 Light absorption and consequent photoexcitation of the electron–hole pairs take place when the energy of the incident photons matches or exceeds the bandgap (Tahir and Amin 2013)

TiO_2 is doped with metal ions, such as Au, Ag, Pt, Pd, Cu, Mn, Ce, Al, Co, and Eu that creates shallow or deep-level states without altering the bandgap of the host material and also able to absorb the light visible region. However, these intra-band states created in a metal doping can act as a recombination center, which will certainly reduce the photocatalytic efficiency. In the case of nonmetal doping, such as N, S, Cl, and C, impregnation of dopant atom leads to engineer the bandgap by forming the local defect states and also extends the absorption region of TiO_2 toward the visible range. These results thus further limit the photocatalytic performance due to their remarkable reduction of active sites in the surface of the host materials (Asahi et al. 2001; In et al. 2007; Hu and Teng 2010; Li et al. 2010, Li et al. 2013).

The fundamental criteria for choosing visible light active photocatalyst with better charge separation efficacy can be achieved by introducing co-catalysts into host metal oxide in the form of nanocomposites (Zhang et al. 1998; Zheng et al. 2011; Papaefthimiou et al. 1998; Qiu et al. 2010). Nanocomposite is a composite matrix with more than two phases, in either any one phase or all the phases can have particulate in the dimensions in the range of nanometer. The physiochemical, thermoelectric, and electrochemical properties of the nanocomposite will vary from the individual component materials. Also, the hybrid nanocomposite materials play a vital role in the production of smart materials in future. It allows incorporating various materials like noble metals, semiconductors, carbon nanotubes, and graphene in a native host material to tune the efficient photocatalytic properties (Rawalekar and Mokari 2013).

Therefore, the rational design and fabrication of mixed metal oxide photocatalyst is a challenging work to develop sustainable materials for solar energy conversion and environmental cleaning applications. This chapter provides an extensive outline on synthesis, characterization, and challenges in the fabrication of mixed metal oxides nanocomposites for visible light induced photocatalytic water remediation applications.

Major Technological Challenges and Chances for Photocatalyst

We know that, nanoparticles have larger surface area, unusual morphologies, higher surface energy, and simple fabrication techniques, which are extremely useful to engineer the properties toward photocatalysis. However, the performances of the nanocomposites, especially in visible light active photocatalysis, have to facilitate and meet the requirements in commercial applications (Zeng et al. 2008; Chen and Ye 2008; Kamat 2007; Chen and Mao 2007). Moreover, the basic need is to develop a new design of the semiconductor photocatalysts, which are cost-effective, reusable, and stable material.

The fundamental property of the efficient photocatalytic activity of a semiconducting material is its electronic band energy configuration (Fig. 5.2). In context, photogenerated electrons have to be effectively separated from the holes in a semiconductor, and react with the specific molecules at the surfaces for oxidation and reduction reactions. Due to this, photoinduced charge transfer process between electrons and holes can be recombining at certain point owing to its instability in the higher state, which is a major cause that restricts the quantum efficiency of photocatalyst (Li et al. 2008; Wang et al. 2008; Maeda and Domen 2010; Yi and Ye 2009). In order to hinder the e–h recombination, energy band engineering is a basic criterion to modify the material design and fabrication of semiconductor photocatalysts.

To engineer the bandgap energy and band-edge positions in a perfect technique to avoid recombination process to decorate both metals and metal oxide nanoparticles onto the host semiconductor (TiO_2) to form a nanocomposites structure, which have been extensively studied by Rakkesh and Balakumar 2013, Chen et al. 2011, Tong et al. 2012, Yang et al. 2008. Moreover, the reconstructions of bandgap energy and bandwidth narrowing have been found recently by interparticle electronic charge coupling of semiconductor nanocomposites. Yet another crucial factor that restricts the photocatalytic efficiency of a semiconductor is its interface structure. The influence of surface/interface interaction plays a critical role in electron transport property and energy transfer process between the mixed metal oxides at the interface due to its higher surface energy and excellent chemisorptions property thus leading to higher catalytic activity. In the recent days, many studies have revealed that the exposed facets in the metal oxides are much more reactive than the thermodynamically stable materials for selective applications. The photocatalytic performance of exposed facets semiconductors can be effectively enhanced by the visible light absorption property and excellent charge carrier separation efficiency (Chen and Ye 2008; Bi et al. 2011; Xi and Ye 2010).

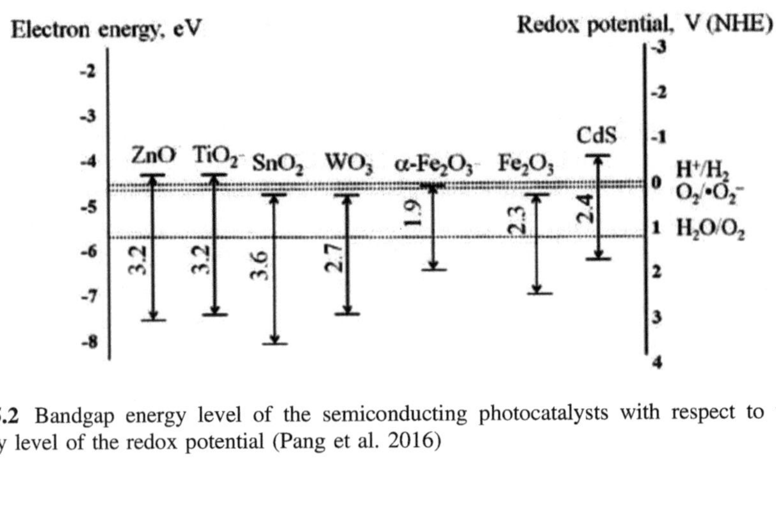

Fig. 5.2 Bandgap energy level of the semiconducting photocatalysts with respect to various energy level of the redox potential (Pang et al. 2016)

Synthesis of Mixed Metal Oxide Photocatalysts

The advancement in the area of mixed metal oxide nanocomposites has promoted many important and noteworthy studies in the field of visible light active photocatalysis. This fascinated results allow to synthesize various potential new hybrid composite materials with different compositions thus creating unique properties for environmental cleaning applications. Herein, we have reviewed and summarized some of the synthesis techniques by which nanocomposites can be fabricated for visible light active photocatalysis.

Sol-Gel Method

The most prominent and potential wet chemical method for preparing homogeneous nanocomposite is the sol-gel approach as it yields particle with uniform size and desired morphology. It is possible to engineer the morphology and size by varying the simple process parameters such as pH, annealing temperature and by adding reaction modifier.

The sol-gel method consists of hydrolysis and polycondensation of alkoxide based metal precursors (MOR) to form an inorganic bridging network. The name "sol" stands for solvated metal precursors forming through a colloidal suspension. The obtained sol turned into a network structure in a continuous liquid phase is called as gel. The precursors required for preparing nanocomposites are usually a metal-based alkoxide element encapsulated by various reactive ligands (Cividanes et al. 2010; Danks et al. 2016; Hench and West 1990). In typical experimental techniques, two different metal precursors are dispersing in an appropriate volume of deionized water to form a sol. However, some metal precursor is chemically unstable in the neutral pH range and hence it can be easily agglomerated as individual nanoparticles. Therefore, this issue can be avoided by adding diluted acids to maintain the pH value in a lower range for higher stability nature. In order to remove the aqueous content in the sol product, gel will be formed under heat treatment. Then, the obtained composite gel slurry should be dried isothermally at 80–100 °C for more than 12 h and then the residue was transferred into the furnace at various temperatures for more than 2 h to crystallize the materials by calcination process (Rakkesh and Balakumar 2013, Rakkesh and Balakumar 2015) (Fig. 5.3).

The chemical reactions involved in the sol-gel process are illustrated below;

$$\text{MOR} + \text{H}_2\text{O} \rightarrow \text{MOH} + \text{ROH (hydrolysis)}$$

$$\text{MOH} + \text{ROM} \rightarrow \text{M} - \text{O} - \text{M} + \text{ROH (condensation)}$$

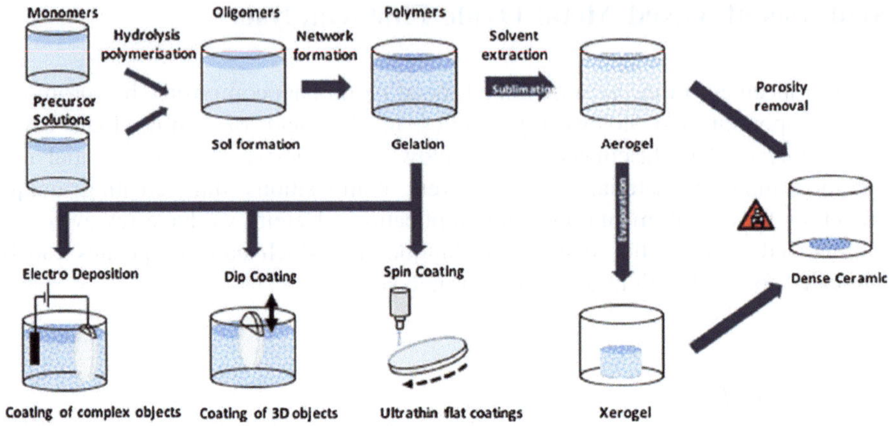

Fig. 5.3 Process involved in the sol-gel process (Owens et al. 2016)

Hydrothermal Method

The name "hydrothermal" came from a regime of high temperature and water pressure involved in the chemical reaction to synthesis of a material. The apparatus used in this method is called "autoclaves" or a "bombs" which is made up of high quality stainless steel lined with Teflon material in the inner core to avoid contamination. Hydrothermal method requires water to synthesize nanocomposites with distinct morphologies; water can act as a catalyst as well as reducing agent at elevated temperature (above 100 °C) and pressure (approximately few atmospheric pressure) (Byrappa and Yoshimura 2001; Lobachev 1973).

In a typical synthesis process, two different precursors (appropriate stoichiometric) are dissolved in double-distilled water and the solution was stirred until it becomes a homogenous solution. If you use anything other than water as a solvent, then it is called as solvothermal method. The saturated solution is poured into Teflon-lined stainless steel autoclave. The autoclave is evenly locked and kept in a furnace maintained at a temperature above 100 °C for different hours. The final product is washed and dried in an oven above room temperature (Rakkesh et al. 2015).

The hydrothermal synthesis route gives more advantages over other wet chemical routes to synthesize nanocomposites especially for photocatalytic applications. Due to its unusual nanomaterial morphologies and desired crystal facets exposed on the surfaces enhances the photocatalytic activity. It is a cost-effective method to fabricate mixed metal oxide nanocomposites and is environmentally friendly than many other methods.

The hydrothermal synthesis route has numerous advantages such as (i) rapid reaction kinetics, (ii) variety of morphologies, (iii) ability to fabricate new hybrid materials at nanoregime, (iv) pollution-free synthesis route—reaction takes place in a closed environment, (v) facile synthesis route which does not need any specific catalyst, seed solution, expensive and hazardous surfactant, and (vi) large-scale production is possible with high purity of materials (Roy 1994) (Fig. 5.4).

Precipitation Method

Precipitation method is one of the wet chemical routes for synthesizing mixed metal oxide nanocomposites. It is based on bottom-up approach. Nanoparticles are formed by the solution under precipitation process; the process thoroughly relies on the presence of preferred nuclei.

Generally, the metal acetate or hydroxide based precursors are used to precipitate in an aqueous solution due to their lower solubility nature. The precipitation reaction can be done by their own precursors or increasing the pH of the solution by adding alkaline precipitating agents (Gupta et al. 2010; Nejati 2012; Schmidt 2001).

The following reaction mechanism occurs during precipitating such as (i) the starting precursors attains supersaturating condition by mixing with aqueous solution, (ii) nucleation, (iii) growth of crystal nuclei, and (iv) aggregation of the crystal nuclei to form a pure crystal upon calcination (Fig. 5.5).

Spray Pyrolysis Techniques

Spray pyrolysis technique is basically a thermal deposition method using metal precursors. The reaction procedure consists of hardly five processes (Fig. 5.6) such as (i) loading of a precursor solution in the tiny needle to form a small droplets (atomizer); (ii) spraying of the droplets on top of the thermal substrate; and (iii) fabrication and deposition of the mixed metal oxides film onto thermal substrate by thermal decomposition process (Widiyastuti et al. 2007).

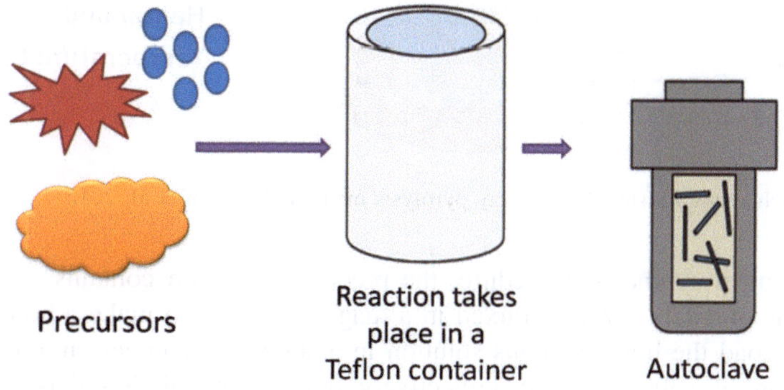

Fig. 5.4 Hydrothermal reaction setup (Rakkesh et al. 2016)

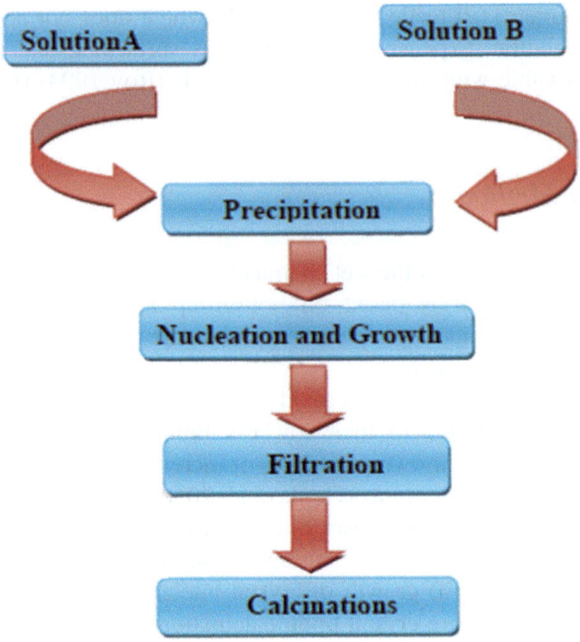

Fig. 5.5 Process involves during precipitation reaction (Sharma et al. 2017)

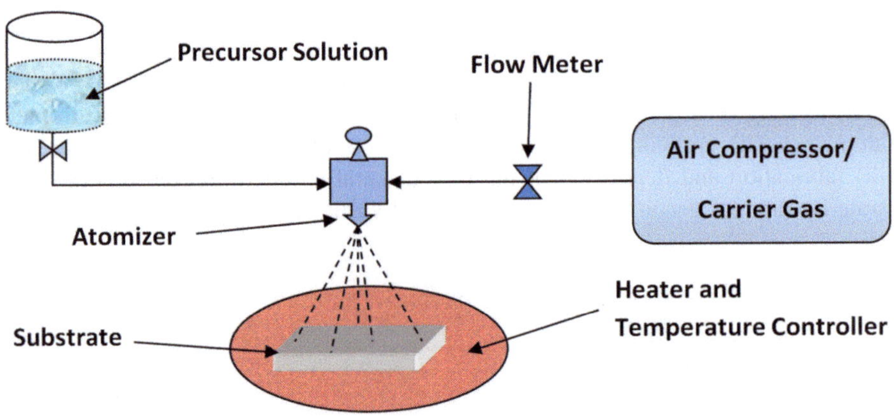

Fig. 5.6 Schematic illustration of spray pyrolysis method (Khatami et al. 2015)

In a typical synthesis procedure, the precursor solution contains various concentrations of $TiCl_3 + ZnCl_2$ mixed in a magnetic stirrer to make a homogeneous solution. Load the homogeneous solution in a spraying chamber; atomization and evaporation of solvent occurs collectively before the solution deposits on the thermal substrate. The nucleation and growth of the mixed metal oxides are formed on the film by chemically induced thermal deposition reaction. The deposition rate and growth of the thin films are monitored by heating rate of the thermal substrate and spray time (Filipovic et al. 2013). The main advantages of using spray pyrolysis for the preparation of mixed metal oxide nanocomposites are smaller particle size, uniform size distribution throughout the film, rapid growth rate, and synthesis of multiphase material with smaller size distribution at very low deposition temperature (Fig. 5.7).

Photocatalytic Applications of Mixed Metal Oxide Nanocomposites

The mixed metal oxides based photocatalysis are highly useful in the field of energy and environment such as decomposition of toxic pollutants, eco-friendly fuel production, water treatment, self-cleaning windows, antifogging coatings, and disinfection of bacterial species. In this chapter, we are going to concentrate on the specific application, which resolves the major issues related to our environmental system.

Degradation of harmful pollutants: Removing the hazardous pollutants in the water bodies remains a huge problem in our environmental system. Harmful organic pollutants are released from various manufacturing firms like textiles, leather factories, automobile industries, etc., which contains highly carcinogenic nitroaromatic dyes that are used as coloring agents finding their way into water bodies (Fig. 5.8). These harmful pollutants will devastate our environment in the

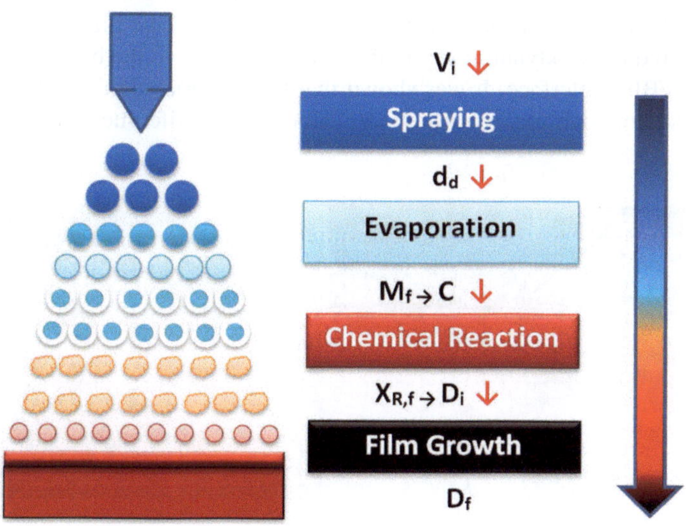

Fig. 5.7 Stages of spray pyrolysis reaction (Khatami et al. 2015)

near future. The photocatalytic activities of the mixed metal oxide based semi-conducting nanocomposites are greatly useful for the degradation of hazardous pollutants from the water bodies (Rakkesh et al. 2015).

Ismail et al. have fabricated and explained the formation of mesoporous WO_3–TiO_2 mixed metal oxide nanocomposites with various WO_3 precursor ratio by sol-gel method. It was found that the homogenous particle size of WO_3–TiO_2 nanocomposite photocatalyst showed a monoclinic and a triclinic WO_3 crystallographic phase. The WO_3–TiO_2 nanocomposites by a sol-gel method exhibited excellent photocatalytic activity toward the degradation of imazapyr organic pollutant under visible light irradiation. The authors also explained the working mechanism of the sol-gel prepared WO_3–TiO_2 nanocomposite photocatalyst under visible light illumination. It is stated that the TiO_2 is photoactive only under UV light; the TiO_2 present in WO_3–Tio_2 nanocomposite can work as the main photocatalyst, while WO_3 can act as a sensitizer for absorbing visible light. Under visible light irradiation, the photogenerated electrons could be excited from the valence band (VB) to the conduction band (CB) of WO_3 and the holes from the VB of WO_3 might be transferred to those of VB of TiO_2. As a result, the holes are accumulated in the VB of TiO_2. This positive holes and negative electrons interact with the water molecules to form hydroxyl radicals and superoxide anions, which are able to destroy organic pollutant effectively in a short period of time (Ismail et al. 2016) (Fig. 5.9).

Humayun et al. demonstrated the wet chemical route to synthesize P-type $BiFeO_3$ (P–BFO) and TiO_2–$BiFeO_3$ (T/P–BFO) nanocomposites and study their photocatalytic performances for the degradation of gas phase acetaldehyde and liquid phase phenol. The author illustrated the charge transfer process of the nanocomposites in Fig. 5.10. It is shown that, when T/P–BFO nanocomposite is illuminated under visible light with the photon energy higher than the CB of P–BFO, the photoinduced charge carriers are generated and the high energy electrons are transferred thermodynamically to the CB of TiO_2 due to the built in electric field at the TiO_2/BFO interface. It was shown that this material significantly suppressed the electron–hole pair recombination rate, increased the lifetime, and also enhanced

Fig. 5.8 Pictorial representation of industrial pollutants released to the water bodies (Rakkesh et al. 2016)

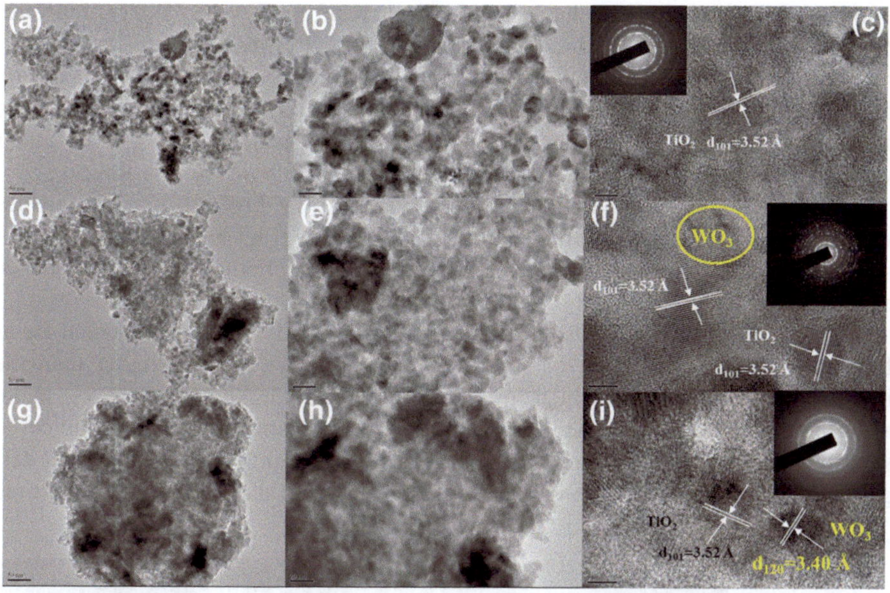

Fig. 5.9 TEM micrographs of mesoporous WO_3–TiO_2 mixed metal oxide nanocomposites (Ismail et al. 2016)

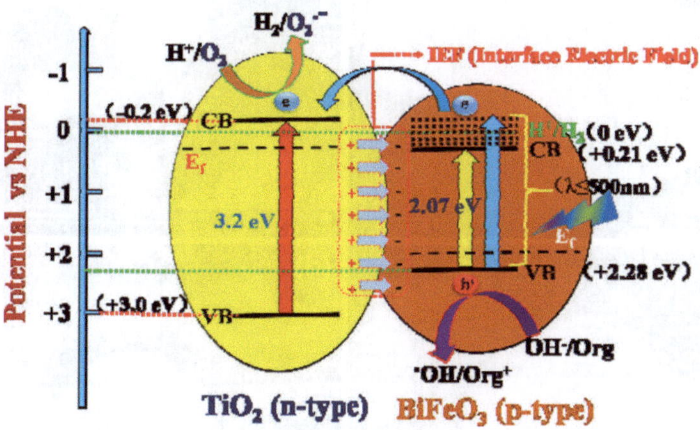

Fig. 5.10 Schematic illustration of charge transfer process occurs during photocatalytic reaction (Humayun et al. 2016)

the separation properties. It showed that TiO_2/BFO nanocomposites play a promising role in the field of visible light active photocatalysis for remediation of pollutants (Humayun et al. 2016).

Chen et al. demonstrated the hydrothermal techniques to synthesize SnO_2–SnS_2 mixed metal oxide heterostructures and explained their photocatalytic properties.

The author depicted the enhancement property of SnO_2–SnS_2 nanocomposites compared with pure SnO_2 and SnS_2 nanoparticles. It was believed that the formation of heterostructure between the SnO_2–SnS_2 nanocomposites plays a vital character to hinder the radial recombination rate of electron and hole pairs and also to facilitate interfacial electron transfer for enhanced photocatalytic activity under visible light range (Chen et al. 2016).

Xie and his co-workers explained the formation of V_2O_5/TiO_2 nanocomposites with core–shell morphology by hydrothermal method (Fig. 5.11). The core–shell structured V_2O_5/TiO_2 nanocomposites exhibit excellent photocatalytic performance over removal of arsenic.

The authors quoted the reasons for superior photocatalytic activities which are as follows: (i) the surface area of core–shell nanostructure is very higher than the individual nanoparticles, which provides larger active sites for heavy metal absorption and leads to efficient photocatalytic performance. (ii) the specific core–shell morphology provides multiple reflections of incident light and locks the targeted molecules between the cavities, thus diffusing in the mean irradiation time

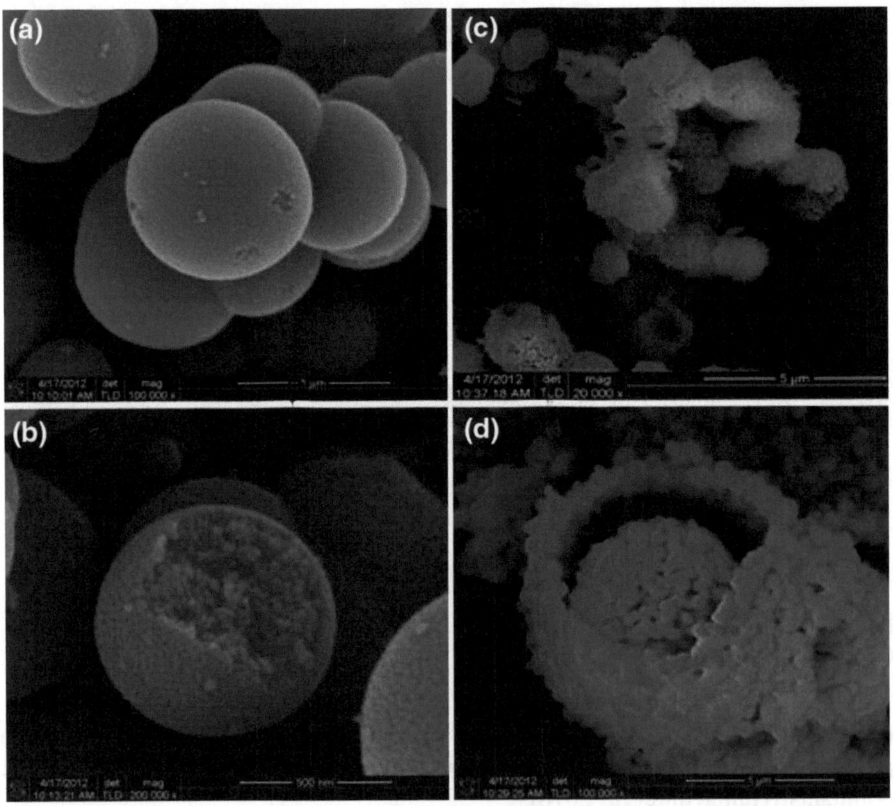

Fig. 5.11 Electron micrographs of V_2O_5/TiO_2 core–shell nanocomposites with different imaging views (Xie et al. 2016)

and (iii) finally, the core–shell–shell morphology can be able to transfer photogenerated carriers effectively to the surface and separate the photogenerated holes and electrons by the creation of local electric field between the interfaces. It is believed that these findings could open the new avenue for preparing multilevel structure materials for specific properties and excellent way to treat the arsenic polluted water bodies in our ecosystem (Xie et al. 2016).

Huo et al. developed a facile two-step wet chemical process to synthesize $ZnWO_4$–CdS nanocomposite photocatalyst for the degradation of antibiotics (ciprofloxacin) in wastewater under the visible light illumination. The nanocomposites photocatalysts displayed the significant advancement in the field of photodegradation processes. The authors demonstrated that the CdS can act as a photosensitizer and also prevent photocorrosion property. It is stated that the photogenerated charge carriers in the VB of CdS can degrade the antibiotics directly and effectively inhibit the photocorrosion ability (Fig. 5.12). Hence, $ZnWO_4$–CdS mixed metal oxide photocatalysts show the superior photocatalytic activity with better stability over various cyclic performances (Huo et al. 2016).

We have developed a simple and scalable wet chemical route to fabricate ZnO–Tio_2 core–shell nanostructures with different shell thickness by increasing the concentration of TiO_2 precursors (Rakkesh and Balakumar 2013). The morphology of the as fabricated nanostructures shows uniform shape and size. We have found that the ZnO–Tio_2 core–shell nanostructures showed higher photocatalytic activity when compared with synthesized pure ZnO and TiO_2 nanomaterials by degrading

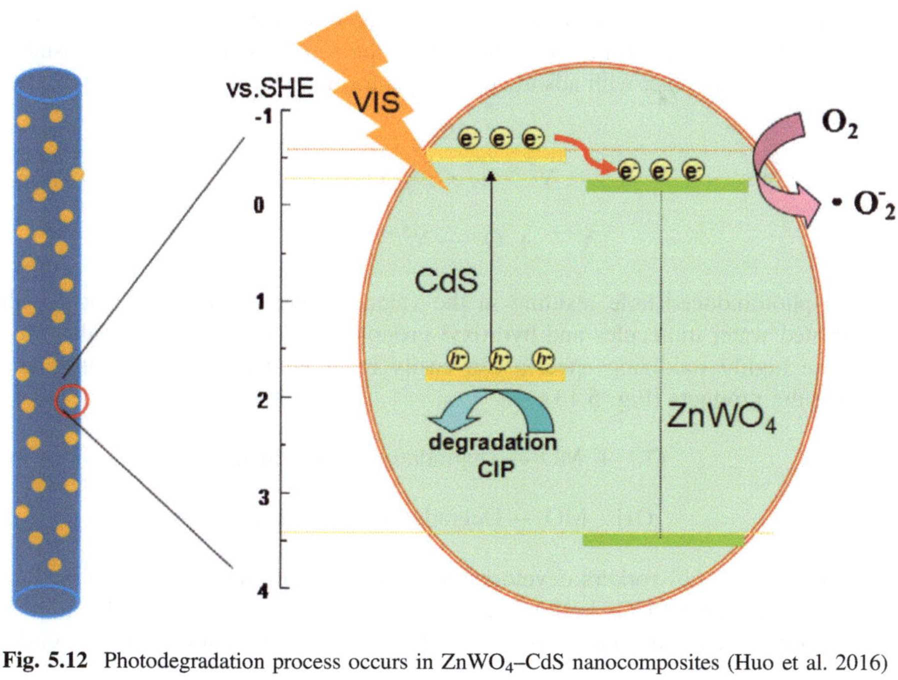

Fig. 5.12 Photodegradation process occurs in $ZnWO_4$–CdS nanocomposites (Huo et al. 2016)

mutagenic acridine orange dye. It is noted that the important factor contributing to the higher photocatalytic activity is the unique core–shell structure and that provides larger specific surface area as compared to the pure nanomaterials. Moreover, it can produce very high density of active site for oxidation/reduction reactions as well as a high interfacial charge carrier transfer processes. Hence, these criteria are greatly believed in the materials design for the superior photocatalytic performance over the degradation harmful pollutants under visible light irradiation.

Zhang and his co-workers have successfully fabricated yttrium-doped anatase TiO_2 nanosheets with exposed {001} facets using a low cost solvothermal method. The as-fabricated composite material exhibits a layer-by-layer self-assembled nanostructure. Surprisingly, the Y-doped TiO_2 nanosheets exhibit excellent photocatalytic efficiency than the commercially available TiO_2 photocatalyst for both oxidation of methylene orange (MO) dye and also reduction of Cr (VI) pollutant. The authors claimed that this higher photocatalytic performance could be due to the synergistic effect of the charge separation and charge transfer process. Furthermore, the authors proposed a mechanism for visible light active photocatalytic degradation activity of MO. The photocatalytic reaction mechanism begins with the creation of electron–hole pairs in TiO_2 under irradiation with photon.

$$TiO_2 + h\nu \rightarrow h^+ + e^-$$

It is well known that, yttrium ions doped with anatase TiO_2 nanosheets and it acts as an electron scavenging agent. The photogenerated electrons and holes can efficiently transfer and separated due to doped Y^{3+} ions in the hybrid system. Consequently, the photoinduced electrons are rapidly trapped by the Y^{3+} ions in yttrium-doped TiO_2 hybrid system, ensuing in the formation of Y^{2+}. The unstable Y^{2+} ions may also react with adsorbed O_2, thus forming highly reactive superoxide ions (O_2^-).

$$Y^{3+} + e^- \rightarrow Y^{2+}$$

$$Y^{2+} + O_2 \rightarrow Y^{3+} + O_2^-$$

The photoinduced hole residing in the valence band of the TiO_2 reacts with dissociated water molecules and hydroxyl groups thus forming•OH radicals.

These highly oxidizing species can easily react with organic molecules and degrade the products (Fig. 5.13).

$$O_2^{\cdot -} + MO \rightarrow \text{Degradation of products}$$

$$\cdot OH \quad MO \rightarrow \text{Degradation of products}$$

Wang and his co-workers developed a single-pot hydrothermal method to fabricate bismuth-doped TiO_2 hollow sheets with {001} exposed facets. The authors demonstrated the photocatalytic performance of as prepared nanocomposites under

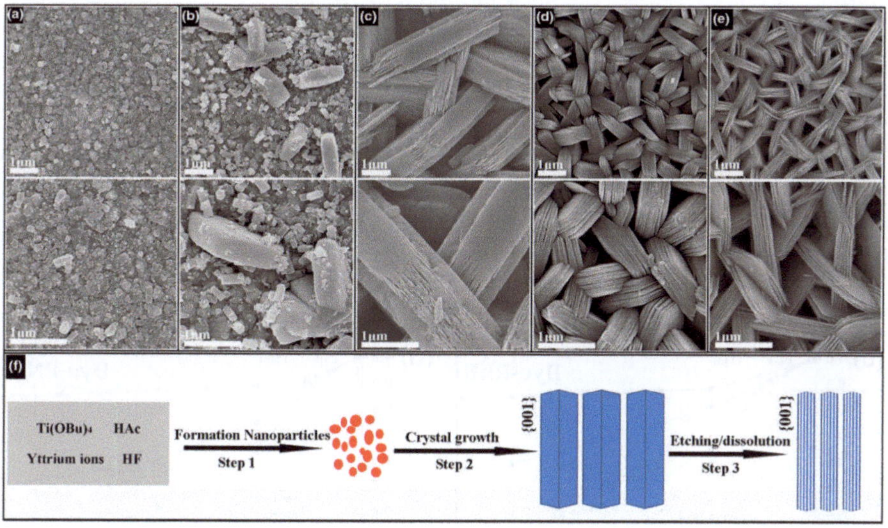

Fig. 5.13 SEM images of yttrium-doped TiO_2 nanosheets fabricated at different reaction times and schematic illustration of the growth mechanism of Y-doped TiO_2 nanosheets (Zhang et al. 2016)

visible light irradiation. Based on the obtained results, it was confirmed that Bi-doped anatase TiO_2 hollow nanocomposite exhibited enhanced photocatalytic activity toward the photodegradation of both cationic and anionic dyes under visible light irradiation. The authors attributed that, the highest photodegradation performance was due to synergistic effect of Bi-doped ions into TiO_2 nanosheets, excellent carrier separation and transfer by the highly active {001} facets (Fig. 5.14).

We have reported the fabrication strategies and comparative analysis on the photocatalytic efficacy of GNS–Tio_2 and GNS–ZnO with chemically bonded interface by modified Hummer's method (Rakkesh et al. 2014). Further, the photodegradation property of the fabricated nanocomposites has been tested toward methylene blue (MB) dye under visible light. It was noted from the set of experimental studies that the photoinduced electrons could not transfer from organic pollutant (MB dye) to the semiconductor, because there was a difference in their energy band levels and only way to flow into the semiconductor through graphene —semiconductor pathway. We have demonstrated the formation of π–π stacking with GNS, which traps and consequently degrades the dye molecules by the creation of reactive radicals. Hence, it was found that the enhanced photodegradation property could be due to increasing the bandgap of GNS and also the π–π interactions. Further, rapid electron mobility, excellent charge separation efficiency, and interfacial charge transfer effect improve the photocatalytic activity.

The design and fabrication of one-dimensional core–shell nanocomposites have been gaining attraction in the recent days due to their unusual morphological

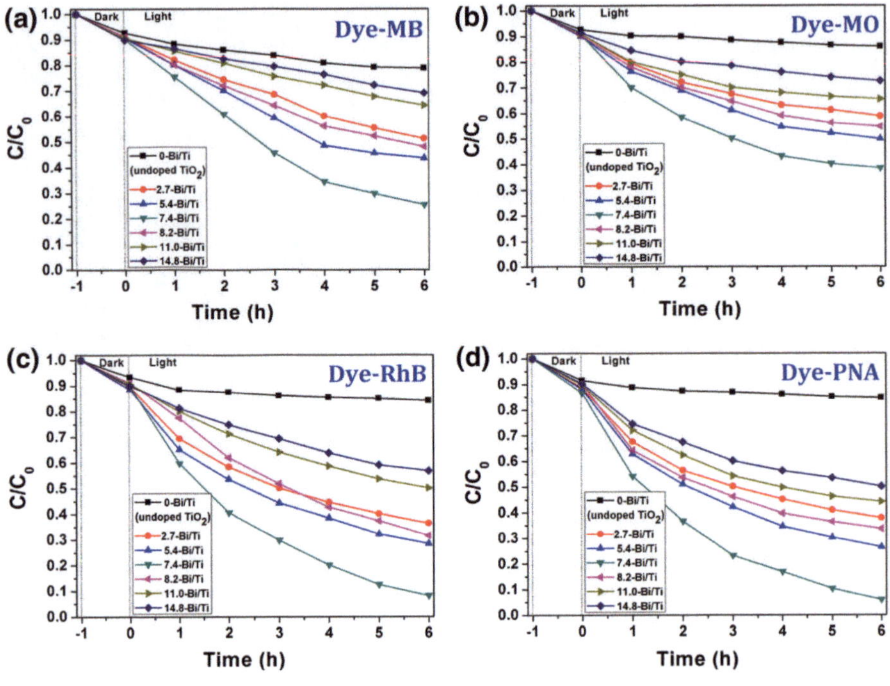

Fig. 5.14 Photocatalytic performance of **a** MB, **b** MO, **c** RhB, and **d** PNA with different Bi doping anatase TiO_2 nanocomposites under visible light irradiation (Wang et al. 2016)

dependent properties. In our earlier work on the fabrication of GNS–V_2O_5–TiO_2 core–shell nanoarchitecture for water remediation applications (Rakkesh et al. 2015). Such nanoarchitecture is fabricated by one-dimensional V_2O_5–TiO_2 core–shell nanorods chemically decorated on GNS with close interfacial contacts. The chemically bonded nanoarchitecture provides excellent photodegradation property than the pure materials. It was demonstrated that work function and ionization energy values greatly attributed the interfacial charge transfer effect. This could be the reason for create good spatial condition for charge transfer from GNS to semiconductors via the chemical bonds, and thus lead to higher photocatalytic activity (Rakkesh 2015).

Recently, we have developed a new smart material, which could simultaneously sense the organic molecules and destroy it under visible light irradiation (Rakkesh et al. 2016). Such smart material was made up of ZnO–Ag–GNS nanoassembly, which showed surface-enhanced Raman spectroscopy (SERS) active and therefore enabled to sense the organic pollutant. It was also explained that the Ag nanomaterials providing a localized surface plasmon resonance effect (LSPR) progressed Raman scattering of organic molecules and thus enabled the sensing through SERS. Since the GNS was impregnated with Ag nanomaterial in this stack, it was expected that the electrons assembled on the CB of the semiconductor could move to the

GNS through a metal–graphene pathway and were completely taken up by oxidation–reduction reactions to destroy the organic pollutants.

Hydrogen Production: Visible light active photocatalytic water splitting produces an eco-friendly, sustainable, clean, and most abundant way of developing hydrogen fuel, which can be used to drive electric devices and vehicles. The following are the chemical reactions involved in the photocatalytic water splitting under visible light irradiation: (i) Water (H_2O) molecules are dissociated into hydrogen (H_2) and oxygen (O_2) by reduction reactions occurred on the surface of the semiconducting photocatalyst. (ii) The photoinduced electrons react with the water molecules and reduce H_2O to H_2 while the photoinduced holes oxidize H_2O to O_2. In such a way, hydrogen fuel was evaluated and collected using visible light induced photocatalytic water splitting processes (Fig. 5.15).

Hou et al. developed the novel hybrid photocatalytic hybrid material with high efficiency and durability for hydrogen generation by a foaming-assisted electrospinning technique. The authors used ternary hybrid nanocomposites of TiO_2/CuO/Cu in a systematic strategy (Fig. 5.16).

The hydrogen evolution studies were demonstrated with TiO_2/CuO/Cu mesoporous system under visible light irradiation. The hybrid material showed highly stable and prominent photocatalytic H_2 evolution efficiency, when compared commercially available photocatalysts. The authors mainly attributed the enhanced photocatalytic behaviors to the formation of heterojunctions among the TiO_2, CuO, and Cu interfaces, which could be the reason for efficient charge separation and transfer of the photoinduced electrons and hole pairs.

Photocatalytic disinfection of bacterial species: Bacterial stains have always endangered the health issues of mankind in our world. Every year, millions lost their lives due to the diseases caused by pathogenic microorganisms present in the drinking water. Hence, proficient techniques have to be identified for limiting the spread of diseases caused by microorganisms in our ecosystem is highly needed. Recently, Jia et al. developed the binanosheet, which is made up of TiO_2–Bi_2WO_6 photocatalysts used to disinfect the *E. Coli* in the waterbodies under visible light

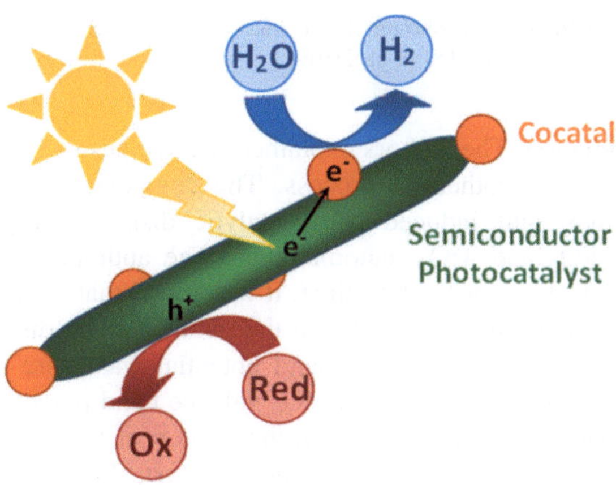

Fig. 5.15 Schematic representation of visible light induced photocatalytic hydrogen production from water (Regulacio and Han 2016)

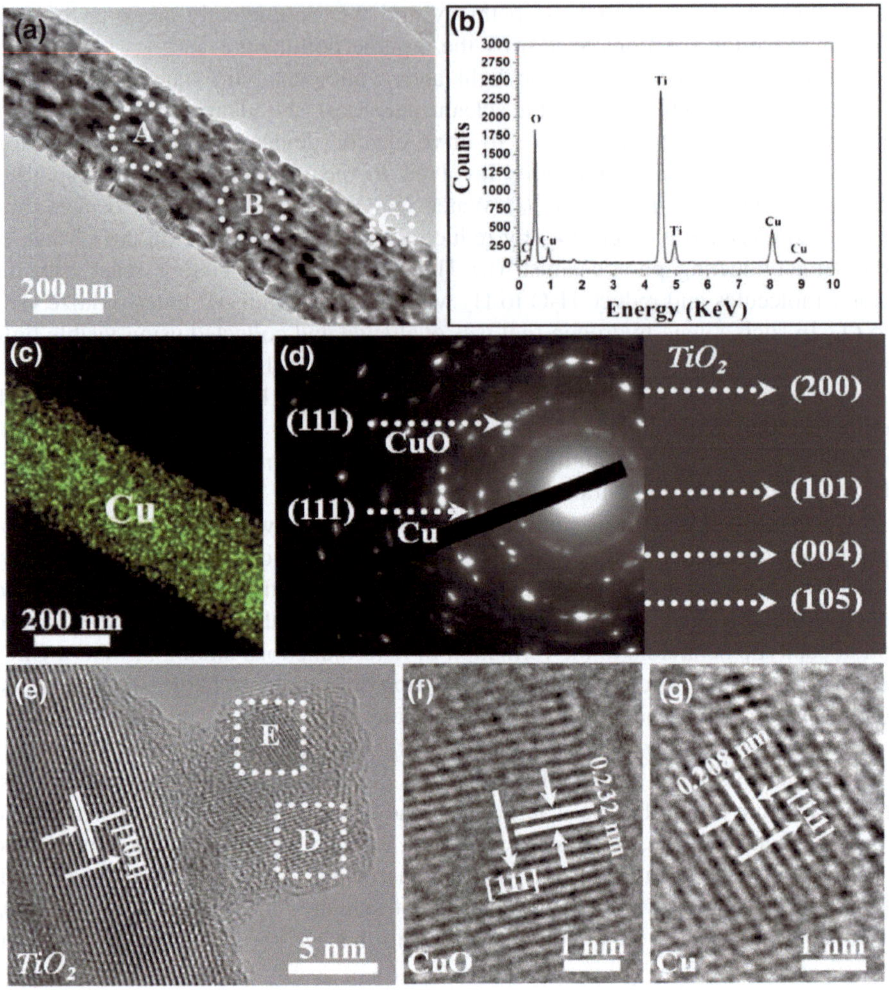

Fig. 5.16 High-resolution TEM micrographs, EDS spectra and mapping of ternary hybrid nanocomposites (Hou et al. 2016)

irradiation. The authors explained the synthesis of TiO_2–Bi_2WO_6 binanosheet by a simple hydrothermal process. The composite photocatalyst showed an efficient visible light induced photocatalytic disinfection property compared with pure Bi_2WO_6 and TiO_2 nanomaterials. The authors also proposed the mechanism of disinfection for the first time; it was found that the active species, including h^+ and e^-, played important roles in the photocatalytic disinfection process.

The TiO_2 conduction band potential was more negative than that of Bi_2WO_6 nanomaterial, and the Bi_2WO_6 valence band potential was more positive than that of TiO_2 nanomaterial. When the photon of visible light radiation impinges on the surface of TiO_2–Bi_2WO_6 photocatalyst, Bi_2WO_6 could be generated to form electron and hole pairs. Then, the potential differences between Bi_2WO_6 and TiO_2 could create a local electric field at their interface of the heterojunction. Finally, the hole in the Bi_2WO_6 valence band gets transferred to the TiO_2 valence band. This charge transfer could successfully increase the charge separation rate and also suppress the electron–hole pair recombination rate, which showed the way to excellent photocatalytic disinfection. Subsequently, the photogenerated electrons would be trapped by O_2 to form $\bullet O_2^-$. After that, the holes could react with H_2O/OH^- to form •OH radicals, which combined to form H_2O_2 and $\bullet HO_2$, the stronger oxidizing reactive species. All of these photogenerated reactive species could attack the bacterial cells, oxidize, and tear the cell membrane, and finally lead to the death of the bacteria (Fig. 5.17).

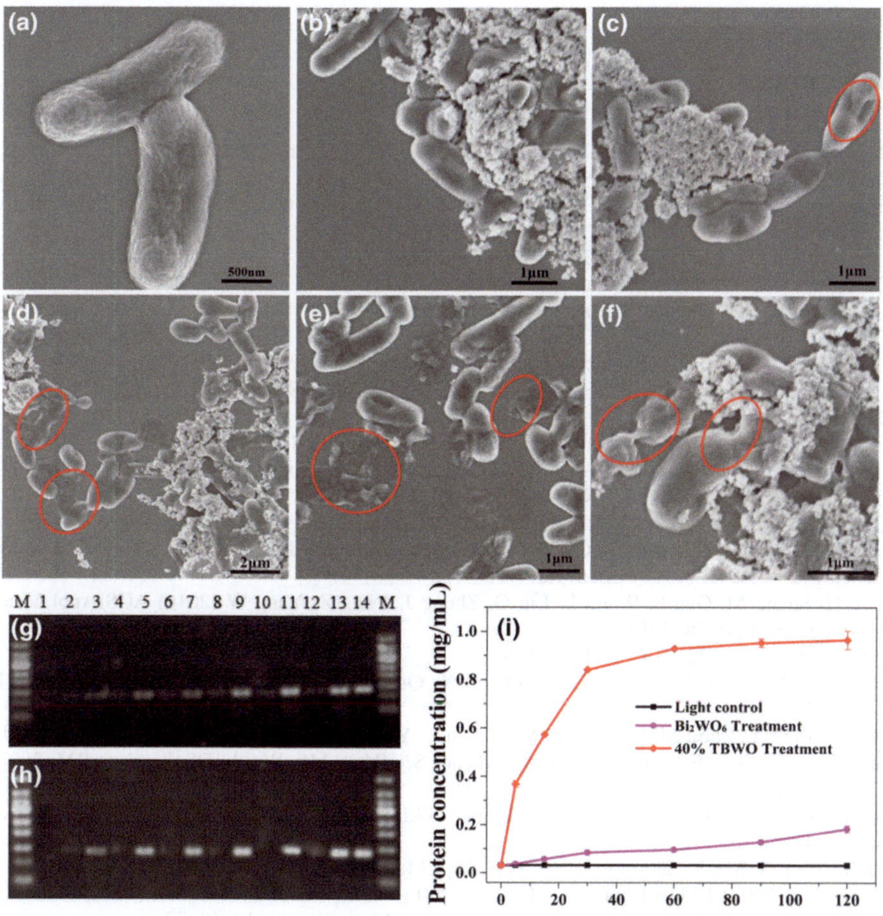

Fig. 5.17 Electron microscopic images, PCR products, and protein leakage in *E. Coli* after photocatalytic disinfection using TiO_2–Bi_2WO_6 photocatalyst under visible light irradiation (Jia et al. 2016)

Conclusion

In this chapter, we have particularly discussed the recent advancement in the field of mixed metal oxide nanocomposites for visible light induced photocatalysis. We have selectively summarized the key issues, scientific and technological challenges in the commercial photocatalysts and also discussed about the opportunities for ongoing and upcoming research on photocatalytic materials. The recent advancement in the design, fabrication, and application of mixed metal oxide nanocomposites are also explained with good reported data. It is expected to develop low-cost, eco-friendly photocatalytic materials for water remediation application which will find a vital role in the healthy ecosystem in the near future.

References

Asahi R, Morikawa T, Ohwaki T, Aoki K, Taga Y (2001) Science 293:269–271

Bi YP, Ouyang SX, Umezawa N, Cao JY, Ye JH (2011) J Am Chem Soc 133:6490–6492

Byrappa K, Yoshimura K (2001) Handbook of hydrothermal technology. Noyes Publications, New Jersey, USA

Chen X, Mao SS (2007) Chem Rev 107:2891–2959

Chen D, Ye JH (2008) Adv Func Mater 18:1922–1928

Chen XQ, Ye JH, Ouyang SX, Kako T, Li ZS, Zou ZG (2011) ACS Nano 5:4310–4318

Chen H, Gu M, Pu X, Zhu J, Cheng L (2016) Fabrication of SnO @SnS heterostructure with enhanced visible light photocatalytic activity. Mater Res Exp 3(6):065002

Cividanes LS, Campos TMB, Rodrigues LA (2010) J Sol-Gel Sci Technol 55:111–125

Danks AE, Hall SR, Schnepp Z (2016) Mater Horiz 3:91–112

Filipovic L, Selberherr S, Mutinati GC, Brunet E, Steinhauer S, Köck A, Schrank F (2013) In: Proceedings of the World Congress on Engineering, vol 2. pp 987–992

Fujishima A, Honda K (1972) Nature 238:37–38

Greene LE, Law M, Yuhas BD, Yang PD (2007) J Phys Chem C 111:18451–18456

Gupta A, Singh P, Shivakumar C (2010) Solid State Commun 150:386–388

Hench LL, West JK (1990) Chem Rev 90:33–72

Hou H, Shang M, Gao F, Wang L, Liu Q, Zheng J, Yang Z, Yang W (2016) ACS Appl Mater Interfaces 8:20128–20137

Hu CC, Teng HS (2010) J Catal 272:1–8

Humayun M, Zada A, Li Z, Xie M, Zhang X, Qu Y, Raziq F, Jing L (2016) Appl Catal B 180:219–226

Huo P, Tang Y, Zhou M, Li J, Ye Z, Ma C, Yu L, Yan Y (2016) J Indus Eng Chem 37:340–346

In S, Orlov A, Berg R, Garcia F, Pedrosa-Jimenez S, Tikhov MS, Wright DS, Lambert RM (2007) J Am Chem Soc 129:13790–13791

Ismail AA, Ahmed IA, Al-Sayari H, Robben SAL, Bahnemann DW (2016) J Hazard Mater 307:43–54

Jia Y, Zhan S, Ma S, Zhou Q (2016) ACS Appl Mater Interfaces 8:6841–6851

Kamat PV (2007) J Phys Chem C 111:2834–2860

Khatami SMN, Ilegbusi OJ, Trakhtenberg LI (2015) Mater Sci Appl 6:68–77

Li XK, Kikugawa N, Ye JH (2008) Adv Mater 20:3816–3820

Li QY, Kako T, Ye JH (2010) Chem Commun 46:5352–5354

Li B, Liu T, Hu L, Wang Y (2013) J Phys Chem Solids 74:635–640

Lobachev AN (ed) (1973) Crystallization processes under hydrothermal conditions. Consultants Bureau, New York
Maeda K, Domen K (2010) Chem Mater 22:612–623
Mills A, Hunte SL (1997) J Photochem Photobiol, A 108:1–3
Nejati K (2012) Cryst Res Technol 47:567–572
Owens GJ, Singh RK, Foroutan F, Alqaysi M, Han CM, Mahapatra C, Kim HW, Knowles JC (2016) Prog Mater Sci 77:1–79
Pang YL, Lima S, Ong HC, Chong WT (2016) Ceram Int 42:9–34
Papaefthimiou P, Ioannides T, Verykios XE (1998) Appl Catal B 15:75–92
Qiu XQ, Miyauchi M, Yu HG, Irie H, Hashimoto K (2010) J Am Chem Soc 132:15259–15267
Rakkesh RA (2015) Graphene Based Nanoassemblies: Role of Interfacial Charge Transfer Effect and Chemical Bonding Processes for Environmental Cleaning Applications, Ph.D. Thesis, NCNSNT, University of Madras, India
Rakkesh RA, Balakumar S (2013) J Nanosci Nanotechnol 13:370–376
Rakkesh RA, Balakumar S (2015) J Nanosci Nanotechnol 15:4316–4324
Rakkesh RA, Durgalakshmi D, Balakumar S (2014) J Mater Chem C 2:6827–6834
Rakkesh RA, Durgalakshmi D, Balakumar S (2015) RSC Adv 5:18633–18641
Rakkesh RA, Durgalakshmi D, Balakumar S (2016) RSC Adv 6:34342–34349
Rawalekar S, Mokari T (2013) Adv Energy Mater 3:12–27
Regulacio MD, Han MY (2016) Acc Chem Res 49:511–519
Roy R (1994) J Solid State Chem 111:11–17
Schmidt H (2001) Appl Organometal Chem 15:331–343
Sharma G, Kumar D, Kumar A, Muhtaseb AH, Pathania D, Naushad M, Mola GT (2017) Mater Sci Eng, C 71:1216–1230
Tahir M, Amin NS (2013) Energy Convers Manage 76:194–214
Tong H, Ouyang S, Bi Y, Umezawa N, Oshikiri M, Ye J (2012) Adv Mater 24:229–251
Wang DF, Kako T, Ye JH (2008) J Am Chem Soc 130:2724–2725
Wang K, Chen JJ, Zeng ZM, Tarr J, Zhou WL, Zhang Y, Yan YF (2010) Appl Phys Lett 96:123105
Wang W, Zhu D, Shen Z, Peng J, Luo J, Liu X (2016) Ind Eng Chem Res 55:6373–6383
Widiyastuti W, Wang WN, Lenggoro IW, Iskandar F, Okuyama K (2007) J Mater Res 22:1888–1898
Xi GC, Ye JH (2010) Chem Commun 46:1893–1895
Xie L, Liu P, Zheng Z, Weng S, Huang J (2016) Appl Cataly B184:347–354
Xu Y, Schoonen MAA (2000) Am Mineral 85:543–556
Yang HG, Sun CH, Qiao SZ, Zou J, Liu G, Smith SC, Cheng HM, Lu GQ (2008) Nature 453:638–641
Yi ZG, Ye JH (2009) J Appl Phys 106:074910–074915
Zeng HB, Cai WP, Liu PS, Xu XX, Zhou HJ, Klingshirn C (2008) ACS Nano 2:1661–1670
Zhang Z, Wang C, Zakaria R, Yin JY (1998) J Phys Chem B 102:10871–10878
Zhang YG, Ma LL, Li JL, Yu Y (2007) Environ Sci Technol 41:6264–6269
Zhang Q, Fu Y, Wu Y, Zhang Y, Zuo T (2016) ACS Sustain Chem Eng 4:1794–1803
Zhao H, Fu W, Yang H, Xu Y, Zhao W, Zhang Y, Chen H, Jing Q, Qi X, Cao J, Zhou X, Li Y (2011) Appl Surf Sci 257:8778–8783
Zheng Z, Huang B, Qin X, Zhang X, Dai Y, Whangbo M (2011) J Mater Chem 21:9079–9087

Chapter 6
Nanoporous Nanocomposite Materials for Photocatalysis

Zahra Hosseini, Samad Sabbaghi and Naghmeh Sadat Mirbagheri

Abstract Aside from chemical modification, increasing the surface area of a photocatalyst material is a very important strategy to improve its photocatalytic activity. Although nanoparticles possess large surface area, problems arise from the aggregation of nanoparticles in solutions and difficulties related to the recycling of nanoparticles limit their practical applications. However, introduction of porosity into a photocatalyst structure not only provides a large surface area for adsorption of organic molecules and their subsequent photodegradation, but also improves the light harvesting by increasing the optical path length of the incident light inside the porous structure. The porous structure of a photocatalyst also provides a medium for better diffusion of reactants and products and can add selectivity to the properties of the photocatalyst material. Moreover, nanoporous photocatalyst materials can be easily recycled and reused which is very important in practical applications. In this chapter, the most common methods for the preparation of nanoporous nanocomposites for photocatalytic applications are presented. The parameters controlling the morphological characteristics of nanoporous structures together with the photocatalytic activity of these structures are discussed.

Keywords Nanoporous · Nanocomposite · Photocatalysis · Synthesis

Introduction

Solid materials interact with their environment only at their surfaces. Surface area and porosity are therefore two important physical properties that highly impact the performance of solids in applications such as sensing (Li 1999), energy storage

Z. Hosseini (✉) · S. Sabbaghi (✉) · N.S. Mirbagheri
Faculty of Advanced Technologies, Shiraz University, Shiraz, Iran
e-mail: zahrahosseini@shirazu.ac.ir

S. Sabbaghi
e-mail: sabbaghi@shirazu.ac.ir

M.M. Khan et al. (eds.), *Nanocomposites for Visible Light-induced Photocatalysis*,
Springer Series on Polymer and Composite Materials,
DOI 10.1007/978-3-319-62446-4_6

(Fröschl et al. 2012), purification (Pan et al. 2010; Ensie and Samad 2014), catalysis (Leofanti et al. 1998; Bell 2003), and photocatalysis (Kubacka et al. 2012).

Photocatalysis as a strategy for cleaning up environmental pollution is of utmost importance. In a photocatalytic process, absorption of photons with energy equal to or greater than the band gap of a semiconductor creates electrons and holes in conduction band and valence band of the semiconductor, respectively. The photogenerated electrons and holes, which get a chance to reach the semiconductor surface before recombining with each other, participate in reduction and oxidation reactions with electron acceptor/donor species and organic pollutants which are adsorbed on the surface of the semiconductor respectively (Linsebigler et al. 1995). The basic principle and detailed mechanism of photocatalysis are presented in Chap. 2.

The first step in any photocatalytic process is the adsorption of (to be decomposed) molecules on the surface of the photocatalyst. This implies that a good photocatalytic architectural design needs to offer a large surface area for the adsorption of molecules. Nanosized semiconductors thus seem to be the best candidates as photocatalytic materials because of their large surface area. However, the problem with nanoparticles is their tendency to aggregate if used in suspension. This leads to the reduction of available surface area for photocatalytic reactions. Separation, recovery, and reuse of nanoparticles together with the biocompatibility and toxicity concerns are other barriers limiting their practical applications in the form of dispersed particles in a solution (Martins et al. 2016; Sabbaghi et al. 2016). Immobilizing nanoparticles as a thin film on a substrate provides the advantage of suppressing the aggregation of nanoparticles as well as facilitating separation and recovery of the photocatalyst material. However, the reduced surface area often limits the photocatalytic efficiency of thin film lower than that of powder material (Scotti et al. 2009). Introduction of porosity into the semiconductor matrix helps to compensate this problem. Moreover, it is well known that chemical reactions are most effective when the transport paths through which molecules move into or out of the nanostructured material are included as an integral part of the architectural design (Rolison 2003).

Since the breakthrough of wet photoelectrochemical (PEC) solar cells by Gratzel and coworkers (O'Regan and Grätzel 1991), a great deal of attention has been paid to nanoporous semiconductors. Nanoporous structures possess a large number of pores with large surface area. The pores define the transport paths for reactants and products and their size determines the feasibility of molecular transport in the porous structure. According to IUPAC (Sing 1985), nanoporous materials can be divided into three categories as microporous materials with pore size 0.2–2 nm, mesoporous materials with pore size 2–50 nm, and macroporous materials with pore size greater than 50 nm. The micropores may hinder the molecular transport in the porous structure but small molecules can transport through meso- and macro-pores up to the diffusion rates in open medium. The arrangement of pores in the nanoporous structure is another factor controlling the molecular diffusion and transport into the structure. Disordered pores with wide pore size distribution may also hamper the molecular transport and it is likewise the problem with ordered mesoporous structures with very small pores. Therefore, ordered mesoporous structures with large pores will present better architectural design for photocatalytic applications. A mesoporous structure may have 1D pores or 3D connected pore network. A clogging problem may occur in a structure with 1D pores as a result of which the walls of the clogged pores will not participate in photocatalytic reaction. Therefore, a 3D connected pore network is favorable for the diffusion of reactants and products.

In a photocatalytic process, conduction band electrons and valence band holes are generated as a result of light absorption by the semiconductor particles. Therefore, light harvesting is another key factor governing the efficiency of a photocatalytic process. Light harvesting improves in a nanoporous structure because the multiple scattering of light from the pore walls increases the optical path length inside the nanoporous structure. Many different semiconductors can be utilized as the photocatalyst material among which TiO_2 is widely used because it is nonphotocorrosive, nontoxic, available, inexpensive and capable of photo-oxidative destruction of most organic pollutants (Linsebigler et al. 1995). TiO_2 has three main crystal phases known as anatase, rutile, and brookite with anatase being the most common structure used in photocatalytic applications. Anatase TiO_2 has a band gap of 3.2 eV, which makes it absorb just the UV photons. On the way to improve the photocatalytic property of TiO_2 many attempts have been done in order to decrease the band gap of TiO_2. This is more important in outdoor applications where the light source is Sun. Application of a visible light active photocatalyst will lead to better utilization of solar spectrum and thereby increase the efficiency of photocatalytic effect. Different strategies have been developed for making visible light active photocatalysts such as nonmetal doping, metal deposition, dye sensitization, and coupling semiconductors. A comprehensive review on visible light active titanium dioxide photocatalysts is provided by Pelaez et al. (2012).

Photocatalytic property is also dependent on the crystallinity of semiconductor particles. Lower recombination of photogenerated electron–hole pair is achieved in semiconductor particles with better crystallinity and lower surface defects. To improve the crystallinity of a photocatalytic structure, heat treatment at high temperature needs to be done. Sometimes heat treatment of a mesoporous structure in high temperature causes the whole structure to collapse as a result of unfavorable growth of grains. Therefore, low thermal stability of mesoporous structures remained as a bottleneck in the application of mesoporous structures until solutions like post-synthesis treatments (Zhou and Fu 2013) and using structural stabilizers (Chen et al. 2009a) developed. The latter can be done by making composites in which one of the components plays the role of structural stabilizer.

The electron–hole recombination can also be hindered or retarded by making composites in which the electrons and holes are scavenged by different components of the composite. Nanocomposites of metal/photocatalyst or semiconductor/photocatalyst have been widely investigated by the intention of hindering electron–hole recombination. Graphene/photocatalyst nanocomposites also show improved photocatalytic effect because of good separation of electron–hole in these nanocomposites. Carbon-based nanostructures, especially graphene, have extraordinary electron mobility (Woan et al. 2009; Tu et al. 2012). Therefore, making nanocomposites of graphene and photocatalyst particles is a good strategy for better separation of electron–hole through the transfer of photogenerated electrons to the graphene (Geng et al. 2013). Different designs of graphene/TiO_2 nanocomposite have been tried including graphene scaffold filled with titania nanoparticles (Ng et al. 2010), TiO_2 thin films doped with titania (Du et al. 2011a), exfoliated graphene into highly ordered mesoporous titania (Malfatti et al. 2014), hierarchically ordered macro-/mesoporous TiO_2/graphene composite (Du et al. 2011a), etc. Carbon-based nanocomposites for visible light induced photocatalysis are comprehensively discussed in Chap. 8.

In summary, photocatalytic activity can be improved through both chemical and physical modifications. Chemical modification is usually done by making nanocomposites. Different compositions improve the photocatalytic effect in different ways. On the other hand, photocatalytic activity is directly affected by the microstructure of the photocatalyst. Therefore, physical modification of photocatalyst material for achieving suitable crystalline structure and large surface area is also necessary. Many design strategies have been developed for the fabrication of nanoporous materials. In this chapter, we focus on the fabrication of nanoporous nanocomposite materials for photocatalytic applications.

Fabrication of Nanoporous Nanocomposite Materials

Nanoporous nanocomposite materials can be made by a wide variety of methods among which wet chemical methods are the most common methods. In the preparation of nanoporous structures, high level of control over morphological and structural properties is desired. In this section, most conventional methods for the preparation of nanoporous structures are presented and the photocatalytic performance of the nanoporous structures fabricated by each method is discussed. The synthesis of zeolite-based and metal–organic framework (MOF) based nanocomposites are also discussed separately.

Template Method

In the template method, the chemical synthesis of nanocomposite material is performed inside a template medium which is eliminated at a later stage. In this approach the template acts as a structure-directing agent to form a porous network and the inner structural arrangements such as pore size and surface area, along with the outer shape and size of the sample can be controlled and tailored easily. This technique is facile and can be applied with a range of methods for the fabrication of materials including polymers, metal oxides, and metals (Huczko 2000; Shchukin and Caruso 2004; Studart et al. 2006). Sol–gel is one of the most common methods, which is used conjunctively with template method for the fabrication of nanoporous structures of inorganic materials. The precursors are introduced into the pores of the template and the material is deposited on the pore walls and inside the pores. The nanoporous structure is obtained after selective removal of the template. The template method is categorized in two general approaches, as hard template method and soft template method, based on the nature of the template materials.

Hard Template Method

Hard template method typically involves mesoporous SiO_2, mesoporous carbon, porous Al_2O_3, or polystyrene spheres (Xiong and Balkus 2005; Alvaro et al. 2006; Yue et al. 2009; Jing et al. 2009; Strandwitz et al. 2010; Daiguji et al. 2012) as the template. Uniform-sized silica nanospheres can be easily assembled into close-packed structures. The close-packed structures of SiO_2 are widely used as a template for the fabrication of other porous structures. Fukasawa et al. used mesoporous SiO_2 structure as a primary template for production of ordered porous graphitic carbon nitride (g-C_3N_4), which was subsequently used as a template to prepare regularly arranged Ta_3N_5 nanoparticles (Fukasawa et al. 2011). The big difference between silica and g-C_3N_4 as template in this process is that silica was not involved in the formation of g-C_3N_4 except serving as the template while in the formation of Ta_3N_5 nanoparticles on g-C_3N_4 template, the g-C_3N_4 served as nitrogen source to convert tantalum precursors into tantalum nitrides (Fukasawa et al. 2011). The full steps of the formation of porous g-C_3N_4 and final porous structure of Ta_3N_5 are shown schematically in Fig. 6.1. Ordered meso- or macro-porous g-C_3N_4 structures were prepared by polymerization of infiltrated cyanamide into the close-packed silica template. The porous g-C_3N_4 structure then served as the template for the formation of regularly arranged Ta_3N_5. Infiltration of Ta source ($TaCl_5$) into the g-C_3N_4 template followed by nitridation at 923–1123 K under a flow of NH_3 led to the formation of regularly arranged Ta_3N_5 nanoparticles. g-C_3N_4 completely decomposes into nitrogen and cyano fragments at temperatures higher than approximately 950 K, even under an inert atmosphere (e.g., N_2 or Ar) (Fischer et al. 2007), then the cyano fragments serve as nitrogen source. The pore size of the porous g-C_3N_4 structure and thereby the pore size of the final porous structure could be tailored easily by changing the size of the silica nanospheres. The average pore sizes in the Ta_3N_5 structure were 13, 20, 55, and 70 nm when the diameters of silica nanospheres in the primary template were 20, 30, 50, and 80 nm, respectively. The structure with 30 nm pore diameter had the largest surface area of about 60 $m^2\ g^{-1}$ and showed the highest rate of photocatalytic H_2 evolution under visible light irradiation which was $\sim$2.6 times higher than that of conventional submicron particles.

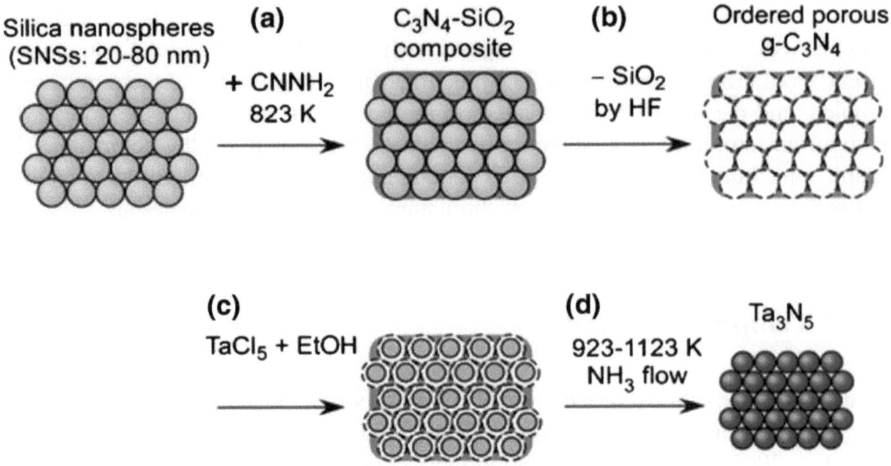

Fig. 6.1 Scheme showing different steps for the synthesis of ordered porous g-C_3N_4 and regularly arranged Ta_3N_5 nanoparticles, using close-packed silica nanospheres as the primary template. **a** Infiltration of cyanamide and polymerization at 823 K, **b** eliminating silica template by etching, **c** infiltration of $TaCl_5$ followed by drying in air, **d** nitridation at 923–1123 K under NH_3 flow. Reprinted from Fukasawa et al. (2011)

Binary and ternary nanocomposites can also be produced by a template method. Sometimes in fabrication of a nanoporous binary nanocomposite, the template is not eliminated and becomes a component of the final nanocomposite. For example, mesoporous Co_3O_4/$BiVO_4$ composite was prepared by deposition of Co_3O_4 particles on $BiVO_4$ template (Dang et al. 2014). $BiVO_4$ template was formed using mesoporous silica KIT-6 as template. Using acidic solution of $Bi(NO_3)_3.5H_2O$ and NH_4VO_3 as precursors and heat treatment at 80 °C for 12 h led to complete filling of KIT-6 pores by the $BiVO_4$ precursor. A second heat treatment at 400 °C for 12 h needed for the formation of $BiVO_4$ walls on the KIT-6 template. The mesoporous $BiVO_4$ was then obtained by elimination of KIT-6 template with a 2 M NaOH aqueous solution. Preparation of Co_3O_4/$BiVO_4$ composites was then completed by the impregnation method from an aqueous solution of $Co(NO_3)_2$. After the evaporation of water under the irradiation of an infrared light, the resulting powder was collected and calcined in air at 300 °C for 2 h. Dang et al. compared the ability of the Co_3O_4/$BiVO_4$ composite in separation of photogenerated electrons and holes for the nonporous and mesoporous structures by fluorescent spectroscopy. Although both nonporous and mesoporous structures of Co_3O_4/$BiVO_4$ composite showed low electron–hole recombination rates, the lower fluorescent intensity of the mesoporous structure clearly showed the better electron–hole separation in mesoporous structure. This was attributed to the larger contact area of Co_3O_4 and $BiVO_4$ in mesoporous composite provided by the larger surface area of mesoporous structure compared to nonporous structure. The larger surface area in mesoporous structure also resulted in an enhanced light absorption ability and adsorption capacity compared to the nonporous $Co_3O_4/BiVO_4$ composite. The collective effect of these characteristics led to an enhanced photocatalytic activity for the mesoporous $Co_3O_4/BiVO_4$ composite. Kornarakis et al. fabricated a ternary nanocomposite, mesoporous polyoxometalate (POM)/Ag_2S/CdS nanocomposite, through a two-step process (Kornarakis et al. 2014). First, they used a silica template (SBA-15) for the growth of CdS–POM heterostructures. Cadmium nitrate, thiourea compounds, and polyoxometalate clusters were the precursors infiltrated into the template voids. Solidification of precursor solution and subsequent etching of silica template led to the formation of mesoporous arrays of POM/CdS composites as a negative replica of the silica template. Kornarakis et al. then used topotactic transformation for the construction of ordered mesoporous POM/Ag_2S/CdS nanocomposite. In this step, they infiltrated $AgNO_3$ aqueous solution into the POM/CdS mesostructure. Partial cation-exchange of Cd^{2+} with Ag^+ ions happened and ordered mesoporous POM/Ag_2S/CdS nanocomposite was formed. The products consisted of hexagonal ternary POM/Ag_2S/CdS nanorod arrays with large internal surface area (35–60 $m^2\ g^{-1}$ for different composites containing different polyoxometalate compounds) and uniform pores (4.3–4.6 nm). These materials displayed increased photoactivity compared to Ag_2S/CdS and CdS samples. Besides the large and accessible pore surface area provided by the mesoporous structure, Ag_2S and POM components served as electron shuttles in photo-oxidation processes. Figure 6.2 schematically shows the stepwise electron transfer from CB of the CdS to the CB of Ag_2S with subsequent electron transfer to the lowest unoccupied molecular orbital (LUMO) of POM in band structure of POM/Ag_2S/CdS composite upon visible light irradiation. Such a band alignment clearly enhances the charge separation efficiency and thereby the photocatalytic activity.

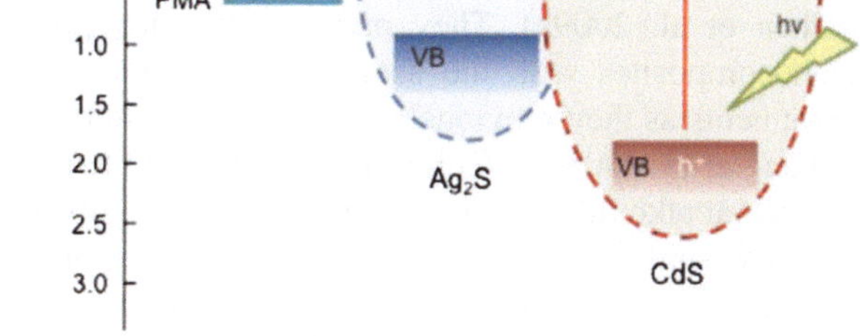

Fig. 6.2 Schematic representation of band alignment in POM/Ag_2S/CdS composite. (*VB* valence band, *CB* conduction band, *STA* $SiW_{12}O_{40}^{4}$, *PTA* $PW_{12}O_{40}^{3-}$, *PMA* $PMo_{12}O_{40}^{3-}$). Reprinted from Kornarakis et al. (2014)

Soft Template Method

Soft template method often uses soft matters like cationic surfactants (e.g., alkyltrimethylammonium), anionic surfactants (e.g., $C_{16}H_{33}SO_3H$), nonionic surfactants (e.g., TX-100), diblock polymers (e.g., Brij 56), amphiphilic triblock polymers (e.g., HO-$(CH_2CH_2O)_{20}(CH_2CH(CH_3)O)_{70}(CH_2CH_2O)_{20}$HP123, HO-$(CH_2CH_2O)_{106}(CH_2CH(CH_3)O)_{70}(CH_2CH_2O)_{106}$H) F127), tetradecyl phosphate, dodecyl phosphate, tetradecylamine, dodecylammonium chloride (DDAC), and cetyltrimethylammonium bromide (CTAB) as template (Kluson et al. 2001; Yoshitake et al. 2002; Crepaldi et al. 2003; Choi et al. 2007; Wang et al. 2008a; Zhao et al. 2008; Zimny et al. 2012).

Porous structure of many different semiconductors and nanocomposites can be prepared through soft template method; mesoporous TiO_2 being one of the simplest. Wang et al. synthesized titania materials with a hierarchical pore structure and studied the effect of sintering temperature on the pore-wall structure and photocatalytic activity (Wang et al. 2005). They used a surfactant template method for the fabrication of mesoporous TiO_2 structure. Titanium tetraisopropoxide (TTIP) and nonionic poly(alkylene oxide)-based surfactant (decaoxyethylene cetyl ether, $C_{16}(EO)_{10}$, Brij56) were used as the titanium source and structural-directing agent, respectively. The aqueous solution of precursors was kept at 80 °C for 24 h, then the product was filtered, dried overnight at 55 °C and calcined in static air at 350–800 °C for 4 h to remove the surfactant and solidify the porous frameworks. Although high-temperature heat treatment is necessary for removing the surfactant and enhancing the crystallization of the TiO_2 network, Wang and coworkers found the hierarchical porous TiO_2 structure collapsed upon sintering above 500 °C and led to the decrease in photocatalytic activity (Wang et al. 2005). A similar phenomenon was also reported by Yu and coworkers for the mesoporous TiO_2 structures prepared by self-assembly method (Yu et al. 2007). They observed an intense decrease in photocatalytic activity after calcination of the mesoporous structure over 500 °C which was attributed to the destruction of the hierarchical macro-/mesoporous structure of the titania and thereby the decrease of specific surface area.

Later, Wang et al. used SiO_2 and ZrO_2 as structural stabilizers for mesoporous TiO_2 structures and made hierarchical macro-/mesoporous TiO_2/SiO_2 and TiO_2/ZrO_2 nanocomposites which were stable under high-temperature thermal treatment (Chen et al. 2009a). They prepared the nanoporous TiO_2/SiO_2 and TiO_2/ZrO_2 nanocomposites with the same surfactant template method with the same heat treatments as they previously used for the synthesis of mesoporous TiO_2 structures (Wang et al. 2005). Zirconium propoxide and tetraethyl orthosilicate (TEOS) were the metal alkoxides used as Zr and Si sources. Investigation on the crystal structure and measurement of surface area and pore volume of the nanoporous TiO_2/SiO_2 and TiO_2/ZrO_2 nanocomposites showed that the presence of metal oxide (SiO_2 or ZrO_2) as a second phase in the TiO_2 mesoporous framework could well improve the stability of the structure. For a pure mesoporous TiO_2, the anatase to rutile phase transformation happens as soon as the calcination temperature moves beyond 500 °C (Wang et al. 2005). In addition, a dramatic decrease in surface area of pure mesoporous TiO_2 observed from 201 to 85 to 2 $m^2\ g^{-1}$ when calcination temperature increased from 350 to 500 to 800 °C, respectively (Wang et al. 2005) which is attributed to the partial and complete collapse of porous framework by increasing the calcination temperature from 350 to 500 °C and from 500 to 800 °C, respectively. Later, by incorporation of a second metal oxide phase into the pure TiO_2 framework Wang et al. achieved a porous nanocomposite framework which was stable under heat treatment at high temperatures (Chen et al. 2009a). For example, TiO_2/SiO_2 mesoporous structure exhibited a high surface area of 298 $m^2\ g^{-1}$ at 350 °C and retained its surface area as large as 92 $m^2\ g^{-1}$ even after calcination at 800 °C for 4 h. The presence of a metal oxide as a secondary phase not only could hinder the excessive crystal growth which was responsible for the instability of the porous framework, but also could inhibit crystallization and anatase to rutile transformation at high temperatures which resulted in the presence of anatase crystalline phase in the porous TiO_2/SiO_2 and TiO_2/ZrO_2 nanocomposites structure after thermal treatment at high temperatures. The mentioned modifications caused by the presence of a second metal oxide phase inside the porous TiO_2 framework led to a better and more stable photocatalytic performance of porous TiO_2/SiO_2 and TiO_2/ZrO_2 nanocomposites compared to the photocatalytic activity of pure TiO_2 porous structure. The photocatalytic activity of nanoporous TiO_2/SiO_2 and TiO_2/ZrO_2 nanocomposites remained as high as 48 and 43% at 500 and 650 °C respectively while the activity of the mesoporous TiO_2 decayed quickly with increasing sintering temperature higher than 350 °C. The better photocatalytic performance of porous TiO_2/SiO_2 and TiO_2/ZrO_2 nanocomposites compared to pure TiO_2 porous structure can also be attributed to the presence of lots of surface hydroxyl groups on the surface of porous nanocomposite. These hydroxyl groups can be converted to active hydroxyl radicals (acidic sites) for oxidizing adsorbed molecules by accepting photogenerated holes. In fact, the nanoporous nanocomposite material provides a large surface area of acidic sites for photocatalytic degradation of organic pollutants.

Molecular selectivity is another superiority of the porous structure over non-porous structure. As an example, Inumaru and coworkers prepared a mesoporous TiO_2/SiO_2 nanocomposite with molecular selective photocatalysis property for organic molecules (Inumaru et al. 2005). Mesoporous silica and the mesoporous TiO_2/SiO_2 nanocomposite were prepared by a surfactant template method. TiO_2 (Degussa P25, 20–30 nm in diameter) was ultrasonically dispersed in a solution of hexadecyltrimethylammonium bromide in hot water. After adjusting the pH to 11.8 by adding ammonia solution, TEOS was quickly added to the mixture, under vigorous stirring. After aging for 1 h, the precipitate was filtered, washed with water, and dried at 80 °C for one day. The materials were calcined at 540 °C for 6 h to obtain the mesoporous composite materials. With this method, they could make TiO_2/SiO_2 nanocomposite photocatalysts with TiO_2 loading up to 60%. The high amount of TiO_2 loading on mesoporous SiO_2 guarantees an excellent photocatalytic performance. These mesoporous nanocomposites had less hexagonally ordered mesopores than mesoporous silica. The detailed characterization of mesoporous silica and mesoporous TiO_2/SiO_2 structures showed that the addition of TiO_2 particles did not interfere with the formation of mesoporous silica, but it did disturb the hexagonal ordering of the pores to some extent. TEM observation (Fig. 6.3a) revealed that the TiO_2 particles were incorporated into the porous silica as illustrated schematically in Fig. 6.3b.

Photocatalytic activity of TiO_2/SiO_2 nanocomposite for the decomposition of a mixed aqueous solution of several alkylphenols was tested and compared with the activity of a mechanical mixture of mesoporous silica and TiO_2 with the same stoichiometric composition as that of mesoporous TiO_2/SiO_2 nanocomposite. The mechanical mixture showed almost the same decomposition rate for all kinds of all alkylphenols in the solution while mesoporous TiO_2/SiO_2 nanocomposite exhibited different decomposition rates for different alkylphenols. This implies that the mesoporous nanocomposite structure shows molecular selective photocatalysis while the photocatalytic activity of TiO_2 nanoparticles is nonselective. This is attributed to the architectural design of the mesoporous TiO_2/SiO_2 nanocomposite structure.

Composites of metal/TiO_2 are of great interest for their enhanced photocatalytic activity compared to pure TiO_2 structures due to the better separation of photogenerated electrons and holes through scavenging the electrons by metal particles. Nanoporous metal/TiO_2 composites can be prepared by template method. Wang and coworkers proposed a number of synthesis routes for the fabrication of porous gold nanoparticle/TiO_2 nanocomposites using a template method with agarose gel as the structure-directing template (Wang et al. 2008b). The basic steps of all routes were as follows: (i) making the agarose gel template by heating agarose in aqueous solution and then cooling to form the gel, (ii) solvent exchange to facilitate sol–gel synthesis of TiO_2 using titanium(IV) isopropoxide as precursor, (iii) TiO_2 precursor infiltration into the template, (iv) hydrolysis and condensation reactions to achieve amorphous TiO_2, and (v) calcination to remove the agarose and crystallize the TiO_2.

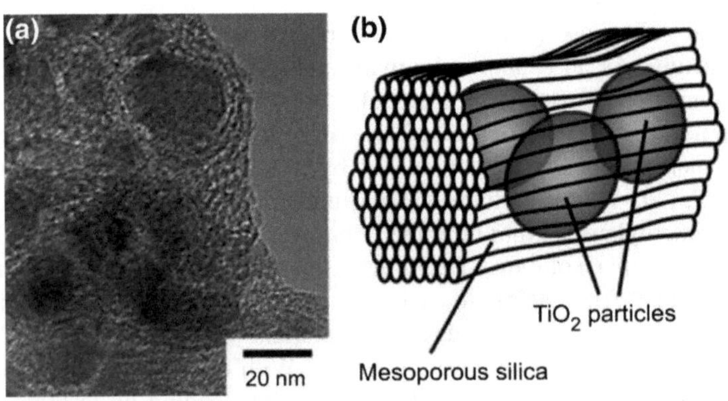

Fig. 6.3 **a** TEM image of mesoporous TiO_2/SiO_2 nanocomposite with 60% TiO_2 loading, and **b** schematic illustration of the nanocomposite material. Reprinted from Inumaru et al. (2005)

They tried five different routes for the incorporation of gold into the porous TiO_2 framework. All these routes are shown schematically in Fig. 6.4. In route I, the gold salt was added during agarose gel template formation. In route II, the gold salt was incorporated into the preformed agarose gel. In route III, the gold salt was added during template infiltration. In route IV, the gold was distributed throughout the structure during the hydrolysis and condensation step. In route V, gold was introduced to the crystallized titania structure using a modified deposition–precipitation method after the agarose template was removed. Different routes led to different gold particle size and distribution. The concentration of gold nanoparticles was found to be another factor affecting both the size and distribution of gold nanoparticles as well as mechanical stability of porous TiO_2 framework. The porosity and mechanical strength of the final structure were decreased with increasing gold content and structure collapsed at maximum gold contents of 0.5, 5, 5, and 16 wt% for the routes I, III, IV, and V respectively. Figure 6.5 shows the effect of gold initial content on the porous structure of Au/TiO_2 nanocomposite prepared by route III (Wang et al. 2008b). Porous structures prepared by route V had the highest stability because in this route porous structure fabricated before the addition of the gold salt and did not get much affected by the gold content. But in other routes, the acidic gold precursor affected either the agarose gel structure (route I) or the sol–gel synthesis of TiO_2 framework (routes III & IV) (Wang et al. 2008b).

The nanoporous Au/TiO_2 composites prepared by different routes were used for the degradation of methylene blue under UV light irradiation. Photocatalytic

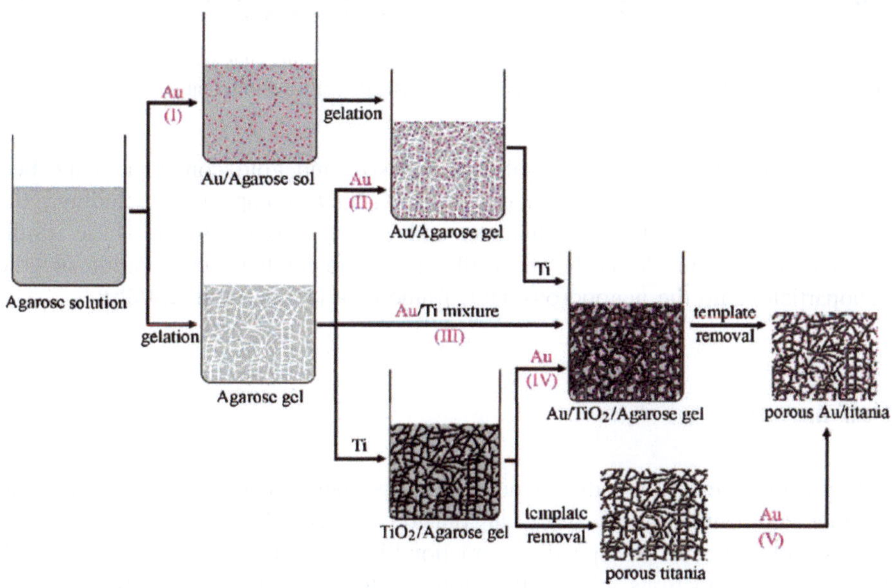

Fig. 6.4 Schematic Representation of five different routes for the incorporation of gold into the porous TiO_2 framework. Reprinted from Wang et al. (2008b)

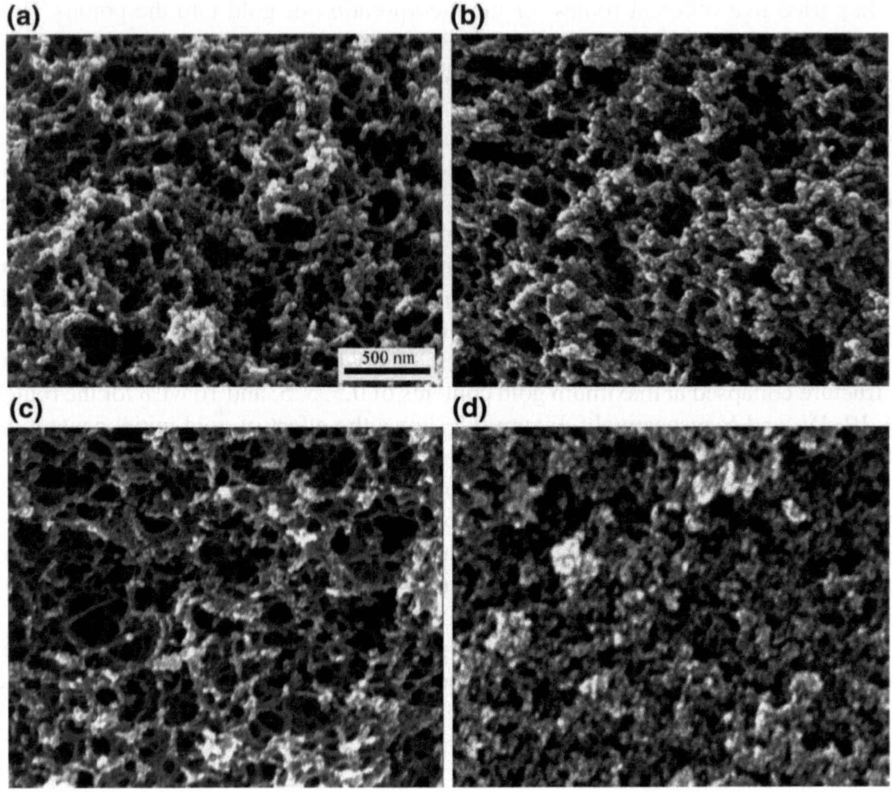

Fig. 6.5 SEM images (all at the same scale) of the Au/TiO_2 composites with **a** 0.25, **b** 0.50, **c** 2, and **d** 5 wt% gold initial content prepared by route III. Reprinted from Wang et al. (2008b)

activity was dependent on the gold particle size and gold content and the best results obtained by using the nanoporous Au/TiO_2 composites including gold nanoparticles of 2 ± 1 nm in diameter and 2.0 wt% gold content. These results also confirmed route V as the most effective route for the incorporation of gold nanoparticles into the nanoporous TiO_2 framework (Wang et al. 2008b).

Self-assembly

Self-assembly generally is the phenomena of spontaneous arrangement of materials without any intervention from an outside source. The driving force in a self-assembly process is a specific interaction like hydrogen bonding, van der Waals forces, π–π interactions, electrostatic forces, etc. Internal characteristics, such as hydrophobicity or hydrophilicity can also direct the assembly of the particles in a system. Nanoparticles can be programmed to self-assemble by changing the functional groups on their surface, taking advantage of the mentioned interactions. Self-assembly has been widely used for the preparation of highly ordered mesoporous TiO_2 films with various mesostructures, such as p6 mm (Grosso et al. 2001; Crepaldi et al. 2003), R3 m (Choi et al. 2006), or Im3 m (Crepaldi et al. 2003). Without a template, self-assembly process often leads to the formation of mesopores with only several nanometers diameter, which makes it difficult for organic molecules to diffuse through the pores easily and limits the accessibility of molecules to the photocatalytic active site (Tang et al. 2004; Choi et al. 2007). A simultaneous application of template and self-assembly process can be used to enforce larger pores into the structure. A more attractive strategy is to construct larger secondary pore channels within a mesoporous structure, which will make the diffusion of reactants and products within the porous structure more easily and increase the availability of the internal active sites (Yang 1998; Sakamoto and Dunn 2002; Zhang and Yu 2003; Fu et al. 2008; Dionigi et al. 2008; Meng et al. 2009; Kimura et al. 2009; Kaune et al. 2009). However, the challenge in this regard is to independently control the porosity at each length scale.

Du et al. successfully produced hierarchically ordered macro-/mesoporous titania/graphene composite films through a confinement self-assembly method Du et al. (2011a). The schematic representation of the method of preparation is shown in Fig. 6.6. Du et al. used Pluronic P123 and polystyrene spheres as a mesostructured template and a macro-structure scaffold, respectively. A glass substrate covered with the polystyrene opal film was immersed in a sol of Pluronic P123, $TiCl_4$, and titanium isopropoxide and graphene oxide in ethanol, water, and THF solvents. After penetration of the sol into the voids of polystyrene opal film, the template took out of the sol and the gelation of sol happened. The withdrawal rate was found to be a critical factor affecting the infiltrated process. Keeping the samples at 35 °C, the hydrolysis and self-assembly of titania particles into a mesoporous structure inside the voids of P123 template happened. The graphene oxide reduction to graphene was finally done under hydrazine vapor at 40 °C for 24 h. A two-step heat treatment at 400 °C under argon for 3 h and at 450 °C for 2 h was done during which the P123 template was eliminated and ordered macro-/mesoporous titania/graphene composite film was achieved (Du et al. 2011a). By this process, Du et al. could produce macro/mesoporous films of anatase titania with well-ordered two-dimensional (2D) hexagonal (p6 mm) mesostructure and well-interconnected periodic macro-pores. The macropore walls consisted of about 3 nm mesopores. Incorporation of interconnected macro-pores in the mesoporous structure improved the mass transport through the structure while a large surface area was available for the catalytic reactions. Evaluation of the photocatalytic activity of these nanoporous nanocomposites was done by degradation of methylene blue in aqueous solution. The dark adsorption of methylene blue on a pure mesoporous titania film and a macro-/mesoporous titania film showed the adsorption of about 17% of dye molecules on macro-/mesoporous titania film compared to less than 1% dye adsorption on pure mesoporous titania film. It clearly confirmed the effect of incorporated macrochannels in the mesoporous structure on the improvement of accessibility of molecules to the internal surface area. In the macro/mesoporous structure, the ordered macrochannels limited the length of mesoporous channels while the mesopores in the pure mesoporous structure were longer with smaller diameters. Besides, the length of the channels in macro-/mesoporous structure can be tuned by changing the size of the polystyrene spheres used as the macroporous template. Smaller polystyrene spheres will make shorter mesoporous channels and more accessible surface area. The increased accessible surface area in macro-/mesoporous titania films, having 170 nm macro-pores, caused an 11-fold enhancement in photocatalytic degradation reaction rate of methylene blue compared to the pure mesoporous structure. Moreover, incorporation of graphene into the macro-/mesoporous titania films, caused the photocatalytic reaction rate constant increased around 2 times, compared to the photocatalytic activity of macro/mesoporous films without graphene. Graphene played the role of electron acceptor in this composite and improved the photocatalytic activity by hindering the electron–hole recombination.

Evaporation induced self-assembly (EISA) method is an extended self-assembly process initiated by Brinker and coworkers for the fabrication of mesoporous silica film (Brinker et al. 1999). In the EISA process, the mesostructure forms as a result of solvent evaporation and thereby enrichment of surfactant and inorganic species in the solution. The solvent is typically ethanol or water. Ordered mesoporous frameworks are then created by subsequent thermal treatment of the mesostructured hybrids. Wei et al. prepared ordered mesoporous carbon-TiO_2 composite through an EISA approach (Wei et al. 2013). The mesoporous structure with oriented and hexagonally arranged pores was obtained by EISA of low-polymerized phenolic resin (carbon source) and titanium tetrachloride (titanium source). Triblock copolymer F127 used as a structure-directing agent. In order to increase the thermal

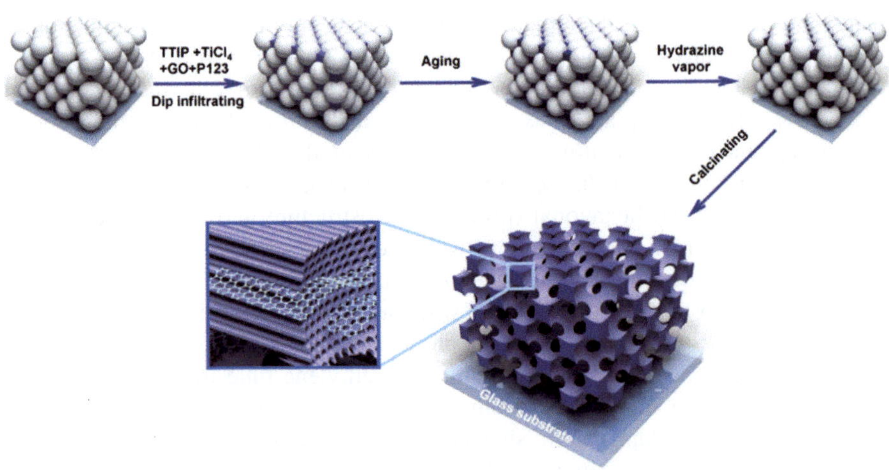

Fig. 6.6 Schematic representation of production of hierarchically ordered macro/mesoporous titania/graphene composite films through a confinement self-assembly method. Reprinted from Du et al. (2011a)

stability of the final mesoporous thin films, a “delayed rapid crystallization” procedure was applied in two steps. First, the nanocomposite was slowly (1 °C min^{-1}) heated to a relatively low temperature (350 °C) and was kept for a long time (10 h) while in the second step rapid heating (5 °C min^{-1}) to the desired temperature (550 °C) was done for a short time (30 min). The first step is necessary for complete condensation of $Ti(OH)_x$, removal of the triblock copolymer, and the release of small molecules from the resin to form a relatively rigid framework. Anatase TiO_2 nanoparticles were mostly formed during the second heat treatment step and the polymeric framework was converted into carbon. The pore walls of carbon framework inhibited the aggregation and thereby increase in the size of TiO_2 nanoparticles. Therefore, small TiO_2 nanoparticles of about 4.2 nm were well dispersed on the pore walls of the carbon framework. In fact, the pore walls of C/TiO_2 nanocomposite were constructed from anatase nanoparticles and amorphous carbon, with the two components being microphase separated at the nanoscale. The specific surface area, pore volume, and micropore volume of the final C/TiO_2 mesoporous structure measured to be 348 $m^2\ g^{-1}$, 0.30, and 0.041 $cm^3\ g^{-1}$, respectively. Mass transportation and access of organic molecules to the photocatalyst active sites was facilitated in this structure because of the relatively large surface area, large pore volume, and uniform mesopore size.

As a result of the mentioned features provided by the mesoporous architecture of the carbon/TiO_2 composite, the composites showed excellent photocatalytic activity. Comparing the activity of C/TiO_2 mesoporous structure with mesoporous TiO_2 and mesoporous TiO_2/SiO_2 structures, higher adsorbing capacity, and better photocatalytic activity in visible light was proved for the mesoporous C/TiO_2 composites. The better photocatalytic activity of C/TiO_2 composite was also attributed to the partial doping of TiO_2 nanoparticles by carbon in this structure. In addition, over 80 adsorption–catalysis cycles was performed without any noticeable loss in photocatalytic activity or the anatase content of the catalyst (Wei et al. 2013).

Hydrothermal Method

Hydrothermal synthesis method is the process of crystallizing substances from aqueous solutions at high temperature and high vapor pressure. It is well known that nanoporous structures of different materials can be produced by this facile and low-cost method. The choice of precursors and internal pressure, which is determined by the temperature and the amount of solution inside the reaction vessel (autoclave), are the factors affecting the architecture of the structure produced by hydrothermal method.

Using the hydrothermal method, it is possible to produce semiconductor nanocomposites with tunable electronic structures in order to enhance the photocatalytic activity. Some ternary metal sulfides like $ZnIn_2S_4$ and $CdIn_2S_4$ with various morphologies have recently been investigated as photocatalyst materials for degradation of dye molecules and H_2 production (Lei et al. 2003; Gou et al. 2006; Kale et al. 2006; Chen et al. 2009b, 2013; Chaudhari et al. 2011; Peng et al. 2011; Cai et al. 2013; Zhou et al. 2013a; Wang et al. 2013b; Hu et al. 2013; Xu et al. 2014a). Although these materials showed attractive photosensitive features, their photocatalytic activity is still away from satisfactory for practical applications. 3D $ZnIn_2S_4/CdIn_2S_4$ (ZIS/CIS) composite nanosheets with different amount of $CdIn_2S_4$(10, 20, and 30%) on carbon cloth (CC) substrates were synthesized to improve the photocatalytic activity (Li et al. 2015). The synthesis performed under hydrothermal conditions at 350 °C for 3 h by using $Zn(CH_3COO)_2{\cdot}2H_2O$, Cd $(CH_3COO)_2{\cdot}2H_2O$, $InCl_3$, and L-cysteine. L-cysteine, having the organic groups of –COOH, $–NH_2$, and –SH, has different roles in the formation of 3D ZIS/CIS composite nanosheets in this hydrothermal process. It acts both as the sulfur source and as the complex to coordinate with the Zn^{2+}, Cd^{2+}, and In^{3+} ions. The coordination of L-cysteine with these ions helps the deposition of composite material on CC substrate instead of precipitation of the bulk compounds. In fact, L-cysteine controls the nucleation rate of ZIS/CIS by slowly releasing S^{2-} ions and prevents the crystals growth due to the pinning effect of impurity atoms. Moreover, L-cysteine plays a crucial role in the formation of the porous structure. Release of gases such as H_2S, CO_2, and NH_3 during the hydrothermal process, which is the result of decomposition and reaction of precursors at high temperatures, leads to the formation of nanopores in the nanosheets. It was observed that no nanopore formed when the reaction temperature was below 250 °C. Different 3D network structures of ZIS, ZIS/CIS(10%), ZIS/CIS(30%) and CIS on CC substrate were produced. SEM and TEM images (Fig. 6.7) show a slight difference in the nanosheets roughness for different materials while similar porous nanosheet structures in a large scale were produced.

The proposed mechanism by Li and coworkers for the formation of 3D network of ZIS/CIS nanosheets is shown schematically in Fig. 6.8 (Li et al. 2015). Zn^{2+}, Cd^{2+}, and In^{3+} were adsorbed on the surface of CC, reacted by H_2S and formed ZIS/CIS nuclei on the CC substrate. As the reaction continued, the nuclei aggregated and formed nanoparticles. These nanoparticles then continued to grow and formed interconnected nanosheets on the CC surface. Finally, the porous nanosheets can be obtained through the high-temperature hydrothermal treatment and the porous nanosheets can aggregate to form the 3D porous architecture.

Photocatalytic degradation of MB dye and photocatalytic H_2 production under visible light irradiation was tested for 3D nanoporous network of ZIS/CIS nanosheets on CC. The result of visible light photocatalytic degradation of MB in the presence of ZIS-CC, ZIS/CIS(10%)-CC, ZIS/CIS(20%)-CC, ZIS/CIS(30%)-CC, CIS-CC and ZIS/CIS(20%) powder confirmed the superior performance ZIS/CIS composite compared to the performance of separate semiconductors, ZIS-CC and CIS-CC. This can be attributed to the increased light harvesting in the ZIS/CIS composite as well as suitable arrangement of band potentials of ZIS and CIS in the heterostructure, which facilitates the transfer of charge carriers and retards the electron–hole recombination. The optimum CIS concentration in ZIS/CIS composite found to be 20% beyond which the photocatalytic activity weakens as the result of

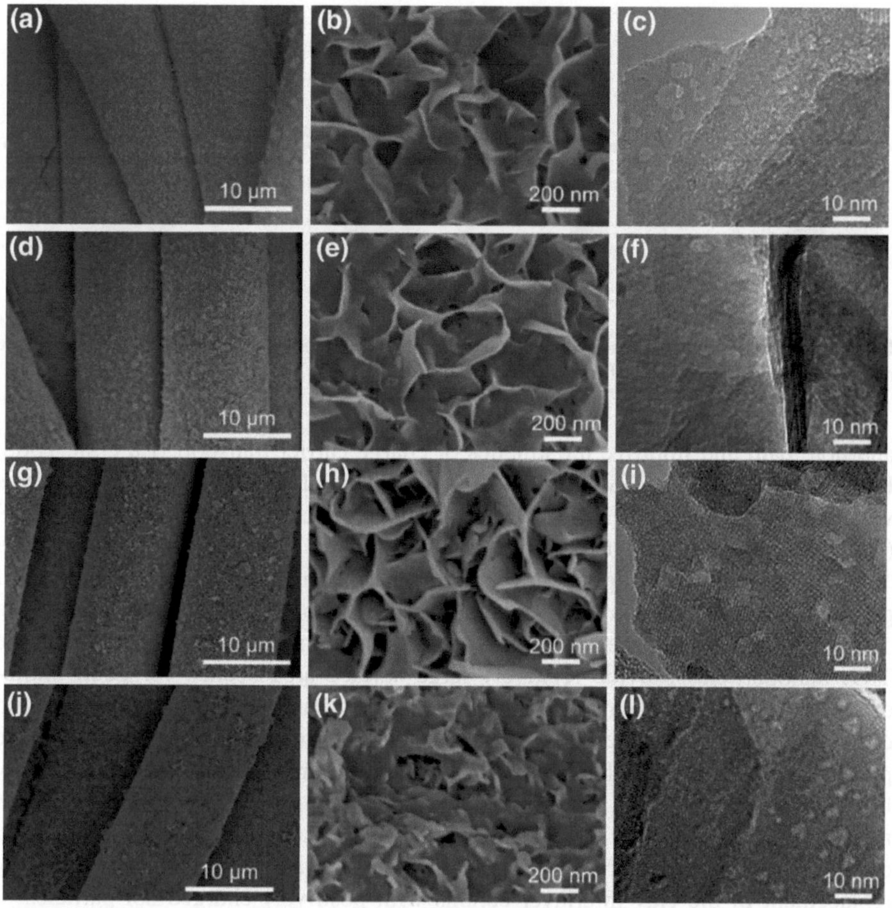

Fig. 6.7 SEM (**a**, **b**, **d**, **e**, **g**, **h**, **j**, and **k**) and TEM (**c**, **f**, **i**, and **l**) images of ZIS-CC (**a**, **b**, **c**), ZIS/CIS(10%)-CC (**d**, **e**, **f**), ZIS/CIS(30%)-CC (**g**, **h**, **i**) and CIS-CC (**j**, **k**, **l**). Reprinted from Li et al. (2015)

recombination of photogenerated electrons and holes on the surface of CIS. Furthermore, ZIS/CIS(20%)-CC showed better photocatalytic activity compared to the ZIS/CIS(20%) powder which was due to the large surface area of CC substrate. Larger surface area of CC provided enough area for the formation of a large number of ZIS/CIS nanosheets, which can increase light harvesting by multiple scattering of light from nanosheets and adsorb a high content of MB dye. Besides providing a large surface area, CC substrate helps to improve the photocatalytic activity by providing fast electronic transmission between the semiconductors and CC, which can lower the recombination rate of photogenerated electrons and holes. Moreover, the porous nanosheets in the 3D porous ZIS/CIS-CC architecture not only provide a large number of active sites for photocatalytic reaction, but also shorten the electron

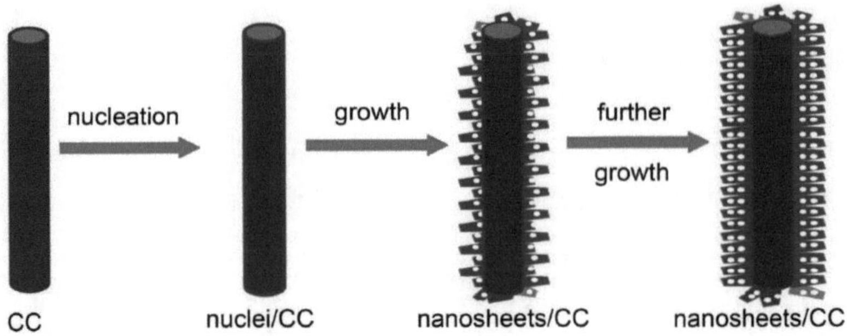

Fig. 6.8 Scheme representing the mechanism for the formation of 3D ZIS/CIS nanocomposite nanosheets on CC. Reprinted from Li et al. (2015)

and hole path and improve the kinetics of charge carrier transport (Lu et al. 2008; Rengaraj et al. 2011). ZIS/CIS(20%)-CC also showed good recyclability which is very important in practical applications. The photocatalytic activity of ZIS/CIS (20%)-CC structure remained almost unchanged even after five successive cycles.

Another example for application of hydrothermal method for the preparation of porous nanocomposites is the preparation of a mesoporous structure with anatase TiO_2 nanoparticles attached on clay layers. As mentioned before, although anatase TiO_2 nanoparticles provide relatively large surface area needed for good photocatalytic activity, problems like agglomeration of nanoparticles in solution and difficulties related to recycling the nanoparticles stimulate the researchers to make porous structures in which nanoparticles embedded in a porous framework. Clay layers are considered to be good support for immobilization of photocatalysts nanoparticles (Ooka et al. 2003; Chae et al. 2004; Yuan et al. 2006). TiO_2-pillared clays first reported by Sterte (1986) and Yamanaka et al. (1987) and were made by replacement of the clay interlayer cations with positively charged sol particles of titanium hydrate during hydrolysis of titanium compounds. Titanium hydrates then converted to titanium oxide pillars by heat treatment at 400 °C. The thickness of the clay layers is about 1 nm while their lateral dimension is between hundreds of nanometers and several microns. Although the TiO_2-pillared clays possessed large surface area, they had some imperfections limiting their application (Sterte 1986; Yamanaka et al. 1987; Yoneyama et al. 1989; Ding et al. 1999). TiO_2 pillars were too small to form crystallites whereas the anatase form of TiO_2 is necessary for an efficient photocatalyst. Moreover, the pore size of TiO_2-pillared clays was too small for the diffusion of large organic molecules before reaching the photocatalytic active sites. Light penetration through many clay layers was also difficult. To circumvent these problems, Xuzhuang and coworkers proposed a new mesoporous structure with size-controllable anatase attached on clay layers (Xuzhuang et al. 2009). In their hydrothermal approach, the reaction of acidic $TiOSO_4$ with laponite clay led to the formation of anatase nanocrystals on the fragments of clay layers. The hydrothermal synthesis was done for 24 h at different temperatures in order to investigate the effect of reaction temperature on the final structure properties.

The final structure of TiO_2/clay composite had large mesopores through which facile diffusion of organic molecules could happen. The ratio of titanium to clay and hydrothermal temperature were found as the two parameters affecting the pore structure and composition of the photocatalysts, as well as the size of anatase nanocrystals. N_2 adsorption/desorption isotherms of samples prepared at hydrothermal temperatures of 100, 150 and 200 °C showed that the meso- and macro-pores volume increased with increasing the hydrothermal temperature. Higher hydrothermal temperature also resulted in higher photoactivity. The 0.5 g/L of sample prepared at 200 °C could fully degrade 25 ppm phenol under UV light irradiation while degradation of only 10% of phenol was achieved by using a TiO_2-pillared clay sample. The superior performance of the new structure was attributed to the presence of well-crystallized anatase nanocrystals and large porosity of the structure, which facilitates the penetration of both light and large organic molecules to access the anatase crystals (Xuzhuang et al. 2009).

Solvothermal Method

Solvothermal method of synthesis is very similar to the hydrothermal method but the precursor solution used in solvothermal synthesis is not aqueous. Different nanoparticles and nanostructures of metals, metal oxides, ceramics, and polymers can be synthesized by the solvothermal method. Easy and precise control over the size, shape distribution, and crystallinity of nanoparticles or nanostructures is attainable in this method by changing reaction temperature, reaction time, solvent, and structure-directing agents, surfactant, and precursors. Moreover, very high reaction temperatures can be applied in solvothermal method by choosing a solvent with high boiling point. Solvothermal synthesis is a facile method for preparation of various geometries and many novel architectural designs of different nanocomposites have been reported using this method. For example, Zhou et al. prepared Cu_2O/Cu nanocomposites by a two-step solvothermal synthesis route without using any templates and additives (Zhou et al. 2011). They used a solution of $Cu(NO_3)_2{\cdot}3H_2O$ in an organic solvent N,N-dimethylformamide (DMF) and did the solvothermal synthesis in two steps at 150 °C for 24 h and at 180 °C for 8–26 h. In their experiments, they found the precursor concentration as the main parameter affecting the morphology of the products (Zhou et al. 2011).

Figure 6.9 shows the microstructure of the Cu_2O/Cu nanocomposites prepared with different precursor concentration by solvothermal synthesis process. Flower-like nanocomposites formed at precursor concentrations less than 0.02 M (Fig. 6.9a, b). The flowers petals were approximately 300–500 nm long and 30–70 nm wide. The petals became shorter and smaller by increasing the precursor concentration to 0.03 M. At 0.05 M concentration, the flower-like morphology was replaced by cubes or octahedra with a few small nanobelts standing on them while these structures converted to large porous spheres and octahedral at the precursor concentration of 0.1 M. These results showed that the precursor concentration has a significant effect on the morphology of the products. Therefore, selective synthesis of nano-flower-like architectures or mesoporous spheres was possible by changing precursor concentration, which is of great importance for designing photocatalysts with large surface area and high activity (Zhou et al. 2011).

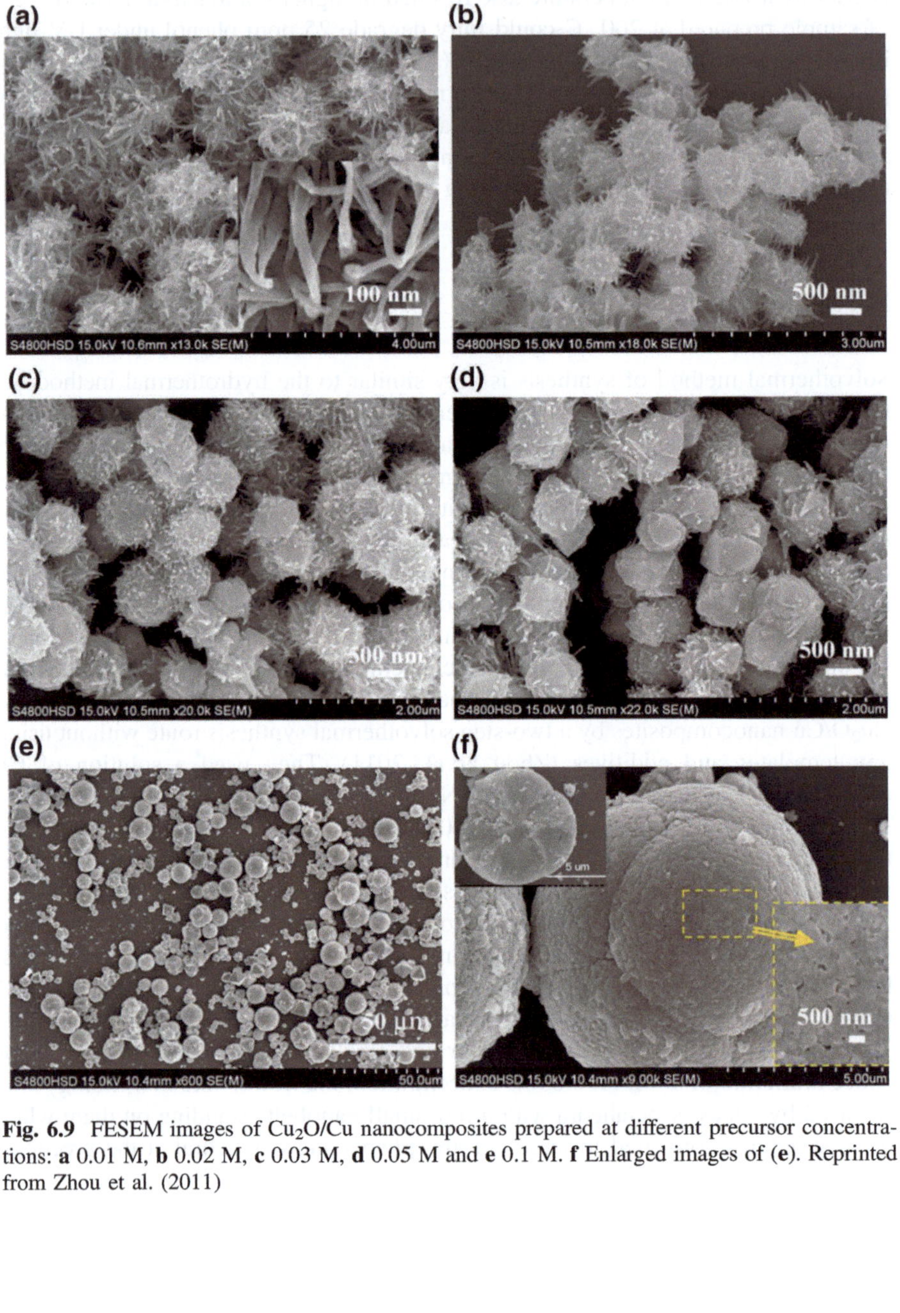

Fig. 6.9 FESEM images of Cu_2O/Cu nanocomposites prepared at different precursor concentrations: **a** 0.01 M, **b** 0.02 M, **c** 0.03 M, **d** 0.05 M and **e** 0.1 M. **f** Enlarged images of (**e**). Reprinted from Zhou et al. (2011)

Later, Wang et al. synthesized pinecone-like $Fe_3O_4@Cu_2O/Cu$ porous nanocomposites via one-pot solvothermal method on Fe_3O_4 (Wang et al. 2013a). The solution used in the solvothermal process was a mixture of $Cu(NO_3)_2{\cdot}3H_2O$ and pre-synthesized Fe_3O_4 nanoparticles in ethylene glycol and ethanol. The pure Fe_3O_4 nanoparticles were of cuboid shape with an average side length of about 19 nm. The solvothermal process was done at 160 °C and in different times from 3 to 7 h. The proposed formation mechanism of the pinecone-like $Fe_3O_4@Cu_2O/Cu$ composites by Wang et al. is shown in Fig. 6.10 (Wang et al. 2013a). First, modified Fe_3O_4 nanoparticles were prepared in a coprecipitation process using $FeCl_3$ as precursor. L-cysteine was used for the modification of magnetic Fe_3O_4 nanoparticles. The coordination of polar groups of L-cysteine with Cu^{2+} ions provided nucleation site for Cu_2O on Fe_3O_4 nanoparticles. The Cu_2O NPs then formed by the reduction of Cu^{2+} using ethylene glycol as the electron acceptor. By increasing the number of small Cu_2O particles around or on Fe_3O_4, aggregation of these nanocrystals happened and microspheres formed because of the high surface energy. During longer times, the inner part of the solid spheres gradually dissolved and recrystallized on the shell of the spheres [Ostwald ripening (Li and Zeng 2007)] to finally form hollow spheres. After the reaction time reached 6 h, partial reduction of Cu_2O nanoparticles to metal Cu happened in ethylene glycol and the nanocomposites resulted in the formation of nanosheets. Finally, pinecone-like $Fe_3O_4@Cu_2O/Cu$ composites formed at 7 h. This mechanism was confirmed by FTIR and XRD data and implies the critical role of both solvent and L-cysteine as linker and reductant in the solvothermal process (Wang et al. 2013a).

Another factor found having a crucial role on the composition and crystal structure of the products obtained by solvothermal process is the reaction time. Detailed investigation of the product obtained after different reaction times by Wang et al. showed that after 3 h of reaction, many Fe_3O_4 nanoparticles with

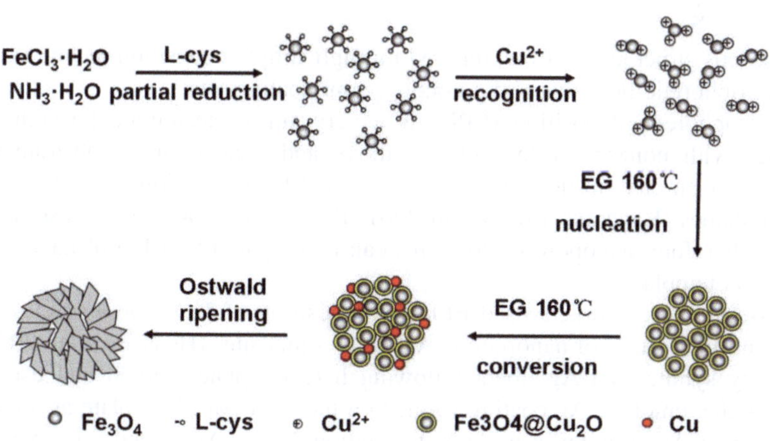

Fig. 6.10 Schematic representation of the proposed growth mechanism for pinecone-like $Fe_3O_4@Cu_2O/Cu$ core-shell nanocomposites. Reprinted from Wang et al. (2013a)

increased average size of 25 nm formed (Wang et al. 2013a). At the reaction time of 5 h, the integrated dispersive porous pompon-like microspheres with an average diameter of 960 nm appeared. The XRD results confirmed the formation of $Fe_3O_4@Cu_2O$ composites after 3 and 5 h with an increased amount of Cu_2O at longer reaction time. Increasing the reaction time to 6 h led to the formation of flower-like spheres with the diameter of about 2 μm which were consisted of many sheets with the average thickness of ~60 nm. Cu crystals partially formed at this stage and pinecone-like $Fe_3O_4@Cu_2O/Cu$ composites formed at 7 h.

The pinecone-like $Fe_3O_4@Cu_2O/Cu$ nanocomposites showed very effective photocatalytic activity of 96% for the degradation of methyl orange under 20 min visible light irradiation. The content of Cu_2O and the presence of Cu crystals proved to be important factors governing the photocatalytic activity. Photocatalytic activities of composites enhanced when the reaction time increased from 3 to 5 h mainly due to the content increase of Cu_2O. In addition, the degradation rate of methyl orange by porous pompon-like $Fe_3O_4@Cu_2O$ obtained at 5 h was lower than pinecone-like $Fe_3O_4@Cu_2O/Cu$ composites. This was attributed to the role of Cu crystals as electron acceptors in $Fe_3O_4@Cu_2O/Cu$ composites, which retarded the recombination of photogenerated electron–hole. Therefore, larger number of electrons and holes can participate in photocatalytic reactions and increase the photocatalytic efficiency. Rapid separation of photogenerated electron–hole in the presence of metal Cu also helped to increase the photostability of the $Fe_3O_4@Cu_2O/Cu$ composites. It is worth mentioning that the presence of Fe_3O_4 in $Fe_3O_4@Cu_2O/Cu$ composites add the magnetic property to the photocatalyst composite which is of great help in recycling the nanocomposite material via magnetic fields in practical applications.

Deposition Method

Nanoporous structures can be prepared through simple deposition techniques such as electrophoretic deposition, dip coating, spin coating, etc.

Electrophoretic deposition (EPD) is a very simple technique through which particles with nonzero surface charge move and deposit on a substrate in the presence of an electric field. EPD is well suited for deposition on complex component shapes (Kaya and Boccaccini 2001; Boccaccini et al. 2004; Besra and Liu 2007). Therefore, nanoporous structures can be prepared by EPD of nanoparticles on porous templates.

Hosseini and coworkers used EPD for deposition of TiO_2 nanoparticles on Ag fibers and fabrication of nanoporous Ag/TiO_2 composites (Hosseini et al. 2008). At first, they synthesized Ag fibers following hard template method and used filter paper as the template. Ag coating formed on the filter paper template as a result of reduction of Ag complexes in $AgNO_3$ solution. Then, Ag-coated filter paper was washed and the template was burned out at 400 °C in air, resulting in the formation of conductive Ag fiber. The SEM images of the Ag fibers before and after burning the template are shown in Fig. 6.11. The structure of Ag fibers follows, to some extent, the structure of the template, where tubular shapes with an average diameter of 10 μm can be observed. During the heat treatment at 400 °C, Ag particles sintered and formed the Ag fibers. This provided good conductivity of the fibers, which is needed for a successful EPD process, as well as good transport of photogenerated electrons of TiO_2.

The Ag fibers were then used as the substrate for EPD of TiO_2 nanoparticles from an aqueous solution of TiO_2 particles with the size of 15–50 nm. EPD was done at low current density of 500 μA/cm^2 and deposition time of 90 s. This duration was long enough for penetration and deposition of TiO_2 nanoparticles onto Ag fiber before agglomeration of particles happened on the surface of fibrous substrate. The SEM images of TiO_2/Ag fibers are shown in Fig. 6.12 (Hosseini et al. 2008).

Verification of photocatalytic activity of TiO_2/Ag fibers in comparison with TiO_2 thin films, prepared with the same method and the same TiO_2 colloidal suspension, was done by testing the degradation of methylene blue. TiO_2/Ag fibers showed about a 1.7 times higher decomposition rate constant in comparison with TiO_2 thin film. High surface area of TiO_2/Ag fibers with large pores that allow easy diffusion of ions and molecules, together with the role of Ag fibers as electron scavenger, which promotes the separation of photogenerated electron–hole pairs, were the main reasons for the better performance of TiO_2/Ag fibers in photocatalytic test (Hosseini et al. 2008).

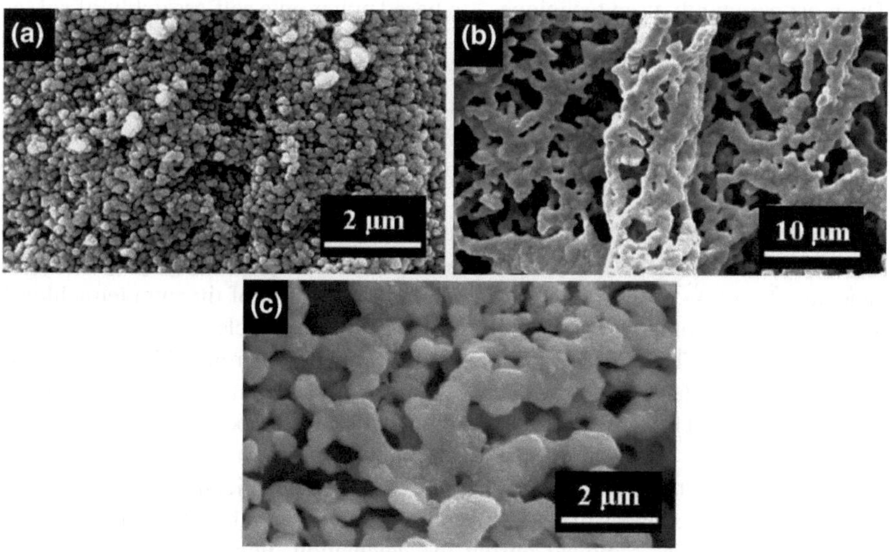

Fig. 6.11 SEM micrograph of **a** Ag coating formed on the surface of filter paper and **b**, **c** Ag fibers. Reprinted from Hosseini et al. (2008)

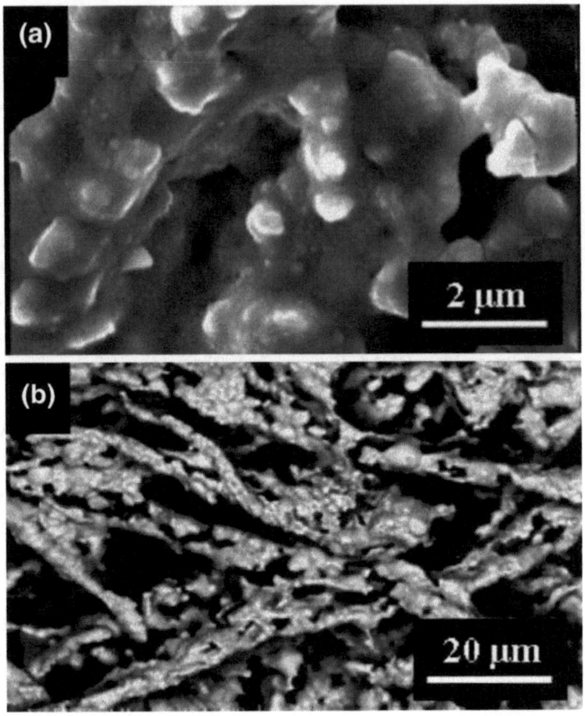

Fig. 6.12 **a** SEM micrograph of TiO_2/Ag fibers. Image **b** is a backscattered electron image. Reprinted from Hosseini et al. (2008)

Preparation of mesoporous TiO_2/SiO_2 nanocomposite is widely investigated. The most commonly used techniques are based either on a self-assembly of sol–gel processed molecular precursors of TiO_2 and SiO_2 with a structure-directing agent (Ogawa et al. 2001; Zhang et al. 2005; Calleja et al. 2008) or on infiltration of titania precursors into the preformed mesoporous silica template (van Grieken et al. 2002; Reddy et al. 2004; López-Muñoz et al. 2005). As an improved approach, Fattakhova-Rohlfing and coworkers proposed a combined method of self-assembly and dip coating for the production of mesoporous TiO_2/SiO_2 composite (Fattakhova-Rohlfing et al. 2009). They used a solution of preprepared TiO_2 crystalline particles and SiO_2 precursors for the dip coating process. TiO_2 nanoparticles of 4–5 nm in size were prepared by reaction of titanium tetrachloride with benzyl alcohol at 60 °C for 24 h. The TiO_2 nanoparticles then redispersed in tetrahydrofuran (THF) in the presence of a suitable Pluronic block-copolymer (P123 or F127). This solution was then mixed with the solution of SiO_2 precursor, which contained TEOS, water, HCl, and THF. The mesoporous TiO_2/SiO_2 films were deposited by dip coating of mixed solution of TiO_2 nanoparticles and SiO_2 precursors at a withdrawal rate of 1.8 mm/s. The final thickness of mesoporous TiO_2/SiO_2 films was measured to be about 300 nm after calcination. The Pluronic block-copolymer was removed during the calcination. Formation of mesoporous TiO_2/SiO_2 nanocomposite films is shown schematically in Fig. 6.13.

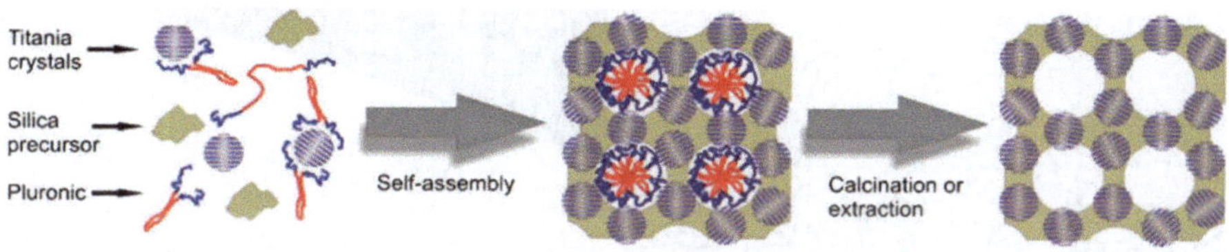

Fig. 6.13 Scheme showing the formation of mesoporous TiO_2/SiO_2 nanocomposite films using preformed titania nanocrystals stabilized by the Pluronic polymer and amorphous sol–gel silica precursor. Reprinted from Fattakhova-Rohlfing et al. (2009)

In this process, SiO_2 acts as both a structure-directing matrix and chemical glue. X-ray diffraction results showed that the addition of nanoparticles to the silica sol had a strong influence on the mesostructure of the final films. In the absence of TiO_2 nanoparticles, SiO_2 particles would assemble in an ordered mesoporous structure while the mesostructure became less organized as TiO_2 particles concentration increased. Addition of more than 10% of particles led to the almost complete disappearance of the mesostructure ordering. Moreover, the size of the mesopores in pure silica films was about 8 nm, which increased to 9–11 nm after incorporation of TiO_2 nanoparticles. This phenomenon was also dependent on the stabilizing agent used in the coating process. Adding the larger and more hydrophilic Pluronic F127 block copolymer provided the possibility of formation of mesoporous TiO_2/SiO_2 nanocomposite films with ordered mesostructure at higher concentration of TiO_2 particles. The higher TiO_2 content is desirable for having a mesoporous composite with good photocatalytic activity. TEM images of the mesoporous TiO_2/SiO_2 nanocomposite films with different particle concentrations calcined at 300 °C are shown in Fig. 6.14. The photocatalytic activity of mesoporous TiO_2/SiO_2 nanocomposite films was tested by photocatalytic oxidation of NO. The activity of mesoporous TiO_2/SiO_2 nanocomposite films increased almost linearly with the TiO_2 content while films of pure silica were inactive. This was predictable since the TiO_2 particles were responsible for the photocatalytic reaction.

Anodization Method

Anodization is an electrochemical process that converts the metal surface into metal oxide. Metal oxide tube arrays or pore arrays can be obtained by anodization process on the surface of metal. Anodic growth of porous Al_2O_3 on aluminum (Keller et al. 1953) and anodic growth of TiO_2 nanotubes (NTs) on titanium (Roy et al. 2011) are the most investigated anodic systems. Zwilling and coworkers reported fabrication of self-organized TiO_2 NTs with tube length of about 500 nm on titanium in 1999 for the first time (Zwilling et al. 1999a, b). After that, a lot of research especially in photocatalytic area devoted to the optimization of anodic growth of TiO_2 NTs (Macak et al. 2007b; Roy et al. 2011; Marien et al. 2016) and a variety of TiO_2 NTs

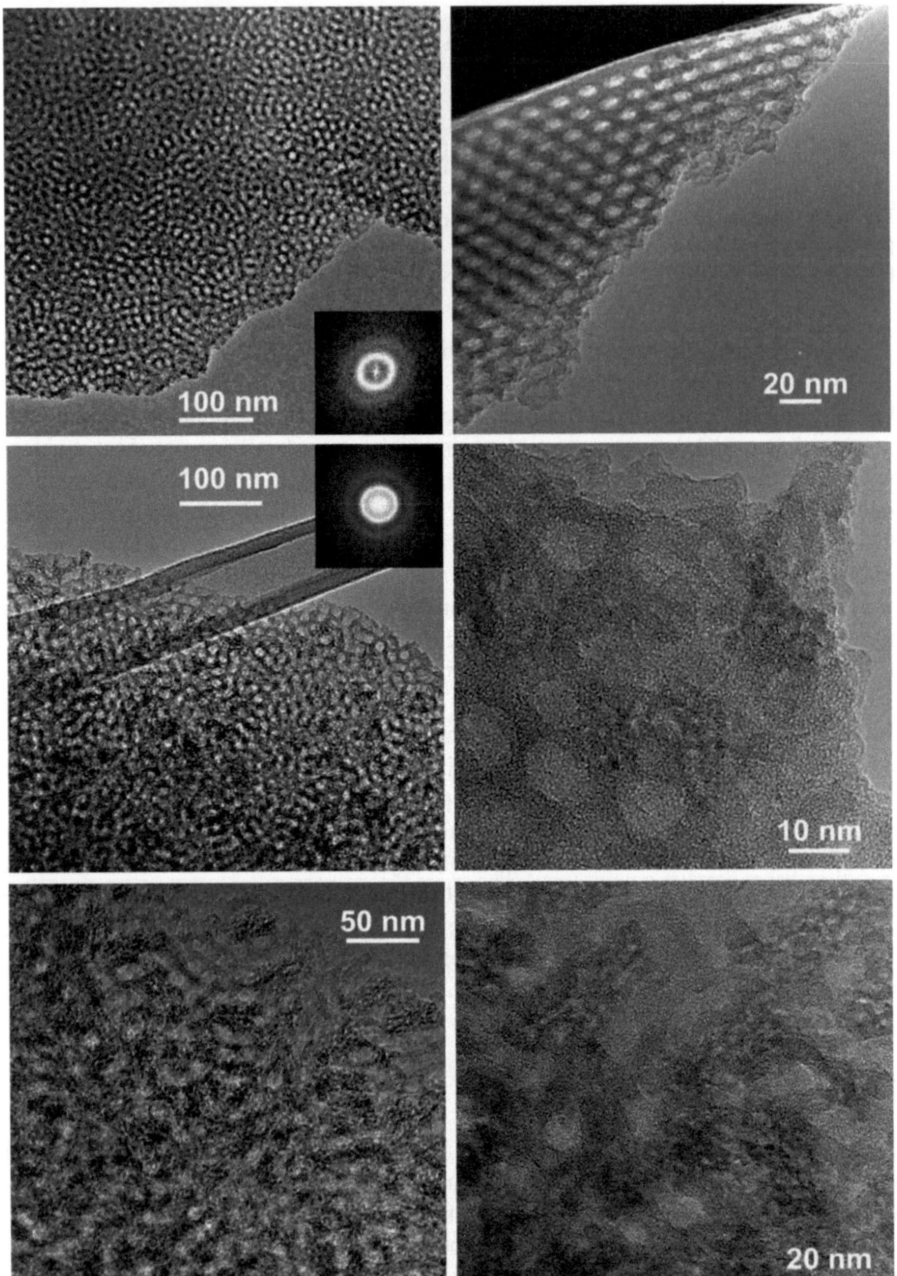
100 nm
20 nm
100 nm
10 nm
50 nm
20 nm

◂**Fig. 6.14** TEM images of mesoporous TiO_2/SiO_2 films containing 0 wt% TiO_2particles (*first row*), 15 wt% TiO_2particles (*second row*), and 30 wt% TiO_2particles (*last row*) after calcination at 300 °C. The *insets* show the Fourier transform of the same picture revealing the periodic mesopore structure. Reprinted from Fattakhova-Rohlfing et al. (2009)

with different aspect ratio and wall thickness reported. Self-organized TiO_2 NTs offering a large internal surface area and more efficient light harvesting soon gained so much attention in photocatalytic applications with superior performance compared to TiO_2 nanoparticles. These characteristics also made self-organized TiO_2 NTs a suitable electrode for photoelectrochemical systems like dye sensitized solar cells (Mor et al. 2006; Zhu et al. 2007; Kuang et al. 2008).

Besides the optimization of NTs characteristics like length and thickness, improvement of photocatalytic activity of TiO_2 NTs can be done by making nanotubes of TiO_2 composites. Kontos et al. fabricated self-organized iron oxide/TiO_2 nanotube arrays (Kontos et al. 2009). First, self-organized TiO_2 NT layers were prepared by electrochemical anodization of titanium foils. Then the surface of TiO_2 nanotubes was covered with iron oxide nanoparticles by dip coating of the TiO_2 NT layers in a dispersion of iron oxide nanoparticles with different concentrations. Iron oxide (Fe_xO_y) nanoparticles coated with dextrin was synthesized in a coprecipitation process using ferric chloride and ferrous chloride as precursors. The final size of iron oxide nanoparticles was 7–15 nm. SEM image of pure TiO_2 NT layers (Fig. 6.15a) showed the formation of a dense array of vertically aligned NTs with tube diameter of ~100 nm and length of around 1 μm. Figure 6.15b shows the SEM image of NTs after dip coating in the iron oxide nanoparticle suspensions. Iron oxide nanoparticles penetrated into the TiO_2 NT pores and aggregates of nanoparticles attached to the pore walls. By increasing the concentration of iron oxide solution, thicker layer of iron oxide was formed on the NT pore walls, which led to a decrease in the diameter of NTs. Figure 6.15c shows the SEM image of a TiO_2 NT array decorated with iron oxide from a solution with high concentration.

To explore the photocatalytic activity of the iron oxide/TiO_2 NT nanocomposite, degradation of methyl orange was tested under UV light irradiation using NT composites functionalized in different concentrations of the iron oxide solution. 5 mg/mL found to be the optimum concentration of the iron oxide solution used for functionalizing the TiO_2 NTs. Below this concentration, the photocatalytic activity decreased because of low concentration of iron oxide nanoparticles on TiO_2 NT walls and beyond this concentration, a very high load of iron oxide nanoparticles led to the blocking of some tubes and reducing the photocatalyst surface area and activity. Iron oxide/TiO_2 NTs prepared in the iron oxide solution with 5 mg/mL concentration showed 30% increase of degradation rate compared to the photocatalytic degradation rate of pure TiO_2 NTs. The better photocatalytic activity of iron oxide/TiO_2 NT nanocomposite was attributed to the enhanced electron–hole separation in the composite due to the hole scavenging ability of the dextrin shell that allows charge transfer between the constituent semiconductor oxides. This also changes the hydrophilic property of the NT surface to some extent.

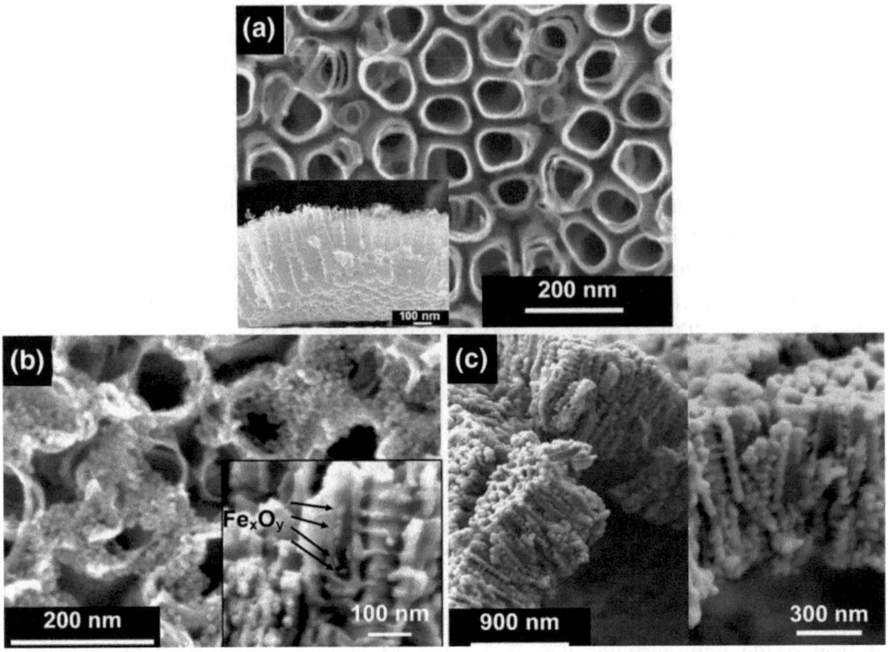

Fig. 6.15 SEM images of **a** pure TiO_2 NT arrays, **b** TiO_2 NTs after deposition of the iron oxide nanoparticles (5 mg/mL) and **c** TiO_2 NTs after deposition of the iron oxide nanoparticles at high concentrations (10 mg/mL). The *insets* show the side view images. Reprinted from Kontos et al. (2009)

In order to improve the light harvesting efficiency and thereby the photocatalytic activity of TiO_2 NTs in visible region, Tsui and Zangari sensitized TiO_2 NT arrays by Cu_2O nanoparticles (Tsui and Zangari 2013b). Sensitization of TiO_2 by Cu_2O nanoparticles can be done by electrodeposition of Cu_2O particles on TiO_2 surface (Tsui et al. 2012; Tsui and Zangari 2013a). But the challenge in this method is how to control the penetration of Cu_2O particles into the TiO_2 NTs and determine the exact deposition point of Cu_2O particles on TiO_2 NTs surface. Tsui et al. tried plasma cleaning of TiO_2 NTs before electrodeposition of Cu_2O on TiO_2 NT walls and obtained a uniform coverage of Cu_2O. They prepared TiO_2 NTs by anodization of Ti foil at voltage of 20 V for 1 h. The solution of NH_4F in ethylene glycol and water was used as the anodization solution. Post heat treatment at 350 °C changed amorphous TiO_2 NT into anatase. Then they deposited Cu_2O particles on TiO_2 NT walls by electrodeposition from a solution containing 0.02 M $Cu(CH_3COO)_2$ and 0.1 M $NaCH_3COO$ (pH 5.7) (Tsui et al. 2012; Tsui and Zangari 2013b). Plasma cleaning of TiO_2 NTs was performed before electrodeposition of Cu_2O. This cleaning step proved to be necessary for the elimination of residual hydrocarbons and better filling and uniform coverage of the TiO_2 NTs. Hydrocarbons can passivate nucleation sites for Cu_2O growth, leading to a less uniform deposit (Berger et al. 2010). The absorption spectra of bare TiO_2 NTs and TiO_2/Cu_2O composite

NTs showed the band gap change from 3.27 eV for TiO_2 to 2.21 eV for TiO_2/Cu_2O composite. This will guarantee higher light harvesting efficiency of TiO_2/Cu_2O composite NTs in photocatalytic and photoelectrochemical applications.

Several other strategies have been developed for the uniform deposition of sensitizer materials such as Cu or Cu_2O on TiO_2 NT walls. Electrochemical self-doping of TiO_2 NTs (Li et al. 2010) resulted in uniform filling of TiO_2 NTs by Cu or Cu_2O due to the enhanced conductivity at the bottom of TiO_2 NTs (Macak et al. 2007a; Li et al. 2011). Application of pulsed voltage during the deposition permits the depleted ion concentration to recover during rest pulses and leads to a uniform coating (Sun et al. 2001; Xiong et al. 2011). Photoreduction of Cu^{2+} to Cu_2O (Hou et al. 2009) and electroless deposition of Cu followed by thermal oxidation of Cu to Cu_2O (Hodgson et al. 2013) are other methods tried for the deposition of Cu_2O on TiO_2 NT walls.

Zeolite-Based Photocatalysts

The problems related to the practical application of nanoparticle photocatalysts, such as aggregation of nanoparticles in slurry and difficulties related to the recycling of nanoparticles, led the researchers to the idea of using well-dispersed photocatalyst species supported or anchored on different supports. Zeolite-based photocatalysts have been widely investigated for different applications in both liquid and gas photocatalysis (Anpo et al. 1997; Anpo and Tekeuchi 2003; Kitano et al. 2007; Dutta and Severance 2011; Kuwahara and Yamashita 2011). It is of special interest to investigate the photocatalytic activity of highly dispersed titanium oxide species on zeolites because of the improved photocatalytic reactivity and selectivity, much higher than that of conventional semiconducting TiO_2 photocatalysts.

Zeolites are microporous, crystalline aluminosilicates with the framework consisting of tetrahedrally coordinated SiO_4^{4-} and AlO_4^{5-} networks and enclosed cages and channels of molecular dimensions. Uniform pore distributions and channel sizes of zeolites provide selectivity to the introduced molecules as well as a large surface area available for adsorption of molecules. In addition to natural zeolites, there are synthetic zeolites, which are usually formed by sol–gel and hydrothermal processes of silica-alumina gel in the presence of alkalis and structure-directing agents (Cundy and Cox 2003). More than 140 frameworks of zeolite are known (Treacy and Higgins 2007).

When using zeolites as supports for TiO_2 photocatalysts, their chemical composition, such as SiO_2/Al_2O_3 ratio and cation type, topology, and morphology (crystallinity, surface area, pore diameter, and particle size) need to be carefully selected (Xu and Langford 1995). To clarify the effect of each of these properties on the photocatalytic activity of TiO_2/zeolite composite, Kuwahara and Yamashita examined photocatalytic activity using a variety of supports including A-zeolite (LTA structure), X- and Y-zeolites (FAU structure) with varying SiO_2/Al_2O_3 ratio from 2.5 to 200 as well as ZSM-5 zeolites (MFI structure) with $SiO_2/Al_2O_3 = 760$

(Kuwahara and Yamashita 2011). Zeolites X and Y had similar pore diameter of about 0.74 nm. Pore diameter of LTA structure and MFI structure were 0.41 nm and 0.51–0.55 nm respectively. Ammonium titanyl oxalate ($(NH_4)_2[TiO(C_2O_4)_2].2H_2O$) was used as the titanium source for the sol–gel preparation of TiO_2 particles within the zeolite micropores with 10 wt% loading amount of TiO_2. The samples were calcined at 823 K in air for crystallization. Photocatalytic activities of TiO_2 on different zeolite supports were evaluated by testing the photocatalytic degradation of 2-propanol diluted in water under UV light irradiation. The results of photocatalytic tests are shown in Fig. 6.16. The specific surface area of the zeolites as well as their water adsorption capacity is also shown on the graph. The numbers below the name of zeolites on horizontal axis are the SiO_2/Al_2O_3 ratio in each zeolite. It is well known that the SiO_2/Al_2O_3 ratio defines the degree of hydrophobicity of the surface of zeolite and it is independent to the topology. From the results shown in Fig. 6.16, it is understood that when using zeolites with similar topologies, the photocatalytic activity improved by increasing the SiO_2/Al_2O_3 ratio and thereby hydrophobicity of the pores surface. Even for zeolites with very low surface area and pore diameter, like H-Y and ZSM-5, the high SiO_2/Al_2O_3 ratio caused an enhanced photocatalytic activity than that of commercial P25 powder.

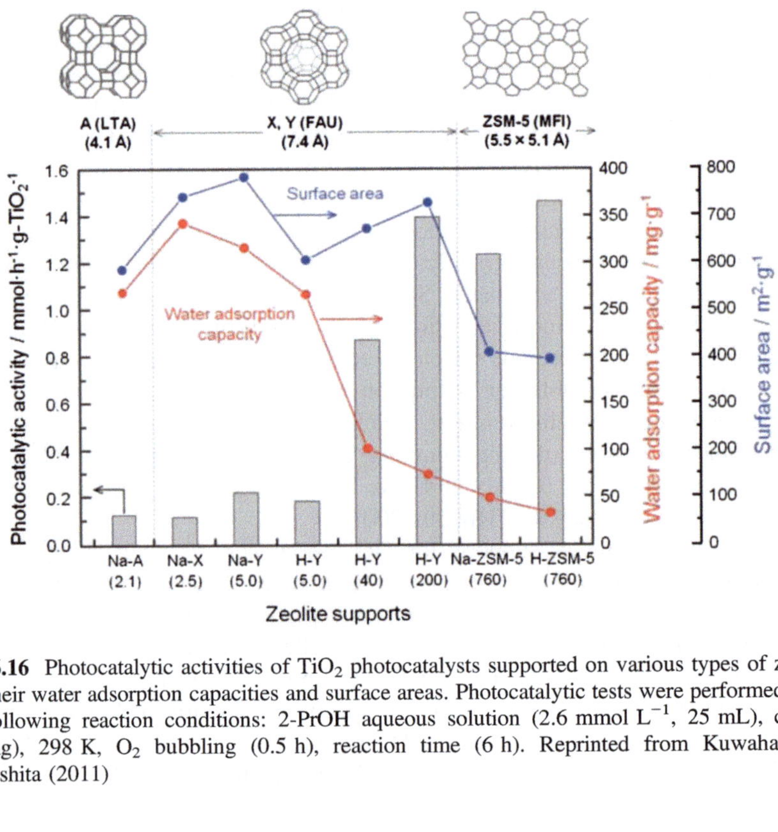

Fig. 6.16 Photocatalytic activities of TiO_2 photocatalysts supported on various types of zeolites and their water adsorption capacities and surface areas. Photocatalytic tests were performed under the following reaction conditions: 2-PrOH aqueous solution (2.6 mmol L^{-1}, 25 mL), catalyst (50 mg), 298 K, O_2 bubbling (0.5 h), reaction time (6 h). Reprinted from Kuwahara and Yamashita (2011)

This implies that when performing photocatalytic tests in an aqueous medium, the photocatalytic activity of TiO_2 supported on zeolite is highly affected by the hydrophobicity of the surface of the pore walls.

Figure 6.17 schematically shows the effect of hydrophilic and hydrophobic pores on the accessibility of organic molecules to the photocatalyst active sites. Hydrophobic pores provide a better medium for the diffusion and adsorption of organic molecules, thus improve the accessibility to the active TiO_2 sites.

Several strategies, including dealumination and surface modification by Silylation, have been employed to improve surface hydrophobicity of zeolite when the intrinsic hydrophobic character of zeolite is insufficient to satisfy the requirements for TiO_2 support completely. Dealumination treatment, in which the Al atoms are expelled from the zeolite framework, can be performed by hydrothermal treatment (Stach et al. 1986; Fleisch 1986), acid leaching (Jones et al. 2001), and chemical treatments with hexafluorosilicate (Lonyi and Lunsford 1992) or silicon tetrachloride (Anderson and Klinowski 1986). This treatment is especially necessary for increasing the Si/Al ratio in zeolites prepared without structure-directing agents (Kuwahara and Yamashita 2011).

The highly dispersed titanium oxide species having fourfold coordination anchored onto the zeolite supports also show unique photocatalytic performance much higher than that of the powdered TiO_2 photocatalysts. These photocatalysts have been investigated especially for the photocatalytic reduction of CO_2 with H_2O (Ikeue et al. 2001, 2002) and NO decomposition into N_2 and O_2 (Zhang et al. 1997; Antes and Thiel 1999; Zhang et al. 2000; Iwakuni et al. 2007).

Ikeue et al. investigated the photocatalytic reduction of CO_2 with H_2O on Ti-ß zeolite photocatalysts (Ikeue et al. 2001). They used OH^- and F^- ion as anions of the structure-directing agents in hydrothermal synthesis of Ti-ß zeolite. Different anions led to the formation of different types of Ti-ß zeolites in both of which the titanium oxide species found to be highly dispersed in their frameworks and existed

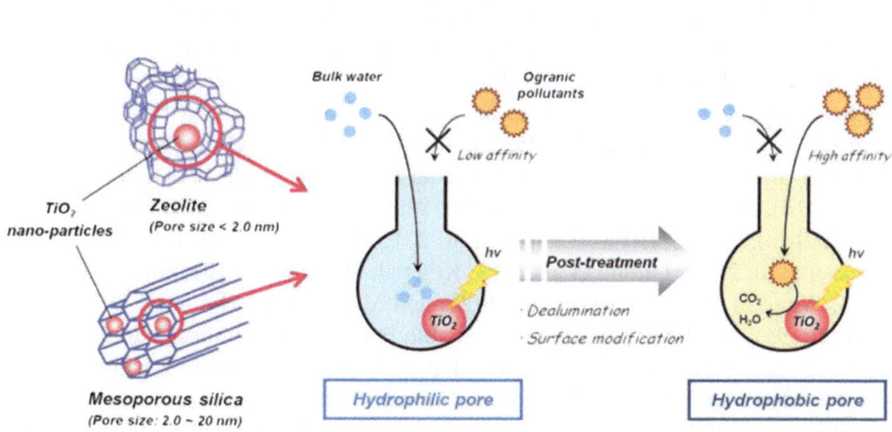

Fig. 6.17 Scheme showing the effect of hydrophilic and hydrophobic pores on the accessibility of organic molecules to the photocatalyst active sites in photodegradation test of organic molecules diluted in water by TiO_2 supported zeolites. Reprinted from Kuwahara and Yamashita (2011)

in tetrahedral coordination states. But the surface properties of the Ti-ß zeolites were highly affected by the anion of the structure-directing agents. Ti-ß zeolites synthesized using OH^- ions (Ti-ß(OH)) exhibited hydrophilic properties and the Ti-ß zeolites synthesized using F^- ions (Ti-ß(F)) showed hydrophobic properties. Therefore, the H_2O molecules were easily able to reach the tetrahedrally coordinated titanium oxide species in the Ti-ß(OH) zeolite and reduction of CO_2 with H_2O happened more efficiently on the surface of Ti-ß(OH) zeolite compared to Ti-ß(F) zeolite with hydrophobic pores.

Comparing this situation (Ikeue et al. 2001) with the experiments of degradation of organic molecules in aqueous medium using TiO_2 supported on zeolite (Kuwahara and Yamashita 2011), one can conclude that different characteristics of the zeolite support such as surface properties need to be carefully selected, taking into account the nature of the support and the situations where the catalyst is used. Degradation of organic molecules in aqueous medium is more efficient using zeolite-supported photocatalyst with hydrophobic pores while photocatalytic reduction of CO_2 with H_2O needs to be done in hydrophilic pores.

MOF-Based Photocatalysts

Metal–organic frameworks (MOFs) are highly porous crystalline materials buildup from metal(oxide) nodes interconnected by organic ligands (Zhou et al. 2012; Zhou and Kitagawa 2014). Due to their diverse and easily tailored structures, as well as high porosity and tunable properties, they gained so much attention in different applications such as gas storage (Millward and Yaghi 2005; Kaye et al. 2007; Getman et al. 2012), catalysisx (Lee et al. 2009; Ma et al. 2009), chemical sensing (Lu and Hupp 2010; Kreno et al. 2012; Hu et al. 2014), drug delivery (Horcajada et al. 2008; Della Rocca et al. 2011), etc. Recently, MOF materials considered as a new class of photocatalysts applied in different areas such as water splitting, CO_2 conversion, and organic materials degradation. To date, several MOFs have been studied as photocatalysts (Wang et al. 2012, 2014a) including MOF-5 (Llabrés i Xamena et al. 2007), MOF-253-Pt (Zhou et al. 2013b), UiO-66 (Sun et al. 2013; Wang et al. 2015b; Zhou et al. 2015), UiO-67 (Wang et al. 2011), NTU-9 (Gao et al. 2014), MIL-53(Fe) (Du et al. 2011b; Wang et al. 2014b), MIL-88(A)(Fe) (Xu et al. 2014b), MIL-88B(Fe) (Laurier et al. 2013; Wang et al. 2014b), MIL-100(Fe) (Laurier et al. 2013; Online et al. 2013; Ke et al. 2015), MIL-101(Fe) (Online et al. 2013; Wang et al. 2014b; Xu et al. 2015) and MIL-125(Ti) (Fu et al. 2012; Wang et al. 2015a), etc. High surface area, tunable pore size and volume, and enhanced light harvesting are the characteristics that make MOFs interesting candidates for photocatalytic applications. The organic linkers in MOF are responsible for light absorption of MOF material. Therefore, band gap of MOFs can be tuned through modification of organic linkers (Gascon et al. 2008) or substitution of linkers with amino groups (Gomes Silva et al. 2010; Fu et al. 2012; Horiuchi et al. 2012) and thus visible light photocatalytic activity can be achieved. MOF materials have been investigated as photocatalyst both in their pure form and in composite form (Petit and Bandosz 2009a, b, 2010, 2011; Petit et al. 2012) with other photocatalysts, magnetic materials or metals. Collective properties of MOF and other components of the composite will provide improved photocatalytic performance.

Online and coworkers fabricated magnetic Fe_3O_4@MIL-100(Fe) core–shell with good photocatalytic activity under both UV and visible light irradiation (Online et al. 2013). Additionally, the magnetic property of the composite material inherited from Fe_3O_4 helped the separation and recycling from the reaction medium without significant loss of photocatalytic activity after used for many times. Fe_3O_4 core nanoparticles were synthesized by a solvothermal method and functionalized with mercaptoacetic acid. The MIL-100(Fe) shell was formed on the Fe_3O_4 cores through a layer by layer strategy using the solutions of $FeCl_3$ and benzenetricarboxylic acid (H_3BTC) (shown schematically in Fig. 6.18).

The photocatalytic degradation of MB under both UV-vis light and visible light irradiation was tested to investigate the efficiency of the Fe_3O_4@MIL-100(Fe) photocatalyst. The results showed that this photocatalyst had remarkable photocatalytic activity for MB degradation both under UV-vis and visible light irradiation, compared to TiO_2 and C_3N_4 photocatalysts. Online et al. studied the MB photodegradation in the presence of H_2O_2 without which the photodegradation efficiency of MB was very low (Online et al. 2013). The photocatalytic efficiency was about 35 and 99% for the degradation of MB under 40 min of UV-vis light irradiation without and with H_2O_2, respectively. This was due to the role of H_2O_2 as the electron acceptor, which suppressed the photogenerated electron–hole recombination, thus enhanced the photocatalytic efficiency. The MB photodegradation experiments also showed that the photocatalytic efficiency of MIL-100(Fe) was almost the same in pure form and in the Fe_3O_4@MIL-100(Fe) composite. This implies that the Fe_3O_4 core did not have any role in photocatalytic activity. However, the Fe_3O_4 make the recycling of Fe_3O_4@MIL-100(Fe) composite easy in practical applications by donating the composite a magnetic property.

Fig. 6.18 Schematic illustration of synthesis of Fe_3O_4@MIL-100(Fe) core–shell microspheres. Reprinted from Online et al. (2013)

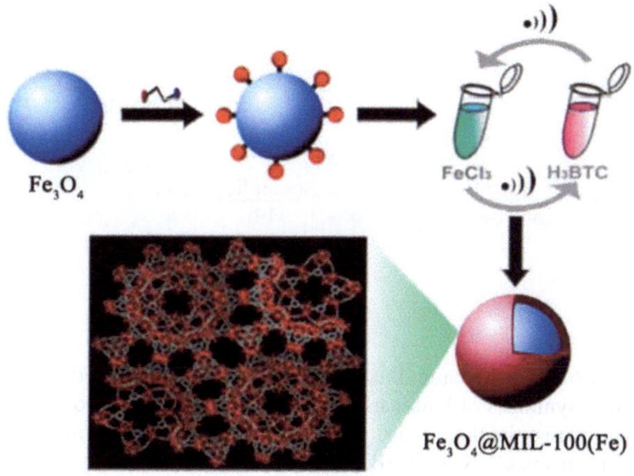

Zeolitic imidazolate frameworks (ZIFs), which are constructed from the tetrahedrally coordinated divalent cations ($M^{2+} = Zn^{2+}$ or Co^{2+}) linked by the uninegative imidazolate ligands (im^{-}), are a subclass of porous MOFs with unusual high thermal and chemical stability (Huang et al. 2006; Tian et al. 2007; Phan et al. 2010). ZIFs usually show very wide band gaps. Therefore, band gap engineering of these materials towards visible light region is necessary if they are considered as photocatalyst materials. Yang et al. tried to tune the gas sorption and photocatalytic properties of ZIF material by both modification of ZIF structural composition and Cu doping (Yang et al. 2012). For this purpose, they chose Co^{-2}-methylimidazole framework ZIF-67 with the SOD topology and synthesized the Cu-doped ZIF-67 material in a solvothermal process by mixing $Cu(COO)_2$, $Co(COO)_2$ and 2-methylimidazole in mixed ethyleneurea hemihydrate (e-urea) and ethanol solvent at 140 °C for 7 days. The band gaps of ZIF-67 and Cu/ZIF-67 were measured to be 1.98 and 1.95 eV, respectively. Although the two materials have almost the same band gap, the photodegradation of methyl orange under the visible light showed a very different photocatalytic activity after Cu doping of ZIF-67. The Cu/ZIF-67 could fully degrade the methyl orange after 25 min visible light illumination while only 10% of methyl orange degraded at this time by using ZIF-67 as photocatalyst. The Cu doping also enhanced the gas uptake capacity of ZIF-67.

Hong et al. reported in situ synthesis of 2D semiconductor/MOF nanocomposite photocatalyst for the first time (Hong et al. 2016). Other reports used the macroscale mixing of preprepared components for the fabrication of composite materials (Zhang et al. 2015) while nanoscale mixing would offer more intimate contact and greater electron contact at the heterojunction. Therefore, Hong and coworkers used an in situ synthesis method for the fabrication of carbon nitride nanosheet/MIL 100 (Fe) nanocomposite hoping to achieve better junctions and improved photocatalytic properties. They synthesized carbon nitride nanosheets (CNNSs) through a two-step calcination–exfoliation method using urea as a precursor. Then the as-prepared CNNSs were added to the MIL-100(Fe) precursor solution for the in situ synthesis of CNNSs-MIL nanocomposites under a non-hydrothermal condition (95 °C). The schematic representation of the synthesis procedure is shown in Fig. 6.19.

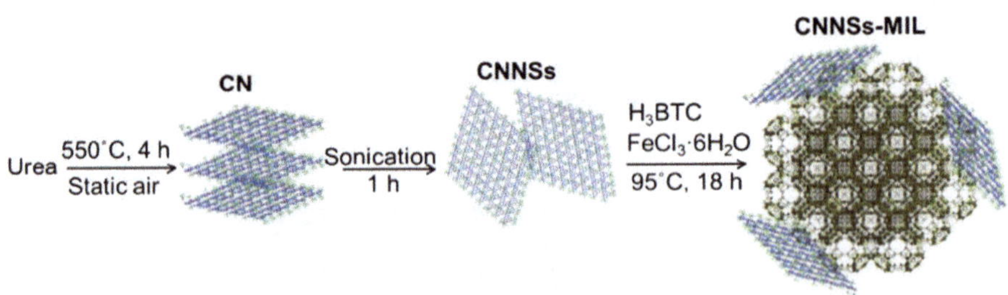

Fig. 6.19 Schematic illustration of the synthesis procedure of CNNSs-MIL nanocomposites. First step: synthesis of bulk carbon nitride(CN) via calcination of urea. Second step: exfoliation of bulk CN via sonication. Third step: synthesis of the MOF in the presence of dispersed carbon nitride nanosheets(CNNSs). H3BTC: benzene-1,3,5-tricarboxylic acid. Reprinted from Hong et al. (2016)

The X-ray diffraction patterns of the nanocomposite materials proved that MOF crystals formed in the presence of up to 5 wt% CNNSs. MOFs have very large surface area which is a desired property in photocatalytic activities. The surface area of the pure MIL 100(Fe) measured to be 1225 m^2 g^{-1}. A slight decrease in surface area happened by making the CNNS/MIL nanocomposite. Nevertheless, the surface area of CNNS/MIL nanocomposites remained higher than 1000 m^2 g^{-1} even at higher concentrations of CNNS up to 5 wt%. To evaluate the photocatalytic performance of the CNNS/MIL nanocomposite, degradation of Rhodamine B (RhB) was tested under visible light irradiation and in the presence of H_2O_2. The photocatalytic degradation results showed that CNNSs-MIL containing 1 wt% CNNS degraded 100% of RhB after 4 h of visible light irradiation. Pure CNNSs and MIL degraded only 55 and 68% of RhB, respectively. The improved photocatalytic performance of CNNSs-MIL nanocomposites compared to the performance of CNNS and MIL was attributed to the better separation of photogenerated electron–hole pair in the nanocomposite. This phenomenon was confirmed by measuring the photoluminescence spectra of the nanocomposites materials in which the fast quenching of PL intensity was obvious by introduction of CNNSs into the MIL framework. Moreover, the photocatalytic tests showed a decrease in the performance of CNNS/MIL nanocomposites as the CNNSs loading increased beyond 1 wt%. This could be due to the presence of stacked CNNSs, which limited the homogeneous dispersion of the nanosheets, thus led to electron–hole recombination and decreased light absorbance and surface area.

Summary and Outlook

The photocatalytic activity of a semiconductor is determined by the amount and accessibility of active surface area, light harvesting efficiency, and recombination rate of electron–hole pairs. Various strategies have been developed for improving the photocatalytic activity of semiconductors. These strategies can be categorized as either chemical modifications or morphological modifications. Chemical modifications have been adopted in order to change the band gap of the photocatalyst material, increase light harvesting efficiency and suppress electron–hole recombination. Metal/nonmetal doping, metal deposition, dye sensitization, and coupling semiconductors are the approaches have been developed in this regard. Attaining large surface area though is possible through morphological modifications leading to nanoparticles or nanoporous materials. Although nanoparticles provide a large surface area, problems related to aggregation and recycling of nanoparticles limit their practical applications. Nanoporous photocatalyst materials not only possess a large surface area but also leave the practical limitations of nanoparticles behind. Moreover, nanoporous photocatalyst materials offer a medium with easy diffusion and accessibility of organic molecules to the active sites and improve the light harvesting efficiency by multiple scattering of light from pore walls. Therefore, nanoporous nanocomposites are considered as the best photocatalytic structures,

Table 6.1 Various synthesis methods and their salient features for fabrication of nanoporous nanocomposite materials

Synthesis method		Description of method	Salient features	Examples
Template method	Hard template	Chemical synthesis of nanocomposite material inside a template medium, which is eliminated at later stage	Easy to control the exact structure, morphology and size of the product pores with high precisian and reproducibility Typical templates: mesoporous SiO_2, mesoporous carbon, porous Al_2O_3 or polystyrene spheres	Porous graphitic carbon nitride (Fukasawa et al. 2011), mesoporous $Co_3O_4/BiVO_4$ (Dang et al. 2014), mesoporous polyoxometalate (POM)/Ag_2S/CdS (Kornarakis et al. 2014)
	Soft template		The reaction has good controllability and the template can be easily formed and removed while the reaction is performed Typical templates: surfactants and biopolymers	mesoporous TiO_2 (Wang et al. 2005); (Yu et al. 2007), macro/mesoporous TiO_2/SiO_2 and TiO_2/ZrO_2 (Chen et al. 2009a), Mesoporous SiO_2 and mesoporous TiO_2/SiO_2 (Inumaru et al. 2005), nanoporous Au/TiO_2 (Wang et al. 2008b)
Self-assembly		Spontaneous assembly and arrangement of the product via specific interactions such as hydrogen bonding and Van der Waals forces, π–π interactions, electrostatic forces, etc	Porous structure can be produced with or without using a template Larger pores and better control on the porous structure are attainable in the presence of a template	mesoporous TiO_2 (Grosso et al. 2001; Crepaldi et al. 2003; Choi et al. 2006; Crepaldi et al. 2003), macro/mesoporous TiO_2/grapheme (Du et al. 2011a), mesoporous carbon-TiO_2 (Wei et al. 2013)
Hydrothermal		Crystallization in an aqueous solution at high temperature and high vapor pressure	Facile tuning of product structure and pore size by changing the internal pressure and the precursor type	$ZnIn_2S_4/CdIn_2S_4$ nanosheets (Li et al. 2015), mesoporous TiO_2/clays (Xuzhuang et al. 2009)
Solvothermal		Crystallization in a non-aqueous solution at high temperature and high vapor pressure	Easy to control the size, shape and crystallinity of the nanostructures by changing the reaction time, reaction temperature, solvent and precursor type	Cu_2O/Cu (Zhou et al. 2011), Pinecone-like $Fe_3O_4@Cu_2O$/Cu (Wang et al. 2013a)

(continued)

Table 6.1 (continued)

Synthesis method	Description of method	Salient features	Examples
Deposition	Preparation of porous structure through simple deposition techniques such as electrophoretic deposition, dip coating, spin coating, etc	Facile preparation of porous films on both flat and complicated substrates	Ag/TiO_2 fibrous electrode (Hosseini et al. 2008), mesoporous TiO_2/SiO_2 (Fattakhova-Rohlfing et al. 2009)
Anodization	Conversion of metal surface into metal oxide through electrochemical process	Formation of metal oxide tube arrays or pore arrays on the surface of metal Facile synthesis of nanotube arrays with different aspect ratio and wall thickness	Al_2O_3/Al (Keller et al. 1953), TiO_2 nanotubes/Ti (Roy et al. 2011; Zwilling et al. 1999a,b; Macak et al. 2007b; Roy et al. 2011; Marien et al. 2016; Mor et al. 2006; Zhu et al. 2007; Kuang et al. 2008), iron oxide/TiO_2 (Kontos et al. 2009), Cu_2O on TiO_2 nanotubes (Tsui et al. 2012; Tsui and Zangari 2013a, 2013b; Li et al. 2010)

which gained the necessary properties through both chemical and morphological modifications.

In this chapter, the most common methods for preparation of nanoporous nanocomposites have been presented. A variety of porous architectural designs can be prepared by template method, self-assembly method, hydrothermal and solvothermal method, anodization and deposition method and used in photocatalytic applications. Table 6.1 presents short description of these methods along with their salient features for fabrication of nanoporous nanocomposite materials. The size and morphology of the pores highly affect the photocatalytic activity. Control over these characteristics is possible by different parameters in each method. Thermal and mechanical stability of the nanoporous structure are other characteristics which are highly desirable for photocatalytic porous structures especially in practical applications. In this chapter we also discussed the nanoporous structures based on zeolites and MOFs for photocatalytic applications. Although zeolites provide frameworks with very large surface area for making nanoporous photocatalysts, the micropores in zeolite-based photocatalyst materials can hinder the diffusion of organic molecules before reaching the photocatalytic active sites. However, the special characteristics of MOFs, arising from the existence of the organic network, such as a 3D order with tunable dimensionality of the cavities, together with their tunable chemical properties, make MOF-based photocatalysts a promising alternative to their zeolite-based counterparts.

References

Alvaro M, Aprile C, Benitez M et al (2006) Photocatalytic activity of structured mesoporous TiO_2 materials. J Phys Chem B 110:6661–6665

Anderson MW, Klinowski J (1986) Zeolites treated with silicon tetrachloride vapour IV. Acidity. Zeolites 6:455–466

Anpo M, Tekeuchi M (2003) The design and development of highly reactive titanium oxide photocatalysts operating under visible light irradiation. J Catal 216:505–516

Anpo M, Zhang SG, Mishima H et al (1997) Design of photocatalysts encapsulated within the zeolite framework and cavities for the decomposition of NO into N_2 and O_2 at normal temperature. Catal Today 39:159–168

Antes I, Thiel W (1999) Adjusted connection atoms for combined quantum mechanical and molecular mechanical methods. J Phys Chem A 103:9290–9295

Bell AT (2003) The impact of nanoscience on heterogeneous catalysis. Science 299:1688–1691

Berger S, Kunze J, Schmuki P et al (2010) Influence of Water content on the growth of anodic TiO_2 nanotubes in fluoride-containing ethylene glycol electrolytes. J Electrochem Soc 157: C18–C23

Besra L, Liu M (2007) A review on fundamentals and applications of electrophoretic deposition (EPD). Prog Mater Sci 52:1–61

Boccaccini AR, Karapappas P, Marijuan JM, Kaya C (2004) TiO_2 coatings on silicon carbide and carbon fibre substrates by electrophoretic deposition. J Mater Sci 39:851–859

Brinker JC, Lu Y, Sellinger A, Fan H (1999) Evaporation-induced self-assembly: nanostructures made easy. Adv Mater 11:579–585

Cai P, Ma D-K, Liu Q-C et al (2013) Conversion of ternary Zn_2SnO_4 octahedrons into binary mesoporous SnO_2 and hollow SnS_2 hierarchical octahedrons by template-mediated selective complex extraction. J Mater Chem A 1:5217–5223

Calleja G, Serrano DP, Sanz R, Pizarro P (2008) Mesostructured SiO_2-doped TiO_2 with enhanced thermal stability prepared by a soft-templating sol–gel route. Microporous Mesoporous Mater 111:429–440

Chae HJ, Nam I-S, Ham S-W, Hong SB (2004) Characteristics of vanadia on the surface of V_2O_5/Ti-PILC catalyst for the reduction of NO_x by NH_3. Appl Catal B 53:117–126

Chaudhari NS, Bhirud AP, Sonawane RS et al (2011) Ecofriendly hydrogen production from abundant hydrogen sulfide using solar light-driven hierarchical nanostructured $ZnIn_2S_4$ photocatalyst. Green Chem 13:2500–2506

Chen X, Wang X, Fu X (2009a) Hierarchical macro/mesoporous TiO_2/SiO_2 and TiO_2/ZrO_2 nanocomposites for environmental photocatalysis. Energy Environ Sci 2:872–877

Chen Y, Ge H, Wei L et al (2013) Reduction degree of reduced graphene oxide (RGO) dependence of photocatalytic hydrogen evolution performance over RGO/$ZnIn_2S_4$ nanocomposites. Catal Sci Technol 3:1712–1717

Chen Z, Li D, Zhang W et al (2009b) Photocatalytic degradation of dyes by $ZnIn_2S_4$ microspheres under visible light irradiation. J Phys Chem C 113:4433–4440

Choi H, Antoniou MG, Pelaez M et al (2007) Mesoporous nitrogen-doped TiO_2 for the photocatalytic destruction of the cyanobacterial toxin microcystin-LR under visible light irradiation. Environ Sci Technol 41:7530–7535

Choi SY, Lee B, Carew DB et al (2006) 3D hexagonal (R-3 m) mesostructured nanocrystalline titania thin films: synthesis and characterization. Adv Funct Mater 16:1731–1738

Crepaldi EL, Soler-Illia GJ de AA, Grosso D, et al (2003) Controlled formation of highly organized mesoporous titania thin films: from mesostructured hybrids to mesoporous nanoanatase TiO_2. J Am Chem Soc 125:9770–9786

Cundy CS, Cox PA (2003) The hydrothermal synthesis of zeolites: history and development from the earliest days to the present time. Chem Rev 103:663–702

Daiguji H, Hwang J, Takahashi A et al (2012) Ion transport in mesoporous silica SBA-16 thin films with 3D cubic structures. Langmuir 28:3671–3677

Dang X, Zhang X, Dong X et al (2014) The p–n heterojunction with porous $BiVO_4$ framework and well-distributed Co_3O_4 as a super visible-light-driven photocatalyst. RSC Adv 4:54655–54661

Della Rocca J, Liu D, Lin W (2011) Nanoscale metal-organic frameworks for biomedical imaging and drug delivery. Acc Chem Res 44:957–968

Ding Z, Zhu HY, Lu GQ, Greenfield PF (1999) Photocatalytic properties of titania pillared clays by different drying methods. J Colloid Interface Sci 209:193–199

Dionigi C, Greco P, Ruani G et al (2008) 3D hierarchical porous TiO_2 films from colloidal composite fluidic deposition. Chem Mater 20:7130–7135

Du J, Lai X, Yang N et al (2011a) Hierarchically ordered macro–mesoporous TiO_2–graphene composite films: improved mass transfer, reduced charge recombination, and their enhanced photocatalytic activities. ACS Nano 5:590–596

Du J-J, Yuan Y-P, Sun J-X et al (2011b) New photocatalysts based on MIL-53 metal–organic frameworks for the decolorization of methylene blue dye. J Hazard Mater 190:945–951

Dutta PK, Severance M (2011) Photoelectron transfer in zeolite cages and its relevance to solar energy conversion. J Phys Chem Lett 2:467–476

Ensie B, Samad S (2014) Removal of nitrate from drinking water using nano SiO_2-FeOOH-Fe core-shell. Desalination 347:1–9

Fattakhova-Rohlfing D, Szeifert JM, Yu Q et al (2009) Low-temperature synthesis of mesoporous titania-silica films with pre-formed anatase nanocrystals. Chem Mater 21:2410–2417

Fischer A, Antonietti M, Thomas A (2007) Growth confined by the nitrogen source: synthesis of pure metal nitride nanoparticles in mesoporous graphitic carbon nitride. Adv Mater 19:264–267

Fleisch T (1986) Hydrothermal dealumination of faujasites. J Catal 99:117–125

Fröschl T, Hörmann U, Kubiak P et al (2012) High surface area crystalline titanium dioxide: potential and limits in electrochemical energy storage and catalysis. Chem Soc Rev 41:5313–5360

Fu Y, Jin Z, Xue W, Ge Z (2008) Ordered macro-mesoporous nc-TiO_2 Films by sol-gel method using polystyrene array and triblock copolymer bitemplate. J Am Ceram Soc 91:2676–2682

Fu Y, Sun D, Chen Y et al (2012) An Amine-functionalized titanium metal-organic framework photocatalyst with visible-light-induced activity for CO_2 reduction. Angew Chemie Int Ed 51:3364–3367

Fukasawa Y, Takanabe K, Shimojima A et al (2011) Synthesis of ordered porous graphitic-C_3N_4 and regularly arranged Ta_3N_5 nanoparticles by using self-assembled silica nanospheres as a primary template. Chem—Asian J 6:103–109

Gao J, Miao J, Li P-Z et al (2014) A p-type Ti(iv)-based metal–organic framework with visible-light photo-response. Chem Commun 50:3786–3788

Gascon J, Hernández-Alonso MD, Almeida AR et al (2008) Isoreticular MOFs as efficient photocatalysts with tunable band gap: an operando FTIR study of the photoinduced oxidation of propylene. Chemsuschem 1:981–983

Geng W, Liu H, Yao X (2013) Enhanced photocatalytic properties of titania–graphene nanocomposites: a density functional theory study. Phys Chem Chem Phys 15:6025–6033

Getman RB, Bae Y-S, Wilmer CE, Snurr RQ (2012) Review and analysis of molecular simulations of methane, hydrogen, and acetylene storage in metal-organic frameworks. Chem Rev 112:703–723

Gomes Silva C, Luz I, Llabrés i Xamena FX, et al (2010) Water stable Zr-benzenedicarboxylate metal-organic frameworks as photocatalysts for hydrogen generation. Chem—A Eur J 16:11133–11138

Gou X, Cheng F, Shi Y et al (2006) Shape-controlled synthesis of ternary chalcogenide $ZnIn_2S_4$ and CuIn(S, $Se)_2$ nano-/microstructures via facile solution route. J Am Chem Soc 128:7222–7229

Grosso D, de A. A. Soler-Illia GJ, Babonneau F, et al (2001) Highly organized mesoporous titania thin films showing mono-oriented 2D hexagonal channels. Adv Mater 13:1085–1090

Hodgson SD, Brooks WSM, Clayton AJ et al (2013) Enhancing blue photoresponse in CdTe photovoltaics by luminescent down-shifting using semiconductor quantum dot/PMMA films. Nano Energy 2:21–27

Hong J, Chen C, Bedoya FE et al (2016) Carbon nitride nanosheet/metal–organic framework nanocomposites with synergistic photocatalytic activities. Catal Sci Technol 6:5042–5051

Horcajada P, Serre C, Maurin G et al (2008) Flexible porous metal-organic frameworks for a controlled drug delivery. J Am Chem Soc 130:6774–6780

Horiuchi Y, Toyao T, Saito M et al (2012) Visible-light-promoted photocatalytic hydrogen production by using an amino-functionalized Ti(IV) metal-organic framework. J Phys Chem C 116:20848–20853

Hosseini Z, Taghavinia N, Sharifi N et al (2008) Fabrication of high conductivity TiO_2/Ag fibrous electrode by the electrophoretic deposition method. J Phys Chem C 112:18686–18689

Hou Y, Li X, Zou X et al (2009) Photoeletrocatalytic activity of a Cu_2O-loaded self-organized highly oriented TiO_2 nanotube array electrode for 4-chlorophenol degradation. Environ Sci Technol 43:858–863

Hu P, Pramana SS, Cao S et al (2013) Ion-induced synthesis of uniform single-crystalline sulphide-based quaternary-alloy hexagonal nanorings for highly efficient photocatalytic hydrogen evolution. Adv Mater 25:2567–2572

Hu Z, Deibert BJ, Li J (2014) Luminescent metal–organic frameworks for chemical sensing and explosive detection. Chem Soc Rev 43:5815–5840

Huang X-C, Lin Y-Y, Zhang J-P, Chen X-M (2006) Ligand-directed strategy for zeolite-type metal-organic frameworks: zinc(II) imidazolates with unusual zeolitic topologies. Angew Chemie Int Ed 45:1557–1559

Huczko A (2000) Template-based synthesis of nanomaterials. Appl Phys A Mater Sci Process 70:365–376

Ikeue K, Nozaki S, Ogawa M, Anpo M (2002) Photocatalytic reduction of CO_2 with H_2O on Ti-containing porous silica thin film photocatalysts. Catal Lett 80:111–114

Ikeue K, Yamashita H, Anpo M, Takewaki T (2001) Photocatalytic reduction of CO_2 with H_2O on Ti-beta zeolite photocatalysts: effect of the hydrophobic and hydrophilic properties. J Phys Chem B 105:8350–8355

Inumaru K, Kasahara T, Yasui M, Yamanaka S (2005) Direct nanocomposite of crystalline TiO_2 particles and mesoporous silica as a molecular selective and highly active photocatalyst. Chem Commun 30:2131–2133

Iwakuni H, Shinmyou Y, Yano H et al (2007) Direct decomposition of NO into N_2 and O_2 on $BaMnO_3$-based perovskite oxides. Appl Catal B 74:299–306

Jing W, Huang W, Xing W et al (2009) Fabrication of supported mesoporous tio_2 membranes: matching the assembled and interparticle pores for an improved ultrafiltration performance. ACS Appl Mater Interfaces 1:1607–1612

Jones CW, Hwang S-J, Okubo T, Davis ME (2001) Synthesis of hydrophobic molecular sieves by hydrothermal treatment with acetic acid. Chem Mater 13:1041–1050

Kale BB, Baeg J-O, Lee SM et al (2006) $CdIn_2S_4$ nanotubes and "Marigold" nanostructures: a visible-light photocatalyst. Adv Funct Mater 16:1349–1354

Kaune G, Memesa M, Meier R et al (2009) Hierarchically structured titania films prepared by polymer/colloidal templating. ACS Appl Mater Interfaces 1:2862–2869

Kaya C, Boccaccini AR (2001) Colloidal processing of complex shape stainless steel woven fiber mat reinforced alumina ceramic matrix composites using electrophoretic deposition. J Mater Sci Lett 20:1465–1467

Kaye SS, Dailly A, Yaghi OM, Long JR (2007) Impact of preparation and handling on the hydrogen storage properties of $Zn_4O(1,4\text{-benzenedicarboxylate})_3$(MOF-5). J Am Chem Soc 129:14176–14177

Ke F, Wang L, Zhu J (2015) Facile fabrication of CdS-metal-organic framework nanocomposites with enhanced visible-light photocatalytic activity for organic transformation. Nano Res 8:1834–1846

Keller F, Hunter MS, Robinson DL (1953) Structural features of oxide coatings on aluminum. J Electrochem Soc 100:411–419

Kimura T, Miyamoto N, Meng X et al (2009) Rapid fabrication of mesoporous titania films with controlled macroporosity to improve photocatalytic property. Chem—Asian J 4:1486–1493

Kitano M, Matsuoka M, Ueshima M, Anpo M (2007) Recent developments in titanium oxide-based photocatalysts. Appl Catal A 325:1–14

Kluson P, Kacer P, Cajthaml T, Kalaji M (2001) Preparation of titania mesoporous materials using a surfactant-mediated sol–gel method. J Mater Chem 11:644–651

Kontos AI, Likodimos V, Stergiopoulos T et al (2009) Self-organized anodic TiO_2 nanotube arrays functionalized by iron oxide nanoparticles. Chem Mater 21:662–672

Kornarakis I, Lykakis IN, Vordos N, Armatas GS (2014) Efficient visible-light photocatalytic activity by band alignment in mesoporous ternary polyoxometalate-Ag_2S-CdS semiconductors. Nanoscale 6:8694–8703

Kreno LE, Leong K, Farha OK et al (2012) Metal-organic framework materials as chemical sensors. Chem Rev 112:1105–1125

Kuang D, Brillet J, Chen P et al (2008) Application of highly ordered TiO_2 nanotube arrays in flexible dye-sensitized solar cells. ACS Nano 2:1113–1116

Kubacka A, Fernández-García M, Colón G (2012) Advanced nanoarchitectures for solar photocatalytic applications. Chem Rev 112:1555–1614

Kuwahara Y, Yamashita H (2011) Efficient photocatalytic degradation of organics diluted in water and air using TiO_2 designed with zeolites and mesoporous silica materials. J Mater Chem 21:2407–2416

Laurier KGM, Vermoortele F, Ameloot R et al (2013) Iron(III)-based metal-organic frameworks as visible light photocatalysts. J Am Chem Soc 135:14488–14491

Lee J, Farha OK, Roberts J et al (2009) Metal–organic framework materials as catalysts. Chem Soc Rev 38:1450–1459

Lei Z, You W, Liu M, et al (2003) Photocatalytic water reduction under visible light on a novel $ZnIn_2S_4$ catalyst synthesized by hydrothermal method. Chem Commun 2142–2143

Leofanti G, Padovan M, Tozzola G, Venturelli B (1998) Surface area and pore texture of catalysts. Catal Today 41:207–219

Li D, Chang P-C, Chien C-J, Lu JG (2010) Applications of tunable TiO_2 nanotubes as nanotemplate and photovoltaic device. Chem Mater 22:5707–5711

Li D, Chien C-J, Deora S et al (2011) Prototype of a scalable core–shell Cu_2O/TiO_2 solar cell. Chem Phys Lett 501:446–450

Li G (1999) Relationships between sensitivity, catalytic activity, and surface areas of SnO_2 gas sensors. Sens Actuators, B 60:64–70

Li J, Zeng HC (2007) Hollowing Sn-doped TiO_2 nanospheres via ostwald ripening. J Am Chem Soc 129:15839–15847

Li L, Peng S, Wang N et al (2015) A general strategy toward carbon cloth-based hierarchical films constructed by porous nanosheets for superior photocatalytic activity. Small 11:2429–2436

Linsebigler AL, Linsebigler AL, Yates JT Jr et al (1995) Photocatalysis on TiO_2 surfaces: principles, mechanisms, and selected results. Chem Rev 95:735–758

Llabrés i Xamena FX, Corma A, Garcia H (2007) Applications for metal–organic frameworks (MOFs) as quantum dot semiconductors. J Phys Chem C 111:80–85

Lonyi F, Lunsford J (1992) The development of strong acidity in hexafluorosilicate-modified Y-type zeolites. J Catal 136:566–577

López-Muñoz M-J, van Grieken R, Aguado J, Marugán J (2005) Role of the support on the activity of silica-supported TiO_2 photocatalysts: Structure of the TiO_2/SBA-15 photocatalysts. Catal Today 101:307–314

Lu F, Cai W, Zhang Y (2008) ZnO hierarchical micro/nanoarchitectures: solvothermal synthesis and structurally enhanced photocatalytic performance. Adv Funct Mater 18:1047–1056

Lu G, Hupp JT (2010) Metal–organic frameworks as sensors: a ZIF-8 based Fabry–Pérot Device as a selective sensor for chemical vapors and gases. J Am Chem Soc 132:7832–7833

Ma L, Abney C, Lin W (2009) Enantioselective catalysis with homochiral metal–organic frameworks. Chem Soc Rev 38:1248–1256

Macak JM, Gong BG, Hueppe M, Schmuki P (2007a) Filling of TiO_2 nanotubes by self-doping and electrodeposition. Adv Mater 19:3027–3031

Macak JM, Zlamal M, Krysa J, Schmuki P (2007b) Self-organized TiO_2 nanotube layers as highly efficient photocatalysts. Small 3:300–304

Malfatti L, Falcaro P, Pinna A et al (2014) Exfoliated graphene into highly ordered mesoporous titania films: highly performing nanocomposites from integrated processing. ACS Appl Mater Interfaces 6:795–802

Marien CBD, Cottineau T, Robert D, Drogui P (2016) TiO_2 nanotube arrays: Influence of tube length on the photocatalytic degradation of Paraquat. Appl Catal B 194:1–6

Martins PM, Miranda R, Marques J et al (2016) Comparative efficiency of TiO_2 nanoparticles in suspension vs. immobilization into P(VDF–TrFE) porous membranes. RSC Adv 6:12708–12716

Meng X, Kimura T, Ohji T, Kato K (2009) Triblock copolymer templated semi-crystalline mesoporous titania films containing emulsion-induced macropores. J Mater Chem 19:1894–1900

Millward AR, Yaghi OM (2005) Metal–organic frameworks with exceptionally high capacity for storage of carbon dioxide at room temperature. J Am Chem Soc 127:17998–17999

Mor GK, Shankar K, Paulose M et al (2006) Use of highly-ordered TiO_2 nanotube arrays in dye-sensitized solar cells. Nano Lett 6:215–218

Ng YH, Lightcap IV, Goodwin K et al (2010) To what extent do graphene scaffolds improve the photovoltaic and photocatalytic response of TiO_2 nanostructured films? J Phys Chem Lett 1:2222–2227

O'Regan B, Grätzel M (1991) A low-cost, high-efficiency solar cell based on dye-sensitized colloidal TiO_2 films. Nature 353:737–740

Ogawa M, Ikeue K, Anpo M (2001) Transparent self-standing films of titanium-containing nanoporous silica. Chem Mater 13:2900–2904

Online VA, Zhang C, Qiu L et al (2013) A novel magnetic recyclable photocatalyst based on a core-shell metal-organic framework Fe_3O_4@MIL-100(Fe) for the decolorization of methylene blue dye. J Mater Chem A 100:14329–14334

Ooka C, Yoshida H, Horio M et al (2003) Adsorptive and photocatalytic performance of TiO_2 pillared montmorillonite in degradation of endocrine disruptors having different hydrophobicity. Appl Catal B 41:313–321

Pan JH, Dou H, Xiong Z et al (2010) Porous photocatalysts for advanced water purifications. J Mater Chem 20:4512–4528

Pelaez M, Nolan NT, Pillai SC et al (2012) A review on the visible light active titanium dioxide photocatalysts for environmental applications. Appl Catal B 125:331–349

Peng S, Wu Y, Zhu P et al (2011) Controlled synthesis and photoelectric application of $ZnIn_2S_4$ nanosheet/TiO_2 nanoparticle composite films. J Mater Chem 21:15718–15726

Petit C, Bandosz TJ (2011) Synthesis, characterization, and ammonia adsorption properties of mesoporous metal-organic framework (MIL(Fe))-graphite oxide composites: exploring the limits of materials fabrication. Adv Funct Mater 21:2108–2117

Petit C, Bandosz TJ (2009a) MOF-graphite oxide composites: combining the uniqueness of graphene layers and metal-organic frameworks. Adv Mater 21:4753–4757

Petit C, Bandosz TJ (2009b) MOF–graphite oxide nanocomposites: surface characterization and evaluation as adsorbents of ammonia. J Mater Chem 19:6521–6528

Petit C, Bandosz TJ (2010) Enhanced adsorption of ammonia on metal-organic framework/graphite oxide composites: analysis of surface interactions. Adv Funct Mater 20:111–118

Petit C, Levasseur B, Mendoza B, Bandosz TJ (2012) Reactive adsorption of acidic gases on MOF/graphite oxide composites. Microporous Mesoporous Mater 154:107–112

Phan A, Doonan CJ, Uribe-Romo FJ et al (2010) Synthesis, structure, and carbon dioxide capture properties of zeolitic imidazolate frameworks. Acc Chem Res 43:58–67

Reddy EP, Sun B, Smirniotis PG (2004) Transition metal modified TiO_2-loaded MCM-41 catalysts for visible- and UV-Light driven photodegradation of aqueous organic pollutants. J Phys Chem B 108:17198–17205

Rengaraj S, Venkataraj S, Tai C et al (2011) Self-assembled mesoporous hierarchical-like In_2S_3 hollow microspheres composed of nanofibers and nanosheets and their photocatalytic activity. Langmuir 27:5534–5541

Rolison DR (2003) Catalytic Nanoarchitectures–the importance of nothing and the unimportance of periodicity. Science (80-) 299:1698–1701

Roy P, Berger S, Schmuki P (2011) TiO_2 nanotubes: synthesis and applications. Angew Chemie—Int Ed 50:2904–2939

Sabbaghi S, Mohammadi M, Ebadi H (2016) Photocatalytic degradation of benzene wastewater using PANI-TiO_2 nanocomposite under UV and solar light radiation. J Environ Eng 142:05015003-1–05015003-7

Sakamoto JS, Dunn B (2002) Hierarchical battery electrodes based on inverted opal structures. J Mater Chem 12:2859–2861

Scotti R, D'Arienzo M, Morazzoni F, Bellobono IR (2009) Immobilization of hydrothermally produced TiO_2 with different phase composition for photocatalytic degradation of phenol. Appl Catal B Environ 88:323–330

Shchukin DG, Caruso RA (2004) Template synthesis and photocatalytic properties of porous metal oxide spheres formed by nanoparticle infiltration. Chem Mater 16:2287–2292

Sing KSW (1985) Reporting physisorption data for gas/solid systems with special reference to the determination of surface area and porosity. Pure Appl Chem 57:603–619

Stach H, Lohse U, Thamm H, Schirmer W (1986) Adsorption equilibria of hydrocarbons on highly dealuminated zeolites. Zeolites 6:74–90

Sterte J (1986) Synthesis and properties of titanium oxide cross-linked montmorillonite. Clays Clay Miner 34:658–664

Strandwitz NC, Nonoguchi Y, Boettcher SW, Stucky GD (2010) In situ photopolymerization of pyrrole in mesoporous TiO_2. Langmuir 26:5319–5322
Studart AR, Gonzenbach UT, Tervoort E, Gauckler LJ (2006) Processing routes to macroporous ceramics: a review. J Am Ceram Soc 89:1771–1789
Sun D, Fu Y, Liu W et al (2013) Studies on photocatalytic CO_2 reduction over NH_2 -Uio-66(Zr) and its derivatives: towards a better understanding of photocatalysis on metal-organic frameworks. Chem—A Eur J 19:14279–14285
Sun M, Zangari G, Shamsuzzoha M, Metzger RM (2001) Electrodeposition of highly uniform magnetic nanoparticle arrays in ordered alumite. Appl Phys Lett 78:2964–2966
Tang J, Wu Y, McFarland EW, Stucky GD (2004) Synthesis and photocatalytic properties of highly crystalline and ordered mesoporous TiO_2 thin films. Chem Commun 1670–1671
Tian Y-Q, Zhao Y-M, Chen Z-X et al (2007) Design and Generation of extended zeolitic metal-organic frameworks (ZMOFs): synthesis and crystal structures of zinc(II) imidazolate polymers with zeolitic topologies. Chem—A Eur J 13:4146–4154
Treacy MMJ, Higgins JB (2007) Powder pattern identification table. In: Collection of simulated XRD powder patterns for zeolites. Elsevier, pp 10–16
Tsui L-k, Wu L, Swami N, Zangari G (2012) Photoelectrochemical performance of electrodeposited Cu_2O on TiO_2 nanotubes. ECS Electrochem Lett 1:D15–D19
Tsui L, Zangari G (2013a) The influence of morphology of electrodeposited Cu_2O and Fe_2O_3 on the conversion efficiency of TiO_2 nanotube photoelectrochemical solar cells. Electrochim Acta 100:220–225
Tsui LK, Zangari G (2013b) Modification of TiO_2 nanotubes by Cu_2O for photoelectrochemical, photocatalytic, and photovoltaic devices. Electrochim Acta 128:341–348
Tu W, Zhou Y, Liu Q et al (2012) Robust hollow spheres consisting of alternating titania nanosheets and graphene nanosheets with high photocatalytic activity for CO_2 conversion into renewable fuels. Adv Funct Mater 22:1215–1221
van Grieken R, Aguado J, López-Muñoz MJ, Marugán J (2002) Synthesis of size-controlled silica-supported TiO_2 photocatalysts. J Photochem Photobiol, A 148:315–322
Wang C, Xie Z, DeKrafft KE, Lin W (2011) Doping metal-organic frameworks for water oxidation, carbon dioxide reduction, and organic photocatalysis. J Am Chem Soc 133:13445–13454
Wang CC, Li JR, Lv XL et al (2014a) Photocatalytic organic pollutants degradation in metal-organic frameworks. Energy Environ Sci 7:2831–2867
Wang D, Choi D, Yang Z et al (2008a) Synthesis and Li-Ion insertion properties of highly crystalline mesoporous rutile TiO_2. Chem Mater 20:3435–3442
Wang D, Huang R, Liu W et al (2014b) Fe-based MOFs for photocatalytic CO_2 reduction: role of coordination unsaturated sites and dual excitation pathways. ACS Catal 4:4254–4260
Wang H, Hu Y, Jiang Y et al (2013a) Facile synthesis and excellent recyclable photocatalytic activity of pine cone-like Fe_3O_4@Cu_2O/Cu porous nanocomposites. Dalt Trans 42:4915–4921
Wang H, Yuan X, Wu Y et al (2015a) Synthesis and applications of novel graphitic carbon nitride/metal-organic frameworks mesoporous photocatalyst for dyes removal. Appl Catal B 174–175:445–454
Wang J-L, Wang C, Lin W (2012) Metal-organic frameworks for light harvesting and photocatalysis. ACS Catal 2:2630–2640
Wang R, Gu L, Zhou J, et al (2015b) Quasi-polymeric metal-organic framework UiO-66/g-C_3N_4 heterojunctions for enhanced photocatalytic hydrogen evolution under visible light irradiation. Adv Mater Interfaces 2:1500037(1–5)
Wang W, Ng TW, Ho WK et al (2013b) $CdIn_2S_4$ microsphere as an efficient visible-light-driven photocatalyst for bacterial inactivation: Synthesis, characterizations and photocatalytic inactivation mechanisms. Appl Catal B 129:482–490
Wang X, Mitchell DRG, Prince K et al (2008b) Gold nanoparticle incorporation into porous titania networks using an agarose gel templating technique for photocatalytic applications. Chem Mater 20:3917–3926
Wang X, Yu JC, Ho C et al (2005) Photocatalytic activity of a hierarchically macro/mesoporous titania. Langmuir 21:2552–2559

Wei W, Yu C, Zhao Q et al (2013) Improvement of the visible-light photocatalytic performance of TiO_2 by carbon mesostructures. Chem—A Eur J 19:566–577

Woan K, Pyrgiotakis G, Sigmund W (2009) Photocatalytic carbon-nanotube-TiO_2 composites. Adv Mater 21:2233–2239

Xiong C, Balkus KJ (2005) Fabrication of TiO_2 nanofibers from a mesoporous silica film. Chem Mater 17:5136–5140

Xiong L, Huang S, Yang X et al (2011) p-Type and n-type Cu_2O semiconductor thin films: controllable preparation by simple solvothermal method and photoelectrochemical properties. Electrochim Acta 56:2735–2739

Xu B, He P, Liu H et al (2014a) A 1D/2D helical CdS/$ZnIn_2S_4$ nano-heterostructure. Angew Chemie Int Ed 53:2339–2343

Xu W-T, Ma L, Ke F et al (2014b) Metal–organic frameworks MIL-88A hexagonal microrods as a new photocatalyst for efficient decolorization of methylene blue dye. Dalt Trans 43:3792–3798

Xu Y, Langford CH (1995) Enhanced photoactivity of a titanium(IV) oxide supported on ZSM5 and zeolite a at low coverage. J Phys Chem 99:11501–11507

Xu Y, Lv M, Yang H et al (2015) $BiVO_4$/MIL-101 composites having the synergistically enhanced visible light photocatalytic activity. RSC Adv 5:43473–43479

Xuzhuang Y, Yang D, Huaiyong Z et al (2009) Mesoporous structure with size controllable anatase attached on silicate layers for efficient photocatalysis. J Phys Chem C 113:8243–8248

Yamanaka S, Nishihara T, Hattori M, Suzuki Y (1987) Preparation and properties of titania pillared clay. Mater Chem Phys 17:87–101

Yang H, He X-W, Wang F et al (2012) Doping copper into ZIF-67 for enhancing gas uptake capacity and visible-light-driven photocatalytic degradation of organic dye. J Mater Chem 22:21849–21851

Yang P (1998) Hierarchically ordered oxides. Science 282:2244–2246

Yoneyama H, Haga S, Yamanaka S (1989) Photocatalytic activities of microcrystalline titania incorporated in sheet silicates of clay. J Phys Chem 93:4833–4837

Yoshitake H, Sugihara T, Tatsumi T (2002) Preparation of wormhole-like mesoporous tio_2 with an extremely large surface area and stabilization of its surface by chemical vapor deposition. Chem Mater 14:1023–1029

Yu J, Su Y, Cheng B (2007) Template-free fabrication and enhanced photocatalytic activity of hierarchical macro-/mesoporous titania. Adv Funct Mater 17:1984–1990

Yuan P, Yin X, He H et al (2006) Investigation on the delaminated-pillared structure of TiO_2-PILC synthesized by $TiCl_4$ hydrolysis method. Microporous Mesoporous Mater 93:240–247

Yue W, Xu X, Irvine JTS et al (2009) Mesoporous monocrystalline TiO_2 and its solid-state electrochemical properties. Chem Mater 21:2540–2546

Zhang F, Shi J, Jin Y et al (2015) Facile synthesis of MIL-100(Fe) under HF-free conditions and its application in the acetalization of aldehydes with diols. Chem Eng J 259:183–190

Zhang J, Minagawa M, Matsuoka M et al (2000) Photocatalytic decomposition of NO on Ti-HMS mesoporous zeolite catalysts. Catal Letters 66:241–243

Zhang L, Yu JC (2003) A sonochemical approach to hierarchical porous titania spheres with enhanced photocatalytic activity. Chem Commun 2078–2079

Zhang SG, Fujii Y, Yamashita H et al (1997) Photocatalytic reduction of CO_2 with H_2O on Ti-MCM-41 and Ti-MCM-48 mesoporous zeolites at 328 K. Chem Lett 26:659–660

Zhang X, Zhang F, Chan K-Y (2005) Synthesis of titania–silica mixed oxide mesoporous materials, characterization and photocatalytic properties. Appl Catal A 284:193–198

Zhao D, Peng T, Lu L et al (2008) Effect of annealing temperature on the photoelectrochemical properties of dye-sensitized solar cells made with mesoporous TiO_2 nanoparticles. J Phys Chem C 112:8486–8494

Zhou B, Wang H, Liu Z et al (2011) Enhanced photocatalytic activity of flowerlike Cu_2O/Cu prepared using solvent-thermal route. Mater Chem Phys 126:847–852

Zhou H-C Joe, Kitagawa S (2014) Metal–Organic Frameworks (MOFs). Chem Soc Rev 43:5415–5418

Zhou H-C, Long JR, Yaghi OM (2012) Introduction to metal-organic frameworks. Chem Rev 112:673–674

Zhou J, Tian G, Chen Y et al (2013a) In situ controlled growth of $ZnIn_2S_4$ nanosheets on reduced graphene oxide for enhanced photocatalytic hydrogen production performance. Chem Commun 49:2237–2239

Zhou J-J, Wang R, Liu X-L et al (2015) In situ growth of CdS nanoparticles on UiO-66 metal-organic framework octahedrons for enhanced photocatalytic hydrogen production under visible light irradiation. Appl Surf Sci 346:278–283

Zhou T, Du Y, Borgna A et al (2013b) Post-synthesis modification of a metal–organic framework to construct a bifunctional photocatalyst for hydrogen production. Energy Environ Sci 6:3229–3234

Zhou W, Fu H (2013) Mesoporous TiO_2: preparation, doping, and as a composite for photocatalysis. ChemCatChem 5:885–894

Zhu K, Neale NR, Miedaner A, Frank AJ (2007) Enhanced charge-collection efficiencies and light scattering in dye-sensitized solar cells using oriented TiO_2 nanotubes arrays. Nano Lett 7:69–74

Zimny K, Roques-Carmes T, Carteret C et al (2012) Synthesis and photoactivity of ordered mesoporous titania with a semicrystalline framework. J Phys Chem C 116:6585–6594

Zwilling V, Aucouturier M, Darque-Ceretti E (1999a) Anodic oxidation of titanium and TA6 V alloy in chromic media. an electrochemical approach. Electrochim Acta 45:921–929

Zwilling V, Darque-Ceretti E, Boutry-Forveille A et al (1999b) Structure and physicochemistry of anodic oxide films on titanium and TA6 V alloy. Surf Interface Anal 27:629–637

Chapter 7
Polymeric Nanocomposites for Visible-Light-Induced Photocatalysis

Chin Wei Lai, Kian Mun Lee and Joon Ching Juan

Abstract TiO_2 photocatalysts have been applied in treating wide range of organic contaminants, ranging from dye effluents to persistent organic pollutants. Discharging of those contaminants has polluted our natural water resources and reduced the quality and quantity of the clean water for our daily usage. Although TiO_2 photocatalysts show high removal efficiency towards most of the pollutants, the fast recombination rate and large bandgap impede its practical use under visible light irradiation. Considerable efforts have been employed to immobilize TiO_2 onto different substrates, particularly on polymer owing to their highly abundance and low cost. This chapter highlighted the various types of polymer-supported TiO_2 photocatalyst in degrading organic pollutants.

Keywords Titanium dioxide · Photocatalysis · Visible light · Polymer substrate · Nanocomposites · Photocatalytic oxidation

Semiconductor Photocatalysis

The term 'photocatalysis' can be defined as acceleration of a chemical reaction in the presence of a light source and a catalyst (Serpone and Pelizzetti 1989). In particular, semiconductor photocatalysis has attracted increased attention due to their destructive nature and can be operated at room temperature by using atmospheric oxygen as the oxidant (Chatterjee and Dasgupta 2005). In these sections, we have discussed the general photocatalytic degradation mechanism and the limitation

C.W. Lai · K.M. Lee · J.C. Juan (✉)
Nanotechnology & Catalysis Research Centre (NANOCAT),
Institute Postgraduates Studies, University of Malaya,
50603 Kuala Lumpur, Malaysia
e-mail: jcjuan@um.edu.my

J.C. Juan
School of Science, Monash University, Sunway Campus,
Jalan Lagoon Selatan, Bandar Sunway, 47500 Subang Jaya, Malaysia

M.M. Khan et al. (eds.), *Nanocomposites for Visible Light-induced Photocatalysis*,
Springer Series on Polymer and Composite Materials,
DOI 10.1007/978-3-319-62446-4_7

of TiO_2 as the photocatalyst. The needs for visible light responsive photocatalyst are also highlighted.

Mechanism of Photocatalytic Degradation

Photocatalysis is an area of study whereby a catalyst is employed to accelerate a photoreaction while the catalyst itself is unchanged at the end of the cycle (Khan et al. 2015). Depending on the mechanism of the photoreaction, the acceleration of the photoreaction by the catalyst may be attributed to the interaction between the catalyst and the substrate in its ground or excited state and/or between the catalyst and the primary photoproduct (Ohama and Van Gemert 2011). Photocatalysis is divided into two different types, which are homogeneous and heterogeneous, with which only the heterogeneous photocatalysis is discussed here. Nursam et al. (2015) stated that heterogeneous photocatalysis involves a solid catalyst and different phases of reactants. The solid catalyst typically refers to a semiconductor such as TiO_2 or ZnO.

Sahoo et al. (2015) asserted that a semiconductor consists of a highest occupied molecular orbital (HOMO) and a lowest unoccupied molecular orbital (LUMO). In this context, HOMO and LUMO refer to the electronically populated valence band (VB) and the largely vacant conduction band (CB), respectively. Both of the bands of interest are often separated from each other by an energy barrier known as the bandgap energy. Due to the existing energy gap, a semiconductor is usually termed as an excellent photocatalyst.

Kharisov et al. (2016) mentioned that a semiconductor undergoes photoexcitation to produce electron donor (reducing) and electron acceptor (oxidizing) sites. With this, redox reactions are feasible, especially in the degradation of organic pollutants present in the environment. When the semiconductor absorbs light of equal or higher energy than its own bandgap energy, the excitation of an electron from the VB to the CB takes place, creating a positively charged hole in the VB and an electron in the CB (Evans et al. 2013). Figure 7.1 illustrates the photoexcitation of a semiconductor under solar light irradiation.

Under the situation where the separation of the charge is still retained, the electron–hole pair may travel to the surface of the photocatalyst and engage in the redox reactions with the adsorbed species (Qamar et al. 2015). Sakar et al. (2016) highlighted that the aforementioned redox reactions refer to photooxidation and photoreduction processes, which occur respectively in the VB and the CB of the photocatalyst. In the photooxidation process, the holes (h^+) may react with OH^- or H_2O to form hydroxyl radicals (•OH), while, on the other hand, in the photoreduction process, the electrons (e^-) may react with O_2 to produce superoxide radical anions ($O_2^{\bullet-}$). It is these two reactive radical species which are mainly responsible for the photocatalytic degradation of organic pollutants. The reaction mechanism can thus be represented by the following Eqs. 7.1–7.5:

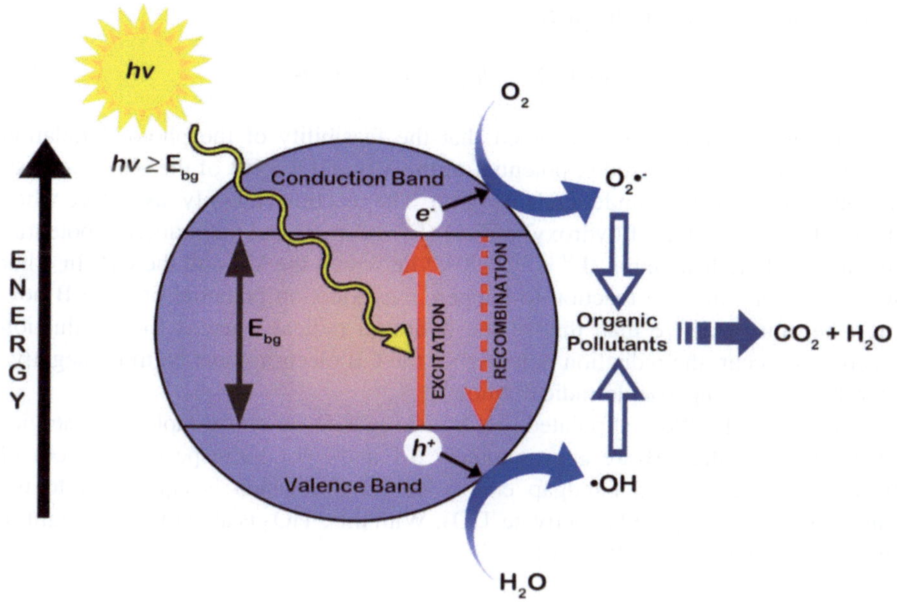

Fig. 7.1 Photoexcitation of a semiconductor under solar light irradiation

Photon absorption by photocatalyst:

$$TiO_2 + hv \rightarrow h^+_{vb} + e^-_{cb} \tag{7.1}$$

Formation of superoxide radical anion:

$$O_2 + e^-_{cb} \rightarrow O_2^{\cdot -} \tag{7.2}$$

Formation of hydroxyl radical:

$$H_2O + h^+_{vb} \rightarrow \cdot OH + H^+ \tag{7.3}$$

Oxidation of organic reactant:

$$R + \cdot OH \rightarrow R'\cdot + H_2O \tag{7.4}$$

Direct oxidation of organic reactant:

$$R + h^+_{vb} \rightarrow R\cdot^+ \rightarrow \text{Degraded products} \tag{7.5}$$

Subsequently, the last process of the reaction can be illustrated by the following Eq. 7.6 (Djokić et al. 2012):

Generation of carbon dioxide:

$$RCOO^- + h_{vb}^+ \rightarrow R\cdot \ + CO_2 \tag{7.6}$$

Vinu and Madras (2012) claimed that the feasibility of the photodegradation reaction depends on the redox potential of the VB and the CB of the photocatalyst, as well as the materials under consideration. The reaction can only take place when the oxidation potential of hydroxyl radical ($E^0 = 2.8$ V) and the reduction potential of superoxide radical anion ($E^0 = -0.28$ V) lie within the VB and the CB. In other words, for an oxidation reaction to happen, the oxidation potential of the VB hole must be more positive than that of the hydroxyl radical whereas for a reduction reaction to occur, the reduction potential of the CB electron must be more negative than that of the superoxide radical anion.

Ghezzar et al. (2007) stipulated that by taking TiO_2 as an example, the energies of the VB and the CB are approximately 3.1 and −0.1 V, respectively. Both of these values result in a bandgap energy of 3.2 eV, which is equivalent to the minimum energy required to activate TiO_2. With this, TiO_2 is able to absorb light in the near UV range ($\lambda < 387$ nm).

Recombination of Electron–Hole

Panda (2009) defined charge carrier recombination as a process whereby the holes in the valence band and the electrons in the conduction band annihilate each other. It is the reverse process of charge carrier generation. The recombination process leads to the disappearance of both charge carriers, and subsequently the dissipation of the absorbed energy in the form of light or heat, depending on the type of recombination process. For light emission, radiative recombination process is involved, whereas for heat evolution, non-radiative recombination process is involved. These two types of recombination process can then be described by the following Eqs. 7.7 and 7.8 (Fulay 2016):

Radiative recombination:

$$h_{vb}^+ + e_{cb}^- \rightarrow \text{photon(light)} \tag{7.7}$$

Non-radiative recombination:

$$h_{vb}^+ + e_{cb}^- \rightarrow \text{phonon(heat)} \tag{7.8}$$

With respect to the usage of TiO_2 as a semiconductor photocatalyst, the photocatalytic activity is relatively low due to the fast recombination of charge carriers (Ibhadon and Fitzpatrick 2013). Despite the fact that it is an undesirable process, recent study has proved its importance in understanding the charge carrier dynamics (Ohtani 2013). Photoluminescence (PL) spectroscopy is a technique used to investigate the recombination of charge carriers (Buha 2013). Apart from finding out the destiny of charge carriers in the photocatalyst, the PL emission spectrum is also useful to determine the efficiency of charge carrier transfer, migration, and trapping (Ma et al. 2011). Li et al. (2002) found that the lower the PL intensity, the lower the recombination rate of charge carriers, and the higher the photocatalytic activity of the photocatalyst.

It well known that recombination rate of charge carriers possesses a strong relationship with photocatalysis, but there are many different type recombination pathways of a photogenerated electron–hole pair (Zhang and Yates 2012). When a semiconductor is irradiated with light of equal or higher energy than its own bandgap energy, the absorption of photons takes place. This leads to the excitation of electrons from the VB to the CB, creating holes in the VB. The photogenerated electron–hole pairs may be trapped in 'trap states'. Due to the presence of these trap states, the recombination process happens through various routes. Trap states can be categorized into shallow and deep trap states, which resulted from the defects or impurities in the semiconductor. As aforementioned, the two main routes are the radiative and non-radiative recombination processes. Figure 7.2 shows the different recombination pathways of a photogenerated electron–hole pair which give rise to the photocatalysis efficiency. The recombination routes in Fig. 7.2b, c are also known as the Shockley–Read–Hall (SRH) recombination (Hall 1952; Shockley and Read Jr. 1952).

Generally, it is believed that TiO_2 has an indirect bandgap (Lin et al. 2006). Therefore, the recombination process of photogenerated charge carriers proceeds through the non-radiative route and results in the loss of energy in the form of heat, which can be detected via the time-resolved photoacoustic spectroscopy (TRPAS) (Schneider et al. 2014). Mendive et al. (2011) reported that the heat energy produced is capable of stimulating the disaggregation of agglomerated TiO_2 nanoparticles. Subsequently, this leads to the increase of the surface area, the adsorption capacity and the degradation rate. In order to study the recombination

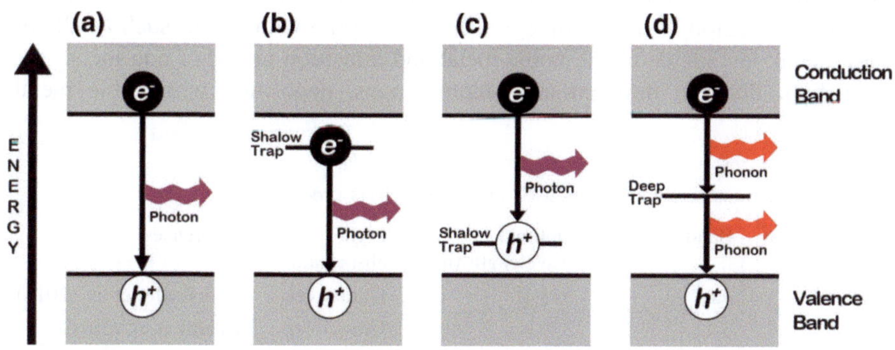

Fig. 7.2 Different recombination pathways of a photogenerated electron–hole pair. **a** Band-to-band radiative recombination, **b** electron-trap state to valence band, **c** conduction band to hole-trap state, and **d** non-radiative recombination via an intermediate state

kinetics of the photogenerated electron–hole pairs, another technique known as the time-resolved absorption spectroscopy (TAS) is utilized. Since photocatalysis activity has a strong relationship with the recombination rate of the charge, several factors such as charge trapping (Skinner et al. 1995), interfacial charge transfer (Colombo and Bowman 1996), and reaction temperature (Katoh and Furube 2011) play important roles.

Visible Light Photocatalysis

Mamba and Mishra (2016) stated that in order to cope with the pollution challenges in the environment, the role of photocatalysts in the aspect of environmental remediation has been studied extensively over the past few years. As compared to other conventional treatment techniques, semiconductor photocatalysts appear to be an effective tool to degrade stubborn organic pollutants in the environment. Table 7.1 summarizes the various semiconductors used in the photocatalysis processes.

Among the semiconductor photocatalysts, TiO_2 remains as one of the predominant photocatalytic materials in degrading a wide variety of organic pollutants because it is stable, efficient, toxic free, inexpensive and environmental friendly (Khan et al. 2014). Sellappan (2013) asserted that nevertheless, TiO_2 suffers from two major drawbacks which lower its quantum efficiency, they are low efficiency of sunlight utilization and fast electron–hole pair recombination rate. Generally, the sunlight is composed of ultraviolet light (λ = 200–400 nm), visible light (λ = 400–800 nm) and infrared light (λ > 800 nm) (Chen et al. 2016). TiO_2 can only absorb high-energy light from the ultraviolet region, which only accounts for 5% of the solar spectrum due to its large bandgap (3.2 eV) (Daghrir et al. 2013).

Wang et al. (2015) explained that with respect to these limitations, several modification techniques have been discovered to maximize the efficiency of electron–hole pair separation, and to extend the photocatalytic response towards the visible light region. Typical strategies include doping with metal (such as alkaline earth metal, lanthanide metal, noble metal and transition metal) or non-metal (such as carbon, fluorine, nitrogen and sulphur) ions, deposition with noble metals,

Table 7.1 Various semiconductors used in photocatalysis processes

Photocatalyst	Bandgap (eV)	Optimal wavelength (nm)	Light absorption	References
SnO_2	3.6	345	Ultraviolet	Abdelkader et al. (2015)
ZnO	3.4	380	Ultraviolet	Ullattil et al. (2016)
TiO_2	3.2	387	Ultraviolet	You and Zhao (2016)
WO_3	2.7	460	Visible light	Ohkura (2013)
V_2O_5	2.2	564	Visible light	Martha et al. (2012)

sensitization with organic dyes and composition/coupling with conducting polymers or other semiconductors. Here, the focus will be on the composition/coupling of large-bandgap TiO_2 with small-bandgap conducting polymers.

Salem et al. (2009) highlighted that conducting polymers are a new class of visible light driven photocatalytic materials that have attracted the interest of a number of researchers. This is mainly due to their high stability in the environment, high visible light absorption coefficients and high electron–hole mobility. The most known conducting polymers include polyaniline (PANI) (Subramanian et al. 2014), polypyrrole (PPy) (Zhang et al. 2014a, b) and polythiophene (PT) (Shahabuddin et al. 2016). Riaz et al. (2015) stipulated that conducting polymers can act as sensitizers for semiconductors with large bandgap. Since they behave as p-type semiconductors, they combine with n-type semiconductors to form p-n junctions. This configuration can later overcome the drawbacks of TiO_2, resulting in the enhancement of visible light response and the reduction of charge carrier recombination rate.

Liu et al. (2015a, b) maintained that in general, when a composite photocatalyst of TiO_2 and a conducting polymer are illuminated with visible light, the conducting polymer absorbs a photon and excites an electron from its HOMO to its LUMO, creating a hole in its HOMO. Then, the excited electron is readily injected into the CB of TiO_2 to take part in the photoreduction process while the hole remains in the HOMO of conducting polymer. Simultaneously, an electron in the VB of TiO_2 is

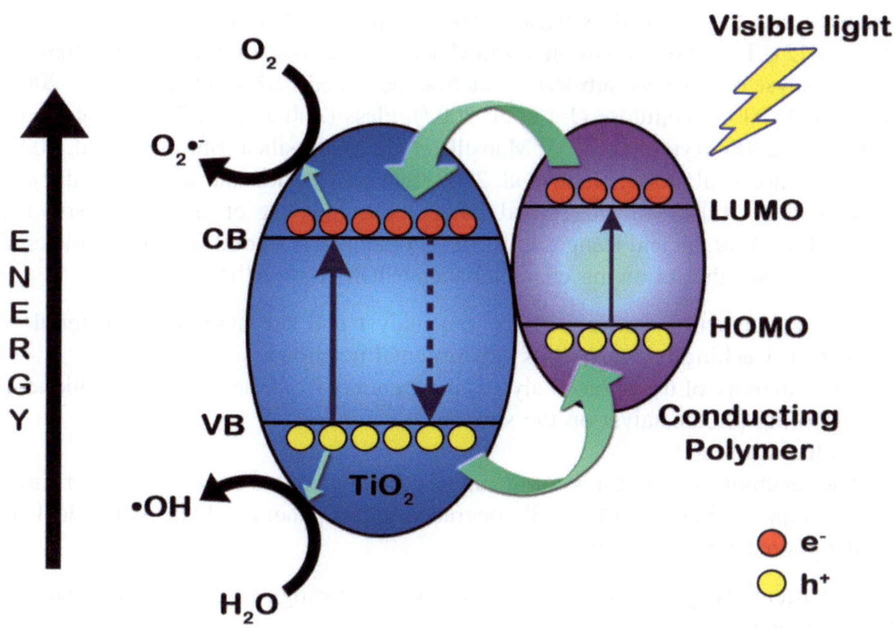

Fig. 7.3 Interfacial charge transfer in conducting polymer/TiO_2 composite photocatalyst under visible light illumination

migrated to the HOMO of conducting polymer to recombine with the hole, which then leads to the generation of a hole in the VB of TiO_2 that is responsible for the photooxidation process. As the redox reaction goes on, the number of electrons injected and the holes produced in TiO_2 gradually increases, yielding more reactive radical species on the surface of TiO_2, thus increasing the photocatalytic activity of the composite photocatalyst. Figure 7.3 depicts the interfacial charge transfer in the conducting polymer/TiO_2 composite photocatalyst under visible light illumination.

Therefore, by coupling TiO_2 semiconductor with conducting polymers, TiO_2 can achieve greater photocatalytic performance under visible light irradiation as the conducting polymers increase the efficiency of sunlight utilization in the visible light range and lower the rate of electron–hole pair recombination in TiO_2, leading to more efficient removals of organic pollutants in the environment (Lee et al. 2012).

Polymer-Supported Photocatalysis

Polymer Support

Most of the photocatalysts are available as powder and have been applied in photocatalysis process in suspended form. Although it offers greater exposed surface area, however, the recovery of photocatalyst and reduction in light utilization efficiency are the major drawbacks (Han and Bai 2010; Krysa et al. 2006; Vaez et al. 2012). The above-mentioned drawback can be overcome by immobilizing a photocatalyst on various substrates, such as activated carbon (Matos et al. 2007; Dona et al. 2007), cellulose (Jin et al. 2007), glass (Zainal et al. 2005; Mahmoodi and Arami 2006; Ryu et al. 2003; Mansilla et al. 2006), silica (Shironita et al. 2008; Lopez-Munoz et al. 2005; Gude et al. 2008) and polymeric materials (Damodar and Swaminathan 2008; Magalhaes and Lago 2009; Zhiyong et al. 2008a; Sriwong et al. 2008; Murugan and Rangasamy 2011). Practically, a good supporting material must possesses the following characteristics (Shan et al. 2010):

- Strong interaction between the photocatalyst and the supporting material to prevent leaching under various experimental reactions;
- The reactivity of the photocatalyst is independent from the selected technique in anchoring photocatalyst on the support
- High surface area
- The resulting photocatalyst on a support must be stable over a period of time
- The support should not be easily degraded by the generated hydroxyl radicals in the photocatalysis process.

In particularly, polymer substrates have been investigated extensively owing to the following advantages:

- Cheap and abundant (Shan et al. 2010)
- Chemically inert and mechanical stable (Han and Bai 2009)
- Stable towards oxidation by highly reactive hydroxyl radicals during photocatalytic reaction (Zhiyong et al. 2008b)
- Hydrophobic nature which enhances adsorption and oxidation capability (Magalhães et al. 2011).

To our best knowledge, the first report on using a polymer-supported TiO_2 photocatalyst was by Tennakone and co-workers in the year 1995 (Tennakone et al. 1995). Since then, numerous polymeric substrates have been utilized as supports, namely polythene (PE) (Baruah et al. 2012; Velásquez et al. 2012), polyethylene terephthalate (PET) (Chan et al. 2014), polyvinyl alcohol (PVA) (Liu et al. 2015a, b), polyaniline (PANI) (Yu et al. 2012) and polymethyl methacrylate (PMMA) (Elfeky and Al-Sherbini 2011a, b).

TiO_2-Polythene (PE) Photocatalyst

Polythene (PE) appears as an excellence candidate as a support for photocatalyst due to its storage and handling properties. Moreover, it is stable against abrasion, caking, crushing and moisture (Omar 1989). Tennakone et al. supported TiO_2 on PE film by thermal treatment method. The surface of the commercial polythene film was evenly spread with TiO_2 powder and rubbed by cotton wool followed by ironing at a temperature of 74 °C. The photocatalytic activity of the TiO_2/PE film was evaluated by photodegrading phenol under solar irradiation with intensity of 0.7 kW/m^2 by monitoring carbon dioxide (CO_2) gas evolution. It was found that more than 75% of phenol was degraded in 5 h. Interestingly, less than 10% of phenol was removed with bare TiO_2 film under identical conditions. Taking their research work further, Tennakone and Kottegoda (1996) investigated the photocatalytic degradation of paraquat by the synthesized TiO_2/PE film. The experimental results were very promising as the paraquat was completely mineralized to CO_2, NH_3, HCI and small quantities of NO_2^-/NO_3^- in the presence of TiO_2/PE film under solar irradiation. They found that photocatalytic reaction was efficient when TiO_2 particles were partially embedded into the PE film, where they remain on the surface for paraquat adsorption.

Elfeky and Al-Sherbini (2011) anchored TiO_2 powders (Degussa P25) onto high-density polyethylene beads (HDPE) by a simple thermal attachment method. The TiO_2/HDPE beads and Au-TiO_2/HDPE beads were then applied in the photocatalytic degradation of Rhodamine 6G (R6G) dye under natural sunlight (Intensity = 0.5–0.6 kW/m^2). The adsorption of R6G dye was ~48% with TiO_2/HDPE beads in the dark conditions. The adsorption efficiency was increased to ~72% in the presence of Au-TiO_2/HDPE beads under the same experimental conditions. The electrostatic attraction between the Au nanoparticles and R6G dye molecules on the TiO_2 photocatalyst surface improved the charge transfer efficiency, which enhanced the percentage of adsorption. Consequently, the photodegradation rate of R6G dye by Au-TiO_2/HDPE beads was twice the photodegradation rate with TiO_2/HDPE beads (9.8×10^{-5} moldm^{-3}s^{-1} vs. 5.1×10^{-5} moldm^{-3}s^{-1}). It is worth to mention that the degradation of R6G only took half the time in the presence of Au nanoparticles, as compared to the absence of Au nanoparticles to achieve same degradation percentage of R6G dye under solar irradiation.

The TiO_2/PE films with different compositions were developed by Mehmood et al. (2015). Briefly, TiO_2 nanoparticles were added in the molten PE solution and ultra-sonicated at 70 °C for an hour. The TiO_2/PE films were then exposed to a 24 W fluorescent lamp and the weight loss of each respective film was studied. The experimental studies revealed that the weight loss increased with increasing TiO_2 concentration and a maximum of 33% weight loss was observed for 20% TiO_2 added. The degraded TiO_2/PE films were then used in the photocatalytic decolorization of Drimarene Brilliant Red (DBR) dye and were compared to fresh TiO_2/PE films. It was found that the degraded TiO_2/PE films showed higher colour removal (81%) than the fresh TiO_2/PE films, which only achieved 32% colour removal. The effectiveness of the degraded TiO_2/PE films could be attributed to the formation of cavities in the PE matrix which enhanced the exposed surfaces for the DBR dye molecules to be adsorbed. In addition, more TiO_2 nanoparticles are readily available in the degraded TiO_2/PE films, which increased the photodegradation efficiency of DBR dye. On the other hand, the fresh TiO_2/PE films exhibited lower photodegradation activity due to less TiO_2 nanoparticles exposed to visible light irradiation. In addition, there was no degradation when the TiO_2/PE films were placed in the dark condition, which further confirm the role of photocatalysis in degrading the DBR dye molecules.

TiO_2-Polyethylene Terephthalate (PET) Photocatalyst

Polyethylene terephthalate (PET) is a cheap and abundant material which could be used as a support for photocatalyst (Hugh et al. 2001). It seems as a viable candidate in removing both organic and inorganic pollutants for large-scale water treatment pilot plant. Moreover, this low cost support is extremely important in many developing countries especially for isolated area, where large population exist (Wegelin et al. 2000). Fostier et al. (2008) coated TiO_2 onto internal surface of PET bottles, by shaking TiO_2 suspension at pH 2.5 ($HClO_4$) for 30 s (Meichtry et al. 2007). The excess suspension was removed before being dried at 55 °C to increase the adhesion of TiO_2 particles on the PET surface. Then, the unfixed TiO_2 particles were removed by washing with distilled water. In the photooxidation of arsenite (As (III)) to arsenate (AS (V)), 200 mL of an As (III) and Fe (II) solution were poured into the TiO_2-coated PET bottles and then exposed to sunlight (Intensity = 1.2 mW/m^2) horizontally. After the reaction was completed, the bottles were kept vertically in the dark for 12 h for precipitation of colloidal material. The precipitates were then filtered before the measurement of As (III) and As (V) concentration in the solution.

In the dark conditions, only 22% of As was removed in 2 h, whereas the removal of As reached 96% when the reaction set-up was irradiated by solar light. This shows that solar light irradiation plays a significant role in the oxidation of As (III). This phenomenon was caused by the generation of large amount of hydroxyl radicals on the TiO_2 surface under irradiation that oxidize As (III) to As (V). On the other hand, small amount of As was removed in the dark which could be attributed to the formation of superoxide anion radicals ($O_2^{\bullet-}$) along with oxidation of Fe(II) (Ryu and Choi 2004). Slight decrease in the As removal was recorded when the reaction was conducted in the presence of humic acid, suggesting this organic matter may compete with the arsenic molecules for the reactive oxidizing species formed in the solution. The study was further investigated by water samples that contaminated with As. It was found that the presence of fluorides (F^-), chlorides (Cl^-), nitrites (NO_2^-), nitrates (NO_3^-) and phosphates (PO_4^{3-}) ions in the natural water samples did not interfere the oxidation of As (III), where over 99% of As removal was observed. This indicated that the irreversible poisoning of the photocatalyst did not occur in this case.

Peerakiatkhajohn et al. (2011) fabricated visible light responsive Ag/TiO_2 on PET film via sol-gel route with different Ag/Ti molar ratio. The Ag/TiO_2 solution was first refluxed at 80 °C for 8 h and then the PET was dipped into the Ag/TiO_2 gel to produce Ag/TiO_2/PET thin films. The resulting thin films were activated by 125 W UV lamp to activate the TiO_2 particles and enhance the adhesion of Ag/TiO_2 onto the PET surface. The photocatalytic activity of Ag/TiO_2/PET thin films was examined by photodecolourization of methylene blue (MB) dye under an 18 W fluorescent lamp. The experimental results revealed that the decolourization was negligible when the reaction was conducted with PET film even after 12 h visible light irradiation. In the presence of TiO_2, the removal efficiency of MB increased to 30%. The photodecolourization of MB was further enhanced by 0.10 Ag/TiO_2 thin film, which achieved 80% removal under identical experiment conditions. Less or higher amount of Ag dopant was not beneficial for the MB removal due to the Ag particle could acts as a recombination centre for electrons and holes (Rengaraj and Li 2006). Upon visible light irradiation, the MB molecules are self-photosensitized, a large number of electrons (e_{CB}) were generated in the conduction band of TiO_2. These e_{CB} were then reacted with oxygen from the atmosphere to produce reactive oxidizing radicals ($O_2^{\bullet-}$, •OOH, •OH) that subsequently degrade the cationic radical ($MB^{\bullet+}$) of MB dye. Similarly, the Ag/TiO_2/PET thin films were used in the photodegradation of benzene, toluene, ethylbenzene and xylene (BTEX) gas under two 8 W fluorescent lamps as visible light source. The results revealed that at least 80% of all gaseous BTEX was decomposed in 4 h. Among the BTEX gases, benzene which is non-polar in nature has the lowest photodegradation efficiency as a result of their low reactivity towards the generated hydroxyl radicals from the TiO_2 photocatalyst.

TiO_2 nanoparticles and Fe-doped TiO_2 nanoparticles were deposited on PET fabric by Dumitrescu and co-workers (2015) by pad-dry-cure and cationization-pad-dry-cure method. The prepared TiO_2/PET fabric and Fe-TiO_2/PET fabric were used to photodecolourize a common textile dye, MB under simulated sunlight.

From the results, the 100% decolourization of MB was achieved by Fe-TiO_2/PET fabric via pad-dry-cure method in 4 h. This is owing to the presence of more TiO_2 nanoparticles on the resulting fabric. The Fe-TiO_2/PET fabric exhibited higher MB removal than TiO_2/PET fabric due to Fe^{3+} trapped the photogenerated holes and electrons, which prolonged the recombination of electron–hole pairs (Luo et al. 2004):

$$\mathrm{Fe(III)} + h^{+}_{VB} \rightarrow \mathrm{Fe(IV)} \quad (7.9)$$

$$\mathrm{Fe(IV)} + OH^{-} \rightarrow \mathrm{Fe(III)} + \cdot OH \quad (7.10)$$

$$\mathrm{Fe(III)} + e^{-}_{CB} \rightarrow \mathrm{Fe(II)} \quad (7.11)$$

$$\mathrm{Fe(II)} + \mathrm{Ti(IV)} \rightarrow \mathrm{Fe(III)} + \mathrm{Ti(III)} \quad (7.12)$$

The Fe^{2+} then reacted with oxygen molecules to form $O_2^{\bullet-}$ that oxidizes MB dye molecules (Zhu et al. 2006):

$$Fe^{2+} + O_2 \rightarrow Fe^{3+} + O_2^{\bullet-} \quad (7.13)$$

TiO_2-Poly (Vinyl Alcohol) (PVA) Photocatalyst

Poly (vinyl alcohol) (PVA) is the first synthetic colloid synthesized by Herrmann and Haehnel in 1924 (Finch 1973). PVA is a water-soluble synthetic polymer, odorless, tasteless, translucent, non-hazardous and safe (Saxena 2004; Thong et al. 2016). PVA are commonly categorized into two types based on their degree of polymerization and hydrolysis. For example, type I is completely hydrolyzed group (98% of acetate groups have been substituted by alcohol groups) and type II is partially hydrolyzed group (ca. 87–89% of acetate groups have been replaced by alcohol groups) (Finch 1973). Type I PVA (Completely hydrolyzed) is very soluble in hot water, has good film forming characteristics, and exhibits good adhesive properties. However, type II PVA (partially hydrolyzed) is soluble in water at room temperature and only slightly soluble in ethanol (Saxena 2004). In term of adhesive, type II is water resistant after cross-linking its linear chains with boric acid, formaldehyde, salts and other insolubility agents (Feldman 1989).

PVA has attracted much attention because of its high visible light photocatalytic activity (Wang et al. 2009) and high reusability (Lei et al. 2012). This phenomenon is due to the formation of Ti-O-C chemical bonds between TiO_2 and PVA during the preparation process (Liu et al. 2015a, b). In other words, the heat treatment process during the preparation of TiO_2-PVA composite is crucial. Upon thermal treatment, the PVA monomer coacervates with TiO_2 and thus formed conjugated structures via Ti–O–C bonds. This formation will act as electron transfer pathway, which will facilitate excited electron from conjugated structure to TiO_2 (Liu et al.

2015a, b). In addition, the high reusability due to this strong Ti–O–C bonds formation gives this composite an advantage for practical application and commercialization. As shown in Table 7.2, the photocatalytic activity of TiO_2-PVA composite is very promising which is able to achieve complete degradation for methyl orange (MO) and phenol after 15 min and 5 h under visible light, respectively. The high photocatalytic activity was also contributed by the swelling ability of PVA when in contact with water which managed to overcome diffusion issue of MO into composite films.

Many techniques have been employed to incorporate TiO_2 into PVA and thermal treatment is the most commonly used techniques. Wang et al. (2009) reported a hydrothermal method to produce conjugated unsaturated D-PVA doped on the surface of TiO_2 (Table 7.1). In their work, $Ti(OH)_4$ precursor and PVA were thermally degraded to produce TiO_2 and D-PVA, respectively. Although this method successfully prepared a highly reusable TiO_2–composite, the photodegradation of MO was only 84% after 40 h under visible light. Similarly, Yang et al. (2015) have prepared TiO_2 into calcined PVA (C-PVA) and then coated on glass substrates. The C-PVA has shown a conjugated C = C bonds and successfully doped with TiO_2 on a glass substrate. In this study, C-PVA/TiO_2 composites photoactivity was due to aggregate states between C-PVA and TiO_2. However, the photodegradation of Rhodamine B (RhB) was 93% after 6 h under visible light.

TiO_2-Polyaniline (PANI) Photocatalyst

Many researchers focus their attention on conducting polymer such as polyaniline (PANI) to be applied as photosensitizer and sensor. Moreover, the application of PANI has gained much momentum in the field of photocatalysis. PANI is a conducting polymer, which exhibits unique electrical, optical and photoelectric properties (Kang et al. 1998; Lei et al. 2012). In general, the formula of aniline polymers is [(–B–NH–B–NH–), (–B–N=Q=N–) 1_Y]x, in which B and Q are denoted as C6H4 rings in the benzenoid and quinonoid forms, respectively. Thereby, PANI are basically poly(p-phenyleneimineamine)s and thus its neutral intrinsic redox states can be varied from the fully oxidized pernigraniline (PNA, y = 0) to that of the fully reduced leucoemeraldine (LM, y = 1) (Kang et al. 1998).

In early study, Li et al. (2008) developed PANI-doped TiO_2 photocatalyst for photodegradation of phenol under visible light irradiation. They managed to prove that PANI-TiO_2 composite has a high photocatalytic activity and reusability. Similarly, PANI-TiO_2 composite has also successfully photodegraded Malachite Green (MG) (Kumar and Sarmah 2011), Rhodamine B (Li et al. 2011), Reactive Red 4 (Razak et al. 2014), Reactive Blue 4 (Masid et al. 2015), Reactive Brilliant Blue KN-R (Yu et al. 2012) and Methylene Blue (MB) (Zhang et al. 2008; Wang et al. 2010) under visible light. This positive enhancement was attributed to the slower electron–hole separation in PANI-TiO_2 nanocomposites which increases

Table 7.2 TiO–PVA composites in various photodegrdation applications

H-PVA/TiO_2	Rhodamine B (RhB)	Visible light (tungsten-halogen lamp, 500 W, 350–2500 nm)	Sol-gel method Thermal treatment	89% degradation of RhB after 6 h under visible light	Song et al. (2014)
TiO_2/D-PVA	Methyl orange (MO)	Visible light (dysprosium lamp, 200 W)	Hydrothermal method	84% degradation of MO after 40 h under visible light	Wang et al. (2009)
PVA/TiO_2	Methyl orange (MO) Phenol	Visible light (xenon lamp, 300 W)	Hydrothermal method	~100% degradation of MO after 15 min under visible light ~100% degradation of phenol after 5 h under visible light	Liu et al. (2015a, b)
TiO_2/ T-PVA/CHC	Rhodamine B (RhB)	Visible light (tungsten-halogen lamp, 500 W)	Sol-gel method Thermal treatment	93% decolourization of RhB after 6 h under visible light	Zhang et al. (2013)
C-PVA/TiO_2	Rhodamine B (RhB)	Visible light (60 mW/cm^2)	Calcination method	92.2% degradation of RhB after 6 h under visible light (mass feed ratio of P/T = 1:6)	Yang et al. (2015)
TiO_2/PVA	Rhodamine B (RhB)	Visible light (xenon lamp, 150 W)	Simple solution method Low temperature heat treatment	Degradation of RhB follows first order kinetic with rate constant of 0.134 min^{-1} under visible light (weight ratio of TiO_2:PVA = 1:0.050)	Filippo et al. (2015)

the oxidative properties of TiO_2 and it was confirmed by photoluminescence spectroscopy which suggests that photo-injected electrons are transferred from TiO_2 to PANI. PANI sensitized visible light and the charge transfer from sensitizer to TiO_2 decrease rate of electron–hole recombination (Wang et al. 2010). Electron spin resonance (ESR) has confirmed that the main oxidative species during photocatalysis of PANI composite are oxygenous radicals ($O_2^{\bullet -}$), hydroxyl radical (•OH) and holes (h^+). When the PANI absorbs light (visible light), the π–π^* transition occurs and then the photogenerated electron transfers to the π^* orbital. Then the photogenerated electron can migrate from PANI to TiO_2 which generates strong hydroxyl and superoxide radical in contact with water. Furthermore, the PANI-TiO_2 composite can be easily separated after use by a fast decantation within 5 min. The ease of separation and highly active under visible light has demonstrated that this composite has high commercial application.

Recently, ternary system which consists of three element of compound in one photocatalyst has been studied widely. Leng et al. (2013) successfully synthesized PANI-modified $CoFe_2O_4$-TiO_2 with hierarchical flower like nanoarchitectures and applied in photocatalytic activity. The composites exhibited excellent photocatalytic activity with rate constant (k) value 0.011/min due to the synergistic effect between PANI, $CoFe_2O_4$ and TiO_2. In another study, ternary system of Ag_3PO_4-PANI-GO composites exhibit high visible light photocatalytic performance and stability (Zhang et al. 2014a, b). The PANI expanded the absorption range in visible light resulting Ag_3PO_4-PANI-GO has 2.1 and 3.1 times better photodegradation rate as compared to that of pure Ag_3PO_4 and Ag_3PO_4-GO, respectively. Chen et al. (2014) have managed to synthesize ternary N-$K_2Ti_4O_9$/$MnFe_2O_4$/PANI composites by an in situ oxidative polymerization method. The uniqueness of these composites is on its excellent ferromagnetic properties of $MnFe_2O_4$ where it can be easily recovered and recycled for next photocatalytic reaction by using magnetic force. PANI was found to have a high adsorption capacity, high utilization of visible light and enhanced charge carrier transfer. PANI-based composites possess better photocatalytic activity than pure materials with photocatalytic rate constant $52.24 \times 10^2\ h^{-1}$. It is obvious that PANI has a vital role for the enhancement of photocatalytic activity. N-$K_2Ti_4O_9$/$MnFe_2O_4$/PANI (7:3) is an optimum ratio for a better photocatalytic activity (Chen et al. 2014).

Ochieng et al. (2017) has synthesized ternary PANI/TiO_2/ZnO composite and applied it for photodegradation of aromatic compounds in abattoir wastewater. They successfully degraded phenol and p-cresol up to 99.2 and 97%, respectively after 10 h under UV light irradiation. Unfortunately, they did not conduct this experiment under the visible light which also expected to give high photocatalytic activity as compared to that of TiO_2 and ZnO. Chitosan-grafted polyaniline/Co_3O_4 nanocube nanocomposites were prepared through an in situ oxidative polymerization technique (Shahabuddin et al. 2015). They incorporated Co_3O_4 into the cross-linked network of the polymeric matrix, and this synergistic effect has led to enhancement of the photocatalytic efficiency of the nanocomposite. The nanocomposite with 2 wt% of Co_3O_4 nanocubes with respect to aniline managed to give 88% photodegradation efficiency after 3 h of irradiation under UV light. Therefore, secondary metal oxide semiconductor is capable to prolong the time of electron at CB and hole at VB which will allow to produce hydroxyl radical and superoxide ion.

TiO_2-Polymethyl Methacrylate (PMMA) Photocatalyst

Polymethyl methacrylate (PMMA) is acrylic with many trade names such as Plexiglas, Perspex, etc. It is a transparent thermoplastic material used in various applications such as shatter-resistant glass, casting resin, coatings and many other usages (Ali et al. 2015). Besides that it is also a relatively stable, economical and hydrophobic polymer which is suitable for food packing. PMMA is an ester of methacrylic acid with the structure $CH_2{=}C[CH_3]CO_2H$, which belongs to acrylic family. The application of PMMA for photocatalysis application has also gained much attraction in recent years because PMMA is transparent to UV–visible light and easily mouldable. It is a good composite for various inorganic particles and it is also proven that the composite comprised of metal oxide fillers exhibit better properties (Demir et al. 2007; Cantarella et al. 2016). The introduction of photocatalyst such as TiO_2 is not only showing high photocatalytic activity but also enhanced the glass transition and thermal stability. Stewart et al. (2015) have successfully synthesized porous PMMA composite with TiO_2 (P25) for photocatalytic degradation of aqueous MO. The prepared porous PMMA-TiO_2 composites demonstrated six times higher photocatalytic activity than that of the non-porous PMMA-TiO_2 composite.

It was reported that composite of PMMA-TiO_2 managed to photodegrade MB and phenol up to ca 78 and 45%, respectively under UV light irradiation (Cantarella et al. 2016). This composite can be recycled up to eight times without any significant changes and there is no leakage occurred after the photodegradation. In order to further enhance the photocatalytic activity under visible light, Au co-doped TiO_2 and PMMA composite was prepared for photodegradation of TrypanBlue (TB) (Elfeky et al. 2011). They managed to completely photodegrade TB under sunlight irradiation at pH 2. The optimum pH was 2 because of strong interaction between photocatalyst ($TiOH^{2+}$) and TB anion. This led to high adsorption of the TB on composite surface and has led to efficient photodegradation. The Au-TiO_2 on PMMA is capable of photodegrading TB up to 90% after 1.75 h under sunlight irradiation. This is better than that of TiO_2 on PMMA which took 2.5 h to photodegrade 87% TB.

Polymer-Supported Buoyant Photocatalysis

Introduction

In the field of photocatalysis, buoyant photocatalysts can offer a cost-effective and easily operating process in promoting effective photocatalytic oxidation reaction. Today, titanium dioxide (TiO_2) has emerged as the leading candidate in water/wastewater treatment, especially water on the Earth's surface due to the rapid attenuation of light or irradiation in water. In general, TiO_2 photocatalyst has been employed as potential substrates for the immobilization of TiO_2 particles on organic fibres, pumice stone, perlite natural materials or glass microspheres in order to form buoyant TiO_2 photocatalyst (Zaleska et al. 2000; Portjanskaja et al. 2004, 2006; Hosseini et al. 2007; Han and Bai 2009). However, an obvious hindrance to the widespread use of those mentioned substrate to form buoyant TiO_2 photocatalyst is its high production cost, fragile and can easily break or spoil, especially in the high-temperature calcination process. Therefore, in the following section, buoyant polymer-supported TiO_2 photocatalyst is reviewed and discussed.

Taking into account of the photocatalytic oxidation reaction on buoyant polymer-supported TiO_2 photocatalyst under solar irradiation, this novel photocatalyst indeed exhibits several advantages including fully utilizing our solar energy without any light attenuation in water/wastewater treatment system; easy and convenient platform to apply in green applications without any complicated installation steps, excellent light-harvesting ability to promote an efficient photocatalytic degradation of suspended insoluble organic contaminants as well as simple self-defense and activation of post-treatment recovery mechanisms (Magalhaes and Lago 2009; Han and Bai 2010; Magalhães et al. 2011). In recent years, polymer-supported substrate (plastics) has received lots of attention and appeared as one of the most promising candidate to form buoyant TiO_2 photocatalyst due to the following reasons, such as cheap, high mechanical strength and high chemical resistance. In this case, many literatures have discussed about the immobilization of TiO_2 photocatalyst onto various polymer-supported substrates, especially polypropylene (PP) and polystyrene (PS) polymers (Dutschke et al. 2003; Yang et al. 2006; Magalhaes and Lago 2009; Han and Bai 2010; Magalhães et al. 2011; Singh et al. 2015). Table 7.3 presented the polymer characteristics of PP and PS in term of density, melting point, UV resistance, acid/base resistance as well as organic solvent resistance.

Table 7.3 Polymer characteristics for PP, and PS (Han and Bai 2010)

Polymer	Density (k/m^3)	Melting point (°C)	UV resistance	Acid/base resistance	Organic solvent resistance
PS	1050	240	Medium	Poor	Poor
PP	855–945	165	Medium	Good	Good

Polypropylene (PP)

Today, polymer became one of the mostly used and studied materials in an enormous and expanding range of products, from paper clips to spaceships, due to their relatively low cost, ease of manufacture, versatility and imperviousness to water. Among many types of polymer, many researchers and scientists preferred PP that acts as the buoyant photocatalyst substrate due to the versatile thermoplastic material, compatible with many processing techniques and used in many commercial applications (Han and Bai 2011). The main advantages for this PP are high chemical and thermal resistance in acid/base organic solvents, excellent moisture and oxygen barrier, high mechanical strength, commercial availability and diversity of shape and structure for automotive equipment, consumer goods, houseware parts, garden furniture and storage containers. In addition, PP leads the way in sustainability, which consumes the least amount of energy during PP production and generates the lowest carbon dioxide (greenhouse gases) emissions as compared to other plastics substrate (Hopewell et al. 2009; Maddah 2016). PP substrate exhibits low density in the range of 855–945 kg/m^3 depending on their crystalline degree, which means that switching to PP reduces the absolute amount of waste as well as making it buoyant on the water surface. Besides, PP substrate can be recycled multiple times before incineration is necessary (Manias et al. 2001; Tokiwa et al. 2009; Maddah 2016). It is a well-known fact that both Starbucks and McDonald's have switched from polyethylene terephthalate (PET) to PP cold cups due to the less plastic consumption and reduction of greenhouse gas emissions. Nevertheless, PP substrate still exhibits poor UV resistance. However, it shows high melting point of about 165 °C and much higher than that of majority group of polymer candidates. As other vinyl polymers, PP cannot be polymerized by radical polymerization due to the presence of allylic carbon. It can be produced only by coordination polymerization like Ziegler–Natta or metallocene catalyst (Shamiri et al. 2014).

As a matter of fact, a Ziegler–Natta catalyst is able to restrict the linking of monomer molecules to a specific regular orientation, either isotactic, when all methyl groups are positioned at the same side with respect to the backbone of the polymer chain, or syndiotactic, when the positions of the methyl groups alternate (Huang and Rempel 1995). In general, commercially available isotactic PP is made with two types of Ziegler–Natta catalysts. The first group of the catalysts encompasses solid (mostly supported) catalysts and certain types of soluble metallocene catalysts. Such isotactic macromolecules coil into a helical shape; these helices then line up next to one another to form the crystals that give commercial isotactic PP with many of its desirable properties (Moore 1996). Another type of metallocene catalysts produces syndiotactic PP (Kaminsky 1998). These macromolecules also coil into helices (of a different type) and form crystalline materials. When the methyl groups in a PP chain exhibit no preferred orientation, the polymers are called atactic. Atactic PP is an amorphous rubbery material. It can be produced commercially either with a special type of supported Ziegler–Natta catalyst or with some metallocene catalysts. The structure of PP includes atactic, isotactic and syndiotactic and the commercial PP products are usually mixtures of mostly isotactic and certain amount of atactic (Claverie and Schaper 2013).

Next, the thermal stability and UV resistance property of the PP polymer have been discussed in literatures. Luzuriaga et al. (2006) found that PP was well stabilized against thermo-oxidation, retaining acceptable mechanical properties and thermo-oxidative stability after 10 months of oven-ageing. Nevertheless, PP was most prone to deterioration, due to structural factors, and a lack of stabilization against UV irradiation. Lipp-Symonowicz et al. (2006) reported that PP exhibited the most degradable under the influence of UV radiation, however, PP able to maintain a high stability of their structure owing to the incorporation of UV stabilizers, such as silver pigment or black pigment. This observation could be reflected in the relatively small change in PP's mechanical properties after addition of UV stabilizers. The authors concluded that coloured PP polymer containing pigments are characterized by a stable structure, and exhibit no significant deterioration in their mechanical properties. Turton and White (2001) claimed that high photodegradation behaviour is observed in PP polymer with the light stabilizer as compared to those without addition of the light stabilizer. In this case, the light stabilizer could act as a radical scavenger. According to the Sachon et al. (2010), incorporation of TiO_2 dopants could further inhibit the photodegradation of PP polymer due to the less penetration of UV irradiation into the PP polymer. Thus, high photodegradation resistance property could be achieved using the PP polymer as the substrate of the buoyant photocatalyst.

In order to make the photocatalytic technology cost-competitive for practical applications in water or wastewater treatment, significant research interest has been conducted to utilize our solar energy as a possible 'free' light source to trigger the photocatalytic process (Han and Bai 2009; Lee et al. 2016). Among several types of photocatalytic processes, it was found that buoyant photocatalysts appeared as a promising candidate to develop an efficient visible light driven photocatalyst for water or wastewater treatment system (Han and Bai 2011; Mukherjee et al. 2014). Over the past few years, buoyant photocatalysts has gained much attention and has been intensively studied because of the unique features of float ability on water surface with strong absorption for solar energy. In fact, buoyant photocatalysts can trigger more powerful hydroxyl radicals and superoxide anions to promote the photocatalytic activity without the light attenuation loss in water medium (Fabiyi and Skelton 2000; Zhong et al. 2014).

TiO_2 has emerged as the leading candidate as buoyant photocatalyst due to its low cost, non-toxicity, self-cleaning property, ready availability and strong photocatalytic activity and high stability against photo-corrosion. In order to further improve the photocatalytic activity, a common approach to achieve this was to modify TiO_2 itself to extend its light activity to the visible light range (400–700 nm) by reducing its bandgap energy (Grimes 2007; Kitano et al. 2007; Kubacka et al. 2012). The main reason of this modification task is to overcome its poor visible light response and rapid recombination rate of charge carriers (Ni et al. 2007; Beranek et al. 2009; Leung et al. 2010). As a matter of fact, TiO_2

photocatalyst can only effectively function under the UV region ($\lambda < 400$ nm) and it only contains about 4–5% of UV rays from our solar energy. Thus, utilization of visible light from our solar energy is essential that leads to the high photocatalytic performance system in water or wastewater treatment system. In general, considerable efforts have been exerted to minimize the recombination losses of charge carriers and extended the spectral response of TiO_2 to visible spectrum by incorporating an optimum amount of cationic, anion doping or transition metal oxide elements into the lattice of TiO_2 (Navarro Yerga et al. 2009; Leung et al. 2010). As a result, bandgap narrowing effects could expand the range of excitation light to the visible region and provide sites that slow down the recombination of charge carriers. Nevertheless, most of these studies involved high calcination temperature process in their preparation method (often >300–400 °C). In order to prepare and immobilize the visible light sensitive TiO_2 photocatalyst on PP substrate, relative low processing temperature is required for the synthesis purpose due to the low melting point of PP substrate (165 °C). Thus, this limitation has restricted the use of conventional synthesis method that required high temperature for crystallization, doping modification or immobilization purpose (Han and Bai 2009).

In order to confine the low processing temperature, synthesis of buoyant TiO_2 photocatalyst on PP substrate can be achieved via three main approaches, including plasma-enhanced chemical vapour deposition (PECVD), liquid phase deposition (LPD) and hydrothermal reaction (Han and Bai 2010; Mukherjee et al. 2014). In general, PECVD method utilizes plasma to enhance chemical reaction rates of the precursors and this processing allows deposition at lower temperatures. The lower temperatures also allow for the deposition of organic coatings, such as plasma polymers, that have been used for nanoparticle surface functionalization. Nevertheless, expensive machine and complicated set-up system are required for this PECVD technique (Dutschke et al. 2003). Literatures have been reported that buoyant TiO_2 photocatalyst on PP substrate using LPD method exhibited the lower crystallinity. In general, low crystallinity of TiO_2 comprises of high concentration of defects such as impurities, dangling bonds and micro-voids, which act as recombination centre and eventually result in a poor photocatalytic performance (Regonini et al. 2010). TiO_2 was modified and immobilized on PP granules via hydrothermal synthesis to prepare buoyant photocatalyst with visible light activity have been reported. The authors introduced a simple one-step process for the simultaneous crystallization and immobilization of the treated TiO_2 nanoparticles on the PP substrate at a low temperature hydrothermal reactor (150 °C). It was found that triethylamine (TEA)-treated TiO_2 nanosol with acetyl acetone as the inhibiting agent led to high visible light activity for the prepared photocatalyst from the low temperature hydrothermal process (Han and Bai 2009). The light absorption edge was up to 800 nm, and the visible light absorption rate reached 32–66%, depending on the TEA-treatment time. Based on the XRD analysis, the XRD patterns showed all the TiO_2-based photocatalysts having the major crystal structure of anatase and minor of brookite. Thus, innovative new approaches and synthesis of a high crystallinity of buoyant TiO_2 photocatalyst on PP substrate are critical and crucial for determining the potential of the material as an efficient candidate in water or wastewater treatment system by utilizing our solar energy.

Conclusion

Photocatalysis is the most promising and viable solution for removing recalcitrant organic pollutants in water. Many efforts have been made to increase the effectiveness TiO_2 photocatalyst under sunlight irradiation which comprised of visible light. Therefore, various modification methods have been developed to increase TiO_2 photocatalyst under visible light and also to ease the recovery of photocatalyst after the treatment. In this chapter, different types of polymer-TiO_2 composite have been summarized and described. Many types of polymer-TiO_2 seem to possess good photocatalytic activity, stability, recyclability, harmless, inexpensive and ease of separation. Nevertheless, there are only proven in the lab scale and thus more is need to be done to bring the lab-scale polymer-TiO_2 to industrial scale for commercialization.

References

Abdelkader E, Nadjia L, Ahmed B (2015) Preparation and characterization of novel $CuBi_2O_4$/SnO_2 p-n heterojunction with enhanced photocatalytic performance under UVA light irradiation. J King Saud Univ Sci 27:76–91

Ali U, Karim KJBA, Buang NA (2015) A review of the properties and applications of poly (methyl methacrylate) (PMMA). Polym Rev 55:678–705

Baruah S, Pal SK, Dutta J (2012) Nanostructured zinc oxide for water treatment. Nanosci Nanotechnol Asia 2:90–102

Beranek R, Macak JM, Gartner M, Meyer K, Schmuki P (2009) Enhanced visible light photocurrent generation at surface-modified TiO_2 nanotube. Electrochim Acta 54:2640–2646

Buha J (2013) Photoluminescence study of carbon doped and hydrogen co-doped TiO_2 thin films. Thin Solid Films 545:234–240

Cantarella M, Sanz R, Buccheri MA, Ruffino F, Rappazzo G, Scalese S, Impellizzeri G, Romano L, Privitera V (2016) Immobilization of nanomaterials in PMMA composites for photocatalytic removal of dyes, phenols and bacteria from water. J Photochem Photobiol A 321:1–11

Chan YL, Pung SY, Sreekantan S (2014) Synthesis of V_2O_5 nanoflakes on PET fiber as visible-light-driven photocatalysts for degradation of RhB dye. J Catal 2014, Article ID: 370696

Chatterjee D, Dasgupta S (2005) Visible light induced photocatalytic degradation of organic pollutants. J Photochem Photobiol B 6:186–205

Chen J, Cen J, Xu X, Li X (2016) The application of heterogeneous visible light photocatalysts in organic synthesis. Catal Sci Technol 6:349–362

Chen Q, He QQ, Lv MM, Liu XT, Wang J, Lv JP (2014) The vital role of PANI for the enhanced photocatalytic activity of magnetically recyclable N-$K_2TiO_4O_9$/$MnFe_2O_4$/PANI composites. Appl Surf Sci 311:230–238

Claverie J, Schaper F (2013) Ziegler-Natta catalysis: 50 years after the Nobel Prize. MRS Bull 38:213–218

Colombo DP, Bowman RM (1996) Does interfacial charge transfer compete with charge carrier recombination? A femtosecond diffuse reflectance investigation of TiO_2 nanoparticles. J Phys Chem 100:18445–18449

Daghrir R, Drogui P, Robert D (2013) Modified TiO_2 for environmental photocatalytic applications: a review. Ind Eng Chem Res 52:3581–3599

Damodar RA, Swaminathan T (2008) Performance evaluation of a continuous flow immobilized rotating tube photocatalytic reactor (IRTPR) immobilized with TiO_2 catalyst for azo dye degradation. Chem Eng J 144:59–66

Demir MM, Castignolles P, Akbey U, Wegner G (2007) In-situ bulk polymerization of dilute particle/MMA dispersions. Macromolecules 40:4190–4198

Dona M, Garriga C, Arãna J, Pérez J, Colón G, Macías M, Navio JA (2007) The effect of dosage on the photocatalytic degradation of organic pollutants. Res Chem Intermed 33:351–358

Djokić V, Vujović J, Marinković A, Petrović R, Janaćković D, Onjia A, Mijin D (2012) A study of the photocatalytic degradation of the textile dye CI Basic Yellow 28 in water using a P160 TiO_2-based catalyst. J Serb Chem Soc 77:1747–1757

Dumitrescu I, Iordache O, Popescu A, Varzaru E, Kim S, Basim B, Ukelge G (2015) The photocatalytic effects of textile materials treated with TiO_2 and Fe/TiO_2. Ind Textila 66:297–305

Dutschke A, Diegelmann C, Lobmann P (2003) Nucleation and growth of TiO_2 thin films on modified polystyrene surfaces. 15:3501–3506

Elfeky SA, Al-Sherbini AA (2011a) Photocatalytic decomposition of Trypan Blue over nanocomposite thin films. Kinet Catal 52:391–396

Elfeky SA, Al-Sherbini AA (2011) Photo-oxidation of Rhodamine-6-G via TiO_2 and Au/TiO_2-bound polythene beads. J Nanomater 2011, Article ID: 570438

Evans RC, Douglas P, Burrow HD (2013) Applied photochemistry. Springer, Dordrecht

Fabiyi ME, Skelton RL (2000) Photocatalytic mineralisation of methylene blue using buoyant TiO_2-coated polystyrene beads. J Photochem Photobiol A 132:121–128

Feldman D (1989) Polymeric building materials. Elsevier Applied Science, London

Filippo E, Carlucci C, Capodilupo AL, Perulli P, Conciauro F, Corrente GA, Ciccarella G (2015) Facile preparation of TiO_2-polyvinyl alcohol hybrid nanoparticles with improved visible light photocatalytic activity. Appl Surf Sci 331:292–298

Finch CA (1973) Polyvinyl alcohol properties and applications. Wiley, London

Fostier AH, Pereira MSS, Rath S, Guimaraes JR (2008) Arsenic removal from water employing heterogeneous photocatalysis with TiO_2 immobilized in PET bottles. Chemosphere 72:319–324

Fulay P (2016) Electronic, magnetic, and optical materials. CRC Press, Bosa Roca

Ghezzar M, Abdelmalek F, Belhadj M, Benderdouche N, Addou A (2007) Gliding arc plasma assisted photocatalytic degradation of anthraquinonic Acid Green 25 in solution with TiO_2. Appl Catal B Environ 72:304–313

Grimes CA (2007) Synthesis and application of highly ordered arrays of TiO_2 nanotubes. J Mater Chem 17:1451–1457

Gude K, Gun'ko VM, Blitz JP (2008) Adsorption and photocatalytic decomposition of methylene blue on surface modified silica and silica-titania. Colloids Surf A 325:17–20

Hall RN (1952) Electron-hole recombination in germanium. Phys Rev 87:387

Han H, Bai R (2009) Buoyant photocatalyst with greatly enhanced visible-light activity prepared through a low temperature hydrothermal method. Ind Eng Chem Res 48:2891–2898

Han H, Bai R (2010) Highly effective buoyant photocatalyst prepared with a novel layered-TiO_2 configuration on polypropylene fabric and the degradation performance for methyl orange dye under UV-Vis and Vis lights. Sep Purif Technol 73:142–150

Han H, Bai RB (2011) The effect of thickness of photocatalyst film immobilized on a buoyant substrate on the degradation of methyl orange dye in aqueous solutions under different light irradiations. Ind Eng Chem Res 50:11922–11929

Hopewell J, Dvorak R, Kosior E (2009) Plastics recycling: Challenges and opportunities. Phil Trans R Soc B 27:2115–2126

Hosseini SN, Borghei SM, Vossoughi M, Taghavinia N (2007) Immobilization of TiO_2 on perlite granules for photocatalytic degradation of phenol. Appl Catal. B Environ 74:53–62

Huang J, Rempel GL (1995) Ziegler-Natta catalysts for olefin polymerization: mechanistic insights from metallocene systems. Prog Polym Sci 20:459–526

Hugh S, Canonica L, Wegelin M, Gechter D, Von Guten U (2001) Solar oxidation and removal of arsenic at circumneutral pH in iron containing waters. Environ Sci Technol 35:2114–2121

Ibhadon AO, Fitzpatrick P (2013) Heterogeneous photocatalysis: recent advances and applications. Catalysts 3:189–218

Jin F, Cao J, Kishida H, Moriya T, Enomoto H (2007) Impact of phenolic compounds on hydrothermal oxidation of cellulose. Carbohydr Res 342:1129–1132

Kaminsky W (1998) Highly active metallocene catalysts for olefin polymerization. J Chem Soc Dalton Trans 9:1413–1418

Kang ET, Neoh KG, Tan KL (1998) Polyaniline: a polymer with many interesting intrinsic redox states. Prog Polym Sci 23:277–324

Katoh R, Furube A (2011) Tunneling-type charge recombination in nanocrystalline TiO_2 films at low temperature. J Phys Chem Lett 2:1888–1891

Khan MM, Adil SF, Al-Mayouf A (2015) Metal oxides as photocatalysts. J Saudi Chem Soc 19:462–464

Khan MM, Ansari SA, Pradhan D, Ansari MO, Lee J, Cho MH (2014) Band gap engineered TiO_2 nanoparticles for visible light induced photoelectrochemical and photocatalytic studies. J Mater Chem A 2:637–644

Kharisov BI, Kharissova OV, Ortiz-Mendez U (2016) CRC concise encyclopedia of nanotechnology. CRC Press, Bosa Roca

Kitano M, Matsuoka M, Ueshima M, Anpo M (2007) Recent developments in titanium oxide-based photocatalysts. Appl Catal A Gen 325:1–14

Krysa J, Waldner G, Mestankova H, Jirkovsky J, Grabner G (2006) Photocatalytic degradation of model organic pollutants on an immobilized particulate TiO_2 layer: roles of adsorption processes and mechanistic complexity. Appl Catal B Environ 64:290–301

Kubacka A, Fernández-García M, Colón G (2012) Advanced nanoarchitectures for solar photocatalytic applications. Chem Rev 112:1555–1614

Kumar A, Sarmah S (2011) Photocatalytic activity of polyaniline-TiO_2 nanocomposites. Indian J Phys 85:713–726

Lee KM, Lai CW, Ngai KS, Juan JC (2016) Recent developments of zinc oxide based photocatalyst in water treatment technology: a review. Water Res 88:428–448

Lee JU, Jung JW, Jo JW, Jo WH (2012) Degradation and stability of polymer-based solar cells. J Mater Chem 22:24265–24283

Lei P, Wang F, Gao X, Ding Y, Zhang S, Zhao J, Liu S, Yang M (2012) Immobilization of TiO_2 nanoparticles in polymeric substrates by chemical bonding for multi-cycle photodegradation of organic pollutants. J Hazard Mater 227–228:185–194

Leng C, Wei J, Liu Z, Xiong R, Pan C, Shi J (2013) Facile synthesis of PANI-modified $CoFe_2O_4$-TiO_2 hierarchical flower-like nanoarchitectures with high photocatalytic activity. J Nanopart Res 15:1643

Leung DYC, Fu XL, Wang CF, Ni M, Leung MKH, Wang X, Fu X (2010) Hydrogen production over titania-based photocatalysts. Chemsuschem 2:681–694

Li X, Li F, Fan C, Sun Y (2002) Photoelectrocatalytic degradation of humic acid in aqueous solution using a Ti/TiO_2 mesh photoelectrode. Water Res 36:2215–2224

Li X, Teng W, Zhao Q, Wang L (2011) Efficient visible light-induced photoelectrocatalytic degradation of rhodamine B by polyaniline-sensitized TiO_2 nanotube arrays. J Nanopart Res 13:6813–6820

Li X, Wang D, Cheng G, Luo Q, An J, Wang Y (2008) Preparation of polyaniline-modified TiO_2 nanoparticles and their photocatalytic activity under visible light illumination. Appl Catal B 81:267–273

Lin H, Huang C, Li W, Ni C, Shah SI, Tseng Y-H (2006) Size dependency of nanocrystalline TiO_2 on its optical property and photocatalytic reactivity exemplified by 2-chlorophenol. Appl Catal B Environ 68:1–11

Lipp-Symonowicz B, Sztainowski S, Kardas I (2006) Influence of UV radiation on the mechanical properties of polyamide and polypropylene fibres in aspect of their restructuring. AUTEX Res J 6:196–203

Liu X, Chen Q, Lv L, Feng X, Meng X (2015a) Preparation of transparent PVA/TiO_2 nanocomposite films with enhanced visible-light photocatalytic activity. Catal Commun 58:30–33

Liu B, Fang Y, Li Z, Xu S (2015b) Visible-light nanostructured photocatalysts: a review. J Nanosci Nanotechnol 15:889–920

Lopez-Munoz MJ, Grieken RV, Aguado J, Marugan J (2005) Role of the support on the activity of silica-supported TiO_2 photocatalysts: structure of the TiO_2/SBA-15 photocatalysts. Catal Today 101:307–314

Luo HM, Takata T, Lee YG, Zhao JF, Domen K, Yan YS (2004) Photocatalytic activity enhancing for titanium dioxide by co-doping with bromine and chlorine. Chem Mater 16:846–849

Luzuriaga S, Kovarova J, Fortelny I (2006) Degradation of pre-aged polymers exposed to simulated recycling: properties and thermal stability 91:1226–1232

Ma CM, Lee YW, Hong GB, Shie JL, Chang CT (2011) Effect of platinum on the photocatalytic degradation of chlorinated organic compound. J Environ Sci 23:687–692

Maddah HA (2016) Polypropylene as a promising plastic: a review. Am J Polym Sci 6:1–11

Magalhaes F, Lago RM (2009) Floating photocatalysts based on TiO_2 grafted on expanded polystyrene beads for the solar degradation of dyes. Sol Energy 83:1521–1526

Magalhães F, Moura FCC, Lago RM (2011) TiO_2/LDPE composites: a new floating photocatalyst for solar degradation of organic contaminants. Desalination 276:266–271

Mahmoodi NM, Arami M (2006) Bulk phase degradation of Acid Red 14 by nanophotocatalysis using immobilized titanium(IV) oxide nanoparticles. J Photochem Photobiol A 182:60–66

Mamba G, Mishra A (2016) Graphitic carbon nitride (g-C_3N_4) nanocomposites: a new and exciting generation of visible light driven photocatalysts for environmental pollution remediation. Appl Catal B Environ 198:347–377

Manias E, Touny A, Wu L, Strawhecker K, Lu B, Chung TC (2001) Polypropylene/montmorillonite nanocomposites. Review of the synthetic routes and materials properties. Chem Mater 13:3516–3523

Mansilla HD, Bravo C, Ferreyra R, Litter MI, Jardim WF, Lizama C, Freer J, Fernandez J (2006) Photocatalytic EDTA degradation on suspended and immobilized TiO_2. J Photochem Photobiol A 181:188–194

Martha S, Das D, Biswal N, Parida K (2012) Facile synthesis of visible light responsive V2O5/N, S-TiO2 composite photocatalyst: enhanced hydrogen production and phenol degradation. J Mater Chem 22:10695–10703

Masid S, Tayade R, Rao NN (2015) Efficient visible light active polyaniline/TiO_2 nanocomposite photocatalyst for degradation of Reactive Blue 4. Int J Photocatalysis Photon 119:190–203

Matos J, Laine J, Herrmann J-M, Uzcategui D, Brito JL (2007) Influence of activated carbon upon titania on aqueous photocatalytic consecutive runs of phenol photodegradation. Appl Catal B Environ 70:461–469

Mehmood CT, Qazi IA, Baig MA, Arshad M, Quddos A (2015) Application of photodegraded polythene films for the treatment of Drimarene Brilliant Red (DBR) dye. Int Biodeterior Biodegradation 102:31–39

Meichtry JM, Lin HJ, de la Fuente L, Levy IK, Gautier EA, Blesa MA, Litter M (2007) Low-cost TiO_2 photocatalytic technology for water potabilization in plastic bottles for isolated regions. Photocatalyst fixation. J Sol Energy Eng 129:119–126

Mendive CB, Hansmann D, Bredow T, Bahnemann D (2011) New insights into the mechanism of TiO_2 photocatalysis: thermal processes beyond the electron-hole creation. J Phys Chem C 115:19676–19685

Moore EP (1996) Polypropylene Handbook. Polymerization, Characterization, Properties, Processing, Applications. Hanser Publishers, New York

Mukherjee D, Barghi S, Ray AK (2014) Preparation and characterization of the TiO_2 immobilized polymeric photocatalyst for degradation of aspirin under UV and solar light. Processes 2:12–23
Murugan E, Rangasamy R (2011) Development of stable pollution free TiO_2/Au nanoparticle immobilized green photo catalyst for degradation of methyl orange. J Biomed Nanotechnol 7:225–228
Navarro Yerga RM, Alvarez Galvan MC, Del Valle F, Villoria de la Mano JA, Fierro JLG (2009) Water splitting on semiconductor catalysts under visible-light irradiation. Chemsuschem 2:471–485
Ni M, Leung KH, Leung DYC, Sumathy K (2007) A review and recent development in photocatalytic water-splitting using TiO_2 for hydrogen production. Renew Sustain Energy Rev 11:401–425
Nursam NM, Wang X, Caruso RA (2015) High-throughput synthesis and screening of titania-based photocatalysts. ACS Comb Sci 17:548–569
Ochieng A, Maurice SO, Thabo B (2017) Photocatalytic degradation of aromatic compounds in abattoir wastewater. Int J Environ Sustain 13:17–30
Ohama Y, Van Gemert D (2011) Application of titanium dioxide photocatalysis to construction materials: State-of-the-art report of the RILEM Technical Committee 194-TDP, vol 5. Springer Science & Business Media, Dordrecht
Ohkura Y (2013) Synthesis and optical ignition of aluminum and silicon-based. Stanford University. Retrieved from http://purl.stanford.edu/yw312xw5478
Ohtani B (2013) Titania photocatalysis beyond recombination: a critical review. Catalysts 3:942–953
Omar AS (1989) Polyethylene-coated urea. 1. Improved storage and handling properties. Ind Eng Chem Res 28:630–632
Panda S (2009) Microelectronics and optoelectronics technology. Laxmi Publications, New Delhi
Peerakiatkhajohn P, Onreabroy W, Chawengkijwanich C, Chiarakorn S (2011) Preparation of visible-light-responsive TiO_2 doped Ag thin film on PET plastic for BTEX treatment. J Sustain Energy Environ 2:121–125
Portjanskaja E, Krichevskaya M, Preis S, Kallas J (2004) Photocatalytic oxidation of humic substances with TiO_2-coated glass micro-spheres. Environ Chem Lett 2:123–127
Portjanskaja E, Preis S, Kallas J (2006) Aqueous photocatalytic oxidation of lignin and humic acids with supported TiO_2. Int J Photoenergy 2006, Article ID: 85927
Qamar M, Drmosh Q, Ahmed MI, Qamaruddin M, Yamani ZH (2015) Enhanced photoelectro-chemical and photocatalytic activity of WO_3-surface modified TiO_2 thin film. Nanoscale Res Lett 10:1–6
Razak S, Nawi MA, Haitham K (2014) Fabrication, characterization and application of a reusable immobilized TiO_2-PANI photocatalyst plate for the removal of reactive red 4 dye. Appl Surf Sci 319:90–98
Regonini D, Jaroenworaluck A, Stevens R, Bowen CR (2010) Effect of heat treatment on the properties and structure of TiO_2 nanotubes: phase composition and chemical composition. Surf Interface Anal 42:139–144
Rengaraj S, Li XZ (2006) Enhanced photocatalytic activity of TiO_2 by doping with Ag for degradation of 2,4,6-trichlorophenol in aqueous suspension. J Mol Catal A Chem 243:60–67
Riaz U, Ashraf S, Kashyap J (2015) Role of conducting polymers in enhancing TiO_2-based photocatalytic dye degradation: a short review. Polym Plast Technol Eng 54:1850–1870
Ryu CS, Kim MS, Kim BW (2003) Photodegradation of alachlor with the TiO_2 film immobilised on the glass tube in aqueous solution. Chemosphere 53:765–771
Ryu J, Choi W (2004) Effects of TiO_2 surface modifications on photocatalytic oxidation of arsenite: the role of superoxides. Environ Sci Technol 38:2928–2933
Sachon E, Matheron L, Clodic G, Blasco T, Bolbach G (2010) MALDI TOF-TOF characterization of a light stabilizer polymer contaminant from polypropylene or polyethylene plastic test tubes. J Mass Spectrom 45:43–50
Sahoo DP, Rath D, Nanda B, Parida K (2015) Transition metal/metal oxide modified MCM-41 for pollutant degradation and hydrogen energy production: a review. RSC Adv 5:83707–83724

Sakar M, Balakumar S, Saravanan P, Bharathkumar S (2016) Particulates vs. fibers: dimension featured magnetic and visible light driven photocatalytic properties of Sc modified multiferroic bismuth ferrite nanostructures. Nanoscale 8:1147–1160
Salem MA, Al-Ghonemiy AF, Zaki AB (2009) Photocatalytic degradation of allura red and quinoline yellow with polyaniline/TiO_2 nanocomposite. Appl Catal B Environ 91:59–66
Saxena SK (2004) Polyvinyl alcohol (PVA). Chem Tech Assess 2004. In: 61st Joint expert committee of food additives, pp 3–5
Schneider J, Matsuoka M, Takeuchi M, Zhang J, Horiuchi Y, Anpo M, Bahnemann DW (2014) Understanding TiO_2 photocatalysis: mechanisms and materials. Chem Rev 114:9919–9986
Sellappan R (2013) Mechanisms of enhanced activity of model TiO_2/carbon and TiO_2/metal nanocomposite photocatalysts. Chalmers University of Technology, Göteborg
Serpone N, Pelizzetti E (1989) Photocatalysis: fundamentals and applications. Wiley, New York
Shahabuddin S, Sarih NM, Ismail FH, Shahid MM, Huang NM (2015) Synthesis of chitosan grafted-polyaniline/Co_3O_4nanocubenanocomposites and their photocatalytic activity toward methylene blue dye degradation. RSC Adv 5:83857–83867
Shahabuddin S, Muhamad Sarih N, Mohamad S, Juan JC (2016) $SrTiO_3$ nanocube-doped polyaniline nanocomposites with enhanced photocatalytic degradation of methylene blue under visible light. Polymers 8:27
Shamiri A, Chakrabarti MH, Jahan S, Hussain MA, Kaminsky W, Aravind PV, Yehye WA (2014) The influence of ziegler-natta and metallocene catalysts on polyolefin structure, properties, and processing ability. Materials 7:5069–5108
Shan AY, Ghazi TIM, Rashid SA (2010) Immobilisation of titanium dioxide onto supporting materials in heterogeneous photocatalysis: a review. Appl Catal A Gen 389:1–8
Shironita S, Mori K, Shimizu T, Ohmichi T, Mimura N, Yamashita H (2008) Preparation of nano-sized platinum metal catalyst using photo-assisted deposition method on mesoporous silica including single-site photocatalyst. Appl Surf Sci 254:7604–7607
Shockley W, Read W Jr (1952) Statistics of the recombinations of holes and electrons. Phys Rev 87:835
Singh S, Singh PK, Mahalingam H (2015) A novel and effective strewn polymer-supported titanium dioxide photocatalyst for environmental remediation. J Mater Environ Sci 6:349–358
Skinner DE, Colombo DP Jr, Cavaleri JJ, Bowman RM (1995) Femtosecond investigation of electron trapping in semiconductor nanoclusters. J Phys Chem 99:7853–7856
Song Y, Zhang J, Yang H, Xu S, Jiang L, Dan Y (2014) Preparation and visible light-induced photo-catalytic activity of H-PVA/TiO_2 composite loaded on glass via sol-gel method. Appl Surf Sci 292:978–985
Sriwong C, Wongnawa S, Patarapaiboolchai O (2008) Photocatalytic activity of rubber sheet impregnated with TiO_2 particles and its recyclability. Catal Commun 9:213–218
Stewart BD, Andrews LG, Pelletier BS, Daly CA, Boyd JE (2015) Porous PMMA-titania composites: a step towards more sustainable photocatalysis. J Water Process Eng 8:179–185
Subramanian E, Subbulakshmi S, Murugan C (2014) Inter-relationship between nanostructures of conducting polyaniline and the photocatalytic methylene blue dye degradation efficiencies of its hybrid composites with anatase TiO_2. Mater Res Bull 51:128–135
Tennakone K, Kottegoda IRM (1996) Photocatalytic mineralization of paraquat dissolved in water by TiO_2 supported on polythene and polypropylene films. J Photochem Photobiol A 93:79–81
Tennakone K, Tilakaratne CTK, Kottegoda IRM (1995) Photocatalytic degradation of organic contaminants in water with TiO_2 supported on polythene films. J Photochem Photobiol A 87:177–179
Tokiwa Y, Calabia BP, Ugwu CU, Aiba S (2009) Biodegradability of plastics. Int J Mol Sci 10:3722–3742
Thong CC, Teo DCL, Ng CK (2016) Application of polyvinyl alcohol (PVA) in cement-based composite materials: a review of its engineering properties and microstructure behavior. Constr Build Mater 107:172–180
Turton TJ, White JR (2001) Effect of stabilizer and pigment on photo-degradation depth profiles in polypropylene. Polym Degrad Stabil 74:559–568

Ullattil SG, Periyat P, Naufal B, Lazar MA (2016) Self-doped ZnO microrods: high temperature stable oxygen deficient platforms for solar photocatalysis. Ind Eng Chem Res 55:6413–6421

Vaez M, Moghaddam AZ, Mahmoodi NM, Alijani S (2012) Decolorization and degradation of acid dye with immobilized titania nanoparticles. Process Saf Environ Prot 90:56–64

Velásquez J, Valencia S, Rios L, Restrepo G, Marín J (2012) Characterization and photocatalytic evaluation of polypropylene and polyethylene pellets coated with P25 TiO_2 using the controlled-temperature embedding method. Chem Eng J 203:398–405

Vinu R, Madras G (2012) Environmental remediation by photocatalysis. J Indian Inst Sci 90:189–230

Wang F, Min S, Han Y, Feng L (2010) Visible-light-induced photocatalytic degradation of methylene blue with polyaniline-sensitized TiO_2 composite photocatalysts. Superlattices Microstruct 48:170–180

Wang W, Huang G, Jimmy CY, Wong PK (2015) Advances in photocatalytic disinfection of bacteria: development of photocatalysts and mechanisms. J Environ Sci 34:232–247

Wang YZ, Zhong MQ, Chen F, Yang JT (2009) Visible light photocatalytic activity of TiO_2/D-PVA for MO degradation. Appl Catal B Environ 90:249–254

Wegelin M, Hug S, Boller M, Gechter D, Vermeal S (2000) Back to the household—also in water treatment. EAWAG News, pp 11–12

Yang H, Zhang J, Song Y, Xu S, Jiang L, Dan Y (2015) Visible light photo-catalytic activity of C-PVA/TiO_2 composites for degrading rhodamine B. Appl Surf Sci 324:645–651

Yang JH, Han YS, Choy JH (2006). TiO_2 thin-films on polymer substrates and their photocatalytic activity. Thin Solid Films 495:266–271

You H, Zhao Y (2016) Synthesis, characterization and visible photocatalytic performance of iron (III) tetracarboxyphthalocyanine-sensitized TiO_2 photocatalyst. J Phys Chem Biophys 5:199

Yu C, Wu R, Fu Y, Dong X, Ma H (2012) Preparation of polyaniline supported TiO_2 photocatalyst and its photocatalytic property. Adv Mater Res 356–360:524–528

Zainal Z, Lee KH, Hussein MZ, Taufiq-Yap YH, Abdullah AH, Ramli I (2005) Removal of dyes using immobilized titanium dioxide illuminated by fluorescent lamps. J Hazard Mater B 125:113–120

Zaleska A, Hupka J, Wiergowski M, Biziuk M (2000) Photocatalytic degradation of lindane, p, p'-DDT and methoxychlor in an aqueous environment. J Photochem Photobiol A Chem 135:213–220

Zhang J, Bi H, He G, Zhou Y (2014a) H. Chen., Fabrication of Ag_3PO_4-PANI-GO composites with high visible light photocatalytic performance and stability. J Environ Chem Eng 2:952–957

Zhang J, Song Y, Yang H, Xu S, Jiang L, Dan Y (2013) TiO_2/T-PVA composites immobilized on cordierite: Structure and photocatalytic activity for degrading RhB under visible light. Water Air Soil Pollut 224:1555–1565

Zhang H, Zong R, Zhao J, Zhu Y (2008) Dramatic visible photocatalytic degradation performances due to synergetic effect of TiO_2 with PANI. Environ Sci Technol 42:3803–3807

Zhang Z, Wang W, Gao E (2014b) Polypyrrole/Bi_2WO_6 composite with high charge separation efficiency and enhanced photocatalytic activity. J Mater Sci 49:7325–7332

Zhang Z, Yates JT Jr (2012) Band bending in semiconductors: chemical and physical consequences at surfaces and interfaces. Chem Rev 112:5520–5551

Zhiyong Y, Keppner H, Laub D, Mielczarski E, Mielczarski J, Kiwi-Minsker L, Renken A, Kiwi J (2008a) Photocatalytic discoloration of methyl orange on innovative parylene-TiO_2 flexible thin films under simulated sunlight. Appl Catal B Environ 79:63–71

Zhiyong Y, Laub D, Bensimon M, Kiwi J (2008b) Flexible polymer TiO_2 modified film photocatalysts active in the photodegradation of azo-dyes in solution. Inorg Chim Acta 361:589–594

Zhong W, Yu Y, Du C, Li W, Wang YZ, He G, Xie Y, He Q (2014) Characterization and high pollutant removal ability of buoyant (C, N)-TiO_2/PTFE flakes prepared by high-energy ball-milling. RSC Adv 4:40019–40028

Zhu J, Deng Z, Chen F, Zhang J, Chen H, Anpo M, Huang J, Zhang L (2006) Hydrothermal doping method for preparation of Cr^{3+}-TiO_2 photocatalysts with concentration gradient distribution of Cr^{3+}. Appl Catal B Environ 62:329–335

[illegible]

Chapter 8
Carbon-Based Nanocomposites for Visible Light-Induced Photocatalysis

Elaheh Kowsari

Abstract This paper gives a brief overview of the progress in the development of carbon-based nanocomposites for visible light-induced photocatalysis, based on graphene, graphene oxide, g-C_3N_4, [60]-fullerenes, and carbon nanotubes nanocomposites. In particular, recent progresses on the emerging strategies for tailoring carbon-based nanocomposites photocatalysts to enhance their photoactivity including elemental doping, heterostructure design and functional architecture assembly are discussed. The reported examples are collected and analyzed; and the reaction mechanism, the influence of various factors on the photocatalytic performance, the challenges involved, and the outlooks of carbon-based nanocomposites as photocatalyst are discussed in detail. Finally, some important applications such as photocatalytic degradation of pollutants, photocatalytic H_2 production, and photocatalytic CO_2 reduction are reviewed.

Keywords Photocatalysis · Graphene · Graphene oxide · Fullerenes · Carbon nanotubes · Nanocomposites

Introduction

In recent years, photocatalytic nanomaterials had been a hot research topic in the field of wastewater treatment, air purification, solar cell, antimicrobial, etc., and have wide application prospects (Wang et al. 2016a, b, c; Deng et al. 2015). Photocatalysis, in which the inexhaustibly abundant and clean solar energy can be harnessed as viable technologies, offers a promising avenue with the tenet of sustainable chemistry toward solar energy conversion. The massive research interest was ignited by the seminal report on photoelectrochemical water splitting to produce H_2 over TiO_2 electrode in 1972 (Fujishima and Honda 1972). Since then, increasing research attention has been given to the development of novel efficient

E. Kowsari (✉)
Department of Chemistry, Amirkabir University of Technology, Tehran, Iran
e-mail: kowsarie@aut.ac.ir

M.M. Khan et al. (eds.), *Nanocomposites for Visible Light-induced Photocatalysis*,
Springer Series on Polymer and Composite Materials,
DOI 10.1007/978-3-319-62446-4_8

photocatalysts and the exploration of different approaches to enhance the performance of semiconductor-based artificial photocatalytic redox processes (Zhang et al. 2015). Typically, these engineering strategies to modify heterogeneous photocatalysts with improved activity can be divided into two types: structural and compositional optimizations. Given that the photocatalytic properties of semiconductors are strongly dependent on their crystal morphologies and structural features at the nanometer level, optimization and control of the structural properties of a given semiconductor is crucial for improving its photocatalytic performance (Zhang et al. 2015).

The development of effective semiconductor photocatalysts has therefore emerged into one of the most important goals in materials science. A similarly dramatic rise in interest has occurred since the mid-1990s with regards to carbonaceous nanomaterials due to their unique properties, and the potential to control these properties via structural and compositional modification. In the past decade, these two fields of interest have merged, with significant attention now being paid to exploring the role that carbonaceous nanomaterials may play in photocatalytic processes (Leary and Westwood 2011).

Photoinduced energy and electron transfer processes involving carbon-based nanocomposite based on fullerenes, carbon nanotubes, and graphenes have been attracting attention in the development of solar-powered systems for environmental remediation and renewable energy generation (Leary and Westwood 2011; Park et al. 2010; Tajima et al. 2011; Zhang et al. 2010; Choi et al. 2010). Graphene, a single layer two-dimensional graphite structure with high specific surface area, shows a number of unique electrical properties (Li et al. 2011). Especially, the high electrical conductivity enables a high speed of transmission of electrons and favors electron transfer from the conduction band of TiO_2 due to its less negative redox potential than the conduction band edge of TiO_2 (Dong et al. 2013; Zhao et al. 2016). As an analog of graphene, graphitic carbon nitrides (g-C_3N_4) nanosheets are developing quickly owing to their good chemical and electronic properties (Du et al. 2012; Wang et al. 2009; Xing et al. 2014; Liang et al. 2015; Shi et al. 2015; Fan et al. 2015; Hong et al. 2014; Yang et al. 2013; Liu et al. 2016a, b, c; Dong et al. 2013). Bulk g-C_3N_4 has a stacked 2D structure and appropriate band gap ($\sim$2.7 eV) for visible light absorption. g-C_3N_4 nanosheets can be obtained by delaminating bulk five-layered g-C_3N_4, which is normally prepared by pyrolysis of nitrogen-rich precursors through bulk reaction or polycondensation. Graphitic carbon nitride (GCN) molecular skeleton is based on tri-s-triazine (C_6N_7, also referred to as melom) building blocks, which presents a layered structure as polycyclic aromatic hydrocarbons (Thomas et al. 2008), and the electrons can transport between the dislocated π–π^* and π–π^* orbitals (Li et al. 2011; Gracia et al. 2009). GCN also has an intrinsic band gap energy at 2.7 eV, which can be activated by visible light to produce electron/hole pairs. Owing to the high reduction ability originated from the high conduction band, GCN has shown effective photoreduction for H_2 production and CO_2 reduction (Wang et al. 2009; Martin et al. 2014; Yan 2012; Xing et al. 2011; Kuriki et al. 2016; Zheng et al. 2014; Niu et al. 2014; Maeda 2014; Liu et al. 2016a, b, c). C_{60} fullerene is a carbonaceous nanocomposites that is photochemically activated under visible light irradiation to produce singlet oxygen (O_2) with high quantum efficiency (Choi et al. 2010; Arbogast et al. 1991; Vileno et al. 2004; Yamakoshi et al. 2003). Carbon nanotubes exhibit the potential to contribute to all three routes of increasing photocatalytic activity already outlined (i.e. high-surface area and high quality active sites, retardation of electron–hole recombination and visible light catalysis by modification of band gap and/or sensitization) (Woan et al. 2009; Tryba 2008; Leary and Westwood 2011).

Carbon nanotubes can be divided essentially into two categories: SWNT and MWNT. Ideally, single-wall carbon nanotube are made of a perfect graphene sheet, i.e. a polyaromatic monoatomic layer made of an hexagonal display of sp^2 hybridized carbon atoms that genuine graphite is built with, rolled into a cylinder and closed by two caps (semi-fullerenes) (Serp et al. 2003). There has been an increasing number of studies over the past few years seeking to develop CNT—photocatalysts mixtures or nanocomposites with improved photocatalytic activity (Yao et al. 2014; Wang et al. 2005; Leary and Westwood 2011; Dai et al. 2009; Xia et al. 2007; Xu et al. 2008; Zhang et al. 2009; Kang et al. 2007). Carbon quantum dots (CQDs), a new class of carbon nanomaterials with sizes below 10 nm, are typically quasi-spherical nanoparticles consisting of amorphous to nanocrystalline cores with predominantly graphitic carbon (sp^2 carbon) or graphene and graphene oxide sheets fused by diamond-like sp^3 hybridized carbon insertions (Di et al. 2015a, b; Lim et al. 2015). Very recently, CQDs have been introduced to photocatalytic applications owing to the superior electron transfer ability. Numerous photocatalytic systems based on the CQDs have been studied and the enhanced activity was obtained (Di et al. 2015a, b), such as CQDs/TiO_2, (Tian et al. 2015; Li et al. 2011; Yu et al. 2014) CQDs/Ag_3PO_4, (Zhang et al. 2012) CQDs/C_3N_4, (Liu et al. 2015) CQDs/Cu_2O, and CQDs/Bi_2WO_6 (Di et al. 2015a, b). This study surveys the literature and highlights recent progress in the development of nanocomposites photocatalysts, covering, carbon nanotubes, [60]-fullerenes, graphene, graphene oxide, g-C_3N_4, carbon nanotubes, and more recently developed morphologies. The reported examples are collected and analyzed; and the reaction mechanism, the influence of various factors on the photocatalytic performance, the challenges involved, and the outlooks of carbon-based nanocomposites as visible light-induced photocatalysis are discussed.

Graphene-Based Semiconductor Nanocomposites as a Visible Light Active Photocatalyst

Even today, the precise electronic band structure of GO has not been clearly elucidated owing to the nanoscale inhomogeneities of the structure (Putri et al. 2016). The functionalization of graphene with oxygen groups involves a complex interplay of phenomena which influences graphene's collective electrical, optical, and chemical attributes. Essentially, the covalent addition of oxygen functionalities on the basal plane of graphene converts the original, unsaturated planar sp^2—of carbon atoms into a tetrahedral sp^3 electronic hybridization (Johns and Hersam 2013). These in turn impose a structural disorder on the lattice, which acts as transport barriers, since they interrupt the continuity of π-network which ordinarily allows classical carrier transport to occur. This removal of π electrons plays a mechanistic role in the change of electric properties by introducing a band gap via the symphonious removal of electronic states and emergence of new energy states from functionalizing oxygen groups. In that context, GO is viewed as an electronically hybrid material containing both conducting π-states from graphitic sp^2-hybridized carbons and a large energy gap between the σ-states from its sp^3-hybridized carbon (Lu et al. 2013). As a result, the variation of these sp^2 and sp^3 carbon fractions in GO by controlling oxidation or reduction parameters can be useful for manipulating band gap energy in GO, and thereby allowing the tunability of its electrical properties to especially cater for photocatalytic application (Matsumoto et al. 2016).

Several models of GO nanosheets have been reported. They consist of two main regions (Matsumoto et al. 2016; Bagri et al. 2010; Kudin et al. 2008; Mkhoyan et al. 2009; Lee and Cho 2009) hydrophobic π-conjugated sp^2 domains and sp^3 domains with hydrophilic oxygen-containing functional groups. In GO nanosheets, the sp^2 domains have high conductivity and are islands surrounded by an insulating matrix of the sp^3 domains. The band gap of sp^2 domains depends on the domain size, (Loh et al. 2010; Eda et al. 2010; Lee and Cho 2009; Kudin et al. 2008; Singh and Yakobson 2009), whereas graphene nanosheets have zero band gap. Thus, the sp^2 domains act as semiconductors and exhibit a photoresponse or photoreactivity when irradiated with light of energy exceeding their band gap. In other words, the sp^2 domains act as a photocatalyst under light (e.g., in the UV region); by photoexcitation, an electron is excited into the π^* conduction band and a hole is created in the π valence band. The electron and hole, respectively, contribute to reduction and oxidation of the GO nanosheets and water molecules (Matsumoto et al. 2016). Figure 8.1 shows a model of defect carbons in relatively small holes with a zigzag edge produced by photoreduction (CH bonds are not shown in this figure for clarity). Consequently, the ferromagnetic properties of the present rGO arise from the many holes formed and therefore the many zigzag edges. No O_2 evolution occurred in the photoreaction (Matsumoto et al. 2016).

The enhancement in photocatalytic activity for Graphene-based photocatalytic materials goes beyond the improved electron transfer provided by the presence of graphene sheets. Another important role of graphene in the GR composite is the electron acceptor and transporter. On the one hand, graphene has been reported to be a competitive candidate for the acceptor material due to its two-dimensional conjugation structure, (Liu et al. 2010) and in the TiO_2–graphene system, the excited electrons of TiO_2 could transfer from the conduction band to graphene via a percolation mechanism (Wang et al. 2009) Thus, in GR, graphene served as an acceptor of the generated electrons of GR and effectively suppressed the charge recombination, leaving more charge carriers to form reactive species and promote the degradation of dyes (Zhang et al. 2010), as shown in Fig. 8.2.

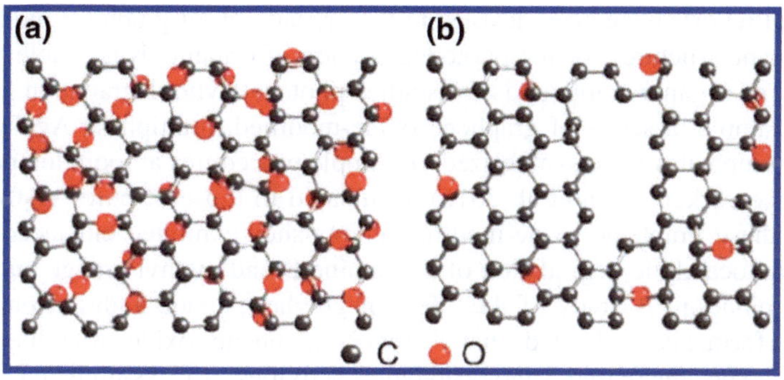

Fig. 8.1 Structural models of nanosheet samples (**a**) before (virgin sample) and (**b**) after the photoreaction. Various oxygen-containing functional groups (especially epoxy C–O–C and hydroxyl COH) were reduced to produce holes (defects) with zigzag edges. Reproduced from Matsumoto et al. (2011), Copyright (2016), with permeation from ACS

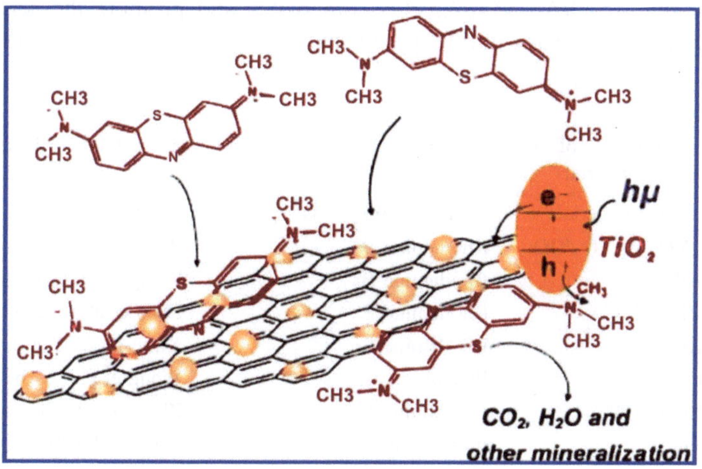

Fig. 8.2 Schematic structure of P25-GR and tentative processes of the photodegradation of methylene blue (MB) over P25-GR. P25 nanoparticles are dispersed on the graphene support, and the carbon platform plays important roles during the photodegradation of MB in three aspects: **i** Increase catalyst adsorptivity. MB molecules could transfer from the solution to the catalysts' surface and be adsorbed with offset face-to-face orientation via π–π conjugation between MB and aromatic regions of the graphene, and therefore, the adsorptivity of dyes increases compared to bare P25. **ii** Extend light absorption. The chemical bonds of Ti–O–C and good transparency of graphene render a red-shift in the photoresponding range and facilitate a more efficient utilization of light for the catalyst. **iii** Suppress charge recombination. Graphene could act as an acceptor of the photogenerated electrons by P25 and ensure fast charge transportation in view of its high conductivity, and therefore, an effective charge separation can be achieved. Reproduced from Zhang et al. (2010), Copyright (2010), with permeation from ACS

Several reports have been devoted to the preparation of graphene modified with ceramic nanostructures for the photodegradation of organic dyes. Table 8.1 shows summary of organic compounds for studies photocatalytic degradation.

For example, a series of graphene oxide-modified, multiphase $Ag_2O/Ag_3VO_4/AgVO_3$ composites were synthesized via simple procedures at room temperature by Ran and coworkers (Ran et al. 2016). Compared to those of pure $Ag_2O/Ag_3VO_4/AgVO_3$, these graphene oxide-modified composites exhibited enhanced activities during photocatalytic degradation of rhodamine B and methyl orange under visible light illumination. A study of the effect of graphene oxide addition on photocatalytic performance indicated that 1.2 wt% graphene oxide was the optimum quantity. The increased photocatalytic activities of as-prepared graphene oxide-modified composites may be attributed to the large surface area possessed by graphene oxide as well as its interactions with other species in the multiphase $Ag_2O/Ag_3VO_4/AgVO_3$ composites during photocatalytic reactions under visible light illumination. From the enhancement in photocatalytic activity, it may be inferred that graphene oxide could improve the adsorption and absorption capabilities of the $Ag_2O/Ag_3VO_4/AgVO_3$ composites and promote the separation of electron-hole (e–h) pairs during photocatalytic reactions compared to those of

Table 8.1 Summary of organic compounds for studies photocatalytic degradation by Graphene-based materials

Graphene-based materials	Organic compounds	References
$Ag_2O/Ag_3VO_4/AgVO_3$/GO	Rhodamine B	Ran et al. (2016)
TiO_2-CdS/RGO	Rhodamine B	Wang et al. (2016a, b, c)
g-C_3N_4 NS/RGO/CA	Rhodamine B	Zhao et al. (2016)
Bi_2WO_6/RGO	Rhodamine B	Xu et al. (2016a, b)
FePc/RGO	Phenol	Wang et al. (2016a, b, c)
Ag_3PO_4/NG/P3HT	Rhodamine B	Zhang et al. (2016a, b, c, d)
BiOBr-GO	direct green	Patil et al. (2016)
Pd-$NiFe_2O_4$/RGO	Rhodamine B	Li et al. (2016a, b, c, d, e, f)
Bi_2S_3 NPs/RGO	Rhodamine B	Chen et al. (2016)
Pd-$NiFe_2O_4$/RGO	Rhodamine B	Li et al. (2016a, b, c, d, e, f)
RGO/Bi_2WO_6	Ciprofloxacin hydrochloride	Li et al. (2016a, b, c, d, e, f)
AgAgX/RGO	*Escherichia coli*	Xia et al. (2016a, b)
GQD/$AgVO_3$	Ibuprofen	Lei et al. (2016)
α-Fe_2O_3/RGO	Rhodamine 6G	Zhang et al. (2016a, b, c, d)
CuTCPP/RGO-TNT	Methylene Blue	Wei et al. (2016)
Ag/$BiOBr_{0.2}I_{0.8}$/RGO	Methylene Blue	Liu et al. (2016a, b, c)
SnO_2/RGO	Methylene Blue	Wei et al. (2016)
ZnO–RGO–TiO_2	Rhodamine B	Nuengmatcha et al. (2016)
ZnO:Cu:RGO	Methylene Blue	Ravichandran et al. (2016)
rGO/C-MoO3	Methylene Blue	Ghaffar et al. (2016)
$Cd_{0.5}Zn_{0.5}S$/rGO	Methylene Blue	Huang et al. (2016)

composites without graphene oxide addition. Moreover, graphene oxide as a modifier was able to partially protect silver species composites from photocorrosion. A possible mechanism was proposed for the photocatalytic degradation of organic dyes on the surface of graphene oxide-modified $Ag_2O/Ag_3VO_4/AgVO_3$.

Reaction mechanism of the photocatalytic degradation of organic dyes RhB and MO) on 1.2 wt% graphene oxide–$Ag_2O/Ag_3VO_4/AgVO_3$ composite (white and hexagonal sheets represented graphene oxide in the diagram) is shown in Fig. 8.3.

Nanocomposites of titanium dioxide (TiO_2) and cadmium sulfide (CdS) were modified by reduced graphene oxide via solvothermal method to synthesize TiO_2–CdS/reduced graphene oxide for the photocatalytic degradation of organic pollutants by Wang and coworkers (Wang et al. 2016a, b, c). Analysis indicated that TiO_2–CdS/reduced graphene oxide had small particle size ($\sim$10 nm), regular crystal structure, and the surface area (109.7 m^2/g) was much enhanced comparing with that of TiO_2–CdS (67 m^2/g) and CdS (53 m^2/g). The photocatalytic efficiency of TiO_2–CdS/reduced graphene oxide has been largely improved, and the degradation ratio of Methylene blue (MB) and Rhodamine B (RhB) reached 97.5 and 93.5% just in 20-min light irradiation, respectively, under suitable conditions. The synthetic process of TiO_2–CdS/reduced graphene oxide composites was shown briefly in Fig. 8.4.

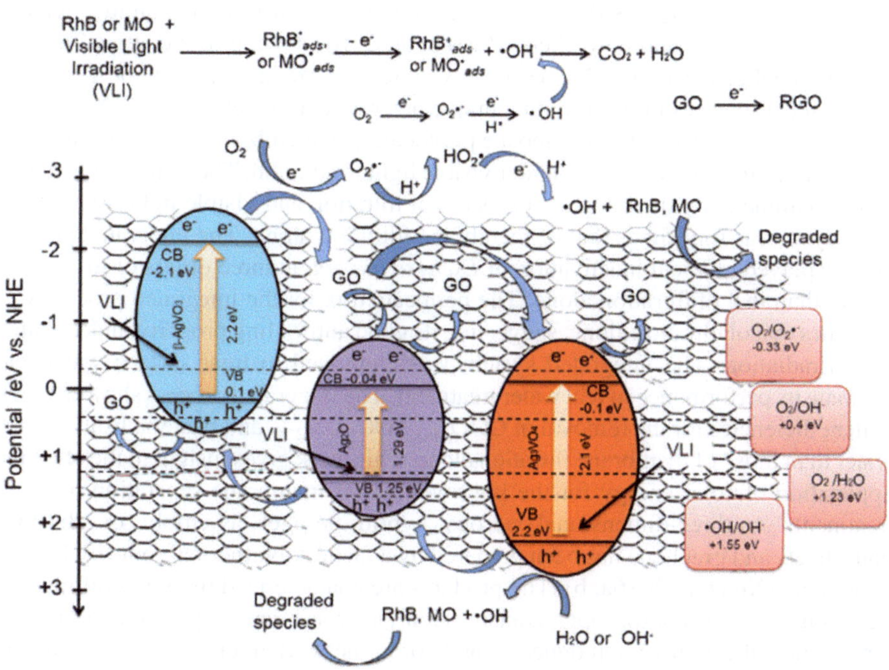

Fig. 8.3 Reaction mechanism of the photocatalytic degradation of organic dyes (RhB and MO) on 1.2 wt% graphene oxide -$Ag_2O/Ag_3VO_4/AgVO_3$ composite (white and hexagonal sheets represented graphene oxide in the diagram). Reproduced from Ran et al. (2016), Copyright (2016), with permeation from Elsevier

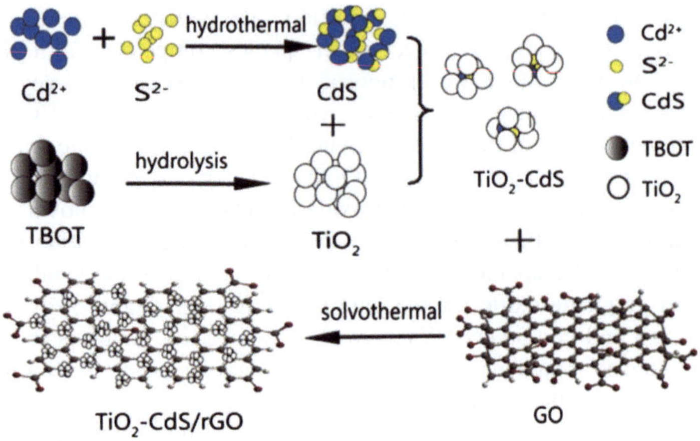

Fig. 8.4 The synthetic process of TiO2-CdS/rGO composites. Reproduced from Wang et al. (2016b), Copyright (2016), with permeation from Elsevier

A graphitic carbon nitride nanosheet/reduced graphene oxide/cellulose acetate composite photocatalytic membrane (g-C_3N_4 NS/RGO/CA) was fabricated by assembling a g-C_3N_4 NS/RGO photocatalyst on the surface of commercial CA membrane by Zhao and coworkers (Zhao et al. 2016). Owing to the attractive photocatalytic efficiency of g-C_3N_4 NS under visible light irradiation and photogenerated charge separation resulting from the unique heterostructure between g-C_3N_4 NS and RGO, g-C_3N_4 NS/RGO/CA composite photocatalytic membranes exhibited superior performance in water treatment under visible light irradiation. The removal efficiency of Rhodamine B by the integrated process of filtration and visible light driven photocatalysis was four times that of membrane filtration alone. The integrated process also displayed efficient inactivation of *Escherichia coli* at three orders of magnitude higher than that of filtration alone. The permeate flux for the integrated process was 3.7 times that of filtration alone, suggesting its good antifouling property under visible light irradiation. The integrated system was employed to treat surface water and evaluate its performance in real water treatment. The integrated process showed much better efficiencies for the removal of COD_{Mn}, TOC, UV_{254}, and bacteria from surface water than those of membrane filtration alone. This work gives insight to the effective application of solar energy for the improvement of membrane separation in water treatment. A three-dimensional porous hybrid aerogel assembly of Bi_2WO_6 nanosheets and graphene has been prepared by a facile solvothermal route by Xu and coworkers (Xu et al. 2016a, b). The products are characterized by X-ray diffraction, Scanning electron microscope, Transmission electron microscopy, and so on. The results show the highly interconnected and porous network microstructure stacked by graphene and Bi_2WO_6 nanosheets in the hybrid aerogel. Figure 8.5 was shown TEM morphologies of Bi_2WO_6. The specific surface area of hybrid aerogel (39.4 $m^2\ g^{-1}$) is 1.6 times higher than that of Bi_2WO_6 nanosheets. The Rhodamine B removal ratio over the composite aerogel reaches up to 99.6% within 45 min, which is higher than that of pure Bi_2WO_6 nansosheets (80%). The excellent photocatalytic performance is mainly owing to the porous structure of aerogel and high electrical conductivity of graphene. The study on the photocatalytic mechanism demonstrates that $O_2^{\bullet-}$ is the main reactive species for Rhodamine B degradation. Figure 8.6 shows schematic of the mechanism for the possible charge separation process by BWGA composites.

Iron(II) phthalocyanine (FePc) is immobilized on graphene sheets by Wang and coworkers (Wang et al. 2016a, b, c) to form a graphene(G)/iron(II) phthalocyanine hybrid (G/FePc) by the π–π stacking method. The result suggests that the interaction between graphene and FePc follows a donor–acceptor mode and the loading of FePc on graphene sheets not only facilitates the dispersion of FePc on graphene but also promotes the exfoliation of the graphene sheets. These samples are tested for photocatalytic degradation of phenol under visible light irradiation ($\lambda \geq 420$ nm). And the photocatalytic activity of graphene, which is not presented when graphene individually acts on phenol, is greatly enhanced because of the π–π stacking interaction when FePc loaded on graphene. The G/FePc hybrid containing 25 wt% FePc (G/FePc-0.25) exhibits the best photoactivity among the different loading content of FePc, whose degradation rate for phenol achieves 77.1% in the presence of H_2O_2 under visible light irradiation for 3 h.

Zhang and coworkers (Zhang et al. 2016a, b, c, d) suggested that the novel Ag_3PO_4/N-doped graphene (NG)/Poly(3-hexylthiophene) (P3HT) composites can

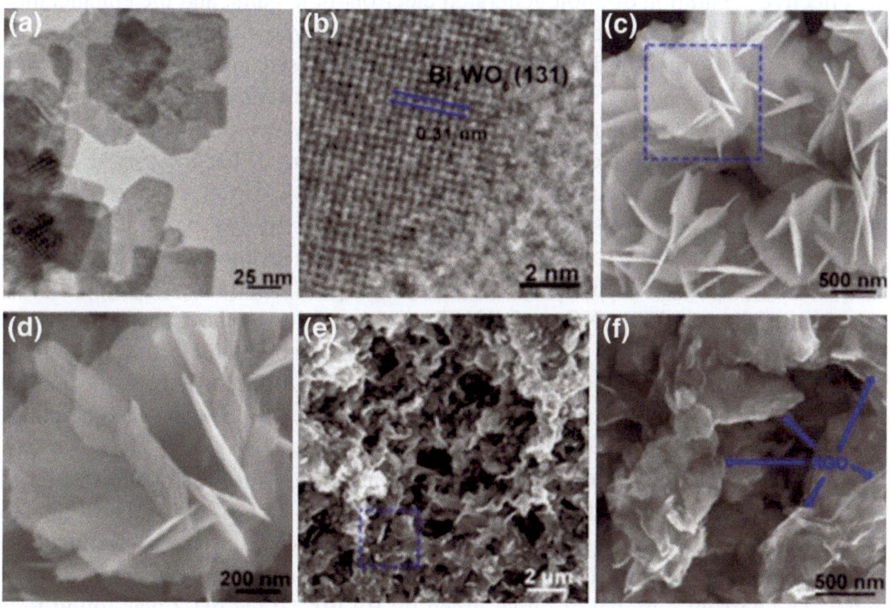

Fig. 8.5 TEM morphologies of Bi_2WO_6 (**a**, **b**); different magnified SEM images of Bi_2WO_6 (**c**, **d**) and BWGA (**e**, **f**). Reproduced from, Xu et al. (2016b), Copyright (2016), with permeation from Elsevier

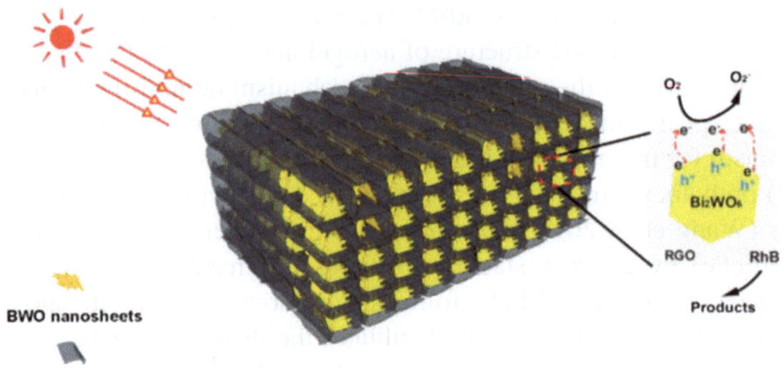

Fig. 8.6 Schematic of the mechanism for the possible charge separation process by BWGA composites. Reproduced from, Xu et al. (2016b), Copyright (2016), with permeation from Elsevier

remove the organic dye Rhodamine B (RhB) from water. The reactive oxygen species trapping experiments indicate that the degradation of RhB over the Ag_3PO_4/NG/P3HT composites mainly results from the holes oxidation and superoxide radical reduction. Besides, Ag_3PO_4/NG/P3HT composites exhibit better recyclability and stability than pristine Ag_3PO_4. Furthermore, the photocatalytic mechanism of Ag_3PO_4/NG/P3HT composites for RhB degradation under visible light was proposed as the synergistic effect of irradiated Ag_3PO_4, P3HT, and NG sheets on the effective separation of photogenerated electron–hole pairs, and the enhancement of visible light absorbance. The possible photocatalytic mechanism of Ag_3PO_4/NG/P3HT composites under visible light was proposed and shown in Fig. 8.5. According to XPS analysis of NG, the doping concentration of nitrogen atom in Zhang and coworkers experiment is 4.1 at percentage. Accordingly, NG can exhibit n-type semiconducting electronic properties. It is inferred that the p-type Ag_3PO_4 particles deposit onto the n-type NG surface forming a heterostructure, which is beneficial to the electron–hole separation. However, further work is needed to figure out the accurate role of p–n heterostructure in enhancing the Ag_3PO_4/NG/P3HT composites' activity. In addition, the specific large π-conjugated structure formed by Ag_3PO_4, NG and P3HT might accelerate the transportation and mobility of photogenerated electrons to effectively separate the electron–hole pairs, which was confirmed by the PL spectra. The proposed photocatalytic mechanism for RhB degradation using Ag_3PO_4/NG/P3HT composites under visible light is shown in Fig. 8.7.

The BiOBr–graphene oxide (BiOBr–GO) nanocomposite was synthesized with sonochemical method by Patil and coworkers (Patil et al. 2016). The effect of incorporation of graphene oxide with BiOBr on photocatalytic performance of BiOBr under exposure of UV–Visible light irradiation was systematically investigated. The percent removal of direct green by BiOBr–GO at pH 7 was found to be 91.9% while by pure BiOBr it was 37%. This exhibits that BiOBr–GO shows enhanced adsorption and photocatalytic performance for removal of direct green under UV–Visible light irradiation. The mineralization efficiency indicates the

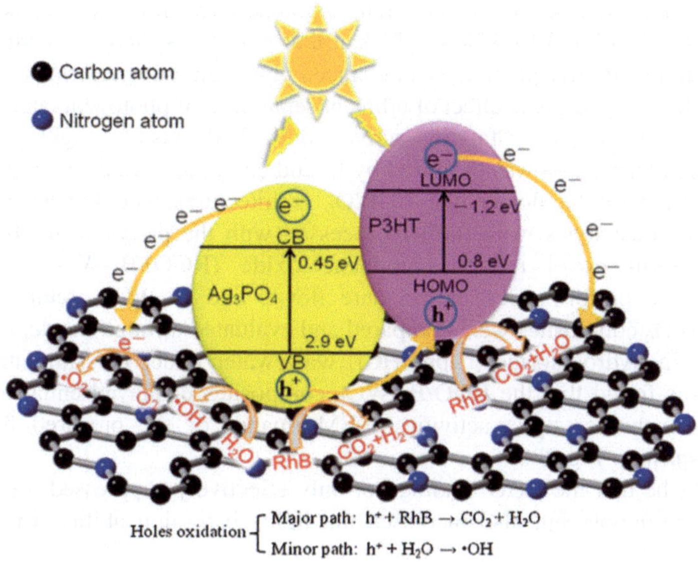

Fig. 8.7 The proposed photocatalytic mechanism for RhB degradation using Ag_3PO_4/NG/P3HT composites under visible light. Reproduced from, Zhang et al. (2016d), Copyright (2016), with permeation from Elsevier

91.7% TOC removal for the degradation of direct green by BiOBr–GO. Magnetically recyclable, multifunctional

Pd–$NiFe_2O_4$/reduced graphene oxide (Pd–$NiFe_2O_4$/rGO) photocatalysts have been prepared by Li and coworkers (Li et al. 2016a, b, c, d, e, f). The integration of $NiFe_2O_4$ with Pd nanoparticles supported on rGO enables efficient harvesting visible light for different catalytic reactions on the heterojunction structures. The Pd–$NiFe_2O_4$/rGO nanocomposites present significantly enhanced photocatalytic activity toward dyes degradation compared to the blank–$NiFe_2O_4$ and the optimum binary $NiFe_2O_4$/rGO, which is ascribed to the added Pd nanoparticles, acting as the mediator on the interfacial layer between $NiFe_2O_4$ and rGO, and transferring the charge carrier across the heterojunction interface. Another important role of Pd nanoparticles is as the electron reservoir, which can directly trap photogenerated electrons. It is interesting that the photogenerated electrons, which transfer to Pd nanoparticles, can increase the electron density of Pd. Chen and coworkers (Chen et al. 2016) present a facile method for the synthesis of highly dispersed Bi_2S_3 nanoparticles (Bi_2S_3 NPs) with an average diameter of ca. 25 ± 3 nm on the surface of reduced graphene oxide (RGO) via a poly(sodium-*p*-styrenesul-fonate) (PSS)-assisted hydrothermal process. Such synthetic strategy can avoid excess aggregates of Bi_2S_3 nanoparticles, meanwhile from effective interfacial contact between Bi_2S_3 nanoparticles and RGO nanosheets, and inhibit the recombination of photogenerated charges. The enhanced charge transfer properties were proved by photoluminescence (PL) measurement. The obtained Bi_2S_3 NPs/RGO composites showed more significant visible light photoactivity for the degradation of 2,4-dichlorophenol and Rhodamine B than that pure Bi_2S_3 and the control sample prepared in the absence of PSS. The enhanced photocatalytic performance could be attributed to the synergistic effect of efficient separation of photogenerated electron–hole pairs, increased catalytic active sites and visible light utilization. A hydrothermal process was proposed by Li and coworkers (Li et al. 2016a, b, c, d, e, f). To prepare the flower-like Bi_2WO_6 architectures, and the as-synthesized Bi_2WO_6 photocatalysts were further processed with the prepared graphene oxide (GO) to form novel reduced graphene oxide (RGO)/Bi_2WO_6 composites. Photocatalytic performances of the pure flower-like Bi_2WO_6 architectures and RGO/Bi_2WO_6 composites were compared and evaluated through the degradation of ciprofloxacin hydrochloride (Cipro HCl) wastewater under the simulated visible light. It was found that the RGO/Bi_2WO_6 composites displayed enhanced visible light-driven photocatalytic activities. SEM images of the prepared flower-like Bi_2WO_6 show in Fig. 8.8.

It might be that the RGO loading not only effectively suppressed the electron–hole recombination, but also increased the light absorption ability. Energy band

Fig. 8.8 SEM images of the prepared flower-like Bi_2WO_6 (**a**, **b**, **c**), 1%RGO/Bi_2WO_6 (**d**), 2% RGO/Bi_2WO_6 (**e**, **f**), 3%RGO/Bi_2WO_6 (**g**), 4%RGO/Bi_2WO_6 (**h**) and 5%RGO/Bi_2WO_6 nanomaterial. Reproduced from, Li et al. (2016f), Copyright (2016), with permeation from Elsevier

diagram and photocatalytic mechanism of Bi_2WO_6 and RGO/Bi_2WO_6 nanocomposites show in Fig. 8.9.

By coupling graphene sheet and plasmonic photocatalysis technologies, a series of Ag-AgX/RGOs (X = Cl, Br, I; RGO = reduced graphene oxide) composites were prepared by Xia and coworkers (Xia et al. 2016a, b) and found to be efficient antimicrobial agents for water disinfection upon visible light. Attributed to the efficient charge transfer by RGO sheets, the optimum Ag-AgBr/0.5% RGO could completely inactivate 2×10^7 cfu mL^{-1} of *Escherichia coli* within 8 min, much faster than bare Ag–AgBr within 35 min. The synergistic antimicrobial mechanism of Ag–AgBr/0.5% RGO was studied by Ag^+ ions release evaluation, radical scavengers study, and radical determination. The enhanced photocatalytic activity of irradiated Ag–AgBr/0.5% RGO originated from the synergistic activities of its three components including Ag, AgBr, and RGO, and the proposed mechanisms contained enhanced attraction by RGO followed by two pathways: primary oxidative stress caused by plasma induced reactive species like H_2O_2 and bactericidal effect of released Ag^+ ions. Furthermore, characterization of *E. coli* cells using SEM, fluorescent microscopy, and cytoplasmic substance leakage illustrated that VL irradiated Ag–AgBr/0.5% RGO could not only cause metabolic dysfunction but also destroy the cell envelope and biomolecular, while irradiated Ag^+ ions play a differential bactericidal action with a limited metabolic injury and no cell membrane damage. This work provides an efficient water disinfection technology and also opens a new idea in studying the antimicrobial mechanism of plasmonic photocatalyst. Proposed synergistic photocatalytic bacterial inactivation mechanism by plasmonic Ag–AgBr/0.5% RGO composite photocatalyst shows in Fig. 8.10.

Single crystalline, nontoxicity, and long-term stability graphene quantum dots (GQDs) were modified onto the $AgVO_3$ nanoribbons by a facile hydrothermal and sintering technique which constructs a unique heterojunction photocatalyst by Lei and coworkers (Lei et al. 2016). Characterization results indicate that GQDs are well dispersed on the surface of $AgVO_3$ nanoribbons and GQD/$AgVO_3$

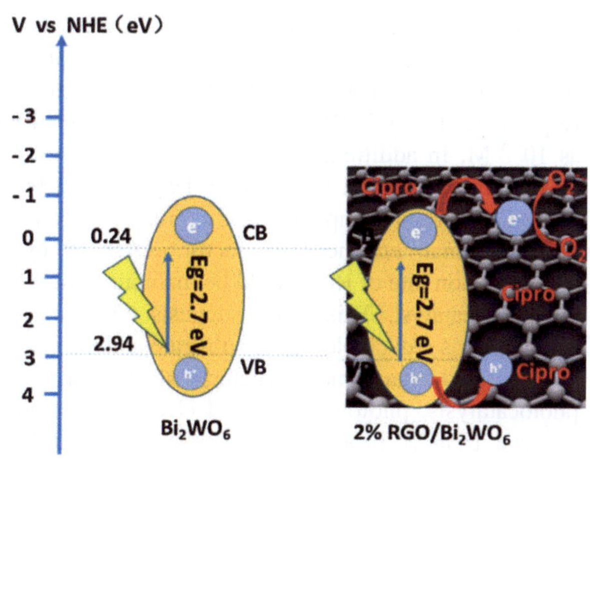

Fig. 8.9 Energy band diagram and photocatalytic mechanism of Bi_2WO_6 and RGO/Bi_2WO_6 nanocomposites. Reproduced from Li et al. (2016f), Copyright (2016), with permeation from Elsevier

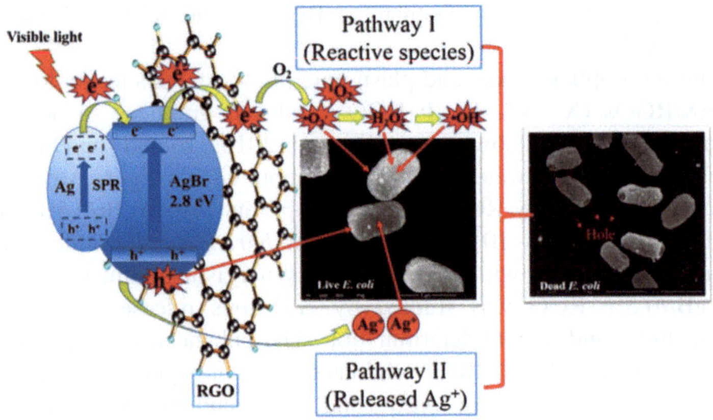

Fig. 8.10 Proposed synergistic photocatalytic bacterial inactivation mechanism by plasmonic Ag-AgBr/0.5% RGO composite photocatalyst. Reproduced from Xia et al. (2016a), Copyright (2016), with permeation from Elsevier

heterojunctions are formed, which can greatly promote the separation efficiency of photogenerated electron–hole pairs under visible light irradiation. By taking advantage of this feature, the GQD/$AgVO_3$ heterojunctions exhibit considerable improvement on the photocatalytic activities for the degradation of ibuprofen (IBP) under visible light irradiation as compared to pure $AgVO_3$. The photocatalytic activity of GQD/$AgVO_3$ heterojunctions is relevant with GQD ratio and the optimal activity is obtained at 3 wt% with the highest separation efficiency of photogenerated electron–hole pairs. Integrating the physicochemical and photocatalytic properties, the factors controlling the photocatalytic activity of GQD/$AgVO_3$ heterojunctions are discussed in detail. Moreover, potential photocatalytic degradation mechanisms of IBP via GQD/$AgVO_3$ heterojunctions under visible light are proposed. α-Fe_2O_3-reduced graphene oxide (α-Fe_2O_3/rGO) composites are rationally designed and prepared Zhang and coworkers (Zhang et al. 2016a, b, c, d) to integrate organic pollutants detection and their photocatalytic degradation. Specifically, the composites are used as the substrate for surface-enhanced Raman scattering (SERS) to detect rhodamine 6G (R6G). Repeatable strong SERS signals could be obtained with R6G concentration as low as 10^{-5} M. In addition, the substrate exhibits self-cleaning properties under solar irradiation. Compared with pure α-Fe_2O_3 and α-Fe_2O_3/rGO mechanical mixtures, the α-Fe_2O_3/rGO composites show much higher photocatalytic activity and much greater Raman enhancement factor. After 10 cycling measurements, the photodegradation rate of R6G could be maintained at 90.5%, indicating high stability of the photocatalyst. This study suggests that the α-Fe_2O_3/rGO composites would serve both as recyclable SERS substrate and as excellent visible light photocatalyst. Well-defined organic nanostructures of porphyrin are promising candidates toward photocatalysis, photovoltaics, and electronics applications where a photoinduced electron transfer process occurs. On the other hand, reduced graphene oxides

(RGO) have attracted much attention in light energy conversion owing to their efficient charge separation property. Bera and coworkers (Bera et al. 2016) have demonstrated a composite of a one-dimensional (1D) nanostructure of 5, 10, 15, and 20-tetrakis (4-carboxyphenyl) porphyrin (TCPP) and RGO for enhancing photoinduced charge separation. The composite was characterized by scanning electron microscopy (SEM), UV–visible spectroscopy, fluorescence spectroscopy, time-correlated single photon counting (TCSPC), and femtosecond fluorescence up-conversion spectroscopy. It is noted that a very fast decay of TCPP NR was observed in the TCPP NR–RGO composite due to the electron transfer process, and the electron transfer rate is found to be 10.0×10^{-4} ps^{-1} for the TCPP NR–RGO system. An increment (1.9 fold) of photocurrent of this composite system under visible light illumination is obtained due to electron transfer from TCPP NR to RGO. This new class of porphyrin-based composite structures opens up new possibilities in solar energy conversion and photocatalytic, photovoltaic, and other new emerging applications. Single crystalline, nontoxicity, and long-term stability graphene quantum dots (GQDs) were modified by Lei and coworkers (Lei et al. 2016) onto the $AgVO_3$ nanoribbons by a facile hydrothermal and sintering technique which constructs a unique heterojunction photocatalyst. Characterization results indicate that GQDs are well dispersed on the surface of $AgVO_3$ nanoribbons and GQD/$AgVO_3$ heterojunctions are formed, which can greatly promote the separation efficiency of photogenerated electron–hole pairs under visible light irradiation. By taking advantage of this feature, the GQD/$AgVO_3$ heterojunctions exhibit considerable improvement on the photocatalytic activities for the degradation of ibuprofen (IBP) under visible light irradiation as compared to pure $AgVO_3$. The photocatalytic activity of GQD/$AgVO_3$ heterojunctions is relevant with GQD ratio and the optimal activity is obtained at 3 wt% with the highest separation efficiency of photogenerated electron–hole pairs. Integrating the physicochemical and photocatalytic properties, the factors controlling the photocatalytic activity of GQD/$AgVO_3$ heterojunctions are discussed in detail. Moreover, potential photocatalytic degradation mechanisms of IBP via GQD/$AgVO_3$ heterojunctions under visible light are proposed. Nickel (Ni) incorporated titanium dioxide (TiO_2)/graphene oxide composite photocatalysts were prepared by anchoring the TiO_2 and Ni onto the surface of graphene oxide (GO) sheets by a straightforward microwave-assisted, one-pot method by Pham and coworkers (Pham et al. 2016). The as-prepared composite photocatalysts with high Ni content (40–50 wt%) showed good adsorption capacity in the dark and high reaction rate constants under visible illumination while the composite photocatalysts with low Ni content (5–10 wt%) exhibited weak activity. An anatase phase, a small amount of rutile phase and Ni metal were detected using X-ray diffraction (XRD) and transmission electron microscopy (TEM). Raman measurements identified a small fraction of $NiTiO_3$ only at high Ni content. The formation of $NiTiO_3$ and the increase in the specific surface area (SSA) for 40 and 50 wt% Ni-loaded catalysts improved the adsorption capacity and photocatalytic activity upon exposure to visible light, resulting in very effective removal of dye contaminants under visible light irradiation. Increasing the Ni content up to 40 and 50 wt% induced not only a structural change affording high porosity but also a narrowing of the band gap to 2.51 eV. Meanwhile, the presence of GO in the composite photocatalysts inhibited the agglomeration of Ni particles even at high Ni content, resulting in similar Ni particle sizes regardless of the Ni content. At the same time, Ni metal accelerated the reduction of the GO sheets, as evidenced by the Raman data. TiO_2 nanotubes (TNT) co-sensitized with copper(II) *meso*-tetra(4-carboxyphenyl) porphyrin (CuTCPP) and reduced graphene oxide nanosheets (rGO), which was fabricated through two-step improved hydrothermal method and heating reflux process by Wei and coworkers (Wei et al. 2016). The effect of rGO and CuTCPP on the co-photocatalytic behavior of TNT for the degradation of Methylene Blue (MB) was measured under visible light irradiation. The results provide a deeper insight into the co-photocatalytic mechanism of CuTCPP/rGO-TNT nanocomposites. The degradation results showed a purification of more than 95% MB in wastewater, which is about five times higher than that of the pure TNT. The results also confirm the prepared CuTCPP/rGO-TNT nanocomposites possess superior co-photocatalytic activities. Yuan and coworkers (Yuan et al. 2016) synthesized MoS_2–graphene composite as a highly efficient cocatalyst to enhance the photocatalytic activity of $ZnIn_2S_4$ under visible light irradiation. Through the optimizing of each composition proportion, the hierarchical MoS_2–graphene/$ZnIn_2S_4$ photocatalyst shows the highest H_2 evolution rate of 4169 μmol h^{-1} g^{-1} under visible light irradiation in presence of Na_2S and Na_2SO_3 as sacrificial reagents when the content of MoS_2–graphene is 1.2 wt% and the weight ratio of MoS_2 to graphene is 10:1, which is almost 22.8 times higher than that of pure $ZnIn_2S_4$. More importantly, the ternary MoS_2–graphene/$ZnIn_2S_4$ composite exhibits much higher photocatalytic activity than Pt-loaded $ZnIn_2S_4$ photocatalyst, suggesting that the MoS_2–graphene composite can act as a more efficient cocatalyst than the commonly used Pt metal. The superior catalytic activity of MoS_2–graphene cocatalyst can be assigned to the positive synergistic effect between MoS_2 and graphene, which act as a hydrogen evolution reaction catalyst and an electron transport bridge, respectively. The effective charge transfer from $ZnIn_2S_4$ to MoS_2 through graphene is demonstrated by the significant enhancement of photocurrent responses in MoS_2–graphene/$ZnIn_2S_4$ composite photoelectrodes. A series of visible light-responsive plasmonic Ag-coated $BiOBr_{0.2}I_{0.8}$ nanosheets are grown by Liu and coworkers (Liu et al. 2016a, b, c) on graphene by a combined solvothermal and photodeposition method. The ternary Ag/$BiOBr_{0.2}I_{0.8}$/graphene nanocomposites exhibit significantly enhanced photocatalytic activity than pristine $BiOBr_{0.2}I_{0.8}$ and the binary $BiOBr_{0.2}I_{0.8}$/graphene composite. When the loading amount of Ag is 1.0 wt%, the Ag/$BiOBr_{0.2}I_{0.8}$/graphene nanocomposite displays the highest photocatalytic activity and 5.57 times as large as those of pristine $BiOBr_{0.2}I_{0.8}$, respectively. The high photocatalytic activity is attributed predominantly to the hybridization of the surface plasmonic resonance (SPR) effect of Ag nanoparticles and the specific electronics effect of graphene, thus enhancing the separation of photogenerated charge carriers of $BiOBr_{0.2}I_{0.8}$. Meanwhile, the excellent adsorption capacity of graphene and the broad absorption in the visible light region also contribute to the enhancement of photocatalytic activity. Both the holes and hydroxyl radicals were the active species in the degradation process.

Suggested mechanism for the photocatalytic enhancement of $Ag/BiOBr_{0.2}I_{0.8}/RGO$ nanocomposites is shown in Fig. 8.11.

Different SnO_2 microspheres like dandelions, silkworm cocoons, and urchins have been synthesized on graphene oxide sheets (GOs) by Wei and coworkers (Wei et al. 2016). The results of XRD revealed that the as-grown SnO_2 microspheres have tetragonal rutile structure. The results of Raman spectra, EDS, XRD, XPS, and SEM showed that the SnO_2 microspheres were grown on GOs and the average diameter of dandelion-like microsphere was about 1.5 μm. The photocatalytic results showed that the dandelion-like SMGs exhibited a much better photocatalytic activity than those of smooth and rough SMGs. Hybrid three-dimensional (3D) structures that were composed of $CeVO_4$ particles and graphene aerogels were fabricated by the electrostatic-driven self-assembly method by Fan and coworkers (Fan et al. 2016). The results showed that graphene nanosheets loading with $CeVO_4$ particles self-assembled into a well-defined and interconnected 3D porous network through strong Van der Waals and π–π interactions. Benefited from the incorporation of $CeVO_4$ particles into graphene nanosheets in such a unique structure, hybrid aerogels exhibited higher photodegradation efficiency toward methylene blue (MB) than that over pure $CeVO_4$ photocatalyst. It is proposed that the efficient physical adsorption of dye molecules and enhanced charge transfer in the composite is account for the improved photocatalytic activity. These findings open a new pathway for the design and fabrication of such functional graphene-based aerogels in water purification and advanced treatment. Visible light-responsive ZnO–graphene–TiO_2 (ZGT) composite catalyst was synthesized via a solvothermal process

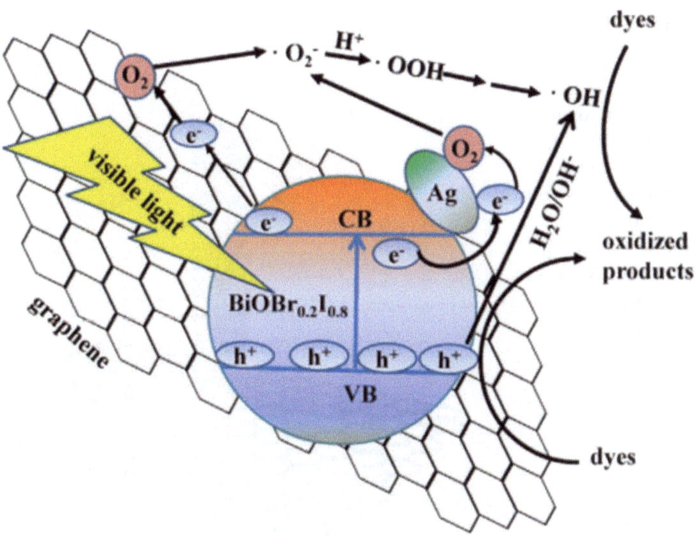

Fig. 8.11 Suggested mechanism for the photocatalytic enhancement of $Ag/BiOBr_{0.2}I_{0.8}/RGO$ nanocomposites. Reproduced from Liu et al. (2016a), Copyright (2016), with permeation from Elsevier

by Nuengmatcha and coworkers (Nuengmatcha et al. 2016). The band gap energies of the samples were evaluated by UV–vis diffuse reflectance spectroscopy (UV-Vis DRS). The photocatalytic activities of as-obtained catalysts were assessed based on the degradation of rhodamine B (RhB), texbrite BAC-L (TBAC), and texbrite NFW-L (TNFW) under visible light irradiation. ZnO, ZnO:Cu and ZnO:Cu: Graphene nanopowders were synthesized by Ravichandran and coworkers (Ravichandran et al. 2016) via a facile wet chemical method. Photoluminescence studies reveal that the incorporation of copper and graphene in ZnO facilitates the efficient photogenerated electron–hole pair separation. It is found that the ZnO:Cu and ZnO:Cu:Graphene nanopowder exhibit improved photocatalytic efficiency for the photodegradation of Methylene Blue (MB) under visible light irradiation. Moreover, improved antibacterial activity of ZnO:Cu:Graphene nanopowder against *Escherichia coli* and *Staphylococcus aureus* bacteria is observed. Ghaffar and coworkers (Ghaffar et al. 2016) reveal that nanoscale carbon layer deposited by hydrothermal process on molybdenum oxide (MoO_3) nanowires surface significantly improve the light absorption range. Furthermore, the graphene–carbon coated MoO_3 nanocopmosite (rGO/C-MoO_3 nanocomposite) exhibits excellent chemical stability and enhanced photocatalytic activity for methylene blue in aqueous solution under visible light irradiation compared to the bare MoO_3 nanowires and carbon coated MoO_3 nanowires (C–MoO_3 nanowires). The enhanced photocatalytic activity of rGO/C–MoO_3 nanocomposite could be attributed to the extended light absorption range, better adsorptivity of dye molecules and efficient separation of photogenerated electrons and holes. Schematic diagram of rGO/C-MoO_3 nanocomposite indicating that mitted electron under visible light irradiation from C-MoO_3 nanowires during photocatalytic process traps in graphene shows in Fig. 8.12.

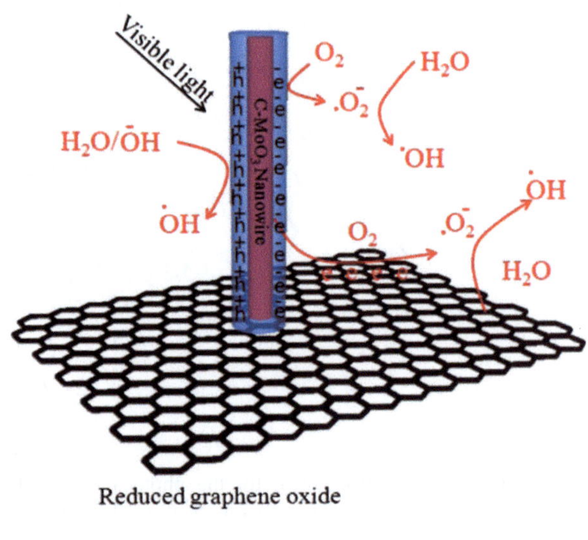

Fig. 8.12 Schematic diagram of rGO/C-MoO_3 nanocomposite indicating that emitted electron under visible light irradiation from C-MoO3 nanowires during photocatalytic process traps in graphene. Reproduced from Ghaffar et al. (2016), Copyright (2016), with permeation from Elsevier

Li and coworkers (Li et al. 2016a, b, c, d, e, f) reported the synthesis of the reduced TiO_2–graphene oxide heterostructure by a facile chemical reduction agent-free one-step laser ablation in liquid (LAL) method, which achieves extended optical response range from ultraviolet to visible and composites TiO_{2-x} (reduced TiO_2) nanoparticle and graphene oxide for promoting charge conducting. 30.64% Ti^{3+} content in the reduced TiO_2 nanoparticles induces the electronic reconstruction of TiO_2, which results in 0.87 eV decrease of the band gap for the visible light absorption. TiO_{2-x}–graphene oxide heterostructure achieved drastically increased photocatalytic H_2 production rate, up to 23 times with respect to the blank experiment. Furthermore, a maximum H_2 production rate was measured to be 16 mmol/h/g using Pt as a cocatalyst under the simulated sunlight irradiation (AM 1.5G, 135 mW/cm^2), the quantum efficiencies were measured to be 5.15% for wavelength $\lambda = 365 \pm 10$ nm and 1.84% for $\lambda = 405 \pm 10$ nm, and overall solar energy conversion efficiency was measured to be 14.3%. These findings provided new insights into the broad applicability of this methodology for accessing fascinate photocatalysts. Schematic illustration of hydrogen evolution mechanism for the strong coupling between TiO_2 and RGO sheets shows in Fig. 8.13.

A series of $Cd_xZn_{1-x}S$ and sulfide/graphene photocatalysts with 3D nanospherical framework have been successfully fabricated by one-pot

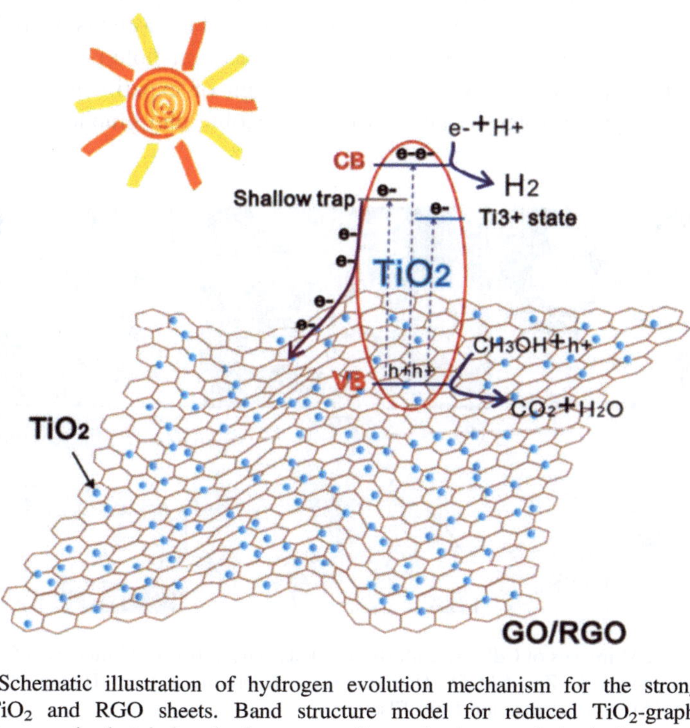

Fig. 8.13 Schematic illustration of hydrogen evolution mechanism for the strong coupling between TiO_2 and RGO sheets. Band structure model for reduced TiO_2-graphene oxide heterostructure and photoinduced charge transfer and photocatalytic hydrogen generation. Reproduced from Li et al. (2016a), Copyright (2016), with permeation from ACS

solvothermal method by Huang and coworkers (Huang et al. 2016). The enhanced photocatalytic activity is mainly attributed to the slow photon enhancement of the 3D structure, and the heterojunction between the 3D nanospherical $Cd_{0.5}Zn_{0.5}S$ solid solutions and a high quality 2D rGO support, which can greatly promote the separation of light-induced electrons and holes. Moreover, the large S_{BET} and extended light absorption range also play an important role for improving the photocatalytic activity. The high photocatalytic stability is due to the successful inhibition of the photocorrosion of $Cd_{0.5}Zn_{0.5}S$/rGO by forming heterojunction between CdS and ZnS, and transferring the photogenerated electrons of $Cd_{0.5}Zn_{0.5}S$ to rGO. This work can provide rational design of graphene-based photocatalysts with large contact interface and strong interaction between the composites for other application. FESEM images of CdS, ZnS, and $Cd_{0.5}Zn_{0.5}S$ and TEM images of CdS/rGO, ZnS/rGO, and $Cd_{0.5}Zn_{0.5}S$/rGO (f) composites are shown in Fig. 8.14.

A series of nitrogen-doped graphene–BiOBr (NG–BiOBr) nanocomposites with different weight addition ratios of nitrogen-doped graphene were prepared via a facile solvothermal method by Li and coworkers (Li et al. 2016a, b, c, d, e, f), and found to possess a higher photocatalytic activity than pure BiOBr toward degradation of methyl orange in water under visible light irradiation. The NG–BiOBr composite with 1.76 wt% NG content exhibited the highest photodegradation efficiency of methyl orange, its degradation rate was about 50, 4.6 and 3.8 times of P25, BiOBr microsphere and 1.76 wt% RGO–BiOBr composite, respectively. The enhanced photocatalytic performance could be ascribed to more visible light harvest and more effective separation of photogenerated electron–hole pairs.

This first-attempt study revealed mixture design of experiments to obtain the most promising composites of TiO_2 loaded on zeolite and graphene for maximal

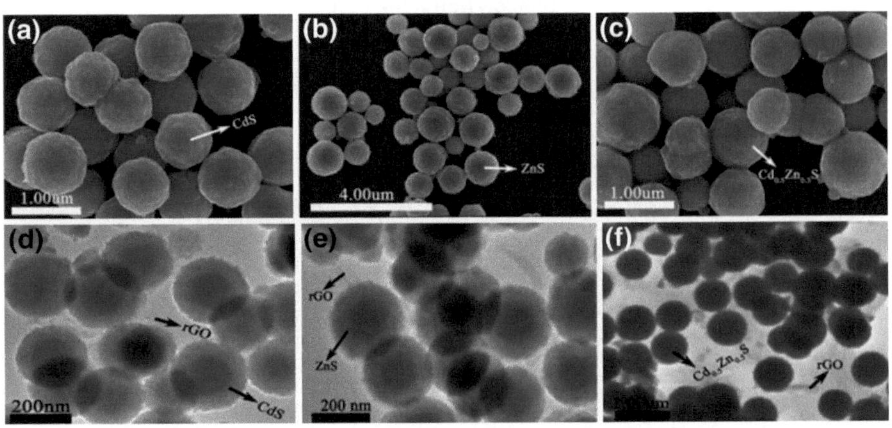

Fig. 8.14 FESEM images of CdS (**a**), ZnS (**b**), and $Cd_{0.5}Zn_{0.5}S$ (**c**); TEM images of CdS/rGO (**d**), ZnS/rGO (**e**), and $Cd_{0.5}Zn_{0.5}S$/rGO (**f**) composites. Reproduced from Huang et al. (2016), Copyright (2016), with permeation from Elsevier

photocatalytic degradation of oxytetracycline (OTC) by Hu and coworkers (Hu et al. 2016a, b). The optimal weight ratio of graphene, titanium dioxide (TiO_2), and zeolite was 1:8:1 determined via experimental design of simplex lattice mixture. In addition, it was uniformly dispersed with promising adsorption characteristics. OTC was used as model toxicant to evaluate the photodegradation efficiency of the GTZ (1:8:1). At optimal operating conditions (i.e., pH 7 and 25 °C), complete degradation (ca. 100%) was achieved in 180 min. The biotoxicity of the degraded intermediates of OTC on cell growth of *Escherichia coli* DH5α were also assayed. After 180 min photocatalytic treatment, OTC solution treated by GTZ (1:8:1) showed insignificant biotoxicity to receptor DH5α cells. Furthermore, EDTA (hole scavengers) and t-BuOH (radical scavengers) were used to detect the main active oxidative species in the system. Reaction mechanism and structure of the GTZ composite are shown in Fig. 8.15.

3D urchin-like TiO_2(UT)/reduced graphene oxide (RGO) composite was fabricated via two processes by Zhou and coworkers (Zhou et al. 2016). Pure UT was first synthesized by solvothermal reaction and UT/RGO(UTG) composite was then prepared by hydrothermal reaction. The urchin-like morphology of as-prepared material was confirmed by SEM and TEM. XRD analysis confirmed the presence of RGO and demonstrated that both pure UT and UTG composite consist of a rutilephase. FTIR, Raman spectroscopy, and XPS characterization demonstrated that graphene oxide was successfully reduced to RGO and RGO well combined with UT. The UV–Visible spectrum of UTG composite showed strong absorption in the visible light region (i.e., a blue shift was observed). The UTG composite exhibited larger specific surface area than pure UT (and much larger than P25)

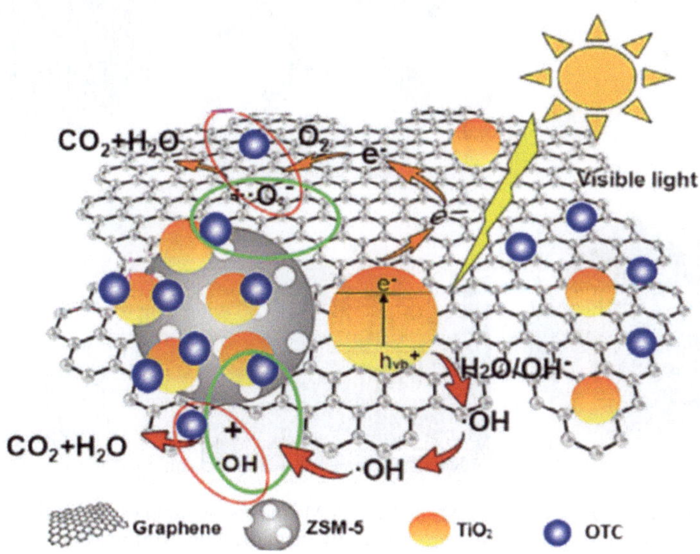

Fig. 8.15 Reaction mechanism and structure of the GTZ composite. Reproduced from Hu et al. (2016b), Copyright (2016), with permeation from Elsevier

owing to the addition of RGO. The formation of Ti–O–C bond possibly separated the photogenerated carriers and transferred the photoelectrons to RGO quickly because of the high electron mobility of RGO, thereby hindering the recombination process. Moreover, experimental results indicated that the UTG composite showed high efficiency in photodegradation of RhB. Representative TEM image of GO and low-magnification SEM image of pure UT, single pure UT, and TEM image of local nanorods grown on single UTG shows in Fig. 8.16.

All these studies consistently reported that graphene sheets played a crucial role in the enhancement of the photocatalytic ability of pristine semicondutor

Photocatalytic Application of G-C_3N_4/Semiconductor (CNS) Nanocomposites

Polymeric graphitic carbon nitride (for simplicity, g-C_3N_4) is a layered material similar to graphene, being composed of only C, N, and some impurity H. Contrary to graphenes, g-C_3N_4 is a medium band gap semiconductor and an effective photocatalyst for a broad variety of reactions, and it possesses a high thermal and chemical stability (Wang et al. 2012). Overall, extensive efforts have been made to

Fig. 8.16 Representative SEM, TEM, and HRTEM images of samples: **a** TEM image of GO; **b** low-magnification SEM image of pure UT; **c** low-magnification SEM image of single pure UT; **d** TEM image of single UTG; **e** TEM image of local nanorods grown on single UTG; **f** HRTEM image of nanorod in UTG. Reproduced from Zhou et al. (2016), Copyright (2016), with permeation from Elsevier

develop graphitic carbon nitride-based materials for application as composite photocatalysts. Wang and coworkers (Wang et al. 2012) described the polycondensation of this structure, how to modify band positions and band gap by doping and copolymerization, and how to texture the organic solid to make it an effective photocatalyst.

Graphitic carbon nitride (GCN) modified by oxygen functional groups was synthesized by a hydrothermal treatment of pristine GCN at different temperatures with H_2O_2 by Liu and coworkers (Liu et al. 2012). Insights into the emerging characteristics of the modified GCN in photocatalysis were obtained by determining the optical properties, band structure, electrochemical activity and pollutant degradation efficiency. It was found that the introduction of GCN with oxygen functional groups can enhance light absorption and accelerate electron transfer so as to improve the photocatalytic reaction efficiency. The photoinduced reactive radicals and the associated photodegradation were investigated by in situ electron paramagnetic resonance (EPR). The reactive radicals, $.O_2^-$ and .OH, were responsible for organic degradation.

Figure 8.17 schematically shows the band structure and associated photocatalysis. For GCN-O-150, a narrow band gap can absorb sufficient visible light to produce excited photoelectrons from VB to CB. With a potential of −1.55 V potential, the electron has an ability to oxidize water-soluble oxygen into O^{2-}. It has been found that the potential energy of valence band (VB) holes (0.7 eV) from g-C_3N_4 is lower that of OH^-/.OH (1.99 eV) and H_2O/.OH (2.37 eV), the holes cannot directly oxidize OH^- or H_2O into .OH. The .OH was generated by the further reaction of photogenerated electron with $.O_2^-$. The reactive radicals ($.O_2^-$ and .OH) can then decompose MB into CO_2 and H_2O.

A composite of graphitic carbon nitride and TiO_2 (g-C_3N_4/TiO_2) with enhanced photocatalytic hydrogen evolution capacity was achieved via calcining melamine and TiO_2 sol-gel precursor by Qu and coworkers (Qu et al. 2016). Compared with the polycondensation of pure melamine, the presence of TiO_2 precursor can promote the formation of melon at a low temperature. The highest photocatalytic activity of g-C_3N_4/TiO_2(400) was achieved when the calcination was performed at 400 °C, exhibiting H_2 production rate of 76.25 μmol/h under UV–vis light irradiation ($\lambda > 320$ nm) and 35.44 μmol/h under visible light irradiation ($\lambda > 420$ nm). The highest photocatalytic performance of g-C_3N_4/TiO_2(400) can be attributed to: (1) the strong UV–vis light absorption due to the narrow band gap caused by synergic effect of TiO_2 and g-C_3N_4, (2) high-surface area and porosity, and (3) the effective separation of photogenerated electron–holes owing to the favorable heterojunction between TiO_2 and g-C_3N_4. Composite photocatalysts with nanoflower-structured MoS_2 grown on pyridine-modified graphitic carbon nitride (g-C_3N_4) have been synthesized through a facile in situ solvothermal approach by Li and coworkers (Li et al. 2016a, b, c, d, e, f). These composites demonstrate greatly enhanced response to visible light, and consequently remarkably enhanced hydrogen evolution performance by photocatalytic water splitting. The addition of 2,5-dibromopyridine during the formation process of g-C_3N_4 can not only enhance the photocatalytic activity but also the durability of the photocatalysts. The MoS_2

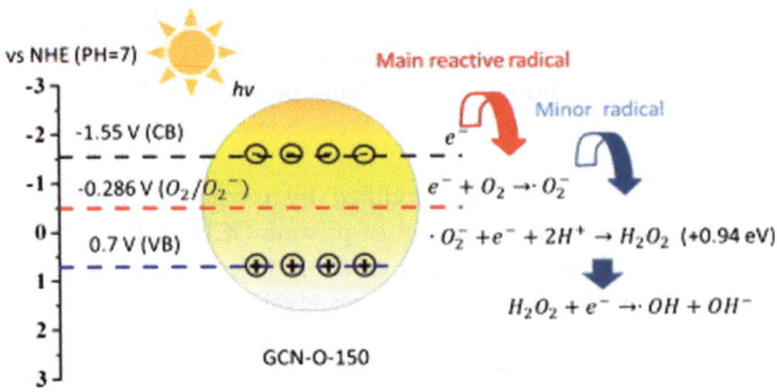

Fig. 8.17 Band structure and the photocatalysis on GCN-O-150. Reproduced from Liu et al. (2016c), Copyright (2016), with permeation from Elsevier

content and the ratio between 2,5-dibromopyridine and g-C_3N_4 in these composites can be well tuned to obtain the optimized photocatalytic activity with a peak H_2 production rate of 25 μmol h^{-1} on 50 mg photocatalyst without adding any noble metal under visible light irradiation at 283 K. A dual synergetic mechanism in MoS_2/pyridine-modified g-C_3N_4 composite, which is featured with significantly promoted separation of photogenerated carriers and stability of S^{2-} and/or $S_2{}^{2-}$ in the composites under visible light irradiation, has been proposed to account for the distinguished hydrogen evolution activity and stability of these composite photocatalysts. Porous polymeric carbon nitride nanosheets were obtained by lithium chloride ions in situ intercalating bulk materials in thermal polycondensation process and followed via liquid exfoliation in water by Ma and coworkers (Ma et al. 2016). The porous nanosheets show two-dimensional layered structure with the thickness of 2–3 nm, a high density in-plane pores with 2–3 nm diameter, a higher surface area (186.3 m^2 g^{-1}), enlarged band gap (by 0.16 eV), prolonged charge carrier lifetime, enhanced electronic transport ability, increased charge carrier density and improved photocurrent responses, which could significantly give rise to photocatalytic activity. The results highlight the crucial role of 2D porous structure, high specific surface area and unique electronic structure on the photocatalytic performance of polymeric carbon nitride materials. The separation free graphitic carbon nitride/SiO_2 (C_3N_4/SiO_2) hybrid hydrogels with three-dimensional (3D) network structures have been prepared via alkali solution and acid-gel process by Zhang and coworkers (Zhang et al. 2016a, b, c, d). The hybrid hydrogels perform enhanced ability to absorb and in situ degrade refractory organic pollutants, such as coking wastewater and phenol. Due to the 3D network structures of C_3N_4/SiO_2 hybrid hydrogels, they show efficient pollutants removal ability by synergistic effect of adsorption and in situ photocatalytic degradation. The 3D network structures guarantee C_3N_4/SiO_2 hybrid hydrogels to be continuously used without adsorption saturation and separation from water, avoiding the photocatalysts aggregation and secondary pollution.

Li and coworkers (Li et al. 2016a, b, c, d, e, f) demonstrated that photocatalytic H_2O_2 production at g-C_3N_4 could be improved by as much as 14 times in the absence of organic scavenger through a carbon vacancy-based strategy. Both the experimental and theoretical calculation results indicated that the creation of carbon vacancies could reduce the symmetry of g-C_3N_4 and produce the effect of electron delocalization. This will allow g-C_3N_4 to possess more excitable electrons and a narrower band gap. On the other hand, carbon vacancies provided more sites to adsorb molecular oxygen and thereby help electrons transfer from g-C_3N_4 to the surface adsorbed O_2. More interestingly, the presence of carbon vacancies changed the H_2O_2 generation pathway from a two-step single-electron indirect reduction to a one-step two-electron direct reduction. This study could not only develop a novel strategy to improve the H_2O_2 production activity of semiconductors, but also shed light on the deep understanding of the role played by surface defect structure on photocatalytic activity of semiconductor photocatalysts.

Lan and coworkers (Lan et al. 2016) present a facile synthesis of bromine doped graphitic carbon nitride (g-C_3N_4) photocatalysts for hydrogen evolution with visible light irradiation. Bromine modification is shown to enhance the optical, conductive, and photocatalytic properties of g-C_3N_4, while still keeping the poly-tri-s(triazine) core structure as the main building blocks of the materials. This modification method can be generally applicable to several precursors of g-C_3N_4, including urea, dicyandiamide, ammonium thiocyanide, and thiourea. The optimal sample CNU-$Br_{0.1}$ shows more than two times higher H_2 evolution rates than pure CNU sample under visible light irradiation, with high stability during the prolonged photocatalytic operation. Results also found that the photocatalytic O_2 evolution activity of CNU-$Br_{0.1}$ was promoted when the sample was subjected to surface kinetic promotion by loading with cobalt oxide as a cocatalyst. This study affords us a feasible modification pathway to rationally design and synthesize g-C_3N_4 based photocatalysts for a variety of advanced applications, including CO_2 photofixation, organic photosynthesis, and environmental remediation.

Carbon nanodots synthesized from rapeseed flower bee pollens were used to couple with ultrathin g-C_3N_4 nanosheets for preparing the 2D/0D type photocatalysts via a hydrothermal method by Liu and coworkers (Liu et al. 2016a, b, c). The g-C_3N_4 nanosheets (UCN) obtained from ammonium chloride blowing dicyandiamide methods exhibit ultrathin two-dimensional structure with a series of hollow spherical structures, and CQDs are well dispersed and uniformly anchored into the UCN network via π–π stacking interactions. The incorporation of CQDs can effectively expand visible light absorption regions via photosensitization and suppress the recombination of photoinduced carriers. Time-resolved fluorescence spectroscopy, electron paramagnetic resonance technology and photoelectrochemical measurements together reveal that CQDs serve as electron transfer mediation to facilitate charge separation and extend the lifetime of photoinduced carriers. The UCN/CQDs-0.2% composite has the optimal H_2 evolution of 88.1 μmol/h, 9.79, 3.02 and 1.91 folds of bulk g-C_3N_4 (BCN), pristine UCN and BCN/CQDs-0.2%, respectively. Consequently, the UCN/CQDs composites exhibit high photocatalytic activity of hydrogen release under the visible light irradiation.

A novel Sb_2S_3/ultrathin g-C_3N_4 sheets heterostructures embedded with g-C_3N_4 quantum dots (CNS) was fabricated via a facile hydrothermal process by Wang and coworkers (Wang et al. 2016a, b, c). It is indicated that the composites have fast electron transport and enhanced solar light absorption. Moreover, the CNS composite exhibits a significant photoelectric conversion property in near-infrared (NIR) wavelength range. Proposed mechanism for the MO degradation on the CNS composites under NIR irradiation is shown in Fig. 8.18.

The band-edge potential levels play a vital role in photoexcited charge carriers in a heterojunction. The minimum CB of Sb_2S_3 is more positive than that of the g-C_3N_4, indicating that the CNS hybrid is beneficial to the separation and transportation of charge carriers. The possible mechanism is exhibited in Fig. 8.16, revealing the utilization of short and long wavelengths in the spectrum. When irradiated by UV and visible light irradiation, both the Sb_2S_3 and utg-C_3N_4 can be excited and generate photoinduced electrons and holes. Meanwhile, the CNQD can absorb NIR light, and then emit the shorter wavelength light due to the up-conversion property, which leads to the subsequent excitation of utg-C_3N_4 and Sb_2S_3. This enables the CNS hybrids to absorb the UV and visible light up-converted from CNQD, and thus producing the e^-/h^+. The electrons are injected from the conduction band of g-C_3N_4 to that of Sb_2S_3 due to the intimate contact between them. Instead, the holes left on the Sb_2S_3 valence band transferred to that of g-C_3N_4. The transfer of charge carrier may allow the charge separation, and then efficiently hindering the recombination of the photogenerated electrons and holes. Subsequently, the excited electrons can be captured by the O_2 adsorbed on the surface of Sb_2S_3 to form O^{2-}, which are one of the main oxidizing species to decompose MO. On the other hand, the MO molecules absorbed on the photocatalyst surface could be degraded by the separated holes via direct holes oxidation.

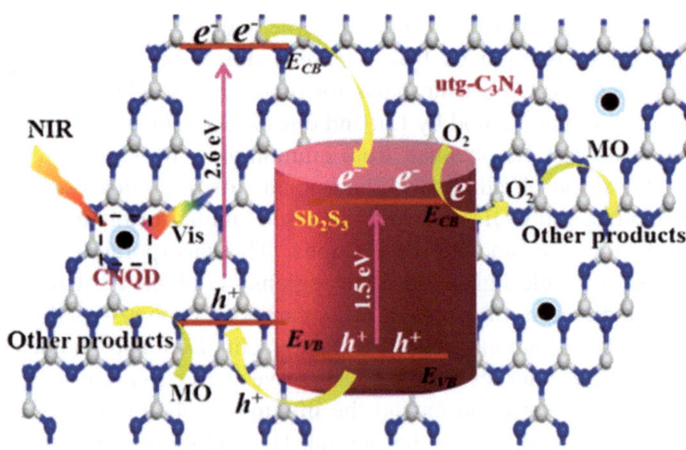

Fig. 8.18 Proposed mechanism for the MO degradation on the CNS composites under NIR irradiation. Reproduced from Wang et al. (2016a), Copyright (2016), with permeation from Elsevier

g-C_3N_4/$Bi_4O_5I_2$ heterojunction was prepared by Bai and coworkers (Bai et al. 2016). The photocatalytic data showed that g-C_3N_4/$Bi_4O_5I_2$ heterojunction had higher activity than pure g-C_3N_4 and $Bi_4O_5I_2$. At an optimal ratio of 1.0 mol% (11.4 wt% of $Bi_4O_5I_2$), g-C_3N_4/$Bi_4O_5I_2$ photocatalyst showed the highest photocatalytic reduction activity for CO_2 conversion with 45.6 μmol h^{-1} g^{-1} CO generation. Photocurrent and electrochemical impedance (EIS) spectroscopy revealed that higher photoinduced carrier separation efficiency of g-C_3N_4/$Bi_4O_5I_2$. Z-scheme charge transfer mode was proved by I_3^-/I^- redox mediator existence and superoxide radical (O_2^-.) and hydroxyl radical (.OH) quantification experiments.

Ag/g-C_3N_4 photocatalysts were synthesized via a rapid microwave-assisted polyol process by Sun and coworkers (Sun et al. 2016). The presence of Ag nanoparticles in Ag/g-C_3N_4 photocatalysts enhanced the visible light absorption and suppressed the recombination of photogenerated electron/hole pairs. The Ag/g-C_3N_4 photocatalysts exhibited the superior visible light-responsive photocatalytic activity for rhodamine B degradation. The mechanism of visible light-induced photocatalysis over Ag/g-C_3N_4 photocatalysts was also discussed.

A photocatalyst based on polyacrylonitrile (PAN)-supported graphitic carbon nitride coupled with zinc phthalocyanine nanofibers (g-C_3N_4/ZnTcPc/PAN nanofibers) was prepared by Xu and coworkers (Xu et al. 2016a, b), where g-C_3N_4/ ZnTcPc was introduced as the catalytic entity and the PAN nanofibers were employed as support to overcome the defects of easy aggregation and difficult recycling. Herein, rhodamine B (RhB), 4-chlorophenol and carbamazepine (CBZ) were selected as the model pollutants. Compared with the typical hydroxyl radical-dominated catalytic system, g-C_3N_4/ZnTcPc/PAN nanofibers displayed the targeted adsorption and degradation of contaminants under visible light or solar irradiation in the presence of high additive concentrations. According to the results of the radical scavenging techniques and the electron paramagnetic resonance technology, the degradation of target substrates was achieved by the attack of active species, including photogenerated hole, singlet oxygen, superoxide radicals, and hydroxyl radicals. Based on the results of ultra-performance liquid chromatography and mass spectrometry, the role of free radicals on the photocatalytic degradation intermediates was identified and the final photocatalytic degradation products of both RhB and CBZ were some biodegradable small molecules.

[60]-Fullerene—Photocatalyst Nanocomposites

C_{60} fullerene is a carbonaceous nanomaterial that is photochemically activated under visible light irradiation to produce singlet oxygen ($^1O_2$) with high quantum efficiency (Choi et al. 2010; Arbogast et al. 1991; Vileno et al. 2004; Yamakoshi et al. 2003), enabling effective sensitized oxidation of organic pollutants and inactivation of viruses with relatively low energy input.

In the presence of dissolved oxygen, energy transfer from photoexcited triplet-state C_{60} ($3C_{60}^*$) to triplet-state oxygen ($3O_2$) results in efficient production of singlet oxygen ($^1O_2$) (Arbogast et al. 1991; Yamakoshi et al. 2003; Orfanopoulos et al. 1995; Lee and Cho 2009). $^3C_{60}^*$ can also be reduced to C_{60} radical anion ($C_{60}^{\bullet -}$) by electron donors such as amines and alcohols, and subsequently reduce oxygen to superoxide radical anion ($O_2^{-\bullet}$) (Yamakoshi et al. 2003). There have been many attempts to enhance the photocatalytic activity using various C_{60} nanocomposites. Arunachalam and coworkers (Arunachalam et al. 2016) described a visible light-responsive photocatalyst of cobalt phthalocyanine (CoPc) and C_{60} nanocomposites synthesized by a reprecipitation process. The full spectrum visible light (420–800 nm) photocatalysis was demonstrated by mineralization under illumination of weak light intensity (in the order of 1 mW/cm^2) for aqueous trimethylamine (TMA) and almost agreed with its absorption spectra. Other aqueous volatile molecules of acetaldehyde (AcH) and 2-mercaptoethanol (ME) were decomposed to CO_2. CoPc/C_{60} nanocomposite exhibited higher photocatalytic activity than independent nanoparticles of CoPc or C_{60} as well as AlPc/C_{60} composite.

Titanium dioxide (TiO_2)/C_{60} hybrid nanocomposite was facilely fabricated by mixing TiO_2 and poly-carboxylic acid functionalized fullerene under an ultrasonication–evaporation method by Zhang and coworkers (Zhang et al. 2016a, b, c, d). It was found that the TiO_2/C_{60} composite could serve as an efficient and reusable photocatalyst for degradation of rhodamine B dye under visible light ($\lambda > 400$ nm). The degradation experiments revealed that the photocatalytic activity strongly depends on the contents of C_{60} from 0.5 to 3% mass ratio. The incorporation of C_{60} into TiO_2 efficiently extended the absorption spectrum of photocatalyst to visible light region, enhanced the adsorption capacity and degradation efficiency, resulting from a synergistic effect of fullerene and TiO_2. The trapping experiments demonstrated that both the photogenerated hole (h^+) and the reactive oxygen species such as superoxide anion radical ($O_2^{-\bullet}$) were involved in the photocatalytic reaction. Figure 8.19 shows schematic illustration of the proposed photodegradation mechanism of the RB dye over the TiO_2/C_{60} nanocomposite photocatalyst.

C_{60} modified $Cr_{2-x}Fe_xO_3$ nanostructure photocatalysts were synthesized by a absorbing process and employed in the photocatalytic H_2 evolution by Song and coworkers (Song et al. 2016). The as-prepared C_{60}–$Cr_{2-x}Fe_xO_3$ nanocomposites exhibits significantly enhanced photocatalytic activity for photocatalytic H_2 evolution without any noble metal. It is shown that the photoinduced electrons achieve high migration efficiency on the interface C_{60} and $Cr_{2-x}Fe_xO_3$, which is generated by the intense interaction of C_{60} and $Cr_{2-x}Fe_xO_3$ with conjugative three-dimensional π-system. A possible mechanism was proposed and supported by the PL emission spectrum technique and transient photocurrent responses.

C_{60} enhanced mesoporous CdS/TiO_2 architectures were fabricated via an evaporation-induced self-assembly route together with an ion-exchanged method by Lian and coworkers (Lian et al. 2016). C_{60} clusters were incorporated into the pore wall of mesoporous CdS/TiO_2 with the formation of C_{60} enhanced CdS/TiO_2 hybrid architectures, for achieving the enhanced photostability and photocatalytic activity in H_2 evolution under visible light irradiation. Such greatly enhanced photocatalytic performance and photostability could be due to the strong

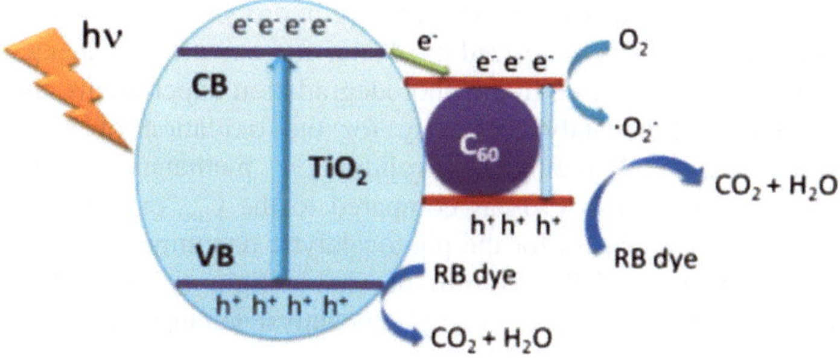

Fig. 8.19 Schematic illustration of the proposed photodegradation mechanism of the RB dye over the TiO_2/C_{60} nanocomposite photocatalyst. Reproduced from Zhang et al. (2016c), Copyright (2016), with permeation from Elsevier

combination and heterojunctions between C_{60} and CdS/TiO_2. The as-formed C_{60} cluster protection layers in the CdS/TiO_2 framework not only improve the light absorption capability, but also greatly accelerated the photogenerated electron transfer to C_{60} clusters for H_2 evolution. C_{60}-enhanced $B_{i2}TiO_4F\ _2$ hierarchical microspheres were prepared by Li and coworkers (Li et al. 2013) via a facile solvothermal method. Compared to the pure $Bi_2TiO_4F_2$ photocatalyst, the C_{60}/ $Bi_2TiO_4F_2$ samples exhibit much stronger photocatalytic performance for degrading Rhodamine B (RhB) and Eosin Y (EY) under visible light irradiation. Such greatly enhanced photocatalytic activity may be ascribed to strong combination and heterojunctions between C_{60} and $Bi_2TiO_4F\ _2$, favorable for charge separation and light adsorption. Loading C_{60} on $Bi_2TiO_4F_2$ results in a new photocatalytic mechanism (based on photogenerated h^+_{vb} and O_2 radicals) different from that of pure $Bi_2TiO_4F_2$ A series of TiO_2–graphene (GR), TiO_2–carbon nanotube (CNT), and TiO_2–fullerene (C_{60}) nanocomposite photocatalysts with different weight addition ratios of carbon contents are synthesized by Yang and coworkers (Yang et al. 2013) via a combination of sol-gel and hydrothermal methods. Photocatalytic selective oxidation of benzyl alcohol to benzaldehyde is employed as a model reaction to evaluate the photocatalytic activity of the TiO_2–carbon (GR, CNT, and C_{60}) nanocomposites under visible light irradiation. The results reveal that incorporating TiO_2 with carbon materials can extend the adsorption edge of all the TiO_2–carbon nanocomposites to the visible light region. For TiO_2–GR, TiO_2–CNT, and TiO_2–C_{60} nanocomposites, the photocatalytic activities of the composites with optimum ratios, TiO_2-0.1% GR, TiO_2-0.5% CNT, and TiO_2-1.0% C_{60}, are very close to each other along with the irradiation time. Furthermore, the underlying reaction mechanism for the photocatalytic selective oxidation of benzyl alcohol to benzaldehyde over TiO_2–carbon nanocomposites has been explored using different radical scavenger techniques, suggesting that TiO_2–carbon photocatalysts follow the analogous oxidation mechanism toward selective oxidation of benzyl alcohol. The addition of different carbon materials has no significant influence on the crystal phase, particle size, and the morphology of TiO_2. Organic semiconductor nanoparticles composed of C_{60} and/or partially hydrolyzed aluminum phthalocyanine chloride (AlPc) were synthesized by Zhang and coworkers (Zhang et al. 2012) using a reprecipitation method. In the photodegradation experiments, the composite showed enhanced photocatalytic activity for the oxidation of various organic compounds (e.g., N-methyl-2-pyrrolidone, methanal, ethanal, and 2-mercaptoethanol (ME)) to CO_2 as compared to the C_{60} or AlPc nanoparticles. Moreover, its action spectrum for the photocatalytic decomposition of ME to CO_2 also covered almost the full spectrum of visible light (<700 nm). The present research demonstrated a novel organic photocatalyst featuring a biphase structure and p–n junction-like characteristics.

CNT—Photocatalysts Nanocomposites

Carbon nanotubes can be divided essentially into two categories: SWNT and MWNT as is played in Fig. 8.20. Ideally, single-wall carbon nanotube are made of a perfect graphene sheet, i.e., a polyaromatic monoatomic layer made of an hexagonal display of sp^2 hybridized carbon atoms that genuine graphite is built up with, rolled up into a cylinder and closed by two caps (semi-fullerenes) (Serp et al. 2003).

CNTs show the potential to contribute to all three of the routes of increasing photocatalytic activity, high-surface area and high quality active sites, retardation of electron–hole recombination, and visible light catalysis by modification of band gap and/or sensitization (Leary and Westwood 2011; Serp et al. 2003). There have therefore been an increasing number of studies over the past decade seeking to develop CNT–photocatalysts mixtures or composites with enhanced photocatalytic activity. Recently for enhanced photocatalytic performance of visible light-responsive CdZnS, a series of $Cd_{0.5}Zn_{0.5}S$ solid solutions were fabricated via different methods by Gong and coworkers (Gong et al. 2016). The enhanced photocatalytic hydrogen production of CZS–PH was probably due to stacking fault formation as well as narrow band gap, a large surface area and a small crystallite size. Based on this, carbon nanotubes modified with Cu^{2+} (CNTs (Cu)) were used as a cocatalyst for CZS–PH. The addition of CNTs (Cu) enhanced notably the absorption of the composites for visible light. The highest photocatalytic hydrogen production rate of the $Cd_{0.5}Zn_{0.5}S$-CNTs (Cu) composite was 2995 μmol h^{-1} g^{-1} with 1.0 wt% of CNTs (Cu). The improvement of the photocatalytic activity by loading of CNTs (Cu) was not due to alteration of band gap energy or surface area, and was probably attributed to suppression of the electron–hole recombination by the CNTs, with Cu^{2+} anchored in the interface optimizing the photogenerated electron transfer pathway between the semiconductor and CNTs.

The nanohybrids TiO_2/CNTs materials were synthesized via hydrolysis method by Nguyen and coworker (Nguyen et al. 2016). The results show that, compared to pure TiO_2 or CNTs nanoparticles, nanohybrids TiO_2/CNTs materials exhibit higher catalytic activities in degradation of methylene blue (MB), or methylene orange

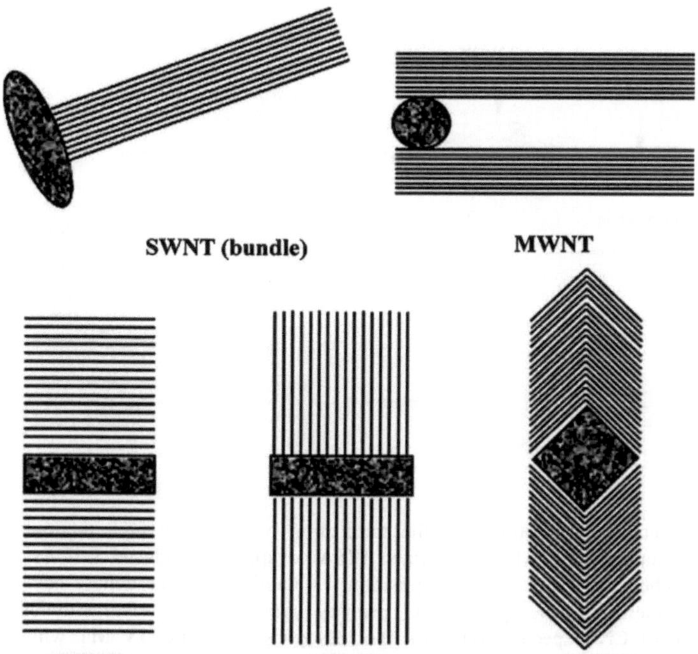

Fig. 8.20 Different types of CNT and GNF. Reproduced from Serp et al. (2003), Copyright (2003), with permeation from Elsevier

(MO). The highest degradation percentages, observed in samples with [TiO_2]/[CNTs] ratio of 5/1, are around 33% for MO and 38% for MB, respectively. This can be attributed to the significantly enhanced visible light absorption of nanohybrids TiO_2/CNTs due to the attachment of the TiO_2 nanoparticles on the sidewall of CNTs. The density functional-theory (DFT) calculations indicate that the stability of nanohybrids TiO_2/CNTs is due to the favored bonding state between nanocrytal TiO_2 and CNTs at the interface.

Pore-hole inorganic hybrid materials: carbon nanotube (CNT)-embedded mesoporous titania (MT) (CNMT) were controlled fabricated in supercritical conditions by deposition of titanium sol containing a liquid crystal template by Liu and coworkers (Liu et al. 2014). The as-prepared hybrid materials were characterized by using modern analytical tools. Embedding of CNTs in the nanophase titania matrix helped protect the mesoporous framework against collapsing, inhibited undesirable grain growth, suppressed transformation of anatase into rutile, and increased the thermal stability of MT during calcination. Schematic illustrations for the fabrication of MT with high thermal stability and improved crystallinity in hybrid materials. EISA: evaporation-induced self-assembly; CNTs shows in Fig. 8.21

Calcination temperature greatly influenced embedding effect of CNTs on the microstructure of MT in hybrid materials. The maximum phenol degradation (99%

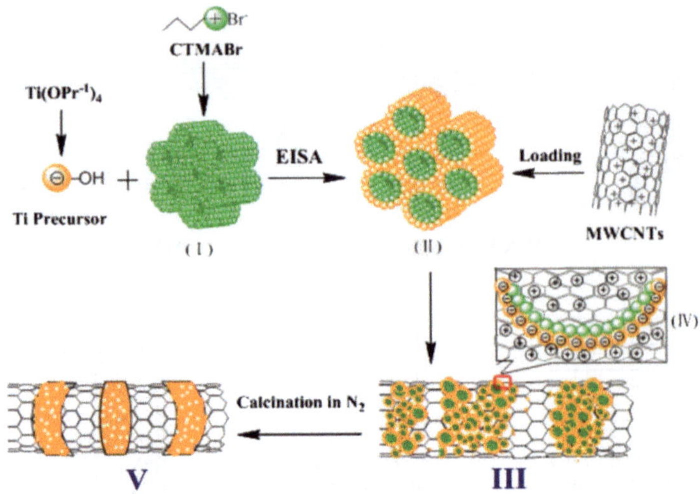

Fig. 8.21 Schematic illustrations for the fabrication of MT with high thermal stability and improved crystallinity in hybrid materials. EISA: evaporation-induced self-assembly; CNTs: carbon nanotubes; SCD:supercritical deposition. **I** Liquid crystal mesophase network; **II** stable mesophase framework of mesoporous TiO_2 primary particles for encircling with CNT species; **III** magnification of CNT species bound on the mesophase structure; **IV** MT with high thermal stability and crystallinity in hybrid materials. Reproduced from Liu et al. (2014), Copyright (2016), with permeation from Elsevier

after 140 min) is observed for CNMT-500 samples. It is attributed to the well mesostructure, which facilitates mass transport, the large surface area that offers more active sites and hydroxyl radicals, and perfect crystallinity that favors the separation of photogenerated electron–hole pairs, confirmed by photoluminescence emission spectrum. Photocatalytic degradation of phenol monitored as the normalized concentration change versus irradiation time in the presence of CNMT shows in Fig. 8.22.

A series of PANi/CNT/TiO_2 photocatalysts immobilized on glass plate irradiated with visible light were presented to degrade DEP by Hung and coworkers (Hung et al. 2016). The PANi/CNT/TiO_2 potocatalysts were fabricated by co-doping with polyaniline (PANi) and two functionalized CNT (CNT–COCl and CNT–COOH) onto TiO_2 followed by a hydrothermal synthesis and a sol–gel hydrolysis. Doping of PANi resulted in the absorption edge of the fabricated potocatalysts shifting to 421–437 nm and the most distinguished red-shift effect was found in hydrothermal synthesized photocatalysts. The best DEP degradation of 41.5–59.0 and 44.5–67.4% was found in the simulated sunlight system irradiated for 120 min for sol-gel hydrolysis PANi/CNT/TiO_2 photocatalysts and hydrothermal synthesized ones, respectively. The optimum pH was determined at 5.0 and 7.0 for the two PANi/CNT/TiO_2 photocatalysts mentioned above, respectively. The reusability of the sol–gel hydrolyzed photocatalysts up to 5 times was observed no decline in the photodegradation efficiency but less photocatalytic stability of the hydrothermal

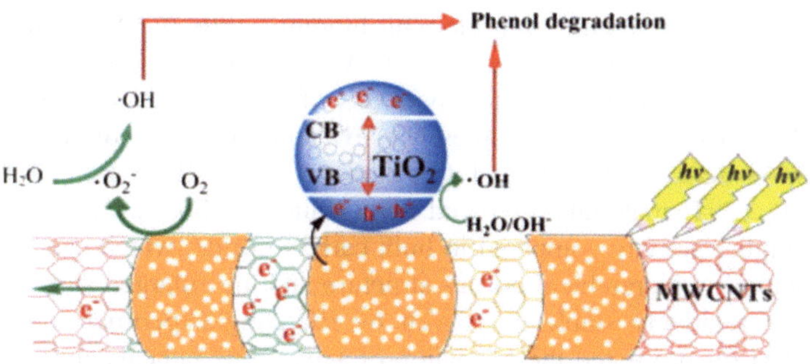

Fig. 8.22 **a** Photocatalytic degradation of phenol monitored as the normalized concentration change versus irradiation time in the presence of CNMT-300 (4), -400 (2), -500 (1), -600 (3), CNTs-MT (5) and CNTs-P25 (6). **b** Schematic showing the photocatalysis of CNMT under visible light irradiation. **c** Respective apparent first-order rate constant determined from the linear graph of $\ln(c/c_0)$ versus time; CNMT-300 (4), -400 (2), -500 (1), -600 (3) and CNTs-MT (5). Reproduced from Liu et al. (2014), Copyright (2016), with permeation from Elsevier

synthesized ones was found. Meanwhile, the active species of OH radicals generated in the DEP degradation system was identified by free radical scavenging experiments, Mohamed and coworkers (Mohamed et al. 2015). TiO_2 synthesized using polymeric template consisting of polyethylene glycol and polyvinyl alcohol (Tev) and loaded with different wt% of Ag (2, 6%) was exploited to create covalent bonds with carboxylate functionalized SWCNT/MWCNT moieties. The synthesized Ag-free Tev–SWCNTs as well as Ag containing Tev–SWCNTs/MWCNTs have been characterized by UV–visible diffuse reflectance, powder XRD, HRTEM, and selected area electron diffraction (SAED), photoluminescence, Raman, FTIR, and N_2 sorptiometry. The materials containing Ag displayed high photocatalytic activity toward degradation of rhodamine B dye under visible irradiation (λ_{max} > 450 nm). Specifically $TevAg_6$–SWCNT has shown the best performance (0.3 g/l catalyst, 20 ppm RhB conc. and 80 min reaction time) due to the synergistic effects derived from TiO_2/Ag°/SWCNT heteroarchitectures. The antibacterial activity of synthesized photocatalysts; under visible light irradiations, towards *Escherichia coli* and *Staphylococcus aureus* was tested by performing bacterial DNA and agar well diffusion method. The results revealed that $TevAg_6$–SWCNT was able to effectively kill both Gram-positive and Gram-negative bacteria. Although $TevAg_6$–SWCNT indicated higher E_g values (1.9 eV) than $TevAg_2$–MWCNT (1.75 cV) and they both exposed not only Ag° nanoparticles but also Ag_2O, the former sample confirmed more lethal action against bacterial growth as well as superior photodegradation activity. This was due to delaying the recombination of electrons and holes, increasing the S_{BET} value as well as decreasing the spherical nanoparticles of Ag° to 3 nm diameter. The mechanisms of the dye degradation and destruction of bacterial cell membranes indicate the efficacy of **.**OH.

Graphitic carbon nitride materials with tri-*s*-triazine- and *s*-triazine-based structures were prepared by thermal condensation of melamine (CNT) and by solution reaction of cyanuric chloride with lithium nitride (CNS), respectively, by Wang and coworkers (Wang et al. 2015). An amphiphilic block copolymer-F68 was used as a soft template for the synthesis of mesoporous carbon nitride. The photocatalytic activity of the samples was evaluated by H_2 evolution from water under visible light irradiation and the degradation of Rose Bengal (RB). The mesoporous CNT materials prepared with Pluronic F68 as template showed markedly higher activity compared to bulk carbon nitride, which can be attributed to its crystallinity, hierarchical porous structure and the enlarged surface area. The catalyst was relatively stable as proven by recycling experiments. Besides, the addition of H_2O_2 promoted the formation of active ·OH radicals and increased the activity of carbon nitride for photodegradation of rose bengal. The carbon nitrides prepared by solution reactions were of very poor activity both in photocatalytic water splitting and rose bengal degradation. The photocatalytic degradation of a cyanobacterial toxin, microcystin-LA (MC-LA), was studied in aqueous solutions under both simulated solar light and visible light irradiation by Sampaio and coworkers (Sampaio et al. 2015). Neat TiO_2 and carbon-based TiO_2 materials, prepared with carbon nanotubes (CNT) or graphene oxide (GO), were compared. The highest photocatalytic activity was obtained with a GO–TiO_2 composite comprising 4 wt% of carbon content (GO–TiO_2-4). Complete conversion of MC–LA was achieved under solar light irradiation in 5 min. GO–TiO_2-4 was also active under visible light illumination, with 88% of MC–LA removal in 2 h. The high photocatalytic activity of GO–TiO_2-4 was attributed to the optimal assembly and interfacial coupling between the TiO_2 nanoparticles and the GO sheets that can effectively inhibit electron/hole recombination. Reaction intermediates of MC–LA photocatalytic degradation were also identified by LC/Q–TOF and LC/MS/MS, most of them resulting from the attack of hydroxyl radicals to the MC–LA molecule under solar light irradiation.

A series of $InVO_4$ incorporated with multiwall carbon nanotubes (CNTs) composite nanofibers were synthesized via an electrospinning technique by Zhang and coworkers (Zhang et al. 2015). The as-collected nanofibers were calcined at 550 C in air to remove polyvinyl pyrrolidone (PVP), which could enable $InVO_4$ to crystallize. $InVO_4$ in the composite illustrated a hollow fibrous morphology and orthorhombic phase, and CNTs were embedded or coated on the $InVO_4$ hollow nanofibers. The photocatalytic performance of the samples was investigated by the degradation rhodamine B (RhB) under visible light irradiation. The CNTs/$InVO_4$ nanofibers in RhB degradation displayed a higher photocatalytic activity than pure $InVO_4$ nanofibers and 10%CNTs/$InVO_4$ nanoparticles. The degradation showed an optimized photocatalytic oxidation for $InVO_4$ nanofibers incorporated with 10 wt% CNTs. The enhanced photocatalytic activity might be ascribed to the role of CNTs as an electron transporter and acceptor in the composites, which could effectively inhibit the charge recombination and facilitate the charge transfer.

Possible mechanism of photocatalytic degradation RhB by CNTs/$InVO_4$ nanofibers under visible light irradiation shows in Fig. 8.23.

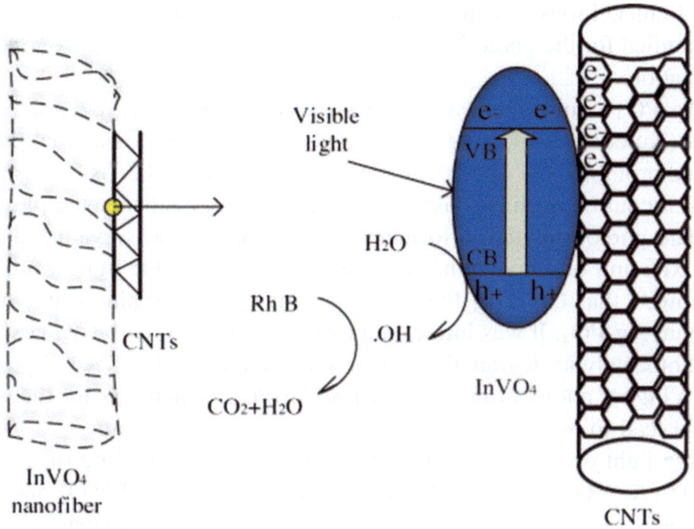

Fig. 8.23 Possible mechanism of photocatalytic degradation RhB by CNTs/$InVO_4$ nanofibers under visible light irradiation. Reproduced from Zhang et al. (2015), Copyright (2015), with permeation from Elsevier

The results of studies on application of TiO_2 and SiO_2 co-precipitated in the presence of CNT to prepare the nanocomposites CNT–TiO_2/SiO_2 are presented by Czech and coworkers (Czech et al. 2015). Nanocomposites were characterized by a large overall surface area (243–252 m^2/g), occurrence of TiO_2 in the anatase form (7–8 nm) and reduced band gap energy (from 3.2 to 2.2 eV). The studies confirmed the role of CNT as a factor not only increasing the overall surface area of the nanocomposites but also allowing for obtaining the systems of uniform distribution of the crystallites TiO_2/SiO_2 on their surface and those activated with visible light. Photocatalytic oxidation carried out for 2 h using the nanocomposites resulted in 75% removal of phenol (UVC irradiation) and a complete loss of methyl orange (UVA irradiation). The nanocomposites containing about 8 wt% of CNT were the most active. Removal of the phenol and methyl orange proceeded according to the first-order kinetics (k_1 for UV decomposition of phenol was 0.4–0.74 × 10^{-2} min^{-1}, and for Vis removal of methyl orange 2–12 × 10^{-2} min^{-1}).

Cobalt-doped nickel ferrite nanoparticles were loaded on the surface of carbon nanotubes using microemulsion method by Singh and coworkers (Czech et al. 2015). In this method, carboxylic group functionalized carbon nanotubes were made to act as a solid support for the beading of ferrite nanoparticles. Sodium dodecyl sulfate was used as a soft template for controlling the shape of nanoparticles. Magnetic studies of $Ni_{1-x}Co_xFe_2O_4$/MWCNTs were carried out using Vibrating Sample Magnetometer (VSM) where all the samples exhibited ferromagnetic behavior. The saturation magnetization increased with increasing concentration of cobalt ion, attributed to the higher magnetic moment of cobalt ions as compared to nickel ions. The photocatalytic activity of ferrite/CNT nanocomposites was also studied for the photodegradation of Rhodamine B (RhB) dye under visible light irradiation.

CNT/Ag_3PO_4 composite was synthesized via a two-step method by Xu and coworkers (Xu et al. 2014). The diffuse-reflection spectra showed that the addition of CNT had promoted the optical absorption of Ag_3PO_4 in the visible region, which meant that it could absorb and use more light energy to enhance the photocatalytic activity. The photocatalytic performance of the samples was evaluated by photocatalytic oxidation of Rhodamine B (RhB) dye under visible light irradiation. The results showed that CNT/Ag_3PO_4 exhibited much higher photocatalytic activity than the pure Ag_3PO_4. It was interesting that CNT/Ag_3PO_4 (0.1 wt%) exhibited the highest photocatalytic degradation efficiency. Moreover, CNT/Ag_3PO_4 (0.1 wt%) exhibited a higher photocurrent than that of Ag_3PO_4, which was for the reason that CNT could lead to good electron transfer between the materials.

A visible light-driven photo fuel cell (PFC) device, consisting of a BiOI-based photoanode (Ag–BiOI/ITO, CNT–BiOI/ITO, and TiO_2–BiOI/ITO) and a Pt photocathode, was used by Shu and coworkers (Shu et al. 2014) to generate electricity by using organic compounds. The photoactivity of BiOI-based photoanode was evaluated in the term of the electricity generation and degradation efficiency for bisphenol A (BPA) organic pollutants. The efficiencies of electricity generation using Ag–BiOI/ITO, CNT–BiOI/ITO, and TiO_2–BiOI/ITO photoanodes were higher in PFC device under visible light compared with BiOI/ITO photoanode. In the mean time, the photocatalytic degradation efficiencies for BPA using Ag–BiOI/ITO, CNT–BiOI/ITO, and TiO_2–BiOI/ITO photoanodes were significantly higher than that of BiOI/ITO. The enhancement in the electricity generation and degradation efficiency for BPA is attributed to the action of additives (Ag, CNT, and TiO_2) on BiOI, which play a key role by driving the electrons to the cathode.

An effective photocatalytic inactivation of *Escherichia coli* K-12 was investigated by Shi and coworkers (Shi et al. 2014) using a series of synthesized Ag/AgX–CNTs (X = Cl, Br, I) composites as photocatalysts under visible light (VL) ($\lambda \geq 400$ nm) irradiation. The results showed that the visible light-driven (VLD) Ag/AgBr–CNTs could completely photocatalytically inactivate 1.5×10^7 cfu mL^{-1} of *E. coli* within 40 min, which was superior to Ag/AgCl–CNTs and Ag/AgI–CNTs. It was found that photocatalytic inactivation of *E. coli* was much more efficient under VL with 435 nm wavelength and the photogenerated holes played an important role in this photocatalytic inactivation system. In addition, the stability and deactivation mechanism of Ag/AgX–CNTs photocatalysts during photocatalytic bacterial inactivation were also studied, and the results showed that the organic debris of decomposed bacteria may be absorbed on the active sites of the photocatalysts leading to the decrease of the photocatalytic activity. Figure 8.24 shows schematic photocatalytic inactivation processes and charge transfer of the Ag/AgBr–CNTs photocatalyst under visible light irradiation.

Carbon nanotube modified $Zn_{0.83}Cd_{0.17}S$ nanocomposite was prepared by Yao and coworkers (Yao et al. 2014) via a solvothermal method. CNTS can efficiently suppress the growth of chalcogenide nanoparticles and improve the dispersity of the

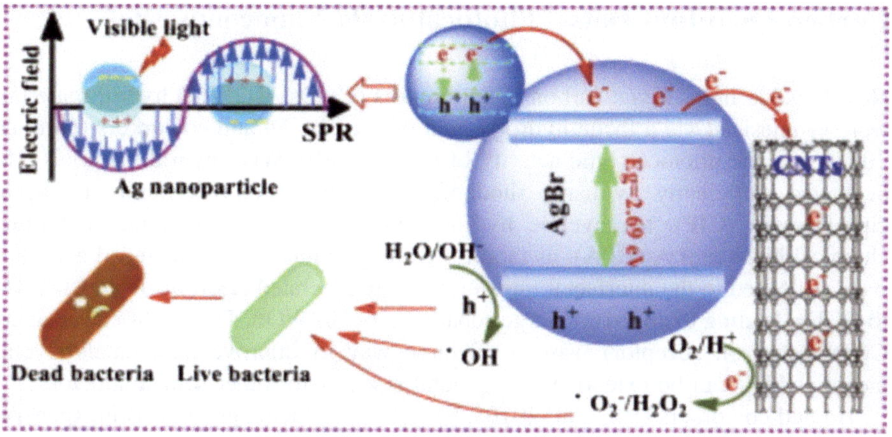

Fig. 8.24 Schematic photocatalytic inactivation processes and charge transfer of the Ag/AgBr-CNTs photocatalyst under visible light irradiation. Reproduced from Shi et al. (2014), Copyright (2014), with permeation from Elsevier

nanocomposite. The absorption edges of $Zn_{0.83}Cd_{0.17}S$/CNTs nanocomposites red-shift and the response of the visible light region (500–800 nm) is strengthened with the increase of CNTs contents in the samples. The prepared $Zn_{0.83}Cd_{0.17}S$/CNTs nanocomposites exhibit an enhanced photocatalytic H_2-production activity and an optimum amount of CNT is determined to be ca. 0.25 wt%, at which the $Zn_{0.83}Cd_{0.17}S$/CNTs displays the highest photocatalytic activity under the irradiation of Xe lamp, with an H_2 production rate of 5.41 mmol h^{-1} g^{-1}. Furthermore, the prepared $Zn_{0.83}Cd_{0.17}S$/CNTs nanocomposite is photostable and no photocorrosion was observed after photocatalytic recycling, compared with pure $Zn_{0.83}Cd_{0.17}S$ photocatalyst.

Pd–ZnO nanocatalyst supported on multiwalled carbon nanotubes was synthesized by via Mohamed and coworkers (Mohamed and Salam 2014) a modified sol–gel method, and the prepared photocatalyst was used for the environmental remediation of aqueous Hg(II) via photocatalytic reduction under visible light. The results showed that both Pd and ZnO nanoparticles were well dispersed over the MWCNTs, and a uniform nanocomposite was formed. The results also illustrated that Pd doping can eliminate the recombination of electron–hole pairs in the catalyst, and the presence of MWCNTs in ZnO composite can change surface properties to achieve sensitivity to visible light. The results demonstrated that optimum mass ratio of CNT:ZnO:Pd were 0.04:1.0:0.08, which resulted in the exceptional performance of the photocatalyst to reduce about 100% of Hg(II) in a 100 mg L solution within 30 min.

Carbon Quantum Dots—Photocatalysts Nanocomposites

Novel visible light-driven carbon quantum dots (CQDs)/Bi2WO6 hybrid materials were synthesized via a facile hydrothermal method by Di and coworkers (Di et al. 2015a, b). The photocatalytic activity of the CQDs/Bi2WO6 hybrid materials was evaluated sufficiently by using rhodamine B (RhB), colorless antibiotic agent ciprofloxacin (CIP), tetracycline hydrochloride (TC), and endocrine disrupting chemical bisphenol A (BPA), as target organic pollutants. The enhanced activities were attributed to the interfacial transfer of photogenerated electrons from Bi2WO6 to CQDs, leading to effective charge separation of Bi2WO6. The modification using CQDs (electron acceptor) was an effective way to improve photocatalytic efficiency, which can be extended to a general strategy for other semiconductors.

Hu and coworkers (Hu et al. 2016a, b) reported a facile, green, and inexpensive top-down strategy towards fluorescent carbon dots (CDs) from coal without incurring the burden of tedious or inefficient postprocessing steps and facing the danger of highly toxic gas liberation. The presented approach shows a high yield and great potential for carbon dot production scale-up using coal, one of our most abundant and low-cost resources. The prepared CDs demonstrate photocatalytic behavior capable of rapidly degrading organic dyes under visible light.

A novel one-step ionic liquid-induced strategy has been reported by Xia (Xia et al. 2016a, b) and coworkers for the controlled synthesis of carbon quantum dots (CQDs)/BiOX (X = Br, Cl) hybrid nanosheets with tunable CQDs loading contents. Such synthetic process allows the CQDs well dispersed on the surface of BiOX nanosheets. Three different types of pollutants, such as phenol rhodamine B (RhB), antibacterial agent ciprofloxacin (CIP), and endocrine disrupting chemical bisphenol A (BPA) were chosen to evaluate the photocatalytic activity of CQDs/BiOX composite nanosheets. They show very interesting CQDs loading content and X composition-dependent photocatalytic activity with 3 wt% CQDs/BiOBr nanosheets showing the highest photocatalytic activity (much better than pure BiOBr nanosheets) for the degradation of RhB, CIP and BPA under visible light irradiation. The results reveal that there are three factors in promoting the photocatalysis of 3 wt% CQD/BiOBr nanosheets: high visible light absorbance, high separation efficiency of photoinduced electrons and holes and lower resistance.

Qian and coworkers (Qian et al. 2016) incorporated highly stable carbon quantum dots (CQDs) with Bi2WO6 to sufficiently photocatalytic removal of gaseous volatile organic compounds (VOCs) utilize solar energy. With the facile decoration of CQDs, the composite photocatalysts of CQDs/Bi2WO6 extend the absorption into visible light region and improve the photoexcited charge separation in comparison with pristine Bi2WO6. The CQDs/Bi2WO6 exhibited higher photocatalytic oxidation activities towards acetone and toluene under both UV–vis and visible light irradiation. In all, CQDs could be a promising candidate for visible light photocatalysts due to their superior ability to extend the visible absorption and suppress the photoexcited charge recombination.

Novel nitrogen-doped carbon quantum dots (N-CQDs)/BiOBr ultrathin nanosheets photocatalysts have been prepared via reactable ionic liquid assisted solvothermal process by Di and coworkers (Di et al. 2016a, b). The one-step formation mechanism of the N-CQDs/BiOBr ultrathin nanosheets was based on the initial formation of strong coupling between the ionic liquid and N-CQDs as well as subsequently results in tight junctions between N-CQDs and BiOBr with homodisperse of N-CQDs. The photocatalytic activity of the as-prepared photocatalysts was evaluated by the degradation of different pollutants under visible light irradiation such as ciprofloxacin (CIP), rhodamine B (RhB), tetracycline hydrochloride (TC), and bisphenol A (BPA). The improved photocatalytic performance of N-CQDs/BiOBr materials was ascribed to the crucial role of N-CQDs, which worked as photocenter for light harvesting, charge separation center for separating the charge carriers, and active center for degrading the pollutants.

Carbon quantum dots (CQDs)-modified BiOCl ultrathin nanosheets photocatalyst was synthesized via a facile solvothermal method by Di and coworkers (Di et al. 2015a, b). The photocatalytic activity of the obtained CQDs modified BiOCl ultrathin nanosheets photocatalyst was evaluated by the degradation of bisphenol A (BPA) and rhodamine B (RhB) under ultraviolet, visible, and near-infrared light irradiation. The CQDs/BiOCl materials exhibited significantly enhanced photocatalytic performance as compared with pure BiOCl and the 5 wt% CQDs/BiOCl materials displayed the best performance, which showed a broad spectrum of photocatalytic degradation activity. The crucial role of CQDs for the improved photocatalytic activity was mainly attributed to the superior electron transfer ability, enhanced light harvesting, and boosted catalytic active sites.

Carbon Quantum Dots-Induced Ultrasmall BiOI Nanosheets with Assembled Hollow Structures for Broad Spectrum Photocatalytic Activity and Mechanism Insight.

Carbon quantum dots (CQDs) induced ultrasmall BiOI nanosheets with assembled hollow microsphere structures were prepared via ionic liquids 1-butyl-3-methylimidazolium iodine ([Bmim]I)-assisted synthesis method at room temperature condition by Di and coworkers (Di et al. 2016a, b). The CQDs/BiOI hollow microspheres structure displayed improved photocatalytic activities than pure BiOI for the degradation of three different kinds of pollutants, such as antibacterial agent tetracycline (TC), endocrine disrupting chemical bisphenol A (BPA), and phenol rhodamine B (RhB) under visible light, light above 580 nm, or light above 700 nm irradiation, which showed the broad spectrum photocatalytic activity. The introduction of CQDs could induce the formation of ultrasmall BiOI nanosheets with assembled hollow microsphere structure, strengthen the light absorption within full spectrum, increase the specific surface areas and improve the separation efficiency of the photogenerated electron–hole pairs. Benefiting from the unique structural features, the CQDs/BiOI microspheres exhibited excellent photoactivity. Figure 8.25 shows schematic drawing illustrating the mechanism of the charge separation and photodegradation process over CQDs/BiOI photocatalysts under visible light irradiation.

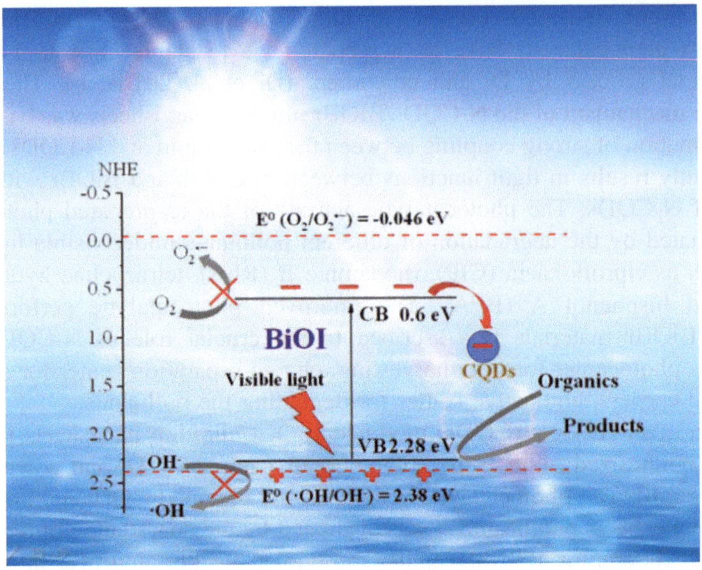

Fig. 8.25 Schematic drawing illustrating the mechanism of the charge separation and photodegradation process over CQDs/BiOI photocatalysts under visible light irradiation. Reproduced from Di et al. (2016b), Copyright (2016), with permeation from ACS

In summary, frontier scientific and technological research into the fields of energy and environmental protection are becoming increasingly challenging because of the need for key materials with high efficiency and functionality. Nanocarbon–photocatalyst systems have been widely investigated and are promising materials for future high activity photocatalysts. Apart from providing a high-surface area support and immobilization for ceramic photocatalyst particles, the presence of the carbonaceous material and nanostructuring may facilitate enhanced photocatalytic activity via one or all of the three primary mechanisms: minimization of electron/hole recombination, band gap tuning/photosensitization and provision of high quality highly adsorptive active sites. It is anticipated that the number of new carbophotocatalyzed conversions will continue to grow and new forms of carbonaceous nanomaterials with engineered morphology or functionalities will emerge as powerful photocatalysts in the near future.

References

Arbogast JW, Darmanyan AP, Foote CS, Diederich FN, Whetten RL, Rubin Y, Alvarez MM, Anz SJ (1991) Photophysical properties of sixty atom carbon molecule (C60) J Phys Chem 95(1):11–12

Arunachalam P, Zhang S, Abe T, Komura M, Iyoda T, Nagai K (2016) Weak visible light ($\sim$mW/cm^2) organophotocatalysis for mineralization of amine, thiol and aldehyde by biphasic cobalt phthalocyanine/fullerene nanocomposites prepared by wet process. Appl Catal B Environ 193:240–247

Bagri A, Mattevi C, Acik M, Chabal YJ, Chhowalla M, Shenoy VB (2010) Structural evolution during the reduction of chemically derived graphene oxide. Nat Chem 29(7):581–587

Bai Y, Ye L, Wang L, Shi X, Wang P, Bai W, Wong PK (2016) g-C_3N_4/$Bi_4O_5I_2$ heterojunction with I_3^-/I^- redox mediator for enhanced photocatalytic CO_2 conversion. Appl Catal B 194:98–104

Bera R, Mandal S, Mondal B, Jana B, Nayak SK, Patra A (2016) Graphene–porphyrin nanorod composites for solar light harvesting. ACS Sustain Chem Eng 4(3):1562–1568

Chen Y, Tian G, Mao G, Li R, Xiao Y, Han Y (2016) Facile synthesis of well-dispersed Bi_2S_3 nanoparticles on reduced graphene oxide and enhanced photocatalytic activity. Appl Surf Sci 378:231–238

Choi SK, Kim S, Lim SK, Park HK (2010) Photocatalytic comparison of TiO_2 nanoparticles and electrospun TiO_2 nanofibers: effects of mesoporosity and interparticle charge transfer. J Phys Chem C 114(39):16475–16480

Czech B, Buda W, Pasieczna-Patkowska S, Oleszczuk P (2015) MWCNT–TiO_2–SiO_2 nanocomposites possessing the photocatalytic activity in UVA and UVC. Appl Catal B 162:564–572

Dai K, Peng T, Ke D, Wei B (2009) Photocatalytic hydrogen generation using a nanocomposite of multi-walled carbon nanotubes and TiO2 nanoparticles under visible light irradiation. Nanotechnol 20(12):125603

Deng X, Zhang Q, Zhao Q, Ma L, Ding M, Xu X (2015) Effects of architectures and H_2O_2 additions on the photocatalytic performance of hierarchical Cu_2O nanostructures. Nanno Res Lett 10:8

Di J, Xia J, Ge Y, Li H, Ji H, Xu H, Zhang Q, Li H, Li M (2015a) Novel visible-light-driven CQDs/Bi_2WO_6 hybrid materials with enhanced photocatalytic activity toward organic pollutants degradation and mechanism insight. Appl Catal B 168–169:51–61

Di J, Xia J, Ji M, Wang B, Yin S, Zhang Q, Chen Z, Li H (2015b) Carbon quantum dots modified biocl ultrathin nanosheets with enhanced molecular oxygen activation ability for broad spectrum photocatalytic properties and mechanism insight. ACS Appl Mater Interfaces 7 (36):20111–20123

Di J, Xia J, Ji M, Wang B, Li X, Zhang Q, Chen Z, Li H (2016a) Nitrogen-doped carbon quantum dots/BiOBr ultrathin nanosheets: in situ strong coupling and improved molecular oxygen activation ability under visible light irradiation. ACS Sustain Chem Eng 4(1):136–146

Di J, Xia J, Ji M, Wang B, Yin S, Xu H, Chen Z, Li H (2016b) Carbon quantum dots induced ultrasmall BiOI nanosheets with assembled hollow structures for broad spectrum photocatalytic activity and mechanism insight. Langmuir 32(8):2075–2084

Dong F, Wang Z, Sun Y, Ho WK, Zhang H (2013) Engineering the nanoarchitecture and texture of polymeric carbon nitride semiconductor for enhanced visible light photocatalytic activity. J Colloid Interface Sci 401:70–79

Du Y, Cheng Z, Yu Z, Do SX, Wang X, Liu L (2012) Hydrothermal synthesized bismuth ferrites particles: thermodynamic, structural, and magnetic properties. J Nanosci Nanotechnol 12 (2):1684–1687

Eda G, Lin YY, Mattevi C, Yamaguchi H, Chen HA, Chen IS, Chen CW, Chhowalla M (2010) Blue photoluminescence from chemically derived graphene oxide. Adv Mater 22(4):505–509

Fan C, Chen C, Wang J, Fu X, Ren Z, Qian G, Wang Z (2015) Black hydroxylated titanium dioxide prepared via ultrasonication with enhanced photocatalytic activity. Sci Rep 4:11712

Fan C, Liu Q, Ma T, Shen J, Yang Y, Tang H, Wang Y, Yang J (2016) Fabrication of 3D $CeVO_4$/graphene aerogels with efficient visible-light photocatalytic activity. Ceram Int 42:10487–10492

Fujishima A, Honda K (1972) Electrochemical photolysis of water at a semiconductor electrode. Nature 238:37–38

Ghaffar I, Warsi MF, Shahid M, Shakir I (2016) Unprecedented photocatalytic activity of carbon coated/MoO3 core–shell nanoheterostructurs under visible light irradiation. Physica E Low-Dimensional Syst Nanostruct 79:1–7

Gong B, Lu Y, Wu P, Huang Z, Zhu Y, Dang Z, Zhu N, Lu G, Huang J (2016) Enhanced photocatalytic activity over $Cd_{0.5}Zn_{0.5}S$ with stacking fault structure combined with Cu^{2+} modified carbon nanotubes. Appl Surf Sci 365:280–290

Hong J, Zhang W, WangY, Zhou T, Xu R (2014) Photocatalytic reduction of carbon dioxide over self-assembled carbon nitride and layered double hydroxide: the role of carbon dioxide enrichment. Chem Cat Chem 6(8):2315–2321

Hu S, Wei Z, Chang Q, Trinchi A, Yang J (2016a) A facile and green method towards coal-based fluorescent carbon dots with photocatalytic activity. Appl Surf Sci 378:402–407

Hu XY, Zhou K, Chen BY, Chang CT (2016b) Graphene/TiO_2/ZSM-5 composites synthesized by mixture design were used for photocatalytic degradation of oxytetracycline under visible light: mechanism and biotoxicity. Appl Surf Sci 362:329–334

Huang M, Yu J, Deng C, Huang Y, Fan M, Li B, Tong Z, Zhang F, Dong L (2016) 3D nanospherical $Cd_xZn_{1-x}S$/reduced graphene oxide composites with superior photocatalytic activity and photocorrosion resistance. Appl Surf Sci 365:227–239

Hung CH, Yuan C, Li HW (2016) Photodegradation of diethyl phthalate with PANi/CNT/TiO_2 immobilized on glass plate irradiated with visible light and simulated sunlight—effect of synthesized method and Ph. J Hazard Mater (In Press, Corrected Proof)

Johns JE, Hersam MC (2013) Atomic covalent functionalization of graphene. Acc Chem Res 46:77–86

Kang Z , Tsang CHA, Wong NB, Zhang Z, Lee ST (2007) Silicon quantum dots: A general photocatalyst for reduction, decomposition, and selective oxidation reactions. J Am Chem Soc129(40):12090–12091

Kudin KN (2008) Zigzag graphene nanoribbons with saturated edges. ACS Nano 2(3):516–522

Kudin KN, Ozbas B, Schniepp HC, Prud'homme RK, Aksay IA, Car R (2008) Raman spectra of graphite oxide and functionalized graphene sheets. Nano Lett 8(1):36–41

Kuriki R, Ishitani O, Maeda K (2016) Unique solvent effects on visible-light CO_2 reduction over ruthenium(II)-complex/carbon nitride hybrid photocatalysts. ACS Appl Mater Interfaces 8 (9):6011–6018

Lan ZA, Zhang G, Wang X (2016) A facile synthesis of Br-modified g-C_3N_4 semiconductors for photoredox water splitting. Appl Catal B Environ 192:116–125

Lim SY, Shen W, Gao Z (2015) Carbon quantum dots and their applications. Chem Soc Rev 1

Liu G, Wang L, Yang HG, Cheng HG, Lu GQ (2010) Titania-based photocatalysts—crystal growth, doping and heterostructuring. J Mater Chem 5

Liu Y, Sun D, Askari S, Patel J, Macias-Montero M, Mitra S, Zhang R, Wen-Feng Lin WF, Mariotti D, Maguire P (2015) Enhanced dispersion of TiO_2 nanoparticles in a TiO_2/PEDOT: PSS hybrid nanocomposite via plasmaliquid interactions Sci Rep 5

Leary R, Westwood A (2011) carbonaceous nanomaterials for the enhancement of TiO_2. Photocatalysis Carbon 49:741–772

Lee G, Cho K (2009) Electronic structures of zigzag graphene nanoribbons with edge hydrogenation and oxidation. Phys Rev B 79:165440

Lei ZD, Wang JJ, Wang L, Yang XY, Xu G, Tang L (2016) Efficient photocatalytic degradation of ibuprofen in aqueous solution using novel visible-light responsive graphene quantum dot/$AgVO_3$ nanoribbons. J Hazard Mater 312:298–306

Li G, Jiang B, Li X, Lian Z, Xiao S, Zhu J, Zhang D, Li H (2013) C_{60}/$Bi_2TiO_4F_2$ heterojunction photocatalysts with enhanced visible-light activity for environmental remediation. ACS Appl Mater Interfaces 5(15):7190–7197

Li L, Yu L, Lin Z, Yang G (2016a) Reduced TiO_2-graphene oxide heterostructure as broad spectrum-driven efficient water-splitting photocatalysts. ACS Appl Mater Interfaces 8 (13):8536–8545

Li Q, Guo B, Yu J, Ran J, Zhang B, Yan H, Gong JR (2011) Highly efficient visible-light-driven photocatalytic hydrogen production of CdS-cluster-decorated graphene nanosheets. J Am Chem Soc 133(28):10878–10884

Li M, Zhang L, Fan X, Wu M, Du Y, Wang M, Kong Q, Zhang L, Shi J (2016b) Dual synergetic effects in MoS_2/pyridine-modified g-C_3N_4 composite for highly active and stable photocatalytic hydrogen evolution under visible light. Appl Catal B 190:36–43

Li S, Dong G, Hailili R, Yang L, Li Y, Wang F, Zeng Y, Wang C (2016c) Effective photocatalytic H_2O_2 production under visible light irradiation at g-C_3N_4 modulated by carbon vacancies. Appl Catal B 190:26–35

Li X, Dong C, Wu KL, Xia SH, Hu Y, Ling M, Liu K, Lu XL, Ye Y, Wei XW (2016d) Synthesis of nitrogen-doped graphene–BiOBr nanocomposites with enhanced visible light photocatalytic activity. Mater Lett 164:502–504

Li Y, Zhang Z, Pei L, Li X, Fan T, Ji J, Shen J, Ye M (2016e) Multifunctional photocatalytic performances of recyclable Pd-$NiFe_2O_4$/reduced graphene oxide nanocomposites via different co-catalyst strategy. Appl Catal B 190:1–11

Li Y, Chen L, Wang Y, Zhu L (2016f) Advanced nanostructured photocatalysts based on reduced graphene oxide-flower-like Bi_2WO_6 composites for an augmented simulated solar photoactivity activity. Mater Sci Eng B 210:29–36

Lian X, Yan B (2016) A post synthetic modified MOF hybrid as heterogeneous photocatalyst for α- phenethyl alcohol and reusable fluorescence sensor. Inorg Chem 55(22):11831–11838

Lian Z, Xu P, Wang W, Zhang D, Xiao S, Li X, Li G (2015) C_{60}-decorated CdS/TiO_2 mesoporous architectures with enhanced photostability and photocatalytic activity for H_2 evolution. ACS Appl Mater Interfaces 7(8):4533–4540

Liu Q, Liu ZF, Zhang XY, Yang LY, Zhang N, Pan GL, Yin SG, Chen YS, Wei J (2009) Polymer photovotaic cells based on solution-processable graphene and P3HT. Adv Funct Mater 19:894–904

Liu G,Wang L,Yang HG, Cheng HG, Lu GQ (2010) Titania-based photocatalysts—crystal growth, doping and heterostructuring. J Mater Chem 5

Liu C, Li Y, Xu P, Li M, Zen M (2014) Carbon nanotube embedded mesoporous titania pore-hole inorganic hybrid materials with high thermal stability, improved crystallinity and visible-light driven photocatalytic performance. Microporous Mesoporous Mater 195:319–329

Liu H, Chen Z, Wang Y (2016a) Plasmonic Ag coated BiOBr 0.2I0.8 nanosheets grown on graphene with excellent visible-light photocatalytic activity. J Photochem Photobiol A 326:30–40

Liu Q, Chen T, Guo Y, Zhang Z, Fang X (2016b) Ultrathin g-C_3N_4 nanosheets coupled with carbon nanodots as 2D/0D composites for efficient photocatalytic H_2 evolution. Appl Catal B 193:248–258

Liu S, Li D, Sun H, Ang HM, Tadé MO, Wang S (2016c) Oxygen functional groups in graphitic carbon nitride for enhanced photocatalysis. J Colloid Interface Sci 468:176–182

Loh KP, Bao Q, Eda G, Chhowalla M (2010) Graphene oxide as a chemically tunable platform for optical applications. Nat Chem 2(12):1015–1024

Lu G, Yu K, Wen Z, Chen J (2013) Semiconducting graphene: converting graphenefrom semimetal to semiconductor. Nanoscale 5:1353–1368

Ma L, Fan H, Wang J, Zhao Y, Tian H, Dong G (2016) Water-assisted ions in situ intercalation for porous polymeric graphitic carbon nitride nanosheets with superior photocatalytic hydrogen evolution performance. Appl Catal B Environ 190:93–102

Matsumoto Y, MiuraY, Takata S (2016) Thickness-dependent flat band potential of anatase TiO_2(001) epitaxial films on Nb:$SrTiO_3$(001) investigated by UHV-electrochemistry approach. J Phys Chem C 120(3):1472–1477

Matsumoto Y, Koinuma M, Ida S, Hayami S, Taniguchi T, Hatakeyama K, Tateishi H, Watanabe Y, Amano S (2011) Photoreaction of graphene oxide nanosheets in water. J Phys Chem C 115(39):19280–19286

Maeda K (2014) Rhodium-doped barium titanate perovskite as a stable p-type semiconductor photocatalyst for hydrogen evolution under visible light. ACS Appl Mater Interfaces 6 (3):2167–2173

Martin DJ, Philip Reardon PJT, Moniz STA, Tang J (2014) Visible light-driven pure water splitting by a nature-inspired organic semiconductor-based system. J Am Chem Soc 136 (36):12568–12571

Mkhoyan KA, Contryman AW, Silcox J, Stewart DA, Eda G, Mattevi C, Miller S, Chhowalla M (2009) Atomic and electronic structure of graphene-oxide. Nano Lett 9(3):1058–1063

Mohamed RM, Salam MA (2014) Photocatalytic reduction of aqueous mercury(II) using multi-walled carbon nanotubes/Pd-ZnO nanocomposite. Mater Res Bull 50:85–90

Mohamed MM, Osman G, Khairou KS (2015) Fabrication of Ag nanoparticles modified TiO_2–CNT heterostructures for enhanced visible light photocatalytic degradation of organic pollutants and bacteria. J Environ Chem Eng 3(3):1847–1859

Nguyen KC, Ngoc MP, Nguyen MV (2016) Enhanced photocatalytic activity of nanohybrids TiO_2/CNTs materials. Mater Lett 165:247–251

Niu P, Yin LC, Yang YQ, Liu G, Cheng HM (2014) Increasing the visible light absorption of graphitic carbon nitride (Melon) photocatalysts by homogeneous self-modification with nitrogen vacancies. Adv Mater 26(47):8046–8052

Nuengmatcha P, Chanthai S, Mahachai R, Oh WC (2016) Visible light-driven photocatalytic degradation of rhodamine B and industrial dyes (texbrite BAC-L and texbrite NFW-L) by ZnO-graphene-TiO_2 composite. J Environ Chem Eng 4(2):2170–2177

Park H, hul Lee YC, Choi BG, Choi YS, Yang JW, Hong WH (2010) Energy transfer in ionic-liquid-functionalized inorganic nanorods for highly efficient hotocatalytic applications. Small 6(2):290–295

Patil SP, Patil RP, Mahajan VK, Sonawane GH, Shrivastava VS, Sonawane S (2016) Facile sonochemical synthesis of BiOBr-graphene oxide nanocomposite with enhanced photocatalytic activity for the degradation of Direct green. Mater Sci Semicond Process 52:55–61

Pham TT, Nguyen-Huy C, Shin EW (2016) Facile one-pot synthesis of nickel-incorporated titanium dioxide/graphene oxide composites: enhancement of photodegradation under visible-irradiation. Appl Surf Sci 377:301–310

Putri LK, Tan LL, Ong WJ, Chang WS, Chai SP (2016) Graphene oxide: exploiting its unique properties toward visible-light-driven photocatalysis. Appl Mater Today 4:9–16

Qian X, Yue D, Tian Z, Reng M, Zhu Y, Kan M, Zhang T, Zhao Y (2016) Carbon quantum dots decorated Bi2WO6 nanocomposite with enhanced photocatalytic oxidation activity for VOCs. Appl Catal B 193:16–21

Qu A, Xinmei Xu X, Xie H, Zhang Y, Li Y, Wang J (2016) Effects of calcining temperature on photocatalysis of g-C3N4/TiO2 composites for hydrogen evolution from water. Mater Res Bull 80:167–176

Ran R, Meng X, Zhang Z (2016) Facile preparation of novel graphene oxide-modified $Ag_2O/Ag_3VO_4/AgVO_3$ composites with high photocatalytic activities under visible light irradiation. Appl Catal B 196:1–5

Ravichandran K, Chidhambaram N, Gobalakrishnan S (2016) Copper and Graphene activated ZnO nanopowders for enhanced photocatalytic and antibacterial activities. J Phys Chem Solids 93:82–90

Sampaio MJ, Silva CG, Silva AMT, Pastrana-Martínez LM, Han C, Morales-Torres S, Figueiredo JL, Dionysiou DD, Faria JL (2015) Carbon-based TiO_2 materials for the degradation of Microcystin-LA. Appl Catal B 170–171:74–82

Serp P, Corrias M, Kalck P (2003) Carbon nanotubes and nanofibers in catalysis. Appl Catal A 253(2):337–358

Shi H, Li G, Sun H, An T, Zhao H, Wong PK (2014) Visible-light-driven photocatalytic inactivation of E. coli by Ag/AgX-CNTs (X = Cl, Br, I) plasmonic photocatalysts: Bacterial performance and deactivation mechanism. Appl Catal B 158–159:301–307

Shi L, Wang T, Zhang H, Chang K, Meng X, Liu H, Ye J (2015) An amine-functionalized iron (III) metal–organic framework as efficient visible-light photocatalyst for Cr(VI) reduction. Adv Sci 2(3)

Shu D, Wu J, Gong Y, Li S, Hu L, Yang Y, He C (2014) BiOI-based photoactivated fuel cell using refractory organic compounds as substrates to generate electricity. Catal Today 224:13–20

Singh AK, Yakobson BI (2009) Electronics and magnetism of patterned graphene nanoroads. Nano Lett 9(4):1540–1543

Singh C, Bansal S, Kumar V, Singhal S (2015) Beading of cobalt substituted nickel ferrite nanoparticles on the surface of carbon nanotubes: a study of their synthesis mechanism, structure, magnetic, optical and their application as photocatalyst. Ceram Int 41(3):3595–3604

Singh AK, Mathew K, Zhuang HL, Hennig RG (2015) Computational Screening of 2D Materials for Photocatalysis. J Phys Chem Lett 196(6):1087–1098

Song T, Huo J, Liao T, Zeng J, Qin J, Zeng J (2016) Fullerene [C_{60}] modified $Cr_{2-x}Fe_xO_3$ nanocomposites for enhanced photocatalytic activity under visible light irradiation. Chem Eng J 287:359–366

Sun T, Jiang HY, Ma CC, Mao F, Xue B (2016) Ag/g-C_3N_4 photocatalysts: microwave-assisted synthesis and enhanced visible-light photocatalytic activity. Catal Commun 79:45–48

Tajima T, Sakata W, Wada T, Tsutsui A, Nishimoto S, Miyake M, Takaguchi Y (2011) Photosensitized hydrogen evolution from water using a single-walled carbon nanotube/fullerodendron/SiO2 coaxial nanohybrid. Mater Adv 23(48):5750–5754

Tryba B (2008) Increase of the photocatalytic activity of TiO by carbon and iron modifications. Int J Photoenergy 2008

Vileno B, Sienkiewicz A, Lekka M, Kulik AJ, Forró, L (2004) In vitro assay of singlet oxygen generation in the presence of water-soluble derivatives of C60. Carbon 42(5–6):1195–1198

Wang X, Zhi LJ, Mullen K (2008) Transparent, conductive graphene electrodes for dye-sensitized solar cells. Nano Lett 8:323–327

Wang X, Blechert S, Antonietti M (2012) Polymeric graphitic carbon nitride for heterogeneous photocatalysis. ACS Catal 2(8):1596–1606

Wang X, Yu JC, Ho C, Hou Y, Fu X (2005) Photocatalytic activity of a hierarchically macro/mesoporous titania. Langmuir 21(6):2552–2559

Wang X, Maeda K, Thomas A, Takanabe K, Gang Xin G, Carlsson JM, Domen K, Antonietti M (2009) A metal-free polymeric photocatalyst for hydrogen production from water under visible light. Nat Mater 8:76–80

Wang Y, Farooq Ibad M, Kosslick H, Harloff J, TBeweries T, Radnik J, Schulz A, Tschierlei S, Lochbrunner S, Guo X (2015) Synthesis and comparative study of the photocatalytic performance of hierarchically porous polymeric carbon nitrides. Microporous Mesoporous Mater 211:182–191

Wang H, Yuan X, Wang H, Chen X, Wu Z, Jiang L, Xiong W, Zeng G (2016a) Facile synthesis of Sb_2S_3/ultrathin g-C_3N_4 sheets heterostructures embedded with g-C_3N_4 quantum dots with enhanced NIR-light photocatalytic performance. Appl Catal B 193:36–46

Wang L, Wen M, Wang W, Momuinou N, Wang Z, Li S (2016b) Photocatalytic degradation of organic pollutants using rGO supported TiO_2-CdS composite under visible light irradiation. J Alloy Compd 683:318–328

Wang Q, Li H, Yang JH, Sun Q, Li Q, Yang J (2016c) Iron phthalocyanine-graphene donor-acceptor hybrids for visible-light-assisted degradation of phenol in the presence of H_2O_2. Appl Catal B 192:182–192

Wei M, Wan J, Hu Z, Peng Z, Wang B (2016) Enhanced photocatalytic degradation activity over TiO_2 nanotubes co-sensitized by reduced graphene oxide and copper(II) meso-tetra (4-carboxyphenyl)porphyrin. Appl Surf Sci 377:149–158

Woan K, Pyrgiotakis G, Sigmund W (2009) Photocatalytic carbon-nanotube–TiO_2 composites. Adv Mate 21(21):2233–2239

Xia HL, Zhuang HS ,Zhang T, Xiao DC (2007) Photocatalytic degradation of acid blue 62 over CuO-SnO_2 nanocomposite photocatalyst under simulated sunlight. J Environ Sci 19(9):1111–1141

Xia D, An T, Li G, Wang W, Zhao H, Wong PK (2016a) Synergistic photocatalytic inactivation mechanisms of bacteria by graphene sheets grafted plasmonic Ag-AgX (X = Cl, Br, I) composite photocatalyst under visible light irradiation. Water Res 99:149–161

Xia J, Di J, Li H, Xu H, Li H, Guo S (2016b) Ionic liquid-induced strategy for carbon quantum dots/BiOX (X = Br, Cl) hybrid nanosheets with superior visible light-driven photocatalysis. Appl Catal B 181:260–269

Xing M, Zhang J, Chen F, Tian B (2011) An economic method to prepare vacuum activated photocatalysts with high photo-activities and photosensitivities. Chem Commun 17

Xing M, Li X, Zhang J (2014) Synergistic effect on the visible light activity of Ti^{3+} doped TiO2 nanorods/boron doped graphene composite. Sci Rep 4

Xu SH, Feng DL, Li DX, Shang G, Wen F (2008) Preparation of magnetic photocatalyst TiO_2 supported on $NiFe_2O_4$ and effect of magnetic carrier on photocatalytic activity. Chin J Chem 26(5):842–846

Xu H, Wang C, Song Y, Zhu J, Xu Y, Yan J, Song Y, Li H (2014) CNT/Ag_3PO_4 composites with highly enhanced visible light photocatalytic activity and stability. Chem Eng J 241:35–42

Xu T, Ni D, Chen X, Wu F, Ge P, Lu W, Hu H, Zhu Z, Chen W (2016a) Self-floating graphitic carbon nitride/zinc phthalocyanine nanofibers for photocatalytic degradation of contaminants. J Hazard Mater 317:17–26

Xu X, Ming F, Hong J, Xie Y, Wang Z (2016b) Three-dimensional porous aerogel constructed by Bi_2WO_6 nanosheets and graphene with excellent visible-light photocatalytic performance. Mater Lett 179:52–56

Yan H (2012) Soft-templating synthesis of mesoporous graphitic carbon nitride with enhanced photocatalytic H_2 evolution under visible light. Chem Comm 28

Yang MQ, Zhang N, Xu YJ (2013) Synthesis of fullerene–, carbon nanotube–, and graphene–TiO_2 nanocomposite photocatalysts for selective oxidation: a comparative study. ACS Appl Mater Interfaces 5(3):1156–1164

Yamakoshi Y, Umezawa N, Ryu A, Arakane K, Miyata N, Goda Y, Masumizu T, Nagano T (2003) Active oxygen species generated from photoexcited fullerene (C_{60}) as potential medicines: O_2^{-*} versus $^1O_2$. J Am Chem Soc 125(42):12803–12309

Yao Z, Wang L, Zhang Y, Yu Z, Jiang Z (2014) Carbon nanotube modified $Zn_{0.83}Cd_{0.17}S$ nanocomposite photocatalyst and its hydrogen production under visible-light. Int J Hydrogen Energy 39(28):15380–15386

Yu J, LowJ, Xiao W, Zhou P, Jaroniec M(2014) Enhanced photocatalytic CO_2–reduction activity of anatase TiO_2 by coexposed {001} and {101} facets. J Am Chem Soc136(25):8839–8842

Yuan YJ, Tu JR, Ye ZJ, Chen DQ, Hu B, Huang YW, Chen TT, Cao DP, Yu ZT, Zou ZJ (2016) MoS_2-graphene/$ZnIn_2S_4$ hierarchical microarchitectures with an electron transport bridge between light-harvesting semiconductor and cocatalyst: a highly efficient photocatalyst for solar hydrogen generation. Appl Catal B 188:13–22

Zhang X, Zhang L, Xie T, Wang D (2009) Low-temperature synthesis and high visible-light-induced photocatalytic activity of BiOI/TiO_2 heterostructures. J Phys Chem C 113(17):7371–7378

Zhang H, Lv X, Li Y, Wang Y, Li J (2010) P25-graphene composite as a high performance photocatalyst. ACS Nano 4(1):380–386

Zhang S, Arunachalam P, Abe T, Iyoda T, Nagai K (2012) Photocatalytic decomposition of N-methyl-2-pyrrolidone, aldehydes, and thiol by biphase and p/n junction-like organic semiconductor composite nanoparticles responsive to nearly full spectrum of visible light. J Photochem Photobiol A 244:18–23

Zhang Y, Ma D, Wu J, Zhang Q, Xin Y, Bao N (2015) One–step preparation of CNTs/InVO4 hollow nanofibers by electrospinning and its photocatalytic performance under visible light. Appl Surf Sci 353:1260–1268

Zhang L, Bao Z, Yu X, Dai P, Zhu J, Wu M, Li G, Liu X, Sun Z, Chen C (2016a) Rational design of α-Fe_2O_3/reduced graphene oxide composites: rapid detection and effective removal of organic pollutants. ACS Appl Mater Interfaces 8(10):6431–6438

Zhang M, Luo W, Wei Z, Jiang W, Liu D, Zhu Y (2016b) Separation free C_3N_4/SiO_2 hybrid hydrogels as high active photocatalysts for TOC removal. Appl Catal B 194:105–110

Zhang X, Qi Wang Q, Zou LH, You JW (2016c) Facile fabrication of titanium dioxide/fullerene nanocomposite and its enhanced visible photocatalytic activity. J Colloid Interface Sci 466:56–61

Zhang Y, Xie C, Gu FL, Wu H, Guo Q (2016d) Significant visible-light photocatalytic enhancement in Rhodamine B degradation of silver orthophosphate via the hybridization of N-doped graphene and poly(3-hexylthiophene). J Hazard Mater 315:23–34

Zhao H, Chen S, Quan X, Yu H, Zhao H (2016) Integration of microfiltration and visible-light-driven photocatalysis on g-C_3N_4 nanosheet/reduced graphene oxide membrane for enhanced water treatment. Appl Catal B 104:134–140

Zheng Y, Lin L, Ye X, Guo F, Wang X (2014) Helical graphitic carbon nitrides with photocatalytic and optical activities. Angew Chem Int Ed Engl 53(44):11926–11930
Zhou Y, Wu Y, Li Y, Liu Y, Yang L, Wang L, Liu H, Li D, Luo Q (2016) The synthesis of 3D urchin-like TiO_2-reduced graphene micro/nano structure composite and its enhanced photocatalytic properties. Ceram Int 42(10):12482–12489

Chapter 9
Nanocomposites of g-C_3N_4 with Carbonaceous π-conjugated/Polymeric Materials Towards Visible Light-Induced Photocatalysts

Sulagna Patnaik, Dipti Prava Sahoo and Kulamani Parida

Abstract The carbonaceous π-conjugated/polymeric materials have been emerging as suitable materials to synthesize nanocomposites because of their attractive nanoporous structure, controllable surface chemistry, mechanical strength and favourable interactions with the semiconducting materials. The photocatalytic performances of the traditional polymeric materials are generally poor. Their performances can be greatly improved by coupling with a host semiconducting material. This is mainly due to their unique crystal structure, stability, high conductivities, nature of formation, efficient catalytic activity, promising electrochemical and optical properties. These polymeric nanocomposites act as photo sensitizers and good visible light absorbers due to π–π* electronic transitions. In this chapter the preparation methods, microstructure analysis and photocatalytic mechanism of graphitic carbon nitride (g-C_3N_4) and various carbonaceous π-conjugated/polymeric material composite catalysts are focused. In particular, modification of g-C_3N_4 by various carbonaceous π-conjugated/polymeric materials result in hybridization owing to strong π–π stacking interaction, which stabilizes the hybrid nanostructure and efficiently utilize the solar spectra by extending the photocatalytic applications in NO removal, CO_2 reduction and oxygen reduction reactions, water splitting to liberate H_2 fuel and degradation of pollutants. The challenges of various π-conjugated/polymeric material modified nanocomposites of g-C_3N_4 in the field of photocatalysis are also highlighted in this chapter to extend their applications in sustainable energy development.

Keywords Polymeric nanocomposite · Carbonaceous materials · Photosensitizer · π–π* transition · Photocatalytic degradation

S. Patnaik · D.P. Sahoo · K. Parida (✉)
Centre for Nano Science & Nano Technology, Institute of Technical Education and Research, Siksha 'O' Anusandhan University, Khandagiri, Bhubaneswar 751030, Odisha, India
e-mail: kulamaniparida@soauniversity.ac.in; paridakulamani@yahoo.com

M.M. Khan et al. (eds.), *Nanocomposites for Visible Light-induced Photocatalysis*, Springer Series on Polymer and Composite Materials,
DOI 10.1007/978-3-319-62446-4_9

Introduction

Environmental problems and energy crisis associated with day-to-day life are now important issues with negative impact on health and ecosystem. In this regard the main goal is to develop novel semiconducting materials for extending photocatalytic activity towards visible region under UV/Vis light irradiation (Chen et al. 2010; Kudo and Miseki 2009; Yang et al. 2014). The most common carbon-based polymeric materials have been emerging as suitable materials to obtain nanocomposites because of their nanoporous structure, controllable surface chemistry, high mechanical strength and favourable interactions with the semiconducting materials. The host semiconducting material also greatly affects the crystal structure, stability and optical properties of the resultant nanocomposites exhibiting different photocatalytic activity (Zhou and Shi 2016). These carbon-based conducting polymeric materials are known as good hole conducting materials and when coupled with other semiconducting nanomaterials, they produce synergistic effects and complementary properties. Moreover, the multifunctional property enhancements in case of these polymeric nanocomposites, may extend their applications in the field of photocatalysis. Various 2D host semiconducting materials like TiO_2, MoS_2, CeO_2, g-C_3N_4, etc. are known to form composites with a number of carbon-based π-conjugated/polymeric materials. This chapter mainly focuses on polymeric nanocomposites utilizing g-C_3N_4 (an organic polymer) as the host and its modifications by various carbonaceous π-conjugated/polymeric materials.

g-C_3N_4 has attracted immense attention as a 2D polymeric semiconducting material owing to its easy preparation from cost-effective starting materials, good chemical and thermal stability, unique electronic structure, high potential for solar energy conversion. Moreover, g-C_3N_4 is also structurally suitable for designing heterojunction photocatalysts because of its 2D planar conjugated structure which provides a better scope for anchoring various composites. Despite of all its goodness, the quantum efficiency is not agreeable owing to low separation rate of photogenerated electron–hole pairs (Thomas et al. 2008; Patnaik et al. 2016a, b, c; Nayak et al. 2015; Martha et al. 2013).

Recently, polymers that can act as photocatalysts have gained much attention from chemists, motivated by the demand for solving pollution problems. This mainly deals with their applications in the reduction of green house gases, degradation of organic dyes and generation of hydrogen fuel by splitting water. Various carbonaceous π-conjugated/polymeric materials owing to their unique electron and hole-transporting nature, high conductivity, suitable redox potential and stability in oxidized state, are compatible to form surface junctions to increase separation of electron–hole pairs. The interface so formed in bulk hetero junction motivates the researchers to form hybrids of these carbon-based conductive polymers with g-C_3N_4 to form organic–organic polymeric heterostructures. The modification of g-C_3N_4 by various carbonaceous π-conjugated/polymeric materials results in hybridization owing to strong π–π stacking interaction, which stabilizes the hybrid nanostructure and efficiently utilize the solar spectra by extending the optical absorption towards visible region. Also several N-doped carbonaceous materials like PANI, PAN, Ppy, Ptp, etc. have been widely used, which upon pyrolysis result N-doped catalysts (Zhou and Shi 2016). The nitrogen atoms doped in the lattice of polymeric carbonaceous materials mainly have four bonding configurations: quaternary N (or graphitic-N), pyrrolic-N, pyridinic-N and pyridine-N-oxide. Out of which pyridinic-N atoms were regarded as the most active sites of the photocatalyst, because of the presence of delocalized π-electrons to favour the adsorption of O_2 molecules to facilitate the degradation processes (Zhou and Shi 2016). Coupling of g-C_3N_4 with various carbonaceous polymeric materials not only compensate the disadvantages of individual semiconductor materials but also induce synergetic effects to improve photostability which ultimately enhances photocatalytic activity (Patnaik et al. 2016a, b; Sahoo et al. 2016). The modification of g-C_3N_4 by various carbonaceous polymeric materials is represented graphically in Fig. 9.1.

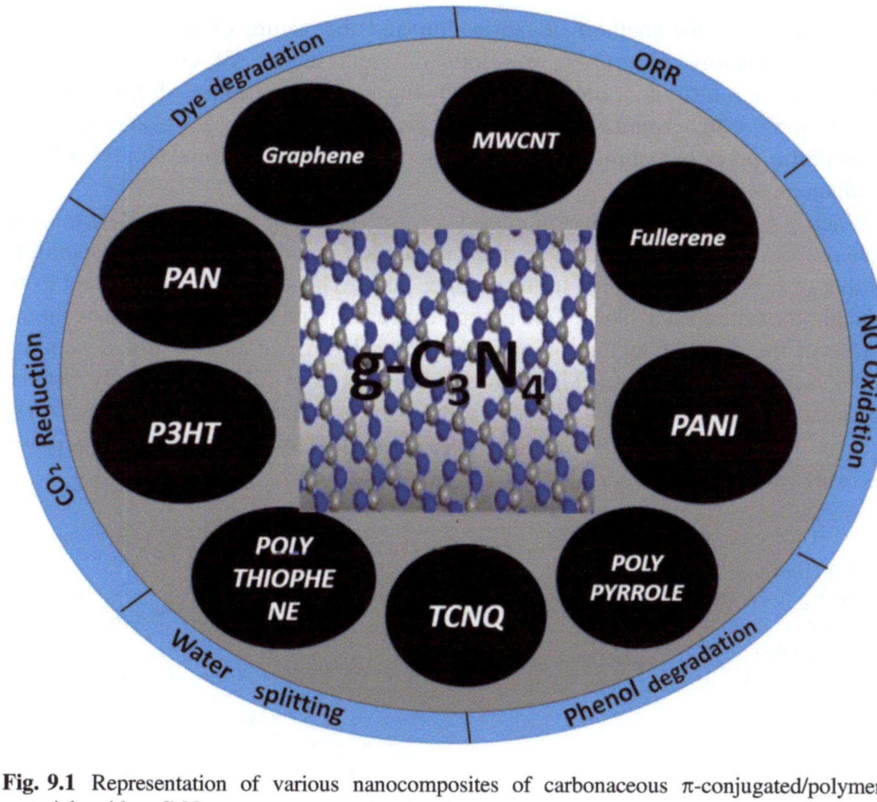

Fig. 9.1 Representation of various nanocomposites of carbonaceous π-conjugated/polymeric materials with g-C_3N_4

A Brief Account of Graphitic Carbon Nitride: A π-conjugated Organic Polymer

One of the oldest synthetic polymer, g-C_3N_4 an analogous of graphene containing alternate C- and N-atoms, was first synthesized by Berzelius and named by Liebig in 1834 as melon. In 1922, Franklin described the structure of this compound which exists in five different allotropic forms α-C_3N_4, β-C_3N_4, graphitic-C_3N_4, Cubic C_3N_4 and pseudo-cubic-C_3N_4. Among different allotropic forms g-C_3N_4 is known to be the most stable one, possessing a graphite-like network containing π-conjugated 2D planar structure, separated by a distance of 0.326 nm, held together by weak van der Waals force of attraction. This polymer is extraordinarily stable due to the presence of strong covalent bonds in the molecular crystal. As the semiconductor possess suitable bandgap energy, upon visible light irradiation it results in spatial charge separation and takes part in subsequent redox reactions with surface adsorbed molecules. The N-atoms with lone pairs of electrons behave as oxidation sites and C-atoms provide the reduction sites (Thomas et al. 2008; Patnaik et al. 2016c).

Depending on the method of fabrication and the nature of the precursor, the polymeric structure of g-C_3N_4 develops different degree of condensation and behaves as a multifunctional catalyst. Owing to the presence of π-conjugated system along with terminal-NH and -NH_2 groups, it exhibits electronic property, nucleophilic property, ability to form hydrogen bond and photocatalytic activity simultaneously (Fig. 9.2).

When g-C_3N_4 receives photon energy $\geq$ Eg (bandgap = 2.7 eV) electrons are excited from the valence band to the conduction band and thus corresponding number of holes are generated in the valence band. The semiconductor is said to be in its photoexcited state. These free charge carriers migrate to the active sites on the surface of the semiconductor and split water into hydrogen and oxygen depending on the sacrificial agent used. The surface properties of g-C_3N_4 intrinsically favours the separation and transfer of charge carriers by generating surface states where electrons and holes are spatially trapped and transferred for subsequent redox reactions (Scheme 9.1).

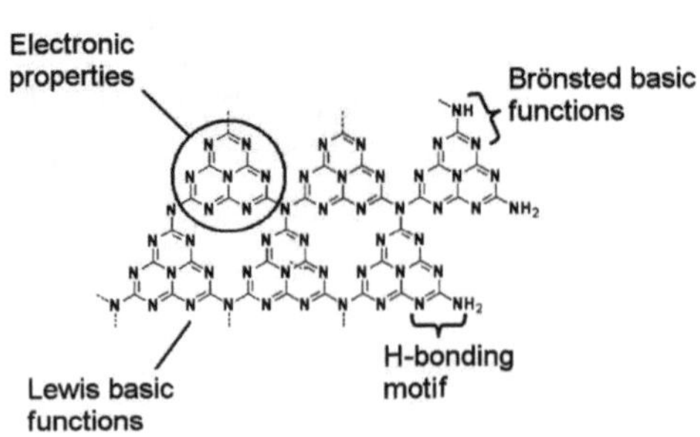

Fig. 9.2 Electronic structure of planar g-C_3N_4 [reproduced from Thomas et al. (2008), License number 4003460342494]

Scheme 9.1 Charge transfer mechanism in neat g-C_3N_4 as photocatalyst

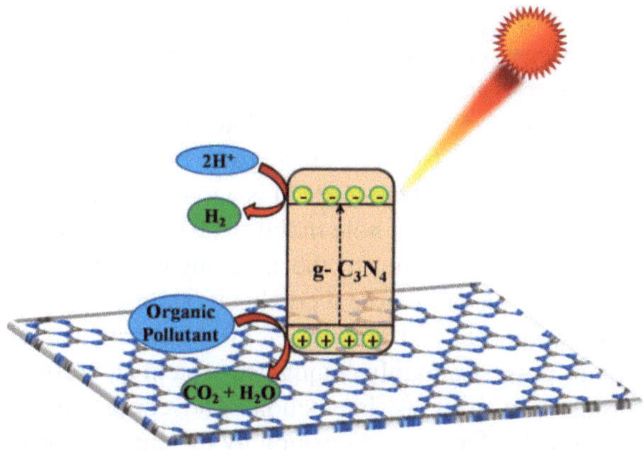

An Insight into the Properties of Carbonaceous π-conjugated/Polymeric Materials

Carbonaceous π-conjugated/polymeric materials are particularly important because of their number of active adsorption sites, efficient electron–hole separation and extended absorption in visible region by sensitization. The electronic integration of g-C_3N_4 with these π-conjugated polymeric materials having unpaired electrons significantly improves delocalization and extends its potential applications. Due to which g-C_3N_4 can be hybridized with various carbon-based π-conjugated/polymeric materials, like graphene, CNT, fullerene, PANI, P3HT, 7,7,8,8-Tetracyanoquinodimethane (TCNQ), polypyrrole (Ppy), polythiophene (Ptp), etc. An optimum amount of polymeric dopant improves the photocatalytic activity due to the extension of the π-conjugated structure towards the carbon dopants. The polymeric nanocomposites show substantial multi-functional property enhancement with lower loading and extended applications. The structural similarity of carbonaceous π-conjugated/polymeric materials with g-C_3N_4 makes it suitable to form 2D-layered junction. Graphene is one of the important carbon-based material known to have a monoatomic thickness. 2D structure of carbon atoms, possessing high chemical/thermal stability, excellent conductivity, superior mechanical flexibility, large surface area (2630 $m^2\ g^{-1}$) and excellent mobility of charge carriers (250,000 $cm^2\ V^{-1}\ s^{-1}$) (Ong et al. 2015). Hence coupling g-C_3N_4 with graphene is an effective strategy to form a large contact area across the interface for charge transfer. CNT also possesses a large specific surface area and controls the morphology and structure of g-C_3N_4. The incorporation of CNT not only enhances the light absorption capacity of the composite but also acts as an acceptor of photogenerated electrons and as a good conductive material increases the efficiency of charge separation (Chen et al. 2014). Another allotrope of carbon, Fullerenes (C_{60}) consisting of 30 bonding molecular orbitals with π-conjugation, favours reduction reaction (Chai et al. 2014). The unique electronic structure acts as an excellent electron acceptor to retard charge recombination. Polyaniline (PANI) is used to functionalize g-C_3N_4

because of its strong photoresponse in visible region. PANI possesses good stability, nontoxicity, corrosion protection and efficient electron–hole transportation ability. Moreover, it involves a facile and low cost method of synthesis and high absorption coefficient ($\sim 5 \times 10^4$) in the visible light range. PANI not only behaves as an electron donor but also behaves as a good hole acceptor under visible light irradiation. Poly-3-hexylthiophene (P3HT) is another p-type semiconducting polymeric material, which possesses high hole mobility (10^{-4}–10^{-3} $cm^2\ V^{-1}\ S^{-1}$) and a suitable bandgap of 1.9–2.1 eV. Owing to the π-conjugation effect, the composites of P3HT greatly reduce the electron–hole recombination process by distribution of electrons and promote photocatalytic activity. 7,7,8,8-Tetracy anoquinodimethane (TCNQ), and its anions possess a highly conjugated system and form charge transfer complexes due to strong π–π stacking interaction. The conjugated π structure of the material and decreased valence band position by the charge transfer interaction are responsible for enhanced photocatalytic performance due to rapid electron–hole separation efficiency. Polyacrylonitrile (PAN) is another polymeric species with conjugated π electronic structures that can be used to fabricate PAN/g-C_3N_4 composites, with high separation efficiency of photogenerated charge carriers in the visible region. The graphite-like aromatic π-conjugated structure of PAN favours effective electron channelization. Although, the introduction of PANI, P3HT and g-PAN is effective to improve the photocatalytic performance, but they are expensive and difficult to fabricate through organic synthesis, which inhibits their practical application. Ppy and Ptp are examples of conductive polymer with high stability in the oxidized state, superior conductivity and a matching bandgap to form heterojunction with g-C_3N_4. The composites increase the photocatalytic activity by increasing number of active sites as its formation involves surfactant-assisted polymerization, assisting well dispersion of Ppy nanoparticles on the surface of g-C_3N_4. The large contact area between the polymer and g-C_3N_4 facilitates interfacial charge transfer to increase the separation of charge carriers.

A Brief Sketch on the Mechanism of Photocatalytic Activity of Nanocomposites of g-C_3N_4 with Carboneous π-conjugated/Polymeric Materials

The photocatalytic activity of g-C_3N_4-based nanocomposites with carbonaceous π-conjugated/polymeric materials is mainly because of their conductivity, stability in the oxidized state and interesting redox properties. Carbonaceous materials with π-conjugated structure are responsible for the channelization of photogenerated electrons to retard recombination of electron–hole pairs. In addition polymers with suitable bandgap help to induce efficient visible light absorption and in the delocalization by forming heterojunction. Most of the polymeric materials act as electron sinks and directly reduce water to liberate H_2, removal of NO and reduction of CO_2. Photogenerated holes also induce degradation, whereas other electrons get channelized to the surface of the cocatalyst and effectively prolong the lifetime of the charge carriers.

Principle of Photocatalytic Degradation of Organic Pollutants

Today's environmental problems include air pollution, water pollution resulting from the contamination of toxic industrial hazardous waste like organic dyes, surfactants, disinfectant by products, insecticides, pesticides, herbicides, volatile organic compounds, heavy metals, green house gases (NO_x, SO_x, CO, CO_2 and NH_3) chlorinated and non-chlorinated aliphatic and aromatic compounds and pathogens (bacteria, fungi and viruses). Depending on the potential of each adsorbate, spontaneous adsorption of these organic pollutants takes place on the surface of the photocatalyst, in the presence of a fluid phase (gas or liquid). During photocatalytic reaction, some active species like hydroxyl radicals ($OH^{\cdot}$), the superoxide radical ($O_2^{\cdot}$), hydroperoxyl radical ($HOO^{\cdot}$) and the holes are formed. Under visible light irradiation in aqueous solutions the hydroxyl radicals ($OH^{\cdot}$) are the primary oxidants and are generated by the direct hole oxidation or photogenerated electron-induced multistep reduction of the dissolved O_2.

Step – I $$O_2 + e^- \rightarrow O_2^{\cdot}$$

Step – II $$O_2 + e^- + 2H^+ \rightarrow H_2O_2 \quad (H^+\text{s are from organic compounds})$$

Step – III $$H_2O_2 + e^- \rightarrow OH^{\cdot} + OH^-$$

The generated hydroperoxyl radicals, also act as scavenger by doubly prolonging the lifetime of photogenerated holes. At the surface of the semiconductor photocatalyst (g-C_3N_4) both the oxidation and reduction can occurs. In the absence of dissolved oxygen electrons and holes recombine, and in their presence the electrons react to form super oxides ($O_2^{\cdot -}$), the hydroperoxyl radical ($HO_2^{\cdot -}$) and subsequently to hydrogen peroxide (H_2O_2). The ions formed reacts to form intermediates and final products. The general mechanism of photodegradation is presented in Scheme 9.2. Depending on the relative potentials of g-C_3N_4 and the polymer, the charge transfer takes either through (2a) or (2b).

Principle of Photocatalytic Water Splitting

Visible light-induced water splitting to generate hydrogen fuel is another important application of photocatalysis. The visible light-induced electrons and holes generated by the semiconductors initiate the reduction and oxidation reactions simultaneously. The electrons and holes react with water molecules to form H_2 and O_2,

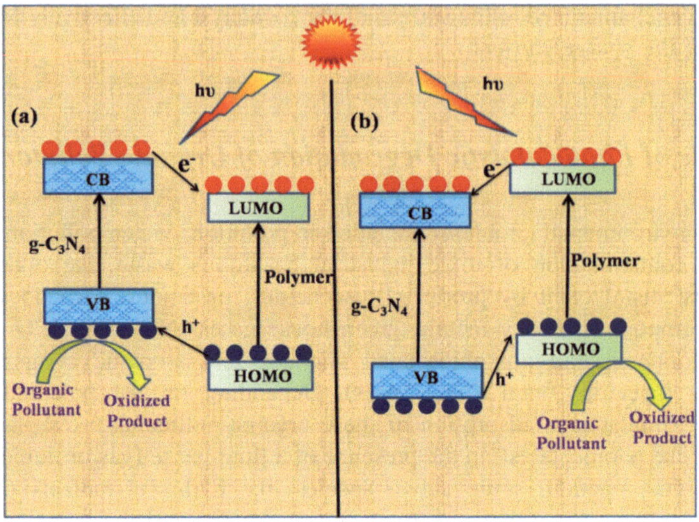

Scheme 9.2 Schematic illustration of charge transport at the g-C_3N_4/polymer heterojunction

respectively. For visible light-induced water splitting to produce H_2 and O_2, the bottom level of the conduction band (CB) of the semiconductor has to be more negative than the redox potential of H^+/H_2 [0 V versus normal hydrogen electrode (NHE)], whereas the valence band (VB) of the semiconductor has to be more positive than the redox potential of O_2/H_2O (1.23 eV). The polymeric semiconductor photocatalyst (g-C_3N_4), is considered to be suitable for water splitting having a bandgap energy greater than 1.23 eV (redox potential of water) and also suitable positioning of the CB and VB levels (Scheme 9.3).

Water splitting reaction is an uphill reaction, $\Delta G = +237.2\,\mathrm{kJ\,mol^{-1}}$
Energy of the light photon is used to overcome this large $(+)\Delta G$.

The electrochemical decomposition of water to H_2 and O_2 is a two step process. The photocatalyst surface is capable of absorbing solar energy to generate electrons and holes that can, respectively, reduce and oxidize H_2O molecule adsorbed on the catalyst (Hu et al. 2015a; Kudo and Miseki 2009; Lu et al. 2014).

$$\text{Oxidation:} \quad H_2O + 2h^+ \rightarrow 2H^+ + 1/2O_2 \tag{9.1}$$

$$\text{Reduction:} \quad 2H^+ + 2e \rightarrow H_2 \tag{9.2}$$

$$\text{Overall reaction:} \quad H_2O \rightarrow H_2 + 1/2O_2 \tag{9.3}$$

Scheme 9.3 The schematic illustration of electron–hole separation and transport at the g-C_3N_4/polymer heterojunction interface in water splitting

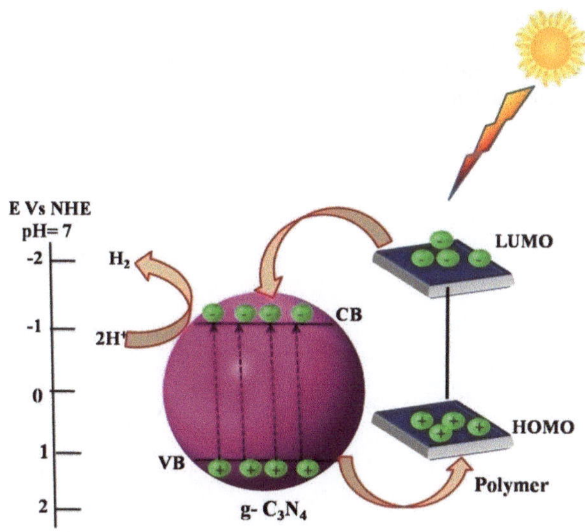

Mechanism of NO (Nitric Oxide) Removal

Nitrogen oxides (NO_x) were found to be a major pollutant responsible for many environmental problems such as acid rain, haze, photochemical smog, etc. For the removal of NO_x at ppb (parts per billion) level the semiconductor photocatalysis is a green technology.

g-C_3N_4 has been used effectively for the photocatalytic removal of NO by increasing its surface area. For which π-conjugated carbonaceous polymeric materials are important to form composite with g-C_3N_4 and provide more surface area for photocatalytic reaction.

The photocatalytic experiments for the removal of NO were carried out at ambient temperature using the prepared photocatalysts in a continuous flow reactor. The concentration of NO, NO_2 and NO_x were measured by a chemiluminescence NO_x analyzer (Ma et al. 2016). The mechanism of NO removal involves several active species like hydroxyl radical ($^{\cdot}OH$), the superoxide radical ($^{\cdot}O^{2-}$), hydrogen peroxide (H_2O_2) and holes. In presence of those free radicals NO is converted into NO_3^- (Scheme 9.4).

Equations:

$$\text{Catalyst} + h\upsilon \rightarrow h^+ + e^- \tag{9.4}$$

Scheme 9.4 The schematic illustration of NO reduction at the g-C_3N_4/polymer heterojunction

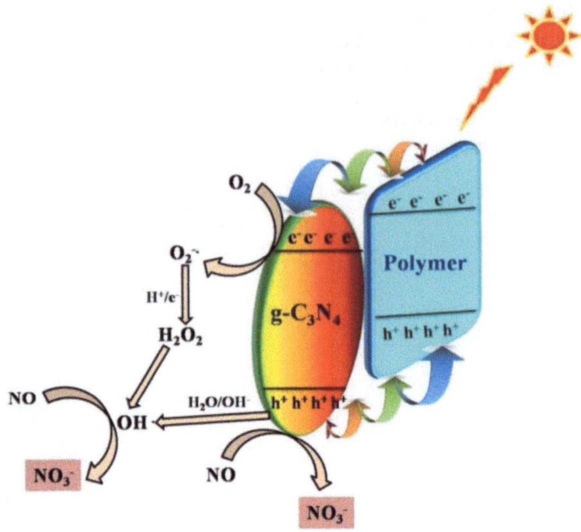

$$e^- + O_2 \rightarrow O_2^{\cdot -} \tag{9.5}$$

$$O_2^{\cdot -} + 2H^+ + e^- \rightarrow H_2O_2 \tag{9.6}$$

$$H_2O_2 + e^- \rightarrow 2OH^{\cdot} \tag{9.7}$$

$$h^+ + H_2O \rightarrow OH^{\cdot} + H^+ \tag{9.8}$$

$$NO + OH^{\cdot} \rightarrow NO_3^- \tag{9.9}$$

$$OH^{\cdot} + h^+ \rightarrow OH^{\cdot} \tag{9.10}$$

$$O_2 + e^- \rightarrow O_2^{\cdot -} \tag{9.11}$$

$$H^+ + O_2^{\cdot -} \rightarrow HO_2^{\cdot}(\text{Peroxy radical}) \tag{9.12}$$

$$NO + HO_2^{\cdot} \rightarrow NO_2 + OH^{\cdot} \tag{9.13}$$

$$NO_2 + OH^{\cdot} \rightarrow HNO_3 \tag{9.14}$$

Mechanism of Oxygen Reduction Reaction

The oxygen reduction reaction (ORR) is a fundamental reaction, which is significantly important during electrochemical energy conversion in fuel cells. However, poor electrochemical conductivity of g-C_3N_4-based cathode is a serious problem for

ORR. Hence, coupling with π-conjugated carbonaceous polymeric material with proper electronic structure helps in ORR by promoting the change transfer. This extends the application of the nano composites of g-C_3N_4 with π-conjugated carbonaceous polymeric materials in energy conversion and storage device (Tian et al. 2014). The nitrogen rich nanostructures of g-C_3N_4 show high performance as ORR catalyst owing to its high surface area, large number of active sites and short diffusion path for electrons and electrolyte/ions in comparisons to bulk or powder carbon nitride (g-C_3N_4). Doping of nitrogen into various carbon-based materials significantly contribute to the better selectivity for ORR since the N-atom with five valence electrons induce polarity in the adjacent carbon atoms and create positively charged sites, for the adsorption and reduction of O_2. Compared to acidic electrolytes N-containing polymers like PANI, Ppy, Ptp, etc. frequently pyrolyzed to form nitrogen-doped carbonaceous polymers. They may be available in four different structural motifs like pyrrolic-N, pyridinic-N, graphitic-N and pyridine-N-oxide, which are regarded as surface active sites and facilitate the adsorption of O_2. The adjacent C-atoms become positively charged and possess strong capacity for ORR (Scheme 9.5).

The kinetics of ORR of a photocatalyst has been investigated by the linear sweep voltammetry (LSV) method.

ORRs in aqueous solution mainly follow two pathways.

- direct 4-electron reduction pathway O_2 to H_2O
- 2-electron reduction pathway O_2 to H_2O

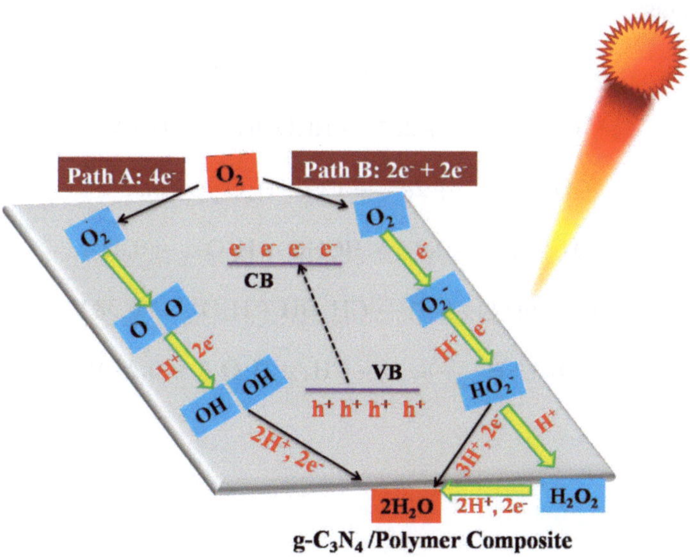

Scheme 9.5 The schematic illustration of electron–hole separation and transport at the g-C_3N_4/polymer heterojunction interface in ORR

Equations:

$$O_2 + 4H^+ + 4e^- \rightarrow 2H_2O \quad 1.22\,V \tag{9.15}$$

$$O_2 + 2H^+ + 2e^- \rightarrow 2H_2O_2 \quad 0.07\,V \tag{9.16}$$

Mechanism of CO_2 Reduction

In order to save the environment from anthropogenic CO_2 emission various strategies have been proposed. Among which, photocatalytic reduction of CO_2 into useful chemicals is one of the alternative. But as the CO_2 molecule is thermodynamically as well as kinetically very stable, a large amount of input energy is required for its conversion. The reduction process is also highly unfavourable and endergonic. For photocatalytic CO_2 reduction, the main requirement is the CB potential which must be higher than that of the reduction potential of CO_2 and should be within the bandgap of the semiconductor. The process of reduction into hydrocarbons also requires more negative potential for the electrons. Moreover, the reduction process involves 2–8 electrons to convert into hydrocarbon fuel. Most of the transformations take place between −0.24 and −0.6 V, whereas at −1.9 V single electron reduction is extremely unfavourable and associated with high overvoltage (Scheme 9.6) (Sultana et al. 2016).

Equations:

$$CO_2 + e^- \rightarrow CO_2^{\cdot -} \quad -1.90\,V \tag{9.17}$$

$$CO_2 + 2H^+ + 2e^- \rightarrow HCOOH \quad -0.61\,V \tag{9.18}$$

$$CO_2 + 2H^+ + 2e^- \rightarrow CO + H_2O \quad -0.53\,V \tag{9.19}$$

$$CO_2 + 4H^+ + 4e^- \rightarrow HCHO + H_2O \quad -0.48\,V \tag{9.20}$$

$$CO_2 + 6H^+ + 6e^- \rightarrow CH_3OH + H_2O \quad -0.38\,V \tag{9.21}$$

$$CO_2 + 8H^+ + 8e^- \rightarrow CH_4 + H_2O \quad -0.24\,V \tag{9.22}$$

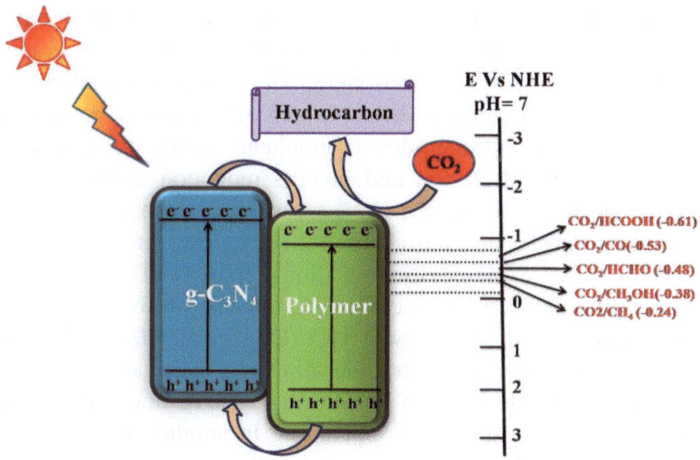

Scheme 9.6 The schematic illustration of electron–hole separation and transport at the g-C_3N_4/polymer heterojunction in CO_2 reduction

An In-depth Study of Various Nanocomposites of g-C_3N_4 with Carbonaceous π-conjugated/Polymeric Materials

Nanocomposites of g-C_3N_4 with Graphene

To tailor the chemistry of g-C_3N_4, coupling with 2D π-conjugated graphene is regarded as an innovative approach. It results a face-to-face contact between g-C_3N_4 and graphene to extend its application due to synergistic coupling interaction. In the process of composite formation, natural graphite can be effectively oxidized to form exfoliated graphene oxide (GO), which possesses a lot of oxygen-containing functional groups on its edges and on the basal planes. These functional groups (OH, COOH, epoxy groups) are important for the covalent interconnection of the GO sheets with g-C_3N_4 and facilitate the subsequent formation of an interface. The structural similarity between GO sheets with g-C_3N_4 also strengthens the chemical bonding through π–π stacking. Various group of researchers reported extensive works related to g-C_3N_4/GO-based composites and their applicability in the field of photocatalysis (Wang et al. 2014; Dai et al. 2014; Li et al. 2014). The synthesis methods mainly include simple calcination method, impregnation method followed by reduction, sonochemical method, hydrothermal method and through cryodesiccation route. A graphene/g-C_3N_4 composite photocatalysts were prepared by an impregnation-chemical reduction strategy involving polymerization of melamine in presence of GO where hydrazine hydrate was used as the reducing agent (Xiang et al. 2011). In a one-step synthesis method a new 3-dimensionally (3D) interconnected carbon nitride tetrapods wrapped with nitrogen-doped graphene was designed which shows significant ORR activity in acidic medium (Unni et al. 2014). Simple dip-coating of melamine foam (MF) with GO results a 3D g-C_3N_4 framework and GO is transformed to graphene when annealed at 900 °C in an inert atmosphere. The nitrogen available in melamine foam is trapped by the graphene sheets, which subsequently get doped into the graphene matrix. These nitrogen-doped graphene matrix thereby creates more active sites for oxygen adsorption and favours reduction reaction. In the photocatalyst, the connectivity between g-C_3N_4 tetrapods and nitrogen-doped graphene helps in fast mass transfer and introduces active centres to trap oxygen molecules and reduce them. A facile one-pot impregnation cum thermal reduction strategy was reported to design sandwich-like metal-free graphene-g-C_3N_4 nanocomposites by the polymerization of urea along with GO (Ong et al. 2015). The presence of graphene oxide plays the role of structure-directing agent at 520 °C. Such a 2D sandwiched nanocomposite helps in efficient light harvesting and provides a short diffusion layer for interfacial charge separation. In another method to address the drawbacks of bulk g-C_3N_4 with low surface area, mesoporous-g-C_3N_4/graphene oxide nanocomposites were fabricated via a facile sonochemical way by using dicyandiamide and SiO_2 as a hard templating agent (Li et al. 2014). This typical procedure involves formation of a porous structure and hybrid with graphene simultaneously. A monolayer g-C_3N_4 dots@graphene nanocomposite was synthesized by depositing monoatomic-thick graphitic carbon nitride (g-C_3N_4) dots on the basal plane of the graphene sheet by hydrothermal treatment (Wang et al. 2015). This g-C_3N_4dots@graphene nanocomposite system possesses a series of advantages over the conventional g-C_3N_4/graphene composites in ORR. It provides an enhanced surface with more number of active catalytic sites, the intimate contact between g-C_3N_4 and the graphene facilitates electron flow through the interface. Porous g-C_3N_4-graphene nanocomposite can also be prepared without using any template following a simple calcinations method from cyanamide where gas bubbles generated during polymerization act as template to form a porous structure in the composite (Yu et al. 2014). Another g-C_3N_4–Pt/graphene heterogeneous structure has been designed by depositing g-C_3N_4 quantum dots on 3D graphene networks by hydrothermal method followed by ultrasonication which acts as an efficient catalyst for the oxidation of methanol (Hu et al. 2015a, b). The crystal structure, interlayer stacking and formation of interface in case of g-C_3N_4-graphene nanocomposites were studied by different physiochemical characterization techniques like XRD, SEM, TEM and FTIR, etc. In almost all g-C_3N_4-graphene nanocomposites it was found that modification with graphene or GO do not alter the lattice structure of g-C_3N_4. The XRD patterns of g-C_3N_4 were maintained showing the same diffraction peaks at 13.1° and 27.4° representing in planar structural packing motif of conjugated aromatic system. However, slightly broad and less intense peaks were observed due to integration of g-C_3N_4 with graphene. In addition, it was also observed that there were no apparent peaks for GO at 9.51° or graphene at 24.51°, which represents the destruction of regular stacking of GO during reduction process and presence of very small amount of graphene in the composite. The enlarged view of 002 peak in graphene/mesoporous-g-C_3N_4 composites indicates that the peak position was slightly shifted to lower angle for

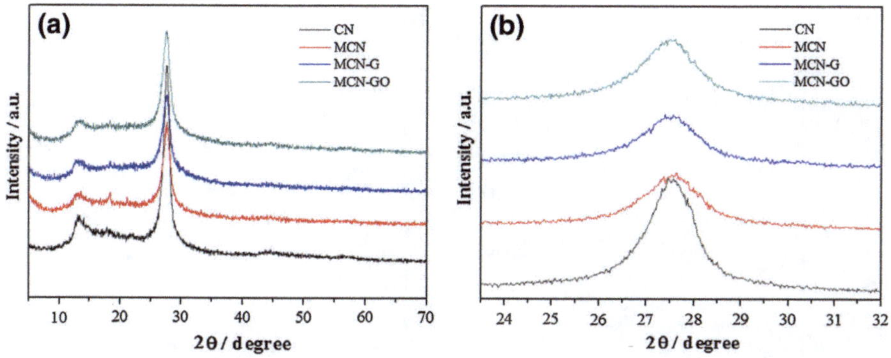

Fig. 9.3 **a** Represents XRD patterns of g-C_3N_4, mesoporous-g-C_3N_4 and graphene and GO modified mesoporous-C_3N_4, **b** represents the enlarged view of 002 peak in graphene/mesoporous-g-C_3N_4 composites [reproduced from Li et al. (2014), License number 4003470036166]

mesoporous-g-C_3N_4 (Fig. 9.3a) but coupling with graphene or graphene oxide shows no noticeable change in peak position (Fig. 9.3b).

However, the presence of graphene sheets in the graphene/g-C_3N_4 composites can be easily evidenced by TEM, FTIR and Raman spectroscopy. From the TEM image, it was observed that nanoparticles of g-C_3N_4 are densely distributed on GO sheets indicating a clear interface and preferential heterogeneous nucleation. In case of sandwiched composite of graphene/g-C_3N_4, a more compact structure was found. After hybridization with grapheme, g-C_3N_4 forms multiple 2D corrugated layers due to π–π stacking and hydrogen bonding interaction (Fig. 9.4).

According to all reported studies, the absorption edge was red shifted and absorption in the visible region increases by increasing the amount of graphene loading (Ong et al. 2015), which may be due to electronic transition between excited g-C_3N_4 and graphene and increase in the surface electric charge of g-C_3N_4 due to hybridization with graphene. In case of mesoporous-g-C_3N_4 and graphene hybrid the light harvesting efficiency is further increased (Li et al. 2014). The porous g-C_3N_4 shows an absorption tail which may be attributed to the porous network of g-C_3N_4. The absorption tail disappeared due to the combination of scattering from porous structure and the strong absorption of graphene in porous g-C_3N_4/graphene hybrid. Incorporation of graphene in g-C_3N_4 matrix modifies its electronic structure and renders the nanocomposite more sensitive to visible light. Upon visible light illumination the electrons get excited from the VB (2p orbital of N) to the CB (2p orbital of C) of g-C_3N_4 and get transferred to graphene due to its lower Fermi level. Owing to enriched electron density of graphene sheets, CO_2, O_2 and NO molecules can be adsorbed onto the surface by conjugation interaction, resulting destabilization and deactivation of the adsorbed molecules. When graphene is hybridized with mesoporous-g-C_3N_4, the enlarged surface area provides more number of surface active sites for photocatalytic reaction and quick diffusion rate of the reactant. The negative shift in CB potential results in high reduction power of photoelectrons,

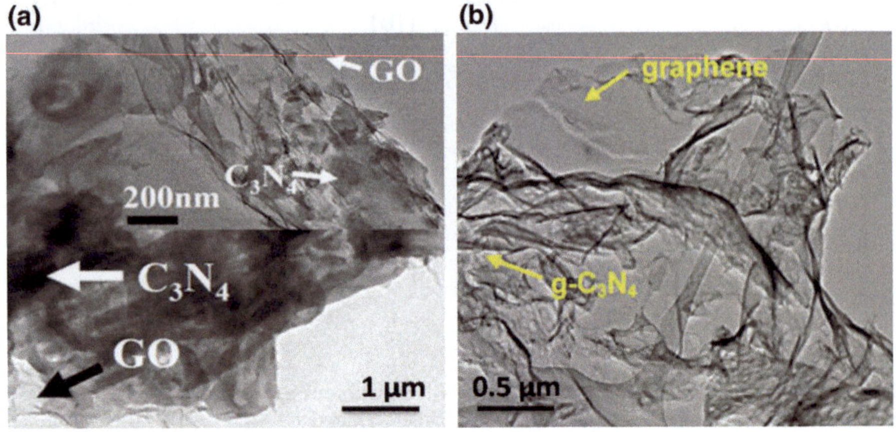

Fig. 9.4 TEM images of **a** g-C_3N_4/GO [reproduced from Dai et al. (2014), License number 4005370379927], and **b** g-C_3N_4/graphene [reproduced from Ong et al. (2015), License number 4003470036166]

which is beneficial for degradation of pollutants. Finally, the electron delocalization favours charge separation and improves the visible light photocatalytic activity significantly as graphene nanosheets behave as electron conducting channels to facilitate electron transfer. For which g-C_3N_4/graphene hybrids are very good catalysts for clean energy conversion and storage systems such as fuel cells, photo electrochemical cells to carry out hydrogen evolution reaction (HER), oxygen evolution reaction (OER) and oxygen reduction reaction (ORR) and lithium storage batteries. As reported, with an optimal graphene content (1.0 wt%) in the composite shows H_2 production rate of 451 μ mol h^{-1} g^{-1}, which was 3.07 times more than that of neat g-C_3N_4 (Xiang et al. 2011). By hybridizing mesoporous-g-C_3N_4 with graphene, photocatalytic NO removal efficiency increases. The NO removal ratio is increased from 16.8% for neat g-C_3N_4 to 64.9% for mesoporous-g-C_3N_4/graphene and 60.7% for mesoporous-g-C_3N_4 graphene oxide (Li et al. 2014). Graphene and GO nanosheets act as electronic conductive channels to improve charge separation and exhibit enhanced photocatalytic activity. The porous g-C_3N_4/graphene composites also show photocatalytic degradation of an organic dye (MB). Under visible light irradiation, g-C_3N_4/graphene composite degrades six times more than that of neat g-C_3N_4. The enhanced activity may be attributed to the good adsorption ability due to interaction between the cationic dye and the negatively charged surface of g-C_3N_4 (Yu et al. 2014). Although g-C_3N_4 shows an impressive photocatalytic activity in hydrogen evolution, pollutant degradation and oxygen reduction reaction, its low electrical conductivity restricts its potential applications. Taking account the superior conductivity of graphene, the graphene modified g-C_3N_4 nanosheets improves the electrical conductivity to enhance ORR. It has been observed that porous nanocomposites of graphene modified g-C_3N_4 nanosheets prepared by using silica template shows enhanced ORR activity in alkaline medium (Wang et al. 2015). The ORR activity of graphene/g-C_3N_4 samples were analysed by rotating ring disk electrode (RRDE) analyser. The composites with graphene show superior ORR activity than neat g-C_3N_4 in terms of onset potential, ORR limiting current density and half-wave potential. In another study it was found that g-C_3N_4 when hybridized with N-doped graphene, the ORR activity of the composite was further improved. This enhanced activity was due to the presence of pyridinic- and pyridine-type nitrogenous hetero atom, which extend the coordination by forming an adduct with dioxygen (Unni et al. 2014). The graphene/g-C_3N_4 composites also show the activity of photocatalytic reduction of CO_2 to CH_4 under visible light irradiation. In the absence of graphene, neat g-C_3N_4 also shows CO_2 reduction activity with a yield of 2.55 μmol g^{-1}; however, upon graphene hybridization the yield of CH_4 increases to 5.87 μmol g^{-1} and the graphene/g-C_3N_4 catalyst displayed good stability even after 10 h maintaining the reactivity of 90% (Ong et al. 2015). Upon 2D graphene modification the photogenerated electrons channelized easily through the conductive network of graphene at the interface as the Fermi level of graphene (−0.08 V) was at lower potential than that of g-C_3N_4 (−1.42 V) to retard the recombination of charge carriers in agreement with the PL spectra. Owing to the π-conjugated structure of CO_2, it gets adsorbed on the surface of graphene forming conjugation interaction and accelerates the reduction of CO_2 to CH_4 involving 8-electron process (Fig. 9.5).

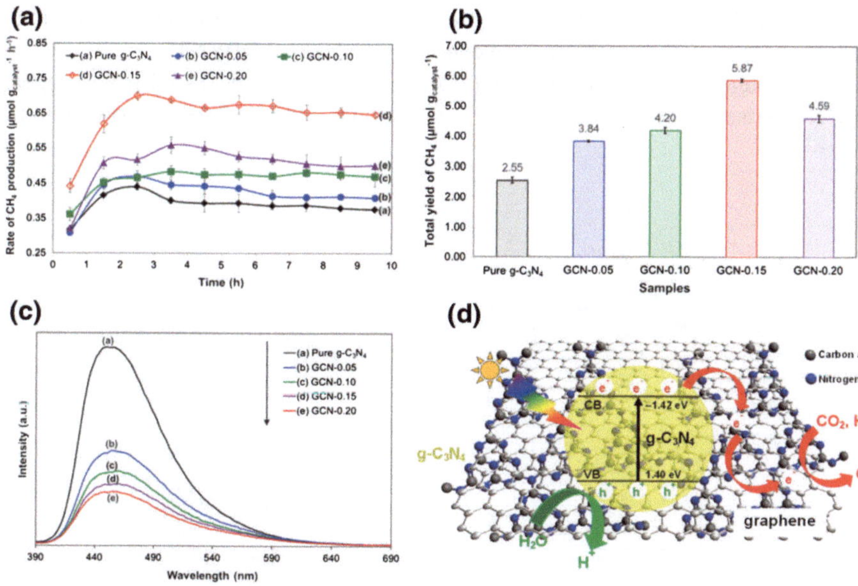

Fig. 9.5 **a** Time dependence rate of CH_4 production, **b** yield of CH_4, **c** PL spectra of the samples, **d** schematic representation of CO_2 reduction by graphene/g-C_3N_4 catalyst in presence of water under visible light irradiation [reproduced from Ong et al. (2015), License number 4005370678462]

Nanocomposites of g-C_3N_4 with Carbon Nanotube (CNT)

Among various carbonaceous nanomaterials, CNT is an important π-conjugated material owing to its special structure and unique conductivity. Functionalized CNT is used to form hybrids with g-C_3N_4 to extend its photocatalytic activity. Different research groups studied the photocatalytic activity of g-C_3N_4/CNT nanocomposites by designing the composites in various methods. The g-C_3N_4/CNT nanocomposites were synthesized by simple heating method using cyanamide and functionalized CNT (Chen et al. 2014). In another method g-C_3N_4/CNT nanocomposites were designed by electrostatically driven self-assembly using white C_3N_4 (protonated g-C_3N_4 by HCl), CNT, and 1-hexadecyl-3-methylimidazoliumnchloride (C16mim) Cl in hydrothermal method (Xu et al. 2013; Scheme 9.7). The synthesis of a π-conjugated nanocomposite of g-C_3N_4 with multiwall carbon nanotubes (MWCNTs) was reported by hydrothermal method, adding functionalized MWCNT to 1.5 M aqueous solution of cyanamide, to maximize its photocatalytic H_2 evolution under visible light (Suryawanshi et al. 2012). There is 100% improvement of its activity for an optimum of 0.5% MWNT/g-C_3N_4 nanocomposite. In another method, a novel MWNT/g-C_3N_4 nanocomposite was synthesized by facile heating method, using cyanamide and MWNT with diameter 20–40 nm and 110 cm^{-1}/g surface area (Ge et al. 2012). The MWNTs owing to their high electrical conductivity and high electron capture capacity help to stimulate electron transfer process from g-C_3N_4 to the surface of MWNT. As the Fermi level of MWNT is at lower potential region, it enhances the charge transfer by forming a Schottky barrier at the interface between the g-C_3N_4 and MWNTs.

The Z-potential (Fig. 9.6) of CNT (−24.0 mV) and g-C_3N_4 (+3.3 mV) suggests that CNT possess –COO groups on the surface, owing to that it gets adsorbed on the

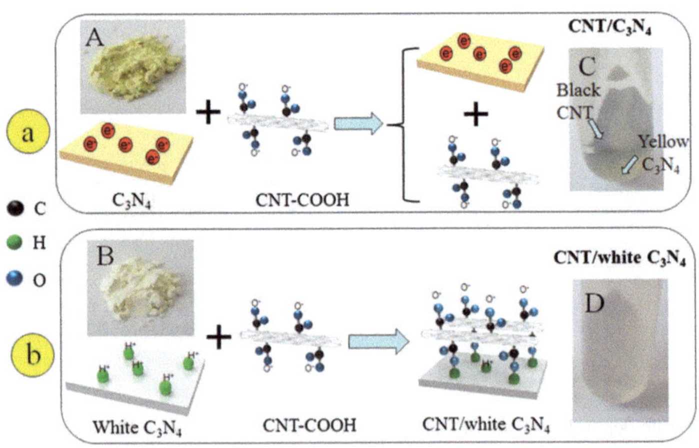

Scheme 9.7 Schematic representation of formation process of CNT/white g-C_3N_4 [reproduced from Xu et al. (2013), License number 4005471106720]

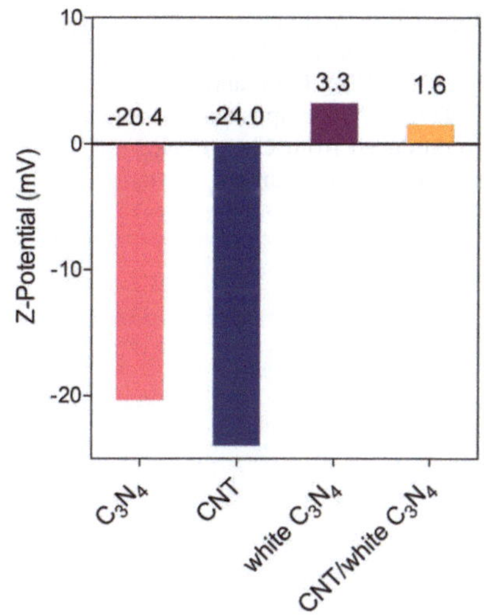

Fig. 9.6 Zeta-potential of the g-C_3N_4, CNT, white C_3N_4 and CNT/white C_3N_4 [reproduced from Xu et al. (2013), License number 4005471106720]

surface of positively charged protonated g-C_3N_4. The electrostatic attraction is beneficial for the formation of well-dispersed and deposited CNT in g-C_3N_4/CNT hybrid.

After CNT modification, the XRD patterns of g-C_3N_4 remain intact which corresponds to inplane distance of nitride pores and graphitic like stacking. No diffraction peaks for CNT were observed due to low loading amount. From which it can be concluded that presence of CNT scarcely alter the structural characteristic of g-C_3N_4 (Fig. 9.7a). The Raman spectra of the composites further support the structure of carbonaceous nanomaterial (CNT). The observed peaks at 1117, 1152, 1212, 1233 and 1460 cm^{-1} show no significant change in position and intensity, which suggests low doping of CNT in the composite for which the lattice structure and molecular skeleton remains unaltered. The Raman spectrum of CNT show three peaks centred at 1310, 1580 and 1609 cm^{-1}, which were assigned to the D, G and D0 bands, respectively, (Chen et al. 2014). The D and D0 bands were attributed to structural defects created due to the presence of oxygenated groups on the basal plane of carbon and the G-band is attributed to the sp2-hybridized carbon networks (Fig. 9.7b). It was also found that the intensity of the CNT bands in the CNT/g-C_3N_4 composites increases with increasing the amount of CNT and a blue shift was observed which confirms charge transfer between g-C_3N_4 and CNT. The peptide bond formed between the terminal amino groups of g-C_3N_4 and carboxyl groups of CNT favours direct charge transfer between two components.

Upon hybridization with CNT, g-C_3N_4 exhibits a stable lamellar morphology with CNT being wrapped by aggregated g-C_3N_4 particles. From the low magnification TEM image, an uniform dyadic structure was observed due to well dispersion of CNT along with intimate interaction. The TEM micrographs of MWCNT/g-C_3N_4 composite show independently coiled MWCNT grown on the surface of g-C_3N_4 (Fig. 9.8).

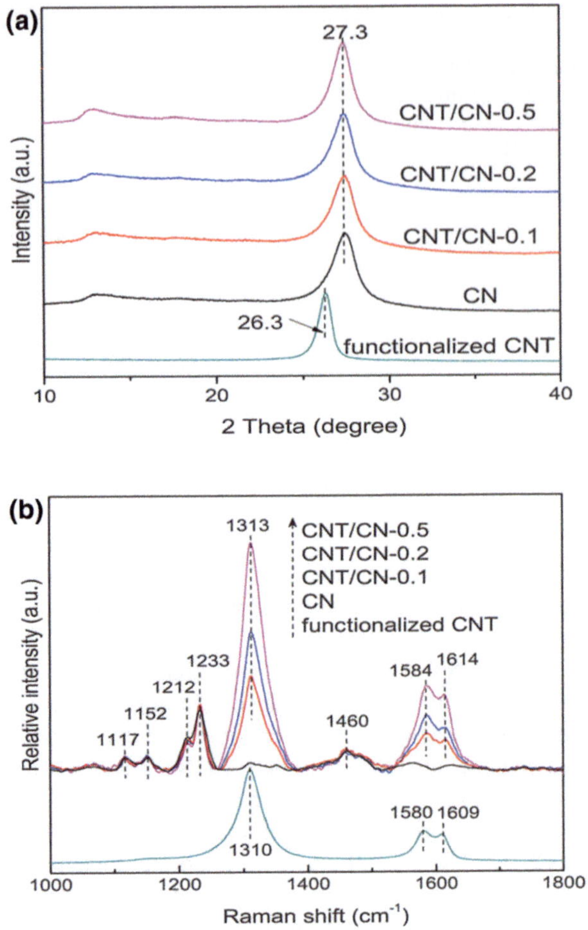

Fig. 9.7 **a** XRD patterns and **b** Raman spectra of the g-C_3N_4, (CN), CNT and CNT/g-C_3N_4 composites [reproduced from Chen et al. (2014), License number 4005371307787]

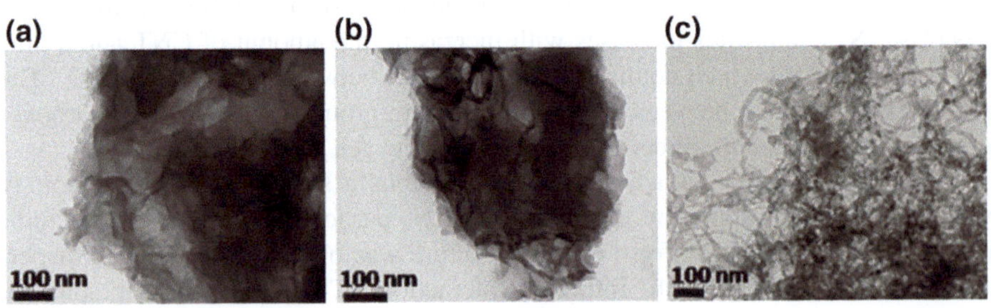

Fig. 9.8 TEM micrographs of the g-C_3N_4 (**a**), and CNT/g-C_3N_4 composites (**b** and **c**) [reproduced from Xu et al. (2013), License number 4004110206676]

In MWNT/g-C_3N_4 composite, incorporation of MWNT has a remarkable effect on the optical properties. Neat g-C_3N_4 shows absorption at 450 nm corresponding to its bandgap 2.7 eV. However, the absorption edge shows a red shift with increase in MWNT content covering a wide range from 200 to 800 nm. The increased absorption is attributed to the enhanced surface electric charge due to addition of MWNT and π–π* electronic transition in the system. The absorption intensity also strengthens the interaction with increasing the amount of MWNT, which is further supported by colour change in case of the composites from yellow to black. However, a blue shift of the absorption edge and decrease in absorption intensity in the CNT/g-C_3N_4 composite was reported, owing to the formation of white g-C_3N_4 upon protonation (Xu et al. 2013; Fig. 9.9a). The same pattern of absorption edge in a physical mixture of g-C_3N_4 and MWNT in the visible region was observed

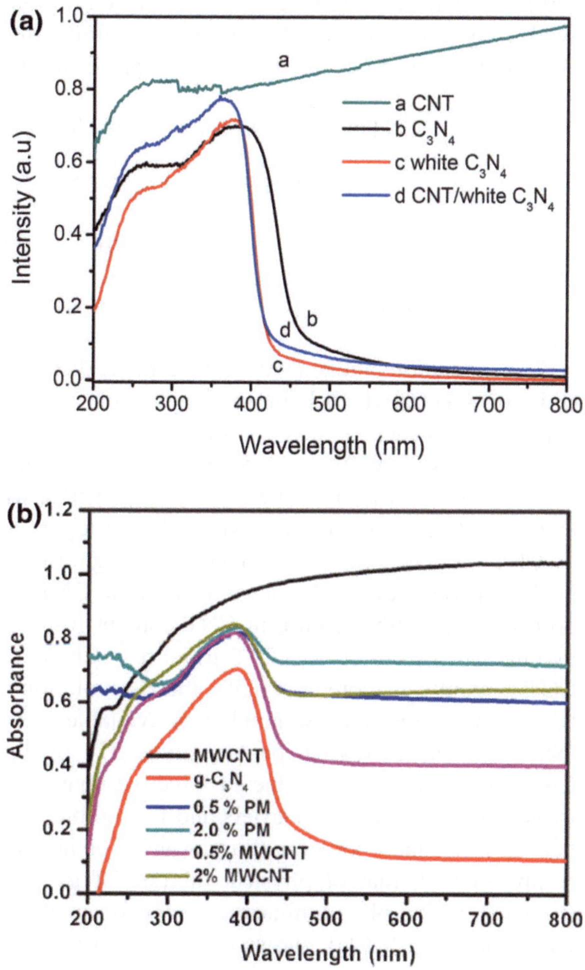

Fig. 9.9 UV–Vis Spectra of **a** g-C_3N_4, and CNT/white-C_3N_4 composites [reproduced from Xu et al. (2013), License number 4006881144455], **b** comparative study with the physical mixture of MWCNT, and g-C_3N_4 [reproduced from Suryawanshi et al. (2012), License number 4004110893670]

although the absorption is suppressed to a greater extent in case of the physical mixture (PM) (Suryawanshi et al. 2012; Fig. 9.9b).

The photocatalytic activity of various g-C_3N_4/CNT nanocomposites were evaluated by degradation of dyes (MB, MO and RhB) and H_2 evolution by water splitting under visible light irradiation. Pure CNT has no photocatalytic activity towards degradation of various dyes. However, upon hybridization with g-C_3N_4 it shows higher photoreactivity. The photocatalytic ability of CNT/white C_3N_4 is about 66.5 and 34.5% higher than that of white g-C_3N_4 and neat g-C_3N_4 in 1.5 h, respectively, (Xu et al. 2013). This may be due to chemically bonded interfacial contact between CNT and the white g-C_3N_4 that improves the electron–hole separation efficiency. The degradation efficiency of g-C_3N_4/CNT nanocomposites towards MO and RhB were found to be 89.7 and 85.4% at 3 h, respectively. But white g-C_3N_4 shows much lower activity owing to its positive polarity which hinders adsorption of the dye on the surface of the catalyst due to electronic repulsion and less light absorption capacity. Functionalized CNT due to the availability of surface–COO–groups exhibit great adsorptivity and favours photo-oxidation of organic dyes due to physical adsorption (Xu et al. 2013).

$$\mathrm{CNT/whiteC_3N_4} + h\vartheta \rightarrow \mathrm{CNT}(e^-)/\mathrm{WhiteC_3N_4}(\mathrm{h}^+) \quad (9.23)$$

$$\mathrm{CNT}(e^-) + \mathrm{O_2} \rightarrow \mathrm{CNT} + {}^{\cdot}\mathrm{O_2}^- \quad (9.24)$$

$${}^{\cdot}\mathrm{O_2}^- + 2\mathrm{CNT}(e^-) + 2\mathrm{H}^+ \rightarrow \mathrm{OH} + \mathrm{OH}^- \quad (9.25)$$

$$\mathrm{whiteC_3N_4}(\mathrm{h}^+)\ \text{or}\ {}^{\cdot}\mathrm{OH} + \mathrm{MB} \rightarrow \text{degrdation product} \quad (9.26)$$

The separation efficiency of photogenerated charge carriers are obtained from PL emission spectra which is in good agreement with the photocatalytic activity. The reactivity order was found to be CNT/white C_3N_4 > neat g-C_3N_4 > CNT/g-C_3N_4 > white g-C_3N_4 > MB photolysis (Xu et al. 2013). The degradation rate was almost 8:1 times higher than white CNT and 2.4 times higher than that of neat g-C_3N_4. A good number of reports were available on H_2 evolution by CNT/g-C_3N_4 composites under visible light irradiation. The activity towards H_2 evolution was found to be dependent on the CNT content in the composite. It was observed that with increase in CNT content from 0 to 0.1 wt%, the H_2-evolution rate increases from 16.4 to 23.5 μmol h^{-1} (Chen et al. 2014). However, when the CNT content was more than 0.5%, the H_2 evolution rate decreases significantly due to agglomeration and less number of available of reactive sites. In another study, there was an increased photocatalytic H_2 evolution of 42 μmol/g by 0.5% CNT/g-C_3N_4 composites due to effective charge transfer from g-C_3N_4 to CNT, longer lifetime of charge carriers and favourable morphological changes in g-C_3N_4 due to CNT incorporation (Suryawanshi et al. 2012). When H_2 evolution was compared with that of physical mixture of g-C_3N_4 and CNT it was much less (6 μmol/g). The photoluminescence spectra of the neat g-C_3N_4 and CNT/g-C_3N_4 composite (Fig. 9.10a), shows quenching in the PL intensity and reveals efficient

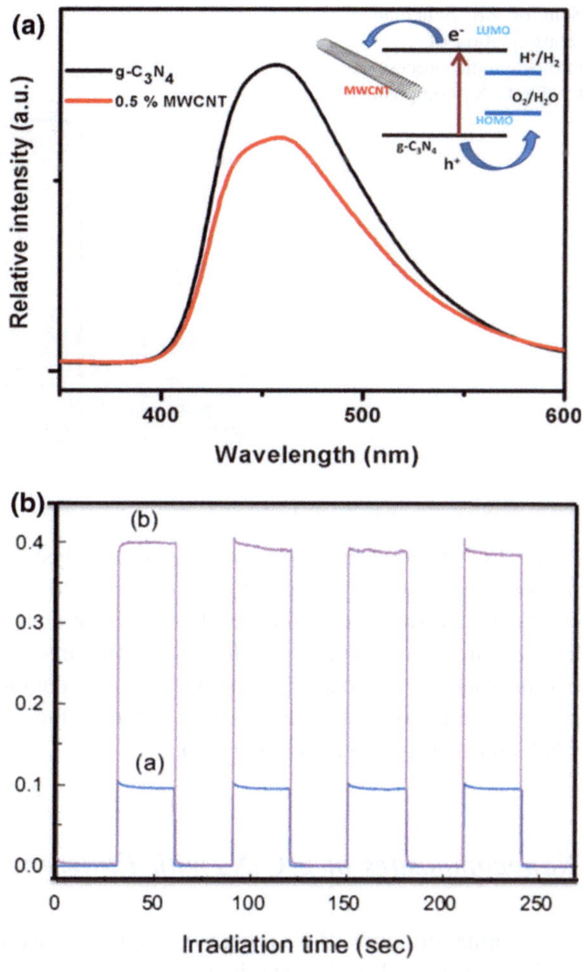

Fig. 9.10 **a** PL Spectra [reproduced from Suryawanshi et al. (2012), License number, 4004110893670], **b** photocurrent response of g-C_3N_4, and CNT/g-C_3N_4 composites [reproduced from Ge et al. (2012), License number, 4004110689017]

charge separation in the composite according to the experimental results. The H_2 evolution of CNT/g-C_3N_4 composites with an optimal CNT loading, i.e. 2.0 wt% was studied and corresponding H_2 evolution rate is 7.58 μmol/g which was about 3.7 times higher than that of neat g-C_3N_4 (Ge et al. 2012). CNTs possess higher electron capture capacity and favour transfer of electron from g-C_3N_4 surface to CNT surface as the Fermi level of CNT is at lower potential. A Schottky barrier was formed at the interface to facilitate charge transfer and helps in water reduction reaction, whereas the holes in the VB can react with methanol as a sacrificial reagent (Ge et al. 2012). All the reported studies established good stability of the catalysts after the photocatalytic reaction. The CNTs/g-C_3N_4 composite photocatalyst shows higher photocurrent intensity than that of neat g-C_3N_4 to further support the H_2 evolution data (Fig. 9.10b).

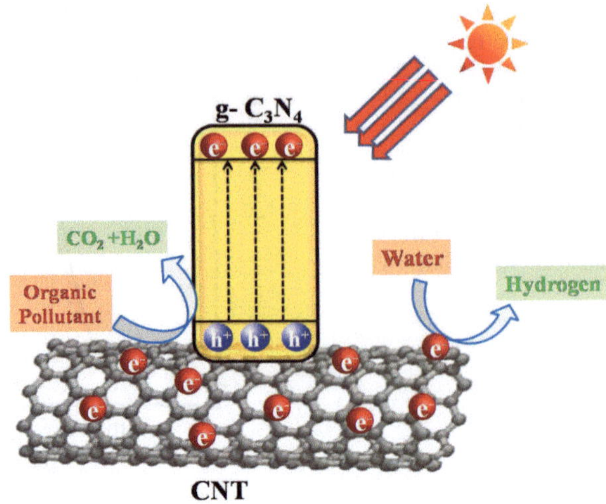

Scheme 9.8 Schematic representation of photocatalytic process of CNT/g-C_3N_4 composite

The CNT modified composite photocatalyst shows significantly enhanced photocatalytic activity owing to a well-matched π-conjugated molecule skeleton, similar lattice structure, enhanced BET surface area, great adsorptivity of dyes and high electron–hole separation efficiency compared to neat g-C_3N_4. The composites exhibit superior photocatalytic activity for dye degradation and for water reduction reaction under visible light irradiation due to the synergetic interactions between CNT and neat g-C_3N_4 (Scheme 9.8).

Nanocomposites of g-C_3N_4 with Fullerene

The unique structure of C_{60}, high exciton mobility (>1.3 $cm^2\ V^{-1}\ S^{-1}$) and large exciton diffusion length make it suitable to act as an excellent electron acceptor to enhance various photocatalytic reactions. C_{60} modified graphitic carbon nitride composite photocatalysts (C_{60}/g-C_3N_4) was fabricated by a facile thermal treatment method involving polymerization of dicyandiamide in the presence of C_{60}. During calcinations g-C_3N_4 network materials result disordered structures, which may contains two different monomeric units; triazine (C_3N_3) and heptazine (C_6N_7) (Bai et al. 2014). The proportion of triazine and heptazine strongly depends on the precursors used and condensation process. Therefore, in the process of calcination, –NH bonds cleaves between C_3N_3 and C_6N_7 by C_{60} clusters and a strong chemical bond (C–N) takes place. In another method C_{60}/g-C_3N_4 nanocomposite was synthesized by incorporating C_{60} into the matrix of g-C_3N_4 by a simple adsorption approach using urea as the precursor of g-C_3N_4 and C_{60}. In which C_{60} molecules are physically adsorbed on the surface of g-C_3N_4. Moreover, due to the evolution of huge amount of gases like NH_3, CO_2 during the condensation process of urea also makes the material porous (Chai et al. 2014). The XRD patterns of C_{60} exhibit diffraction peaks of the (111), (220), (311) and (222) planes at 2θ = 10.7°, 17.7°, 20.7° and 21.7°, which represent cubic phase of C_{60}. By preparing C_{60}/g-C_3N_4 composite the position of the diffraction peaks of g-C_3N_4 and C_{60} show no significant change indicating that C_{60} does not influence the lattice structure of g-C_3N_4. In 1 wt% C_{60}/g-C_3N_4 even no characteristic diffraction peaks are found due to small amount of C_{60} and proper dispersion in the composite (Fig. 9.11).

The SEM and TEM images of the composite reveal a two-dimensional lamellar structure possessing 10–20 nm thickness layer. The 1 wt% C_{60}/g-C_3N_4 composite also represents a porous flaky structure similar to neat g-C_3N_4, suggesting that the introduction of C_{60} has no effect on the morphology of neat g-C_3N_4. Furthermore, the XPS and FTIR results provide information on the chemical interaction between g-C_3N_4 and C_{60} in the composite. When the composite was synthesized by thermal treatment, the intensity of C1s XPS peak represents, formation of a new C–N bond between C_{60} and g-C_3N_4 (Fig. 9.12a). However, when the composite was synthesized by simple adsorption approach due to the presence of physical interaction there was no change in the C1s peak intensity confirming weak interaction, which was further supported by FTIR analysis (Fig. 9.12b). The peak at 1253 cm^{-1} in case of g-C_3N_4 slightly shifted to lower wave number, i.e. to 1245 cm^{-1} in the C_{60}/g-C_3N_4 composite indicates weak C–N bond confirming the interaction between g-C_3N_4 and C_{60}. Existence of these interactions facilitates electron transfer and increases the efficiency of photocatalytic reaction.

The absorption of nanocomposites of g-C_3N_4 with fullerene shows a red shift of about 20 nm along with enhanced absorption intensity. By preparing C_{60}/g-C_3N_4 composites the bandgap energy is decreased showing strong absorption in visible region. The PL emission spectra further confirm the decrease in recombination tendency of photogenerated charge carriers. The main emission peak for neat g-C_3N_4 centred at 455 nm was mainly due to band–band transition. The intensity of the PL peaks for C_{60}/g-C_3N_4 composites was found to be much lower than g-C_3N_4 and shows a blue shift confirming C_{60} as a good electron acceptor material having

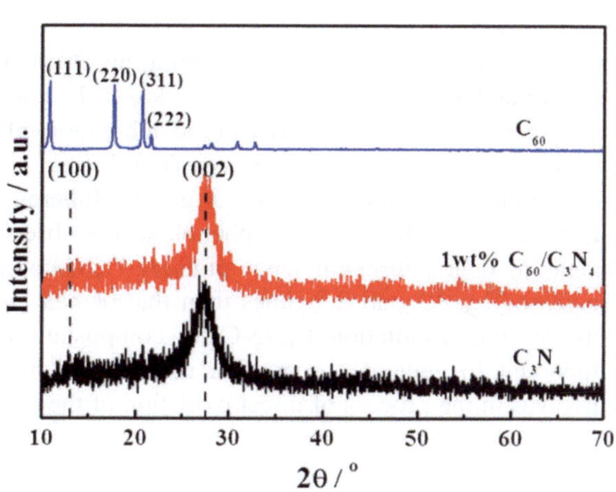

Fig. 9.11 XRD patterns of neat g-C_3N_4, C_{60}, and 1 wt% C_{60}/g-C_3N_4 composite [reproduced from Chai et al. (2014), License number 4004130394222]

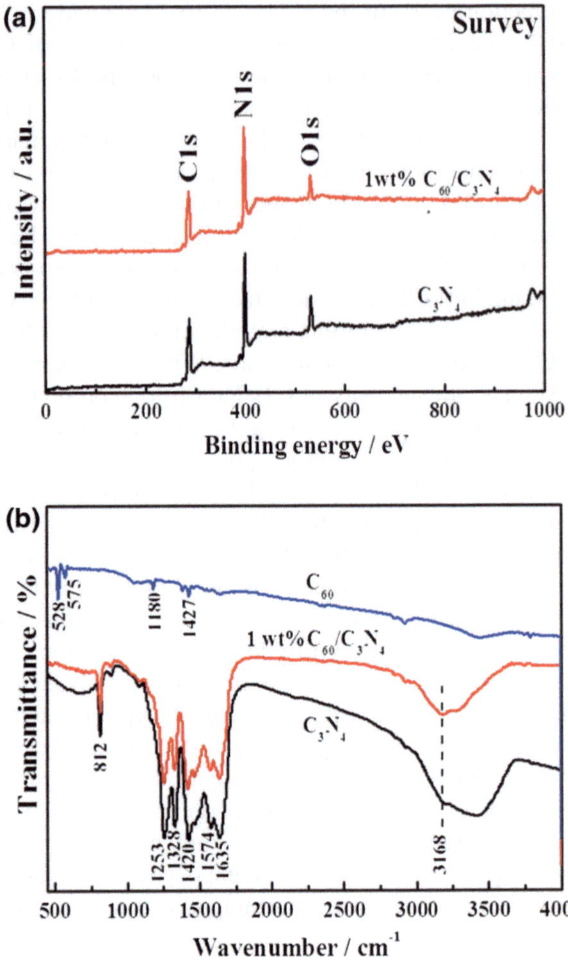

Fig. 9.12 **a** XPS Spectra, **b** FTIR Spectra of g-C_3N_4, and C_{60}/g-C_3N_4 composites [reproduced from Chai et al. (2014), License number 4004121185499]

π-conjugated structure which becoming an efficient separation centre of the photogenerated electrons and holes. A lower PL intensity in case of C_{60}/g-C_3N_4 composites suggests a decrease in recombination rate of charge carriers which is in agreement with the bandgap energy (Fig. 9.13).

The photocatalytic behaviour of g-C_3N_4/fullerene composite was studied by the degradation of MB, RhB and phenol. It was observed that after modified by fullerene the degradation efficiency of C_{60}/g-C_3N_4 composite towards MB and phenol increased by 3.2 and 2.9 times than that of neat g-C_3N_4 (Bai et al. 2014). Upon visible light irradiation C_{60}/g-C_3N_4 composite system O_2 molecules are more important to generate superoxide radicals on the surface of the catalyst which involves in the process of direct oxidation of the dyes. Holes and hydroxyl radicals

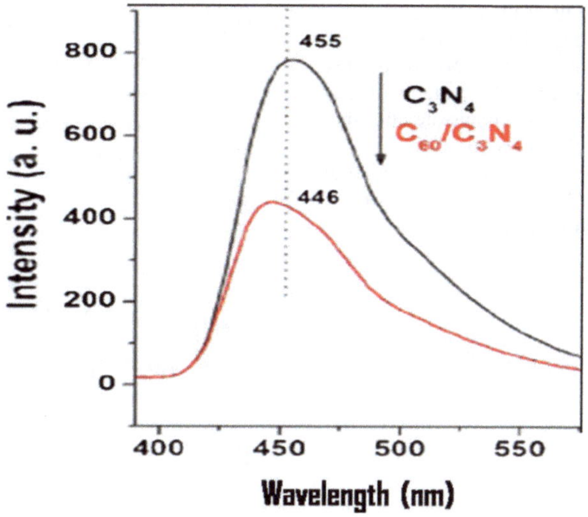

Fig. 9.13 PL emission spectra of neat g-C_3N_4, and C_{60}/g-C_3N_4 composites [reproduced from Bai et al. (2014), License number, 4004121485535]

also show improved mineralization and help in conjugated ring opening. The improved rate of photocatalytic reactions is further supported by electrochemical impedance spectroscopy. In the Nyquist plot of neat g-C_3N_4 and the composite, the high frequency arc represents the resistance offered by the electrodes at the contact interface between the electrode and the electrolyte solution. The smaller arc radius in case of the C_{60}/g-C_3N_4 composite than that of neat g-C_3N_4 represents superior conductivity and higher reaction rate.

Further to confirm the mechanism trapping experiments were carried out which suggests hydroxyl radical as the main oxidative species to carry out the reaction. In the composite when the photogenerated electrons get transferred to C_{60} particles and help to capture the adsorbed O_2 to form oxidative species for degradation purpose. Due to accumulation of electrons the valence band of g-C_3N_4 shifts to lower energy position and gives strong photo-oxidation ability. The photocatalytic activity of C_{60}/g-C_3N_4 composites was further studied by investigating the degradation rate constants of RhB (Chai et al. 2014). Neat g-C_3N_4 has a very low (54%) degradation rate for RhB under visible light irradiation for 60 min, whereas the 1% C_{60}/g-C_3N_4 composite shows degradation rate of 97% in 60 min. The rate constants of different samples (neat g-C_3N_4, 0.5, 1, 2 wt% C_{60}/g-C_3N_4 composites) are 0.00998, 0.03349, 0.05818, 0.03308 and 0.00041 min^{-1}, respectively, showing higher rate in 1% C_{60}/g-C_3N_4 composite (Fig. 9.14a). Moreover, from the stability and reusability tests it was found that the high photocatalytic degradation efficiency of RhB was maintained even after five recycling runs and there is no obvious deactivation. The XRD pattern (Fig. 9.14b) of the recycled C_{60}/g-C_3N_4 composites after five runs of photoreaction is found similar to that of the original one with the same intensity, suggesting considerable photostability of C_{60}/g-C_3N_4 composites.

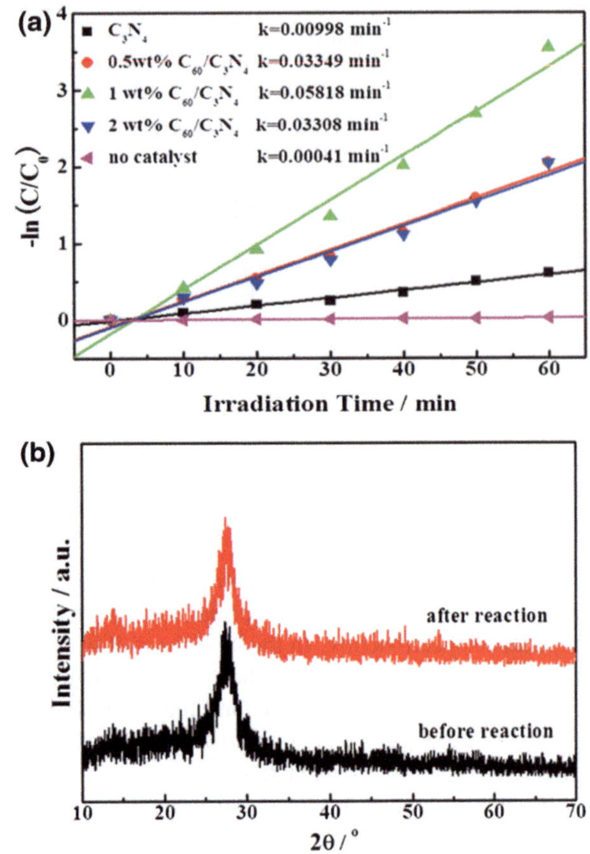

Fig. 9.14 **a** Rates of degradation of neat g-C_3N_4, and C_{60}/g-C_3N_4 composites, and **b** XRD patterns of C_{60}/g-C_3N_4 composite after the photocatalytic run [reproduced from Chai et al. (2014), License number 4004130394222]

Nanocomposites of g-C_3N_4 with Polyaniline (PANI)

In the recent years, the conductive polymer polyaniline (PANI) has been extensively studied as a photocatalyst because of good stability, nontoxicity, corrosion protection and low cost synthesis. In the field of photocatalysis PANI has shown great potential due to its high absorption coefficient (5×10^4) and greater mobility of charge carriers. A novel PANI–g-C_3N_4 nanocomposite was fabricated with different PANI:g-C_3N_4 ratios by "in situ" deposition and oxidative polymerization of aniline monomer in the presence of g-C_3N_4 powder in an ice bath (Ge et al. 2012). A hierarchical nanocomposites of polyaniline (PANI) nanorod arrays on g-C_3N_4 sheets was designed by polymerization under −20 °C by using g-C_3N_4, 1 M $HClO_4$ ethanol solution and aniline monomer and ammonium persulphate was used as the polymerizing agent (Zhang et al. 2013). A gold nanoflower decorated g-C_3N_4 polymer nanosheet–PANI hybrids (AuNF@ g-C_3N_4–PANI) for the detection of dopamine was designed by an in situ synthesis method from PANI, gold nano flowers and g-C_3N_4 nanosheets (Lu et al. 2014). The prepared hybrids were then deposited on carbon electrode to act as an electro-chemiluminescence (ECL) sensor. In another method an interfacial polymerization method was used to synthesize highly dispersed PANI nanorods on porous g-C_3N_4 nanosheets (Yu et al. 2015). A mixture of the oxidant (ammonium persulfate) and porous g-C_3N_4 were mixed in aqueous phase and aniline in the organic phase which upon polymerization were formed at the interface and diffused to the aqueous phase. Formation of the composite mainly involves the interaction between conjugated π-electrons of aniline and the basal plane of g-C_3N_4. This conjugate interaction also increases the activation energy of thermal degradation process of g-C_3N_4/PANI composites and confirms the higher thermal stability of the composite.

The interaction between PANI and porous g-C_3N_4 was clearly seen in the SEM images (Fig. 9.15). In case of g-C_3N_4/PANI composite the hierarchical structures were found showing a lot of PANI nanorods uniformly distributed on the rough surface of g-C_3N_4 sheets (Zhang et al. 2013). This confirms that the nucleation and growth processes only take place on the surface of g-C_3N_4 sheets. The rough surface of the composite was beneficial for the propagation of visible light. Moreover, the hierarchical structure also allowed multiple reflections or scattering of light within the interior void, which could lead to more efficient use of the visible light. PANI nanorod with a diameter of 40–60 nm arrayed with a random stacking are found in clubbed morphology on porous g-C_3N_4 nanosheets. When PANI was coupled with porous g-C_3N_4 a powerful interfacial bonding between porous g-C_3N_4 and PANI was revealed (Ge et al. 2012).

The efficiency of the photocatalyst and more efficient absorption in the visible region was further supported by photoluminescence (PL) spectral analysis. The separation efficiency of photogenerated electrons and holes was responsible for higher rate of photocatalysis. It was found that when excited at the wave length of 365 nm, the position of the emission peak was similar to that of neat g-C_3N_4 sample but the emission intensity was significantly decreased, which indicates much lower recombination tendency of electron–hole pairs in the composite. The pure PANI molecule not only absorbs UV light, but also shows strong absorption both in visible light and near infrared regions, which can be ascribed to transitions in the PANI molecules. The PANI/g-C_3N_4 composite compared to that of neat g-C_3N_4, shows strong absorption in the visible region and with increase in PANI contents the absorption intensity becomes stronger and a red shift was observed. The UV–Vis spectra results indicated that with enhanced light absorption more number of electron–hole pairs were formed under visible light irradiation, and expected to activate the photocatalytic performance.

Various groups of researcher studied the photocatalytic activities of g-C_3N_4/PANI composites towards the degradation of organic dyes (MB, MO, etc.). When the photodegradation of MB was studied, it was observed that 92.8% of MB degradation occurs at an optimum PANI concentration in PANI/g-C_3N_4 composite, whereas neat g-C_3N_4 only degrades 41.2% in 120 min (Ge et al. 2012). In another study, with an optimum PANI loading 78.6% for MB and 99.8% for MO undergo degradation under visible light (Zhang et al. 2014a, b). When the amount of PANI loading is more, the photocatalytic capability of the composite was decreased to

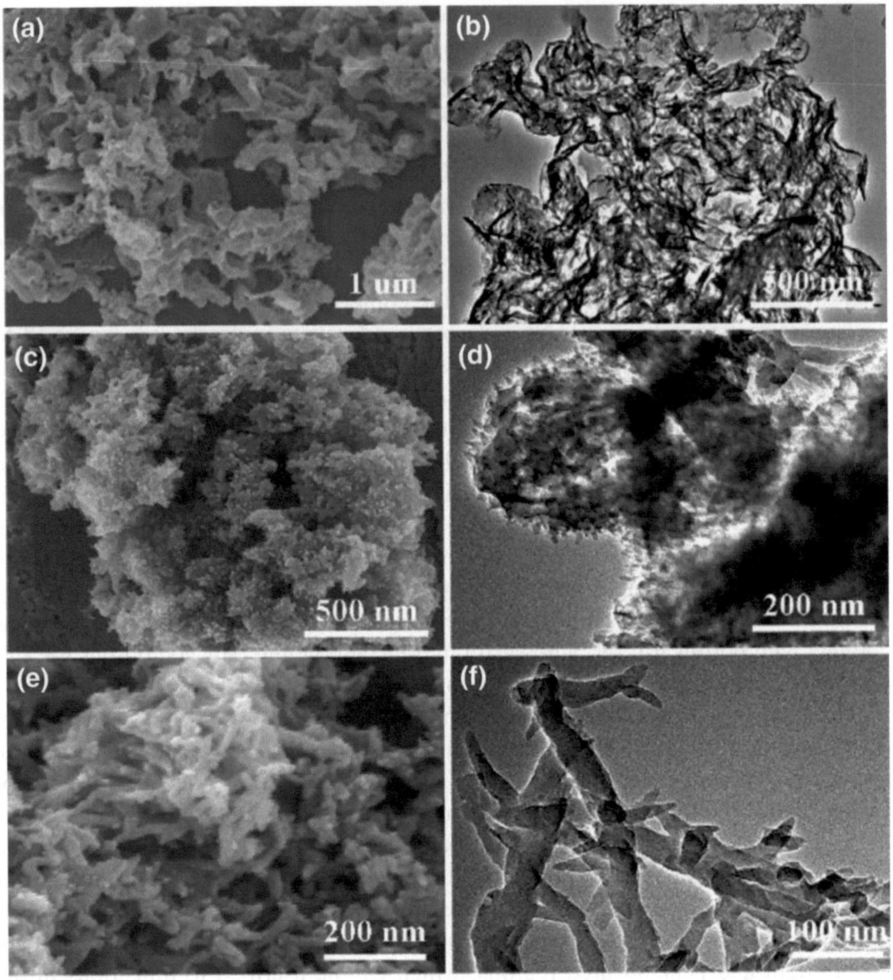

Fig. 9.15 Typical SEM images of porous g-C_3N_4 (**a**), PANI (**b**), and porous g-C_3N_4/PANI composite (**c**) [reproduced from Zhang et al. (2014), License number 4004150262468]

67.7% for MB and 91.7% for MO. That may be due to excessive agglomeration of PANI on the surface of g-C_3N_4, which hinders the transfer of the photoinduced charge carriers. The degradation efficiency was more in case of MO due to greater adsorption capacity of MO than that of MB. Because MO is an anionic dye whereas MB is a cationic dye. Owing to which the positively charged sites of PANI attract the negatively charged MO molecules through strong electrostatic attraction. On the other hand the positively charged MB molecules get adsorbed less effectively due to electrostatic repulsion. As reported by Zhang the specific surface area of PANI/g-C_3N_4 composite (95.4 m^2/g) is also increased after PANI modification which was almost three times higher than that of neat g-C_3N_4. This increase in specific surface area favours the adsorption process by providing more number of active sites. The photocatalytic activity of novel AuNF@ g-C_3N_4-PANI hybrid photocatalyst was investigated as an electro generated chemiluminescence (ECL) biosensor for the determination of DA (dopamine), which was helpful to extend the application of luminophore g-C_3N_4 in the field of analytical chemistry (Lu et al. 2014). The photodegradation efficiency of MB under visible light irradiation without any catalyst, the rate of MB decomposition was found to be only 5% after 180 min. However, in porous g-C_3N_4 photocatalyst, the rate of degradation was about 30%. But in presence of porous g-C_3N_4/PANI composites it was around 70% (Yu et al. 2015). The improved photocatalytic performance was due to the synergistic effect of PANI and g-C_3N_4, oxidizing power and electron transport property of PANI. It was also supported by the transient photocurrent density of the composite which was about 3.58 times higher than that of porous g-C_3N_4 (Fig. 9.16).

In the photocatalytic process upon visible light illumination the absorbed photons induce π–π* transition in PANI molecules. The band edge potentials of PANI were determined to be −2.14 and +0.62 eV and the CB and VB potentials of neat g-C_3N_4 are at −1.13 and +1.57 eV, respectively, (Ge and Han 2012; Ge et al. 2012). On the

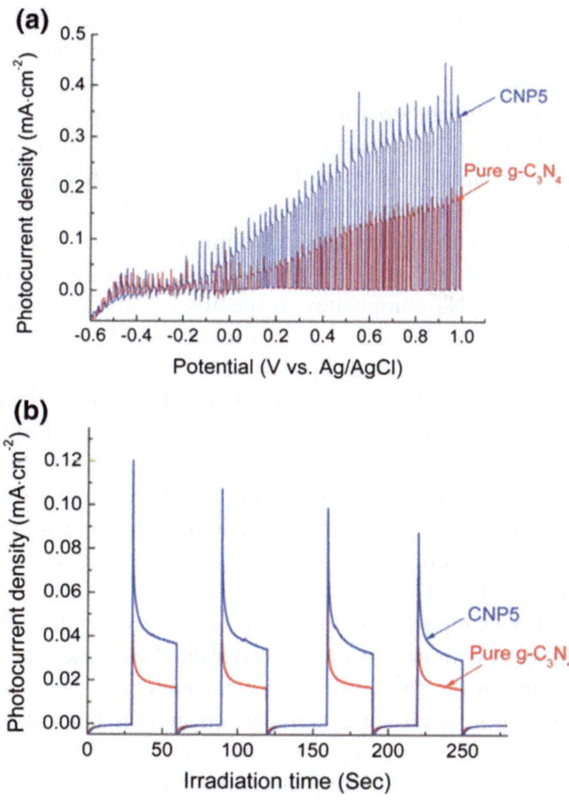

Fig. 9.16 Photocurrent densities versus potential and photocurrent densities versus irradiation time graph of g-C_3N_4/PANI composite (CNP5 represents 0.5:10 PANI:g-C_3N_4 ratio) [reproduced from Ge et al. (2012), License number, 4010071349089]

basis the relative band edge potentials photogenerated electrons get transferred from the CB of PANI to that of g-C_3N_4, and trigger the formation of superoxide radical ion and hydroxyl radical ion via the multistep reduction of adsorbed O_2. Simultaneously photogenerated holes migrate from the VB of g-C_3N_4 to the HOMO of PANI to promote the charge separation process and also photodegrade the adsorbed dye molecules directly.

Nanocomposites of g-C_3N_4 with Polypyrrole (Ppy)

A typical conductive polymer polypyrrole (Ppy) having bandgap of 2.2–2.4 eV, high stability and interesting redox properties, is found suitable to enhance the photocatalytic properties of g-C_3N_4. It can also be synthesized easily at nanolevel through surfactant-assisted microemulsion polymerization method in aqueous media. The modification of g-C_3N_4 by Ppy nanoparticles from surface junction to facilitate transfer of photogenerated electrons and the holes oxidized Ppy to improve conductivity. Different research groups separately tried to synthesize g-C_3N_4/polypyrrole nanocomposites by sonochemical method. The nanocomposite of g-C_3N_4/polypyrrole was designed by loading Ppy on the surface of g-C_3N_4 by ultrasonication method (Sui et al. 2013; Hu et al. 2015a, b). To compare the activity of the composite with other polymeric nanocomposites, g-C_3N_4/polythiophene composites was also synthesized by introducing dispersively distributed polythiophene (Ptp) nanoparticles on g-C_3N_4 surface.

The Ppy nanoparticles were found to have no significant effects on the absorption edge of Ppy/g-C_3N_4 composite. However, the composites show stronger background absorption due to the black colour of Ppy. The absorption edge shows a red shift with increasing the loading amount of Ppy. The observed absorption edges were 539, 563 and 590 nm for Ppy (1.5%)-g-C_3N_4, Ppy(2.5%)-g-C_3N_4 and Ppy (4%)-g-C_3N_4 composites, respectively. The corresponding bandgap energies also decreases to 2.3, 2.2, 2.1 eV for series of Ppy/g-C_3N_4 composites. The PL spectra of neat g-C_3N_4 and Ppy/g-C_3N_4 composite further provide information about separation tendency of photogenerated charge carriers. The broad PL emission peak indicates complicated transitions of the excited states in g-C_3N_4 and decreased intensity of the emission peak in case of Ppy/g-C_3N_4 suggests higher separation of electron–hole pairs (Fig. 9.17). The PL intensity of Ptp/g-C_3N_4 was much lower than that of Ppy/g-C_3N_4, which may be due to stronger interaction in case of Ptp/g-C_3N_4 composite (between g-C_3N_4 and electronegative O-atom in C=O of Ptp) (Hu et al. 2015a, b).

It was reported by various research groups that when g-C_3N_4 was electronically coupled with polymers the proper band alignment between g-C_3N_4 and the polymer results in the formation of a heterojunction. This facilitates the transportation and separation of photogenerated electron–hole pairs at the interface. Upon visible light irradiation, electrons get excited from the VB to the CB of both the semiconducting material and tend to migrate easily from g-C_3N_4 to Ppy (Ptp) having CB potential of

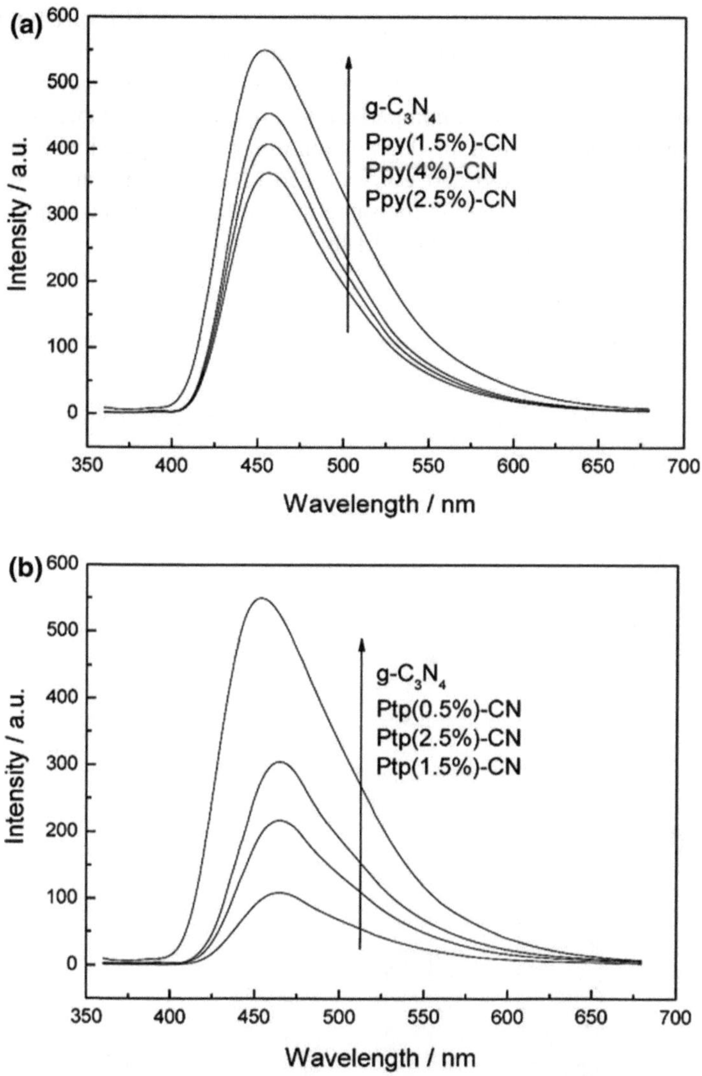

Fig. 9.17 PL Spectra of neat g-C_3N_4, Ppy/g-C_3N_4 composite, and Ptp/g-C_3N_4 composite [reproduced from Hu et al. (2015a, b), License number, 4010080339963]

0.94 eV (1.0 eV), whereas the photogenerated holes transfer from Ppy (Ptp) to g-C_3N_4 having VB potential of 1.1 eV (0.31 eV). To enhance the surface reaction this potential difference acts as the main driving force for overcoming the high dissociation barrier of the Frenkel exciton and electron–hole recombination. The wave length dependent photodegradation activity of RhB by g-C_3N_4/Ppy composite was investigated and compared with that of g-C_3N_4/Ptp composite (Hu et al. 2015a, b; Fig. 9.18a). The activity of g-C_3N_4/Ptp composite was found to be more than that of g-C_3N_4/Ppy composite. Also the photostability of g-C_3N_4/Ptp composite was more than that of g-C_3N_4/Ppy composite, which was attributed to the strength of interaction between g-C_3N_4 and polymers, owing to greater electronegativity of oxygen atom of C=O in Ptp in comparison to Ppy and more stable structure with good adsorption ability. When the degradation rate was compared with that of other three polymers, P3HT, PANI and g-PAN, used to form composites with g-C_3N_4, prepared in the same mass ratio in the same method g-C_3N_4/Ptp composite showed the highest activity as shown in Fig. 9.18b.

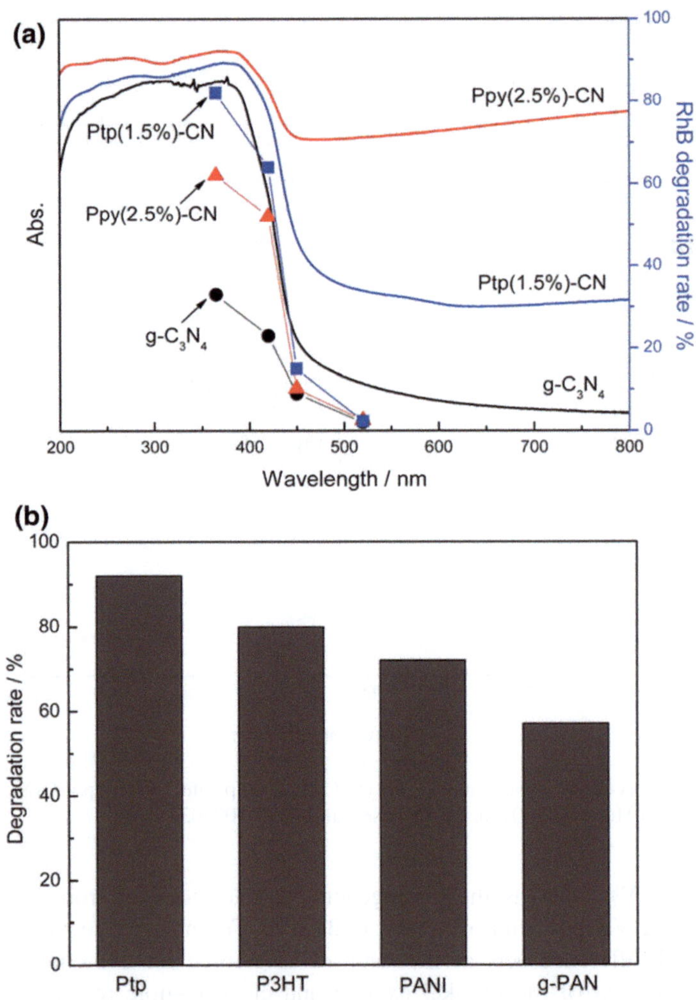

Fig. 9.18 **a** Wavelength-dependent RhB degradation rate of g-C_3N_4, Ppy(2.5%)-CN and Ptp (1.5%)-CN, **b** degradation efficiency of Ptp, P3HT, PANI and g-PAN modified g-C_3N_4 [reproduced from Hu et al. (2015a, b), License number, 4010080339963]

The hydrogen evolution activity of g-C_3N_4/Ppy composite was also studied under visible light irradiation (Sui et al. 2013). Polypyrrole as a polymer has no tendency to liberate hydrogen as a photocatalyst under visible light irradiation. After loading Ppy nanoparticles on the surface of g-C_3N_4 the rate of hydrogen evolution increases by 1.43 times. However, in case of Pt/g-C_3N_4-Ppy composite, the H_2 evolution rate increased up to 49.3 times. Coupling two organic semiconducting materials improves the charge transfer process by forming a surface junction between g-C_3N_4 and Ppy nanoparticles. Based on the relative band edge potentials photoelectrons get transferred easily from Ppy to g-C_3N_4 and favour the water reduction reaction thermodynamically. The photoinduced holes can oxidize Ppy nanoparticles to form p-doping Ppy nanoparticles. These p-doping Ppy nanoparticles possess high electronic conductivity and positive charge in the composite, which induce hydrolysis of water to form H^+ ions. The photoinduced holes in the VB of g-C_3N_4 help to oxidize water to hydrogen peroxide. The H_2 evolution rate of neat g-C_3N_4, Ppy, g-C_3N_4-Ppy composite, Pt/g-C_3N_4-Ppy composites are as in Fig. 9.19a. With an optimum Ppy content 1.5% g-C_3N_4-Ppy composite (PC-1.5) the H_2 evolution was maximum and then with further loading the activity is decreased as in Fig. 9.19b.

Nanocomposites of g-C_3N_4 with Polyacrylonitrile (g-PAN)

Polyacrylonitrile (PAN) is one of the important polymeric materials which upon thermal treatment above 600 °C undergoes graphitization to form a conjugated structure. The graphitized PAN possesses a sheet like structure and facilitates fast transfer of electrons. When coupled with g-C_3N_4 it shows excellent photocatalytic activity. A facile one-pot synthesis method was reported for the fabrication of g-PAN/g-C_3N_4 photocatalysts by thermal treatment (He et al. 2014). A mixture of melamine and PAN are grinded to mix in the solid state properly and then subjected to calcination at 650 °C in an inert atmosphere. Upon heat treatment, g-C_3N_4 and PAN undergo polymerization and get integrated to form conjugated networks for proper electron delocalization and efficient photocatalytic activity.

When PAN was graphitized by heating at 650 °C for 2 h in an inert atmosphere, a broad diffraction peak was found at $2\theta = 26.0°$ instead of at 17.5° in case of pure PAN. Whereas g-C_3N_4 maintains the crystallinity in the composite. In the FTIR spectral analysis also new peaks were observed at 1573 cm^{-1} due to stretching vibrations of C=C and C=N. The aromatic conjugated structure of g-PAN after graphitization was confirmed by the stretching vibrational modes at 805 cm^{-1}. The aromatic conjugation of g-PAN acts as an efficient electron delocalization channel to improve separation efficiency of photogenerated electron–hole pairs. The TEM microgram clearly shows big layered sheets decorated with small nanosheets of g-PAN, which are in intimate contact in the composite (Fig. 9.20).

From UV–Vis DRS study it was observed that the absorption edge of the composite get strengthened and red shifted with increase in the amount of g-PAN.

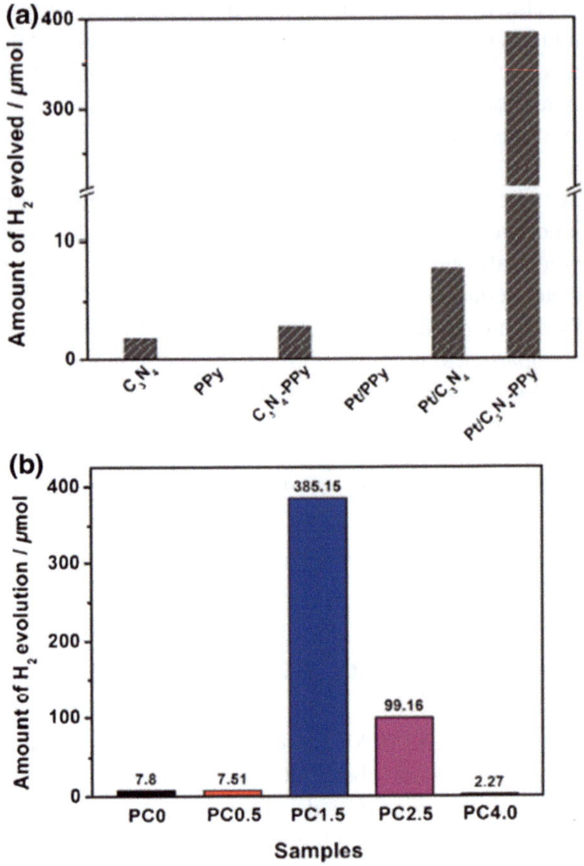

Fig. 9.19 **a** Rate of hydrogen evolution by g-C_3N_4, Ppy, Ppy(1.5%)-CN and Pt/Ptp (1.5%)-CN, **b** H_2 evolution by g-C_3N_4/Ppy composites with varying wt% of Ppy [reproduced from Sui et al. (2013), License number 4004711364744]

The decrease in bandgap also supports the enhanced visible light absorption in case of the composite. Further, the PL emission spectra show the role of g-PAN to increase charge separation. In g-PAN/g-C_3N_4 composites the PL intensity was significantly decreased with increase in PAN content till optimum. The photocatalytic activity of g-PAN/g-C_3N_4 composite under visible light irradiation was investigated for hydrogen evolution in presence of 1.5% Pt using TEOA as the sacrificial agent. When the experiment was carried out only with g-PAN, it was found inactive to generate H_2, but 5 wt% g-PAN/g-C_3N_4 composite improves rate of H_2 evolution by 4.4 times and by using 1.5% Pt the H_2 evolution further increases by 52.8 times compared to that of neat g-C_3N_4 (He et al. 2014). The suitable band edge positions of g-PAN and g-C_3N_4 with CB potential (0.42 and 1.27 eV) favours the transfer of electrons from g-C_3N_4 to g-PAN. The conjugated network promotes charge separation and increases the efficiency on the composite photocatalyst. The proper band alignment in case of g-PAN/g-C_3N_4 composite and enhanced specific surface area (36 $m^2\ g^{-1}$) were found important for optimizing the photocatalytic activity. The composite improves interfacial charge transfer at the interface and helps

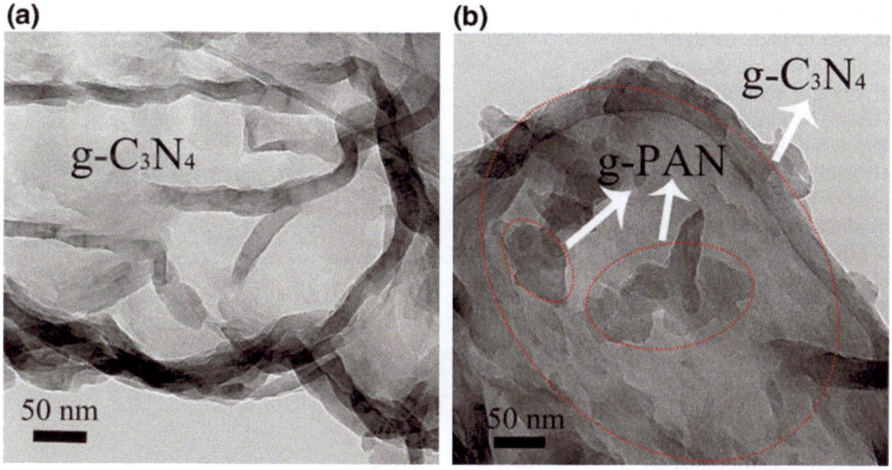

Fig. 9.20 TEM images of g-C_3N_4 (**a**), and 5 wt% g-PAN/g-C_3N_4 composite (**b**) [reprinted with permission from He et al. Copyright 2014]

in water reduction reaction on the surface of g-PAN sheets. The electron separation efficiency was further supported by the electrochemical impedance spectroscopy suggesting lower resistance and faster interfacial charge transfer (Fig. 9.21).

Nanocomposites of g-C_3N_4 with Poly-3-hexylthiophene (P3HT)

P3HT is a p-type semiconducting material having bandgap energy of 1.9–2.1 eV, high solubility, high hole carrier mobility (10^{-4}–10^{-3} cm^2 V^{-1} s^{-1}) and easy processability (Bai et al. 2015). The electronic factor responsible to increase the charge transfer at the interface and the structure of the donor molecule in extending conjugation are responsible for enhancing the photocatalytic activity. P3HT/g-C_3N_4 photocatalysts were synthesized by a ball milling method, which was reported to be an effective way to design polymer modified materials in the field of chemistry. The synthesis method involves calcinations of a mixture of g-C_3N_4 and P3HT at 120 °C for 4 h after mixing properly in the solid state in order to achieve the maximum hole carrier mobility. The –NH_2 and –NH groups on the surface of g-C_3N_4 were mainly involved in polymer processing (Bai et al. 2015). The XRD pattern of the catalyst confirms intimate interaction between g-C_3N_4 and P3HT to favour charge transfer. The TEM microgram shows regular spheres of P3HT and thin sheet like structure of g-C_3N_4 which are coupled together to form a heterojunction. These results revealed that polymeric P3HT adsorbs strongly on to the surface of g-C_3N_4 which results various types of structures ranging from lamellar assemblies to more disordered, bundled conformations. The typical Raman spectra of the samples excited at

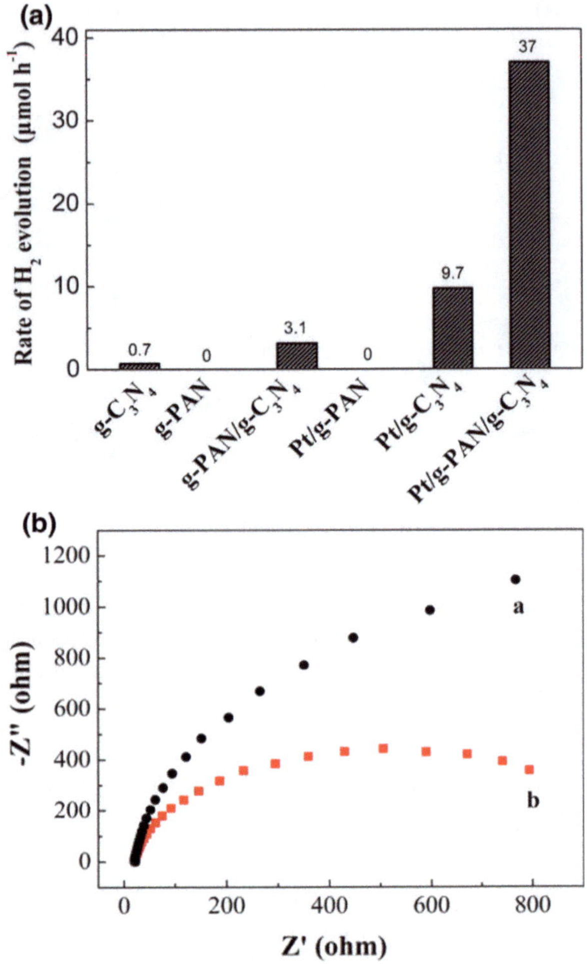

Fig. 9.21 **a** Comparison of the rate of photocatalytic H_2 evolution for g-C_3N_4, g-PAN, 5 wt% g-PAN/g-C_3N_4, Pt/g-PAN, Pt/g-C_3N_4, and Pt/5 wt% g-PAN/g-C_3N_4 in 10 vol% TEOA aqueous solution under visible light irradiation ($\lambda > 400$ nm), **b** EIS Nyquist plots obtained at an AC voltage with amplitude of 5 mV over the frequency range of 1×10^5 to 1×10^{-1} Hz for **a** g-C_3N_4, and **b** 5 wt% g-PAN/g-C_3N_4 electrodes in 0.5 mol L^{-1} Na_2SO_4 aqueous [reprinted with permission from He et al. Copyright 2014]

514 nm show various Raman signals at 400–2000 cm^{-1}; however, the signals due to g-C_3N_4 are weak. Among all, the main in plane ring skeleton modes at 1455 and 1376 cm^{-1} are sensitive to delocalization of π-electrons and conjugation length in case of P3HT molecule. The intensity of C=C bond in the composite increases by three times compared to that of pure P3HT indicating the presence of more ordered and longer conjugated segments in the composite. This increasing trend in Raman intensity is in consistent with its photocatalytic activity due to higher degree of molecular ordering in P3HT/g-C_3N_4 composite (Fig. 9.22). Further in FTIR spectral analysis the band at 1407 cm^{-1} of P3HT/g-C_3N_4 composite was slightly shifted to 1392 cm^{-1} (towards shorter wavelength) suggesting enhanced conjugation between P3HT and g-C_3N_4.

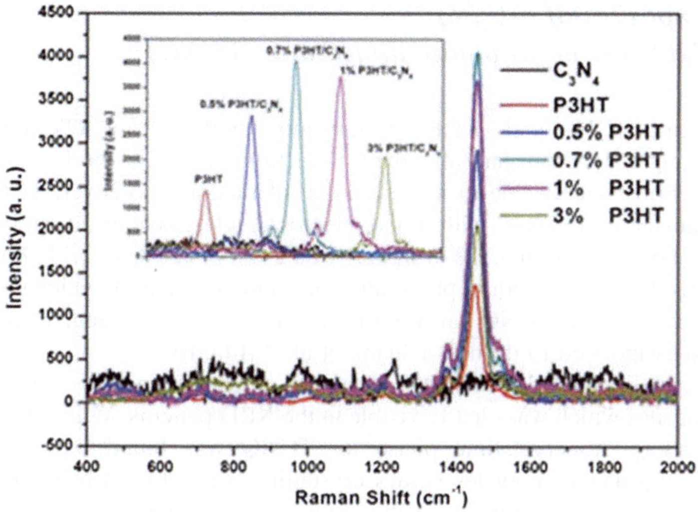

Fig. 9.22 Raman spectrum of the P3HT/g-C_3N_4 composite under 514 nm excitation [reproduced from Bai et al. (2015), License number, 4004740591216]

The optical absorption and absorption edge of the composite can be known from UV–visible diffuse reflectance spectra. Neat g-C_3N_4 shows absorption at 450 nm; however, the absorption edge of the P3HT-g-C_3N_4 composite shifts remarkably to 704 nm with increase in the amount of P3HT. Further the PL emission spectrum show a peak centred at 451 nm. The decrease in PL intensity originated from recombination of charge carriers indicates improved separation of charge carriers by constructing heterostructures. The suppressed PL intensity was due to poor radiative recombination of charge carriers supporting better performance of the photocatalyst. The tight coupling between g-C_3N_4 and P3HT is favourable for charge transfer and promotes the separation of photogenerated electron–hole pairs, subsequently improving the photocatalytic activity.

The photocatalytic activity of g-C_3N_4/P3HT composite was studied by the photodegradation rate of MB with different amounts of P3HT loadings. With increased P3HT loading the activity increased by two times than that of neat g-C_3N_4. The P3HT-g-C_3N_4 sample also shows good efficiency for the decomposition of phenol than that of neat g-C_3N_4. The degradation rate slightly decreases after annealing owing to the change in molecular arrangement. As expected the enhanced photocatalytic activity was also supported by the overall photocurrent generation by the composite photocatalyst. The suitable alignment of the band edge positions allow the photogenerated electrons to migrate to the CB of g-C_3N_4 whereas the holes are effectively transferred to the VB of P3HT through the interface of the type-II p-n junction. The improved efficiency was attributed to the role of P3HT in increasing the π-conjugation between n-type g-C_3N_4 as acceptor and p-type P3HT as donor system.

Nanocomposites of g-C_3N_4 with 7,7,8,8-Tetracyanoquinodimethane (TCNQ)

Both graphitic-C_3N_4 and 7,7,8,8-Tetracyanoquinodimethane (TCNQ) having highly conjugated system and abundant π-electrons, result in the formation of an efficient hybrid material due to π–π stacking interaction. Moreover, TCNQ forms charge transfer complexes with good electrical, electrochemical and magnetic properties. For the first time, the designing of a g-C_3N_4/TCNQ hybrid photocatalyst was reported by a simple liquid phase ultrasonication method. In which g-C_3N_4 was dispersed in DI water, TCNQ was dispersed in DMF and sonicated. Then, both are mixed and evaporated to dryness (Zhang et al. 2014a, b).

The crystal phase of g-C_3N_4 remains unaltered in the g-C_3N_4/TCNQ nanocomposites which was clearly visible in the XRD patterns. With low amount of TCNQ loading, no crystalline phase for TCNQ was found; however, (>5%) g-C_3N_4/TCNQ nanocomposites shows crystalline peaks for TCNQ and the peak intensity increases with increase in the loading amount but presence of excessive TCNQ fails to adhere to the surface of g-C_3N_4 and causes agglomeration. From the TEM image, it was observed that TCNQ possess a layered film morphology of about 10–20 nm dispersed over bulk g-C_3N_4. In addition FTIR spectral analysis also gives evidence in support of TCNQ/g-C_3N_4 composite photocatalysts. The stretching vibrational bands of neat g-C_3N_4 observed at 1628 and 1231 cm^{-1} were shifted to 1632 and 1238 cm^{-1} in the composite with increase in loading amount. Which proves the existence of interaction between g-C_3N_4 and TCNQ to favour charge transfer between them.

The UV–Vis DRS spectra of the TCNQ/g-C_3N_4 composite show two charge transfer bands at 510 and 690 nm. Also the absorption increases remarkably in the visible region with increase in TCNQ loading, indicating enhanced absorption in the composite (Fig. 9.23a).

The typical Raman bands of the composite further gives evidences about the increase in conjugation length due to coupling between π-conjugated compounds. Bands at 1355 and 1558 cm^{-1} for neat g-C_3N_4 were owing to the presence of disorder in graphitic structure (D-band) and the bond stretching motion of sp2 C-atoms (G-band). The Raman bands due to the vibrational modes of TCNQ were found to shift by about 16 and 4 cm^{-1}, respectively, confirming increase in conjugation length. In addition to that the v4 Raman band of TCNQ were observed at around 1456 and 1388 cm^{-1} was due to change in the degree of the charge transfer (Fig. 9.23b).

The suitable VB electronic structures of neat g-C_3N_4 and TCNQ are in good agreement to enhance photocatalytic activity of the nanocomposite. Zhang and group studied the phenol degradation activity of TCNQ/g-C_3N_4 composite and reported that the rate constants of TCNQ/g-C_3N_4 composite was 3.4 and 2.3 times higher than that of neat g-C_3N_4 for the degradation of 2,4-dichlorophenol and bisphenol A, respectively. This increase in activity was a function of charge separation efficiency and light absorption capacity. The mineralization activity of

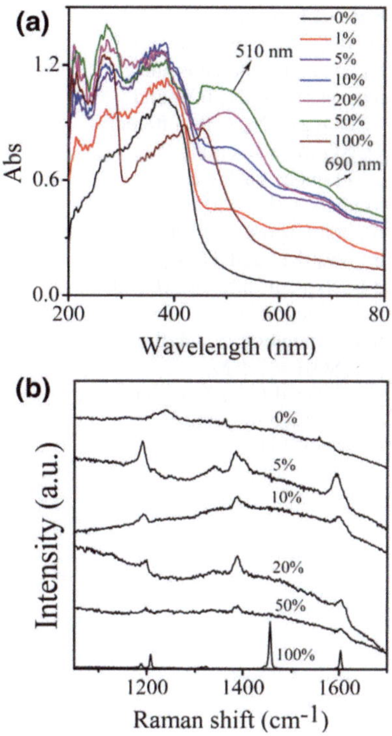

Fig. 9.23 **a** UV–Vis diffuse reflectance spectra of neat g-C_3N_4, pure TCNQ and TCNQ–g-C_3N_4 composites with different mass fractions of TCNQ (1–50%), **b** Raman spectra of g-C_3N_4, pure TCNQ and TCNQ/g-C_3N_4 materials [reproduced from Zhang et al. (2014), License number, 4004750713157]

phenol when investigated at 275 nm for 4 h, it was found that it proceeds through several intermediates such as dihydroxybenzene, 4,4-dihydroxybiphenyl and maleic anhydride, etc. which undergo complete degradation by ring cleavage to form CO_2 and H_2O. When the degradation was studied in presence of different scavengers it was observed that the holes were the main oxidative species in case of 10%-TCNQ/g-C3N4 composite (Fig. 9.24). The electronic interaction between g-C_3N_4 and TCNQ was further confirmed from the photocurrent measurement. Under visible light irradiation the TCNQ/g-C_3N_4 composite shows 23 times more photocurrent than that of neat g-C_3N_4. The flat band potential of neat g-C_3N_4 and TCNQ were calculated to be -1.09 and -0.62 V, respectively. In the composite the shift in flat band potential towards positive side upon TCNQ loading favours charge transfer between g-C_3N_4 and TCNQ, where TCNQ act as an organic acceptor.

Summary

In summary, we can conclude that innovative 2D interface engineering approach to develop polymeric nanocomposites is regarded as a versatile route for tailoring the photocatalytic activity of g-C_3N_4. When the metal-free organic polymer g-C_3N_4

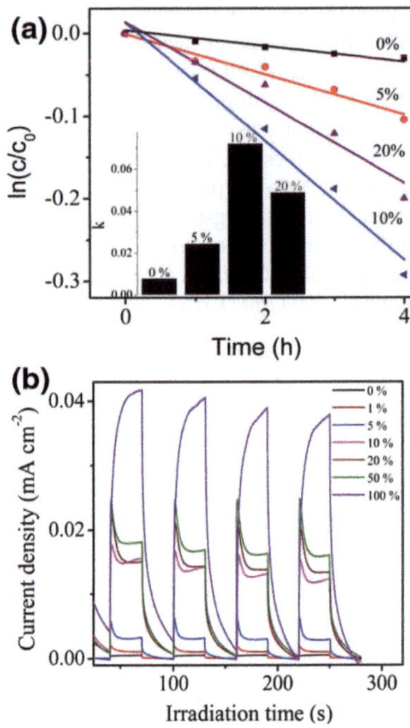

Fig. 9.24 **a** Photocatalytic degradation of phenol, *(inset)* the apparent rate constants over pure g-C_3N_4(0%) and TCNQ–g-C_3N_4 with different TCNQ mass fractions (5–20%) under visible light irradiation (l > 420 nm), **b** photoresponses of pure g-C_3N_4(0%) and TCNQ–g-C_3N_4 with different TCNQ mass fractions under visible light irradiation (l > 420 nm, [Na_2SO_4] = 0.1 M) [reproduced from Zhang et al. (2014), License number 4004750713157]

was electronically coupled with various π-conjugated/carbonaceous polymeric materials, the proper band alignment between g-C_3N_4 and the polymer results in the formation of a heterojunction. This dispersion of conductive polymers even at very low weight ratio, facilitates the transportation and separation of photogenerated electron–hole pairs at the interface. In case of some polymeric nanocomposites (polypyrrole) photoinduced holes can oxidize Ppy nanoparticles to form p-doping Ppy nanoparticles. These p-doping Ppy nanoparticles possess high electronic conductivity and positive charge in the composite, which induce hydrolysis of water to form H^+ ions. This class of polymeric nanocomposites provide processable routes for developing efficient 2D platforms for visible light absorption and various applications not only for photocatalytic hydrogen evolution from pure water system but also for CO_2 reduction, NO removal and pollutant degradation.

References

Bai X, Wang L, Wang Y, Yao W, Zhu Y (2014) Enhanced oxidation ability of g-C_3N_4 photocatalyst via C60 modification. Appl Catal B: Environ 152–153:262–270

Bai X, Sun C, Wub S, Zhu Y (2015) Enhancement of photocatalytic performance via a P3HT-g-C_3N_4 heterojunction. J Mater Chem A 3:2741–2747

Chen X, Shen S, Guo L, Mao SS (2010) Semiconductor-based photocatalytic hydrogen generation. Chem Rev 110:6503–6570

Chen Y, Jianghua L, Hong Z, Shen B, Lin B, Gao B (2014) Origin of the enhanced visible-light photocatalytic activity of CNT modified g-C_3N_4 for H_2 production. Phys Chem Chem Phys 16:8106–8113

Chai B, Liao X, Song F, Zhou H (2014) Fullerene modified C_3N_4 composites with enhanced photocatalytic activity under visible light irradiation. Dalton Trans 43:982–989

Dai K, Lu L, Liu Q, Zhu G, Wei X, Bai J, Xuana L, Wang H (2014) Sonication assisted preparation of graphene oxide/graphitic-C_3N_4 nanosheet hybrid with reinforced photocurrent for photocatalyst applications. Dalton Trans 43:6295–6299

Ge L, Han C (2012) Synthesis of MWNTs/g-C_3N_4 composite photocatalysts with efficient visible light photocatalytic hydrogen evolution activity. Appl Catal B: Environ 117–118:268–274

Ge L, Han C, Liu J (2012) In situ synthesis and enhanced visible light photocatalytic activities of novel PANI–g-C_3N_4 composite photocatalysts. J Mater Chem 22:11843–11850

He F, Chen G, Yu Y, Hao S, Zhou Y, Zheng Y (2014) Facile approach to synthesize g-PAN/g-C3N4 composites with enhanced photocatalytic H2 evolution activity. ACS Appl Mater Interfaces 6:7171–7179

Hu C, Han Q, Zhao F, Yuan Z, Chen N, Qu L (2015a) Graphitic C_3N_4–Pt nanohybrids supported on a graphene network for highly efficient methanol oxidation. Sci China Mater 58:21–27

Hu S, Ma L, Wang H, Zhang L, Zhaoa Y, Wu G (2015b) Properties and photocatalytic Performance of polypyrrole and polythiophene modified g-C_3N_4 nanocomposites. RSC Adv 5:31947–31953

Kudo A, Miseki Y (2009) Heterogeneous photocatalyst materials for water splitting. Chem Soc Rev 38:253–278

Li Y, Sun Y, Dong F, Ho WK (2014) Enhancing the photocatalytic activity of bulk g-C_3N_4 by introducing mesoporous structure and hybridizing with grapheme. J Colloid Interface Sci 436:29–36

Lu Q, Zhang J, Liu X, Wu Y, Yuan R, Chen S (2014) Enhanced electrochemiluminescence sensor for detecting dopamine based on gold nanoflower@graphitic carbon nitride polymer nanosheet–polyaniline hybrids. Analyst 139:6556–6562

Ma J, Wang C, He H (2016) Enhanced photocatalytic oxidation of NO over g-C_3N_4-TiO_2 under UV and visible light. Appl Catal B: Environ 184:28–34

Martha S, Nashim A, Parida KM (2013) Facile synthesis of highly active g-C_3N_4 for efficient hydrogen production under visible light. J Mater Chem A 1:7816–7824

Nayak S, Mohapatra L, Parida KM (2015) Visible light-driven novel g-C_3N_4/NiFe-LDH composite photocatalyst with enhanced photocatalytic activity towards water oxidation and reduction reaction. J Mater Chem A 36:18622–18635

Ong WJ, Tan LL, Chai SP, Yong ST (2015) Graphene oxide as a structure-directing agent for the two-dimensional interface engineering of sandwich-like graphene–g-C_3N_4 hybrid nanostructures with enhanced visible-light photoreduction of CO_2 to methane. Chem Commun 51:858–861

Patnaik S, Martha S, Acharya S, Parida KM (2016a) An overview of the modification of g-C_3N_4 with high carbon containing materials for photocatalytic applications. Inorg Chem Front 3:336–347

Patnaik S, Martha S, Parida KM (2016b) An overview of the structural, textural and morphological modulations of g-C_3N_4 towards photocatalytic hydrogen production. RSC Adv 6:46929–46951

Patnaik S, Martha S, Madras G, Parida KM (2016c) Effect of sulfate pre-treatment to improve deposition of Au-nanoparticles in Gold-modified sulphated g-C_3N_4 plasmonic photocatalyst towards visible light induced water reduction reaction. Phys Chem Chem Phys 18:28502–28514

Sahoo DP, Patnaik S, Rath D, Nanda B, Parida KM (2016) Cu@CuO promoted g-C_3N_4/MCM-41: an efficient photocatalyst with tunable valence transition for visible light induced hydrogen generation. RSC Adv 6:112602–112613

Sui Y, Liu J, Zhanga Y, Zhoua L, Xike T, Chen W (2013) Dispersed conductive polymer nanoparticles on graphitic carbon nitride for enhanced solar-driven hydrogen evolution from pure water. Nanoscale 5:9150–9155

Sultana S, Sahoo PC, Martha S, Parida KM (2016) A review of harvesting clean fuels from enzymatic CO_2 reduction. RSC Adv 6:44170–44194

Suryawanshi A, Dhanasekaran P, Mhamane D, Kelkar S, Patil S, Gupta N, Ogale S (2012) Doubling of photocatalytic H_2 evolution from g-C3N4 via its nanocomposite formation with multiwall carbon nanotubes: electronic and morphological effects. Int J Hydrogen Energy 37:9584–9589

Thomas A, Fischer A, Goettmann F, Antonietti M, Muller JO, Schloglb R, Carlssonc JM (2008) Graphitic carbon nitride materials: variation of structure and morphology and their use as metal-free catalysts. J Mater Chem 18:4893–4908

Tian J, Liu Q, Asiri AM, Alamry KA, Sun X (2014) Ultrathin graphitic C_3N_4 nanosheets/graphene composites: efficient organic electrocatalyst for oxygen evolution reaction. ChemSusChem 7:2125–2130

Unni SM, Illathvalappil R, Gangadharan PK, Bhange SN, Kurungot S (2014) Layer-separated distribution of nitrogen doped graphene by wrapping on carbon nitride tetrapods for enhanced oxygen reduction reactions in acidic medium. Chem Commun 50:13769–13772

Wang H, Mingshi X, Larissa T, Fisher A, Wang X (2014) Strategies on the design of nitrogen-doped graphene. J Phys Chem Lett 5:119–125

Wang X, Wang L, Zhao F, Hu C, Zhao Y, Zhang Z, Chen S, Shib G, Qu L (2015) Monoatomic-thick graphitic carbon nitride dots on graphene sheets as an efficient catalyst in the oxygen reduction reaction. Nanoscale 7:3035–3042

Xiang Q, Yu J, Jaroniec M (2011) Preparation and enhanced visible-light photo catalytic H_2-production activity of graphene/C_3N_4 composites. J Phys Chem C 115:7355–7363

Xu Y, Xu H, Wang L, Yan J, Li Huaming, Song Y, Huangb L, Caib G (2013) The CNT modified white C_3N_4 composite photo catalyst with enhanced visible-light response photoactivity. Dalton Trans 42:7604–7613

Yang L, Zhou H, Fan T, Zhang D (2014) Semiconductor photo catalysts for water oxidation: current status and challenges. Phys Chem Chem Phys 16:6810–6828

Yu Q, Guo S, Li X, Zhang M (2014) Template free fabrication of porous g-C_3N_4/graphene hybrid with enhanced photocatalytic capability under visible light. Mater Technol: Adv Perform Mater 29:172–178

Yu Q, Li X, Zhang L, Wang X, Tao Y, Zhang M (2015) Significantly improving the performance and dispersion morphology of porous g-C_3N_4/PANI composites by an interfacial polymerization method. e-Polymers 2:95–101

Zhang M, Yao W, Lv Y, Bai X, Liu Y, Jiang W, Zhu Y (2014a) Enhancement of mineralization ability of C_3N_4 via a lower valence position by a tetra-cyanoquinodimethane organic semiconductor. J Mater Chem A 2:11432–11438

Zhang S, Zhaoc L, Zenga M, Lib J, Xua J, Wang X (2014b) Hierarchical nanocomposites of polyaniline nanorods arrays on graphitic carbon nitride sheets with synergistic effect for photocatalysis. Catal Today 224:114–121

Zhang Y, Pan Q, Chai G, Liang M, Dong G, Zhang Q, Qiu J (2013) Synthesis and luminescence mechanism of multicolor-emitting g-C_3N_4 nanopowders by low temperature thermal condensation of melamine. Sci Rep 3:1943

Zhou Q, Shi G (2016) Conducting polymer-based catalysts. J Am Chem Soc 138:2868–2876

Chapter 10
Titanium-Based Mixed Metal Oxide Nanocomposites for Visible Light-Induced Photocatalysis

Soumyashree Pany, Amtul Nashim and Kulamani Parida

Abstract The concept of photocatalysis is not new, but the photocatalyst used for the process of photocatalysis is improving day by day. To take the concept of photocatalysis in advanced manner, titanium-based mixed metal oxide nanocomposites photocatalyst has been introduced in the field of photocatalysis. A brief study on the photocatalytic activity of the titanium-based mixed metal oxide nanocomposites (by categorizing blockwise into s, p, d, f groups) has been given in this chapter. The mechanism behind the improved photoactivity of the nanocomposite, due to the efficient charge separation at the heterojunction interface, is summarized. Various structures adopted by titanium-based mixed metal oxides like perovskite, pyrochlore, ilmenite, etc. by considering their ionic radii are reviewed here. Morphology, surface area, lattice and energy level matching etc. are some of the key factors responsible for the improved photoactivity with examples are discussed briefly. The photocatalytic activity of mixed metal oxide nanocomposites beyond titanium is also reviewed here in the last section. This book chapter may give a new insight for the development of research on nanocomposite in the field of photocatalysis as well as other fields such as supercapacitor and sensors.

Keywords Titanium · Mixed metal oxide · Nanocomposite · Visible light · Photocatalysis

All authors contributed equally to this work.

S. Pany · A. Nashim · K. Parida (✉)
Centre for Nano Science and Nano Technology, Institute of Technical Education and Research, Siksha 'O' Anusandhan University, Khandagiri, Bhubaneswar 751030, Odisha, India
e-mail: kulamaniparida@soauniversity.ac.in; paridakulamani@yahoo.com

M.M. Khan et al. (eds.), *Nanocomposites for Visible Light-induced Photocatalysis*, Springer Series on Polymer and Composite Materials,
DOI 10.1007/978-3-319-62446-4_10

Introduction

The depletion of fossil fuels, the growing energy demand, and the environmental issues have triggered research efforts to develop technology for energy generation and environmental remediation. These issues can be addressed by approaching the technology "Photocatalysis" the one of the most considered Holy Grail of chemistry by using the ultimate source of renewable energy resources sun and water; and this leads to the development of new, clean, green, safe, and a sustainable photocatalyst. Earlier in 1972 the term "photocatalysis" has been implemented for the first time by Fujishima and Honda for the photolysis of water by using TiO_2 electrode (Fujishima and Honda 1972). Afterward, many more photocatalysts have been developed and investigated toward photocatalytic performances. But, still the importance of metal oxide is unique and applied widely as an active photocatalyst or as support material because of their acid–base and redox properties (Misono 2013; Gawande et al. 2012). Among the metal oxide-based photocatalysts, a predominant place has been acquired by transition metal because of its innocuous, corrosion resistant, abundant, cheap, low cost of production, and easy regeneration features which make them quite ideal. Further, the metal oxide can be utilized in an effective way by forming mixed metal oxides, as mixed metal oxides lead to the enhancement in the optical absorption properties as well as performance toward photocatalysis. Properties of mixed metal oxides like acid–base, redox properties, and thermal, chemical, and mechanical stability make them versatile catalyst and considered as an important tool for the environmental and energy applications (Fig. 10.1) (Misono 2013). According to the fundamental concept, mixed metal oxides consist of oxygen and two or more metallic ions in the proportion that may vary or follow a strict stoichiometry. Depending upon the presence of number of

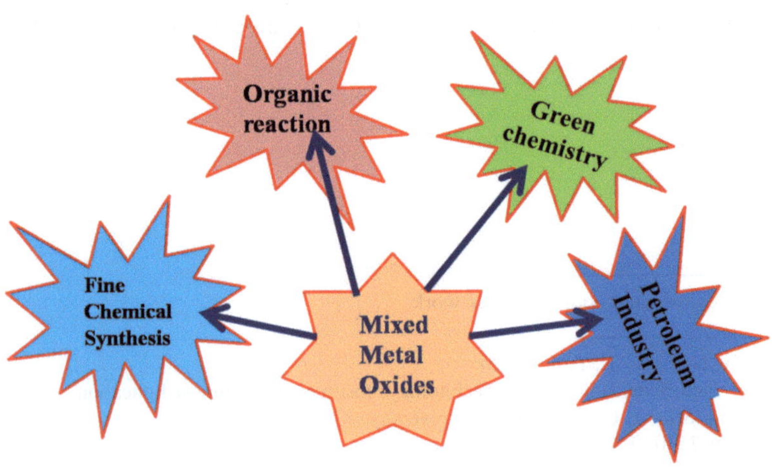

Fig. 10.1 Application of mixed metal oxides

different metal cations; mixed metal oxide may be binary, ternary, quaternary, and so on (Gawande et al. 2012).

Mixed metal oxides give us the opportunity to develop single-phase visible active materials for a photocatalytic activity like $FeTiO_3$, $CoTiO_3$, $NiTaO_3$, etc. (Chen et al. 2010). In comparison to doped ones, corresponding single-phase materials are more stable. They possess less defect sites; therefore, less photoexcitons recombination occurs. Further, the activity of mixed metal oxides can be improved by adopting various methods like metal doping, nonmetal doping, co-doping, forming solid solution, dye sensitization, cocatalyst loading, composite formation, etc. Out of the above-described methods, combining two or more semiconductors together is an effective method for the separation of photogenerated charge carriers by forming composite via heterojunction structure. Moreover, with growing concern about nanotechnology (Fig. 10.2), nanocomposites have much more advantages over composites. A nanocomposite is a multiphase solid material, in which one of the phases has dimensions in nanoscale range, i.e., less than 100 nm or repeated distance of separation between the different phases is in nanoscale range. In comparison to conventional composite, nanocomposites have exceptionally high surface area-to-volume ratio (Fig. 10.3), and may result in enhanced optical, electrical, and dielectric properties. These properties of a material are directly linked with the photoactivity of that material. Metal oxide-based nanocomposites consist of metal oxide matrix and metal oxide nanoparticles, nanowires, etc., as filler. Mixed metal oxide based nanocomposites have attracted the attention of many researchers owing to their synergistic and hybrid properties. For the synthesis of mixed metal oxide nanocomposite, only the simple mixing of two materials does not support the formation of nanocomposite; therefore, in order to establish a chemical bond between two or more semiconductors, some chemical treatment is necessary. The methods such as sol–gel, sol–gel auto-combustion, hydrothermal, solid-state reaction, solvothermal process, soft chemical route, milling annealing, evaporation-induced self-assembly (EISA), wet impregnation, sonochemical route, etc. have been used for the synthesis of mixed metal oxides nanocomposites. There are many literatures which emphasize on the synthesis of mixed metal oxide nanocomposites by taking d^0 (In^{3+}, Ga^{3+}, Ge^{4+}, Sn^{4+}) or d^{10} (Ti^{4+}, Ta^{5+}, Zr^{4+}, W^{6+}, Nb^{5+}) metals (Chen et al. 2010). Here, more attention has been paid to the survey of photocatalytic activity of titanium-based mixed metal oxide nanocomposites and the parameter that related to its activity. Other mixed metal oxide based nanocomposites with their photocatalytic properties are also explained briefly.

Titanium-Based Mixed Metal Oxide Nanocomposites

Titanium-based mixed metal oxide can adopt different crystal structures and accordingly they have different oxide compositions like perovskite (ABO_3), layered perovskite, scheelites (ABO_4), spinels (AB_2O_4), palmeirites ($A_3B_2O_8$), ilmenite,

Fig. 10.2 Application of nanotechnology

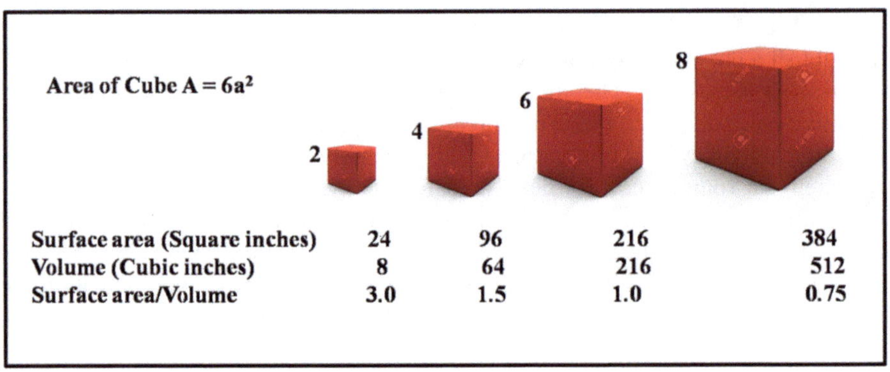

Fig. 10.3 Change in surface area-to-volume ratio with size

and pyrochlore ($A_2B_2O_7$) (Gawande et al. 2012). Most of the titanium-based mixed metal oxides are perovskite and these are tremendously studied because of its good stability as well as simple and flexible structural properties (Zhang et al. 2016a). As a consequence, many more work in the field of photocatalysis has been carried out and they show efficient photocatalytic performances. The general formula of perovskite is ABO_3, where B is the small transition metal cation and A is the larger s, d, f block cations (Kanhere and Chen 2014). In the perovskite structure (Fig. 10.4), the unit cell is built from corner-sharing BO_6 octahedra that are connected through B–O–B linkages and the B-site cation resides in the interstitial site of an octahedron of oxygen anions. At the center of eight corner-sharing BO_6 octahedra, A-site cation is present (Zhang et al. 2016a).

The ideal perovskite is normally cubic crystal structure having tolerance factor (t) = 1, which is defined in Eq. 1 (Pena and Fierro 2001):

$$t = \frac{(r_A + r_o)}{\sqrt{2}(r_B + r_o)}, \quad (1)$$

where $r_{A,}$ $r_{B,}$ and r_O are the ionic radii of A, B, and oxygen elements, respectively. For stable perovskite, the tolerance factor must be lie in between 0.75 and 1.0. When the "t" value is less than 1 ($t < 1$), it produces slightly distorted perovskite structure with orthorhombic or rhombohedral symmetry. The ideal perovskite structure exists in very limited cases; when t is very close to 1 and it formed at high temperature (Zhang et al. 2016a). Apart from perovskite, there also exists different types of structures like pyrochlore, ilmenite, etc. Rare earth titanate, having formula $Ln_2Ti_2O_7$, is considered as one of the versatile materials because of its novel crystal structure. Moreover, based on their dielectric, ferroelectric, piezoelectric, and ionic conductivity, they have several technological applications. For rare earth titanates, there is a linear correlation between the unit cell parameter "a" and ionic radius. Depending upon the radius of the cation, (Shao et al. 2012) $Ln_2Ti_2O_7$ adopts two types of structure, whose stability depends upon the ionic radii of cations,

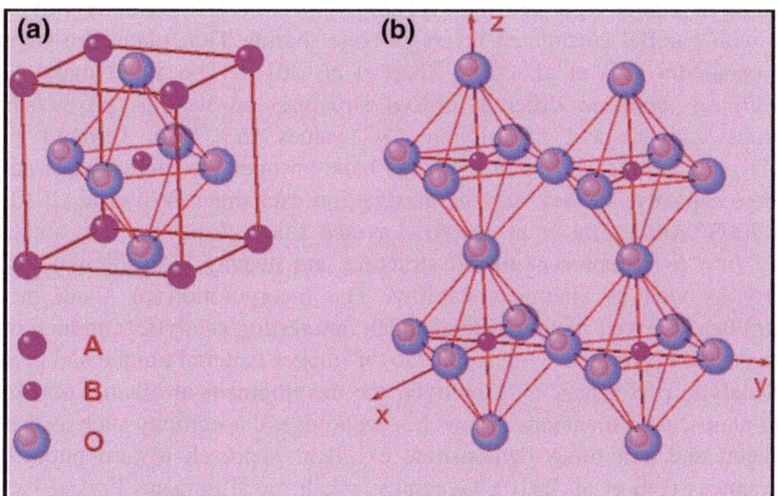

Fig. 10.4 Perovskite structure

i.e., Ln^{3+}/Ti^{4+} (i) Pyrochlore (Ln^{3+}/Ti^{4+} = 1.46–1.78), i.e., Ln = Sm to Lu acquires pyrochlore structure and (ii) Layered perovskite (Ln^{3+}/Ti^{4+} > 1.78), i.e., Ln = La to Nd. Similar to perovskite and pyrochlore another investigated structure of mixed metal oxide is ilmenite. The tolerance factor (t) and the difference in electronegativity value (e) of ABO_3 system for ilmenite formation are statistically analyzed (Liu et al. 2009). The results depict that tolerance factors $t > 0.80$, $e > 1.465$, and $Rm/Ro^{2-} \geq 0.48$ for octahedral factor should be satisfied for the formation of ilmenite structure. For ilmenite structure ($A^{2+}B^{4+}O_3$), both A^{2+} and B^{4+} should be small or intermediate size cation. Unlike perovskite, ilmenites have face shared octahedral. Titanium-based mixed metal oxides having different structures can be significantly modified by forming nanocomposite with other materials. The nanocomposite materials have their unique approach in the field of photocatalysis as these have tremendous features to suppress the photogenerated charge carriers which are beneficial for effective photocatalysis. Here in this section, we categorize titanium-based mixed metal oxide by considering one of the cations from s, p, d, f block elements and focus on their nanocomposite as well as various properties which influence the photocatalytic performances.

S Block Mixed Metal Oxide Titanate Nanocomposites

Titanium-based mixed metal oxide with s block elements such as alkali or alkaline earth metal is one of the most investigated forms of research in photocatalysis. The alkali metal titanates have the general formula $A_2Ti_nO_{2n+1}$ or A_2O. $nTiO_2$ (A = alkali metal ion or proton; where n = 1–9), having their own crystal structures of layered or tunnel type. Mostly, $A_2Ti_nO_{2n+1}$ (n = 3, 4, 5) exhibits a monoclinic lattice with parallel corrugated layers of edge-sharing TiO_6 octahedra stepped by every octahedra (Ma et al. 2005; Zhao et al. 2013). The alkali metal titanates ($A_2Ti_nO_{2n+1}$) represent different crystal structures as well as physicochemical properties with respect to different "n" values (n = 1–9). Layered titanates $A_2Ti_nO_{2n+1}$ (A = Na or K and n = 2–4) have an open structure and large cation exchange capacity, so they may be used as ion exchanger (Wu et al. 2012; Yang et al. 2008; Armstrong et al. 2004). Layered titanates having low alkali metal content (n = 6–8) represent tunnel structure and display high thermal insulating property as well as chemical stability. The incorporation of alkali metals to titanium-based mixed metal oxide elucidate interesting catalytic, conductivity, and its intercalation properties with respect to "n" makes material unique and pivotal for photocatalytic performances. Moreover, the development of alkali-based titanates having nanoscale dimensions and with morphological specificity such as nanofiber, nanosheet, and nanotubes demonstrate excellent approach toward photocatalytic performances (Liu et al. 2010). Generally, alkali metal titanates having nanowire, nanotube-based morphology were prepared through hydrothermal techniques by using different TiO_2-based precursors material in a highly alkaline medium. But still the information regarding its structure is not clear. It has been reported by various research groups that the obtained nanotube and nanowire have different composition and crystal structures such as hydrogen trititanate ($H_2Ti_3O_7$), tetratitanate ($H_2Ti_4O_9{\cdot}H_2O$), lepidocrocite titanate ($Na_xH_{2-x}Ti_3O_7$), and $H_2Ti_2O_4(OH)_2$ (Razali et al. 2012). After thermal treatment, titanates nanotube or nanowire with formula $Na_xH_{2-x}Ti_3O_7$ leads to form different titanates having general formula $Na_2Ti_nO_{2n+1}$ (Sun and Li 2003). Crystal structure and different compositions of the product strongly depend upon the heat treatment during the synthesis process and the condition that has been carried out during the washing period. According to the previous reports, the formation mechanism of nanotube/nanowire is unclear, whether it will be formed during hydrothermal reaction or during washing with HCl and distilled water. But the latter study reveals that the nanotube/nanowire is formed during the hydrothermal synthesis process (Lim et al. 2005). During the reaction period, sodium cations (Na^+) reside in between the edge-shared TiO_6 octahedral layers at high temperature and gradually replaced by H_2O molecules. The size of the intercalated H_2O molecule is larger than Na^+ ions, so expansion in interlayer distance occurs and the static interaction between neighboring (TiO_6) octahedral sheets became weak. Consequently, the layered titanate particles exfoliate to form nanosheets. Moreover, to release the strain energy nanosheets curl up from the edges to form TiO_2 nanotubes. Many more work has been carried out toward alkali-based titanates and implemented toward photocatalytic performances. But for further enhancement in the photocatalytic performance of alkali metal titanates, current research emphasizes on the formation of nanocomposite. $Na_2Ti_3O_7$ nanotube films successfully synthesized by adopting hydrothermal method using Ti foils in NaOH solution (Liu et al. 2013). Then Pt and In_2S_3 were deposited on the surface of $Na_2Ti_3O_7$ nanotube through photochemical reduction and SILAR (successive ionic layer adsorption and reaction) method. The authors studied the photocatalytic performances of the composite (In_2S_3–$Na_2Ti_3O_7$) toward water splitting under visible light irradiation. The authors investigated the effects of $InCl_3$ concentration over the photocatalytic performances of $In_2S_3/Na_2Ti_3O_7$. The size of In_2S_3 nanoparticles strongly depends upon the concentration of $InCl_3$. The observation finding reveals that the optimum coverage of In_2S_3 nanoparticle over the surface of $Na_2Ti_3O_7$ nanotube approaches effective e^-/h^+ pair separation; as the consequences, it effectively enhances the photocatalytic performances. Furthermore, the photocatalytic performance of $In_2S_3/Na_2Ti_3O_7$ has been enhanced by Pt incorporation. In the three-component system (In_2S_3–Pt–$Na_2Ti_3O_7$) the photogenerated electrons that migrated from In_2S_3 to $Na_2Ti_3O_7$ nanotubes will immediately transfer to Pt; as a result, it effectively separates the photogenerated charge carriers as well as enhances the photocatalytic performances.

Similar to alkali metal titanates, alkaline earth metal titanates $MTiO_3$ (where M = Ca, Sr, and Ba with a cubic perovskite structure) have been studied extensively because of their unique features like dielectric, piezoelectric, and ferroelectric properties which have potential approach in various fields like capacitors, transducers, actuators, nonvolatile random access memory devices (Park et al. 1999; Alexe et al. 1999). Ferroelectric materials have internal dipolar fields that separate photogenerated carriers and that draw attention to investigate these materials for photocatalytic applications (Giocondi and Rohrer 2001; Burbure et al. 2010). Heterostructure $SrTiO_3/TiO_2$ nanofiber is fabricated by following the in situ hydrothermal synthesis process where TiO_2 nanofibers were used both as template and reactant (Cao et al. 2011). The heterostructure composite material ($SrTiO_3/TiO_2$ nanofiber) contains nanocubes or nanoparticles of $SrTiO_3$, which accumulate uniformly over the surface of TiO_2 nanofibers. The adopted synthesis method not only ensures successful growth of $SrTiO_3$ nanostructures over the TiO_2 nanofiber substrate but also the $SrTiO_3$ nanostructures are highly dispersed on TiO_2 nanofiber without aggregation. The author investigated the density as well as morphology of $SrTiO_3$ nanostructure by adjusting the alkaline $Sr(OH)_2$ concentration, reaction time, and temperature. The heterostructure $SrTiO_3/TiO_2$ is formed by following the dissolution and precipitation mechanism according to the well-known Ostwald ripening process.

The detailed structural characteristics of the heteroarchitecture is well confirmed from FESEM and TEM study (Fig. 10.5). From these studies, the author observed $SrTiO_3$ nanostructures (nanocubes or nanoparticles), which are strongly bitten into the TiO_2 nanofiber. The author evaluated its photocatalytic performances toward Rhodamine B degradation under UV irradiation. The enhanced photocatalytic performance for $SrTiO_3/TiO_2$ nanofiber heterostructure has been explained on the basis of charge separation because of the coupling effect of TiO_2 and $SrTiO_3$. Two-dimensional $SrTiO_3/TiO_2$ heterostructure nanosheets have been fabricated by following in situ hydrothermal techniques (Yue et al. 2014). They used (001) facet-dominated anatase TiO_2 as a template as well as an initial reactant. The author demonstrates the particle size, morphology, and the content of $SrTiO_3$ attached to the surface of TiO_2 which are controlled by adjusting the reaction time. The photocatalytic performances of $SrTiO_3/TiO_2$ heterostructure nanosheets for Rhodamine B degradation have been observed under UV irradiation. From the finding results, an enhanced photocatalytic performance has been observed in $SrTiO_3/TiO_2$ heterostructure only when TiO_2 nanosheet coupled with $SrTiO_3$ and that form nano–nano-heterojunction which substantially accelerates separation of charge carriers in $SrTiO_3/TiO_2$ heterostructure. Many more work has been carried out for $SrTiO_3$ and modified $SrTiO_3$ which basically approaches for photocatalytic performances under UV irradiation. But in concern to visible light absorption, researcher focuses on modified $SrTiO_3$ which will acquire visible light as well as earn enhanced photocatalytic performances.

Ag–$SrTiO_3$ nanocomposite has been synthesized by adopting one-pot solvothermal method and its photocatalytic performances evaluated for NO degradation (Zhang et al. 2016b). The authors reported that during the synthesis process of Ag–$SrTiO_3$ the mineralizing agent NaOH plays a bifunctional role as it helps to promote the growth of $SrTiO_3$ and Ag nanocrystallites. Furthermore, this technique is beneficial for further fabrication of different Ag–titanates (Ca, Ba) based on composite photocatalysts. From the optical characterization study (Fig. 10.6), the author revealed that the $SrTiO_3$ harvested majority of light below 400 nm and calculated its band gap energy 3.2 eV by using the Kubelka–Munk function. But the Ag–$SrTiO_3$ nanocomposite showed the characteristics LSPR

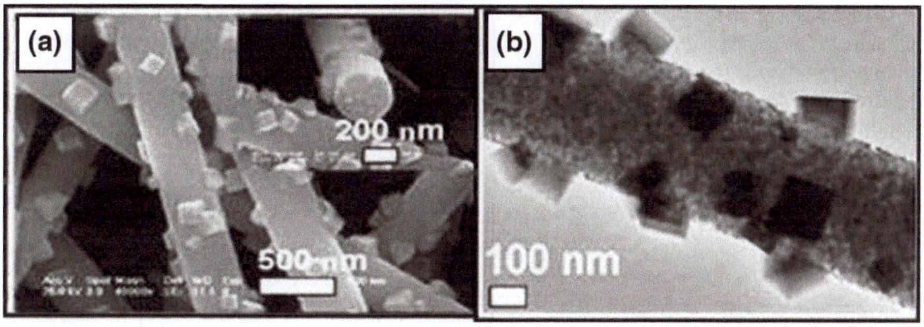

Fig. 10.5 FESEM and TEM micrographs of $SrTiO_3/TiO_2$. Reprinted with permission from Cao et al. (2011). Copyright (2011) American Chemical Society

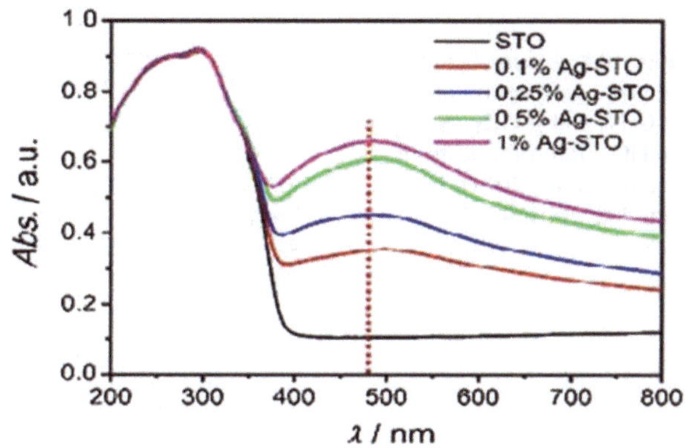

Fig. 10.6 UV–Vis absorbance spectra of bare $SrTiO_3$ and Ag–$SrTiO_3$ nanocomposites at different Ag loading ratios. Reprinted with permission from Zhang et al. (2016b). Copyright (2016) American Chemical Society

(Localized Surface Plasmon Resonance) absorption within the range 400–600 nm confirming the formation of plasmonic photocatalyst. The observed broad plasmon peak is due to the wide size distribution of Ag nanoparticle over the surface of $SrTiO_3$.

The author correlates LSPR intensity and photocatalytic efficiency improvement and demonstrates a quasi-linear relationship existed in the range of 0.1–0.5% of Ag loading ratio (Fig. 10.7). Further increase in Ag loading, absorbance does not lead to a continued increase in NO removal rate. This might be due to recombination of charge carriers in aggregated Ag nanoparticle and that predominantly found at high loading ratios. The improved photocatalytic activity for NO degradation through Ag–$SrTiO_3$ nanocomposite has been explained by the author on the basis of enhanced visible light harvesting property because of the plasmon resonance of Ag nanostructures, and reinforced separation of photogenerated charge carrier at the interface of Ag–$SrTiO_3$ nanocomposite.

Visible light active novel and stable g-C_3N_4/N-doped $SrTiO_3$ hybrid nanocomposites were fabricated via a facile and reproducible polymeric citrate and thermal exfoliation method (Kumar et al. 2014). The authors believe the special

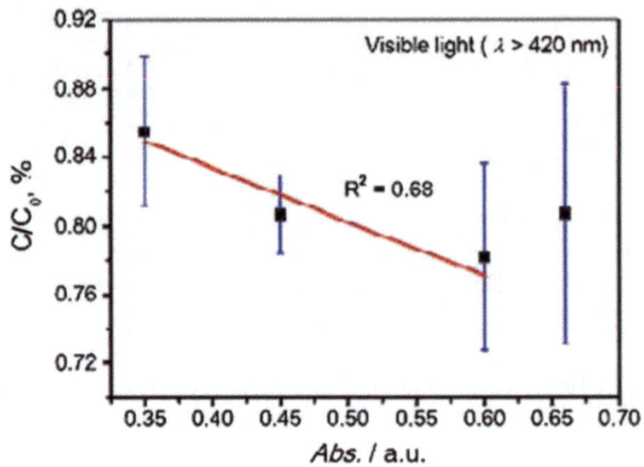

Fig. 10.7 NO removal efficiency (C/C_0) against relative absorbance (a.u.) of Ag–$SrTiO_3$ nanocomposites. Reprinted with permission from Zhang et al. (2016b). Copyright (2016) American Chemical Society

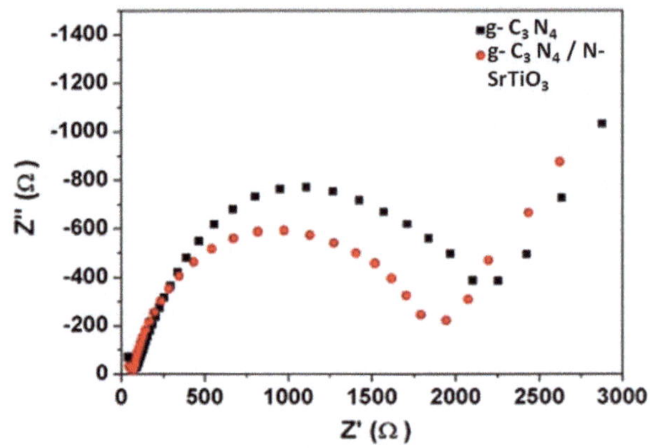

Fig. 10.8 Elecrochemical impedance spectroscopy of g-C_3N_4 and g-C_3N_4/N-doped $SrTiO_3$. Reproduced from Kumar et al. (2014), license no. 4007500354264

structure of the nanocomposite, i.e., N-doped $SrTiO_3$ nanoparticles wrapped in the layers of g-C_3N_4 nanosheets have potentially beneficial for charge separation. To confirm the charge transfer resistance and the separation efficiency of photoinduced electron–hole pairs, they performed electrochemical impedance spectroscopy (EIS) (Fig. 10.8). From the EIS spectra, the authors revealed the presence of N-doped $SrTiO_3$ nanoparticles in g-C_3N_4/N-doped $SrTiO_3$ exhibit much lower resistance than pure g-C_3N_4 nanosheet, which is confirmed from the reduced diameter of the semicircle at high-frequency region. These findings demonstrate that g-C_3N_4/N-doped $SrTiO_3$ nano-heterojunction enhances the separation and transfer efficiency of photogenerated charge carriers; as a consequence, it effectively improves the photocatalytic efficiency.

Also, author carried out PL study, and found a decrease in recombination of charge carriers in g-C_3N_4/N-doped $SrTiO_3$ heterojunction than bare g-C_3N_4. As per the result findings, they concluded that the improved photocatalytic efficiency of g-C_3N_4/N-doped $SrTiO_3$ heterojunction is because of high charge separation efficiency, effective band energy matching, and the decrease recombination factor of g-C_3N_4/N-doped $SrTiO_3$ heterojunction; and these factors make it promising material toward degradation of organic pollutant. The improved separation efficiency of photogenerated charge carriers for $Fe_2O_3/SrTiO_3$ heterojunction has been investigated (Zhang et al. 2013). The author explained the improvement of the charge separation on the basis of calculation of conduction band offset (CBO) between $SrTiO_3$ and Fe_2O_3 and studied the recombination of photogenerated charge carrier. The separation efficiency of photogenerated charge carriers has been confirmed by using the surface photovoltage (SPV) spectroscopy techniques, which is considered as one of the most appropriate direct methods to study the behavior of photogenerated charge carriers under different wavelength irradiations. The SPV intensity of $Fe_2O_3/SrTiO_3$ increases largely and responses toward the visible region in comparison to Fe_2O_3 and $SrTiO_3$; this signifies the improved separation efficiency of photogenerated charge carriers in $Fe_2O_3/SrTiO_3$. Further to better understand the separation efficiency, the author determined energy band diagram of Fe_2O_3 and $SrTiO_3$ from VB region X-ray photoelectron spectroscopy (VB-XPS), and conduction band offset (CBO) between Fe_2O_3 and $SrTiO_3$ was quantified to be 1.26 ± 0.03 eV. Based on the calculation, the band alignment depicting $Fe_2O_3/SrTiO_3$ is a staggered-type heterojunction (Fig. 10.9). The band alignment signifies that Fe_2O_3 is a n-type semiconductor and $SrTiO_3$ is a p-type semiconductor. When n-type Fe_2O_3 and p-type $SrTiO_3$ integrate followed by grinding and sintering, there form an interface between Fe_2O_3 and $SrTiO_3$ as well as at the same time inner electric field is developed at the interface. At equilibrium, the developed inner electric field makes $SrTiO_3$ region negatively charged whereas Fe_2O_3 is positive. The effect of inner electric field has a potential approach toward separation of photogenerated charge carriers. So as per the findings, the author reveals that separation efficiency is basically due to the energy difference between the conduction band (CB) edges of Fe_2O_3 and $SrTiO_3$, and this results in a decrease in recombination for $Fe_2O_3/SrTiO_3$ heterojunction.

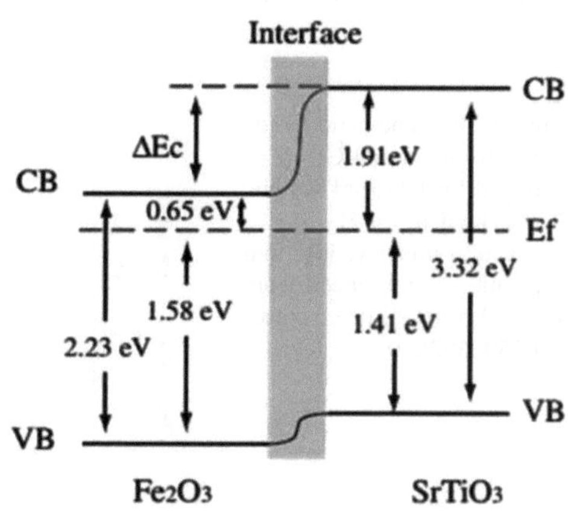

Fig. 10.9 Energy band diagram of Fe_2O_3 and $SrTiO_3$ in $Fe_2O_3/SrTiO_3$ heterojunction semiconductor. Reproduced from Zhang et al. (2013), license no. 4010720653106

$Sr_2TiO_4/SrTiO_3$(La,Cr) heterojunction photocatalyst synthesized by following the simple in situ polymerized complex method and evaluated its photocatalytic performances toward hydrogen evaluation under visible light irradiation (Jia et al. 2013). They also studied the effect of Pt cocatalyst loading and found that $Sr_2TiO_4/SrTiO_3$(La,Cr) heterojunction photocatalyst has higher photocatalytic activity than $SrTiO_3$(La,Cr) and Sr_2TiO_4(La,Cr) in the presence of methanol sacrificial reagent. The formation of heterojunction in $Sr_2TiO_4/SrTiO_3$(La,Cr) photocatalyst is confirmed through matching of lattice fringes of $SrTiO_3$(La,Cr) and Sr_2TiO_4(La,Cr). The combined analysis of absorption spectroscopy and Mott–schottky plots demonstrates that in the composite $Sr_2TiO_4/SrTiO_3$(La,Cr) photocatalyst the photogenerated electrons and holes more readily transfer from $SrTiO_3$(La, Cr) to Sr_2TiO_4(La,Cr) and from Sr_2TiO_4(La,Cr) to $SrTiO_3$(La, Cr) (Fig. 10.10).

The minor potential difference between the two component systems helps to drive the charge transportation and separation. Furthermore, the charge separation in the composite is well achieved by the co-doping of La and Cr in $Sr_2TiO_4/SrTiO_3$(La,Cr) photocatalyst, as in the composite the doped Cr mainly exists in the form of Cr^{3+} rather than high valent Cr^{6+} which behaves as an electron trapping site. Also the authors claim that $Sr_2TiO_4/SrTiO_3$(La,Cr) composite photocatalyst reveals high photocurrent generation in photo-electrochemical measurement, as well from the time-resolved FT-IR observation they noticed long-lived electrons that suggest efficient charge separation in the composite. Furthermore, to facilitate the photocatalytic activity, the author loaded Pt nanoparticles over the surface of electron accepting component Sr_2TiO_4(La,Cr) and this is highly favorable toward hydrogen evolution.

Among the alkaline earth titanates, $BaTiO_3$ is one of the most versatile semiconductor material having strong dielectric and ferroelectric properties. $BaTiO_3/TiO_2$ heterostructure nanotube array has been fabricated through the in situ hydrothermal synthesis process where TiO_2 was used both as template and reactant (Li et al. 2013a). The photocatalytic performances of $BaTiO_3/TiO_2$ heterostructure nanotube arrays under UV irradiation showed enhanced photocatalytic performance toward methylene blue degradation than TiO_2 nanotube. The heterostructure $BaTiO_3/TiO_2$ generates stronger photocurrent and smaller impedance arc radius in

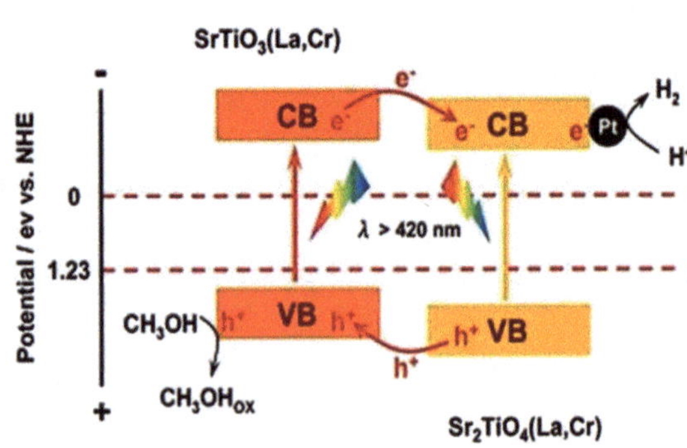

Fig. 10.10 Schematic band structure of La and Cr codoped $Sr_2TiO_4/SrTiO_3$ and its mechanism for H_2 production under visible light irradiation. Reproduced from Jia et al. (2013), license no. 4007500692850

comparison to pristine TiO_2, which demonstrates improved charge carrier separation property. Many more works have been carried out on $BaTiO_3$ and its composite and evaluated its photocatalytic activity under UV irradiation. In concern to the pivotal property of $BaTiO_3$, research focuses to make $BaTiO_3$ active under visible light by forming nanocomposite with visible light active material. By approaching the simple calculation method $BaTiO_3$@g-C_3N_4 nanocomposite has been synthesized (Xian et al. 2015). The photocatalytic performances have been studied through methyl orange (MO) degradation under simulated sunlight irradiation and an enhanced performance has been noticed for $BaTiO_3$@g-C_3N_4 nanocomposite than individual $BaTiO_3$ and g-C_3N_4. The author explained the enhanced photocatalytic performance on the basis of well-matched overlapping band structures of g-C_3N_4 and $BaTiO_3$ which is beneficial to separate photogenerated charge carriers as migration of electron and hole occurs between g-C_3N_4 and $BaTiO_3$, and as a consequence there found increase availability of electrons and holes for photocatalytic reaction.

By using a facile one-pot hydrothermal synthesis method $BaTiO_3$/graphene nanocomposite has been fabricated and evaluated its photocatalytic performances toward methylene blue degradation under visible light irradiation (Wang et al. 2015). The author claims in $BaTiO_3$/graphene nanocomposite; the incorporated graphene acts as an organic dye-like macromolecular photosensitizer for large band gap $BaTiO_3$. The suitable content of graphene loading to $BaTiO_3$ acts vital for the photocatalytic performances of $BaTiO_3$/graphene nanocomposite. The author demonstrates that the integration of graphene with large band gap $BaTiO_3$ reveals visible light absorption toward the red end as well as it offers high absorption intensity in comparison to pristine $BaTiO_3$. This study confirms that graphene has a crucial role toward visible light absorption. Therefore, the author suggests that the photocatalytic process of $BaTiO_3$/graphene is quite similar to the strategy of photosensitization of semiconductor material by matching adsorption of organic dyes through which a wide band gap semiconductor material extends its absorption toward visible region. Furthermore, they identified the major active species that participates during the degradation process through radical trapping experiment technique by using disodium ethylenediaminetetraacetate (Na_2-EDTA) as a hole scavenger. From the observation they confirmed that the hole scavenger does not significantly deactivate the photocatalyst, which suggests that the photogenerated holes do not participate in the photocatalytic degradation. So in the entire reaction process, the photoexcited state of graphene injects an electron into the CB of $BaTiO_3$ and that electron trapped by molecular oxygen species which further generate $^{\cdot}O_2^-$ and other reactive oxygen species which are beneficial for photocatalytic dye degradation. Also from the perovskite family $CaTiO_3$ is one of the best-known oxide materials whose band gap is 3.5 eV and it behaves as insulator (Zhang et al. 2010). As it is UV active, careful donor doping could make the material conductive and visible light active (Ueda et al. 1998; Xian et al. 2016). $CaTiO_3$–graphene nanocomposite has been fabricated through two-step method and its photocatalytic activity has been tested for MO degradation under UV irradiation (Xian et al. 2014). As per the findings, author demonstrates that $CaTiO_3$–grapheme nanocomposite shows enhanced photocatalytic performance than $CaTiO_3$. The enhanced performances have been explained on the basis of effective separation of charge carrier due to the transfer of an electron from CB of $CaTiO_3$ to graphene sheets. Recently, $CaTiO_3$ is fabricated by using different synthesis processes such as solid-state, sol–gel, and hydrothermal methods and tested its performances toward methylene blue degradation (Han et al. 2016). The author demonstrates that the $CaTiO_3$ prepared through hydrothermal synthesis process shows enhanced performances than other $CaTiO_3$ samples (prepared from solid-state and sol–gel process), but lower to TiO_2. In comparison to other perovskite systems, still the research on $CaTiO_3$ for the application of photocatalysis is in a fundamental stage.

Si/$MgTiO_3$ porous heterostructure has been fabricated through Mg thermal reduction of SiO_2/TiO_2 composites followed by HF treatment and tested its photocatalytic performances toward H_2 evolution from pure water without using any sacrificial reagent (Zhu et al. 2016). From this observation, $MgTiO_3$ was considered as host photocatalyst material for the first time. In the composite Si/$MgTiO_3$, the combined Si nanoparticle is a narrow band gap semiconductor with more –ve CB with respect to $MgTiO_3$ and also $MgTiO_3$ possesses more +ve VB. Here, in this study through Z scheme approach the composite shows easy separation of photogenerated charge carriers during the photocatalytic H_2 generation process.

P Block Mixed Metal Oxide Titanate Nanocomposites

Titanium-based mixed metal oxide with cation from p block element has both perovskite and layered structures. Microcrystalline $PbTiO_3$ (mc-$PbTiO_3$) cores coated with nanostructured TiO_2 (ns-TiO_2) shells were fabricated (Li et al. 2012) by adopting sol–gel techniques. The photocatalytic performances for mc-$PbTiO_3$–ns-TiO_2 were studied for methylene blue degradation which is 4.8 times higher than $PbTiO_3$, TiO_2, or mechanical mixture of the phases. The typical particle morphology and the high-resolution image from the TEM observation (Fig. 10.11) confirm the existence of interface between mc-$PbTiO_3$ and ns-TiO_2.

The $PbTiO_3$ core is electron opaque and appears black in image where the nanocrystalline TiO_2 shell appears semi-transparent layer surrounded to the $PbTiO_3$ core. From the diffuse reflectance spectroscopy (DRS) observations, the author reveals that the absorption edge of $PbTiO_3$–TiO_2 heterostructure is very close to absorption edge of $PbTiO_3$ core material and indicates that the core is the primary photon absorbing medium in the visible region of the spectrum. The enhanced performance of heterostructured mc-$PbTiO_3$–ns-TiO_2 toward methylene blue degradation is because of visible light harvesting property of $PbTiO_3$ core, separation of photogenerated charge carriers by the internal fields at the interface, and the reaction on the surface of nanostructured TiO_2 shell. Single-crystal heterostructured $PbTiO_3$/CdS nanorods were fabricated (Jiang et al. 2015) through hydrothermal synthesis process. The author investigated SEM (Fig. 10.12) to study the morphology of the heterostructured composite $PbTiO_3$/CdS and

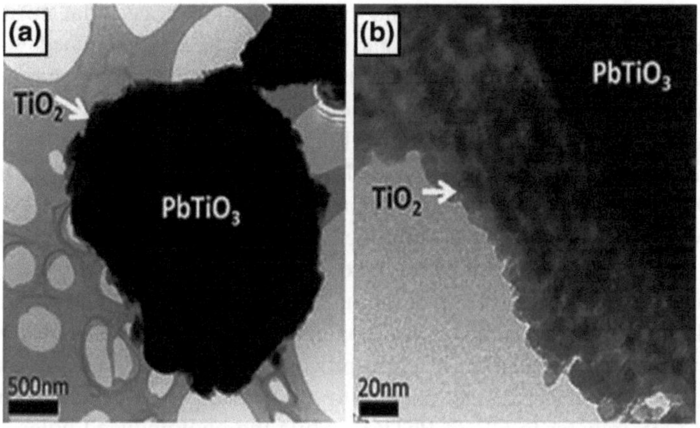

Fig. 10.11 TEM micrographs for $PbTiO_3$–TiO_2. Reproduced from Li et al. (2012), license no. 4007501358432

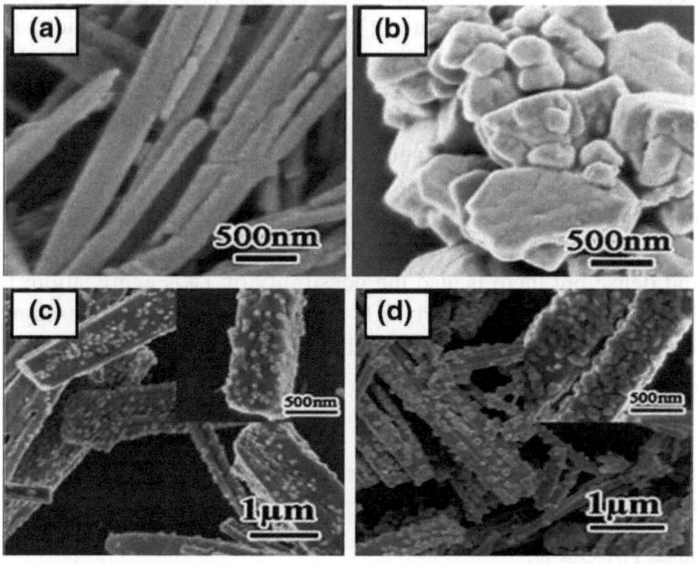

Fig. 10.12 SEM images of **a** $PbTiO_3$ nanorods, **b** CdS nanoparticles, heterostructured $PbTiO_3$/CdS nanorods with mass ratio of **c** 15 wt% and **d** 25 wt%. Reproduced from Jiang et al. (2015), license no. 4007501037063

revealed CdS nanoparticles developed on the surface of $PbTiO_3$ nanorod. From the study they found $PbTiO_3$ has a clean surface having diameter of about 200 nm and CdS nanoparticles aggregates in an irregular manner with particle size distribution from 50 nm to 1 μm. In case of $PbTiO_3$/CdS nanocomposite, smaller CdS nanoparticle was grown uniformly and tightly on the surface of $PbTiO_3$ rod.

From the optical study they found two absorption edges that correspond to $PbTiO_3$ and CdS as well as ensure that the composite has enhanced visible light absorption ability. To study the separation efficiency of photogenerated charge carrier and suppression of recombination in the heterostructured $PbTiO_3$/CdS composite, author carried out PL study (Fig. 10.13).

No emission peak for $PbTiO_3$ reveals low content of surface defects. In the case of $PbTiO_3$/CdS nanocomposite, significant band edge emission for CdS has been noticed when CdS nanoparticle combined to $PbTiO_3$ nanorod and the decreased intensity in $PbTiO_3$/CdS nanocomposite confirms most of the photogenerated charge carriers separated before recombination at the interface. The author deduces reasonable quenching because of the proper band alignment between $PbTiO_3$ and CdS. During the photocatalytic reaction process under visible light irradiation, VB electron of $PbTiO_3$ and CdS get excited to the respective CB, then electrons from the CB of CdS transfer to CB of $PbTiO_3$, while the photoinduced holes on the VB of the $PbTiO_3$ would transfer to the VB of CdS. Therefore, in this heterotype $PbTiO_3$/CdS composite, effective separation of charge carrier and suppression of recombination because of proper band alignment favor enhanced photocatalytic performances. Similar to this work, single-crystal TiO_2/$PbTiO_3$ heterostructured nanofiber composites were fabricated (Yu et al. 2015) by hydrothermal synthesis process by using precursor material perovskite $PbTiO_3$ (nanofiber) and tetra butyl titanate. The author reveals that the composite TiO_2/$PbTiO_3$ consists of tetragonal perovskite $PbTiO_3$ nanofiber and anatase TiO_2 nanorods and they confirm from the SEM study (Fig. 10.14) that the nanorods of TiO_2 grow on the surface of $PbTiO_3$ nanofiber. From the SEM observation, the TiO_2 nanorod grows on the surface of $PbTiO_3$ nanofiber sparsely when the molar ratio of TiO_2/$PbTiO_3$ is 26.7%. As the molar ratio increases from 26.7 to 36.5% the distribution of TiO_2 in $PbTiO_3$ is uniform and dense but there is also the possibility of TiO_2 aggregation. But further increase in molar ratio of TiO_2/$PbTiO_3$ extensively increases the aggregation of TiO_2 nanorod over the surface of nanofiber.

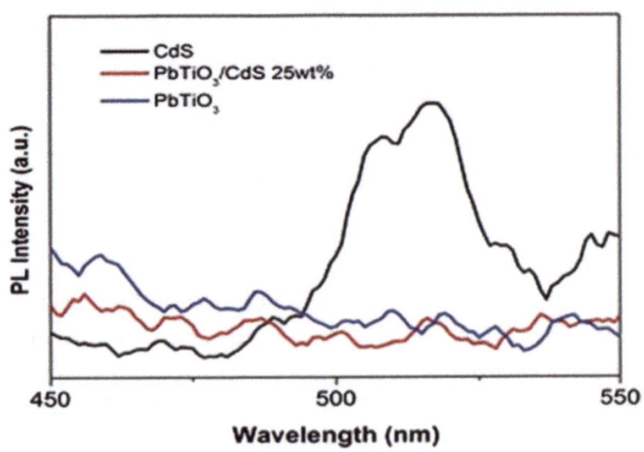

Fig. 10.13 PL spectra of the $PbTiO_3$ nanorods, the CdS nanoparticles, and heterostructured $PbTiO_3$/CdS 25 wt% nanorods. Reproduced from Jiang et al. (2015), license no. 4007510038263

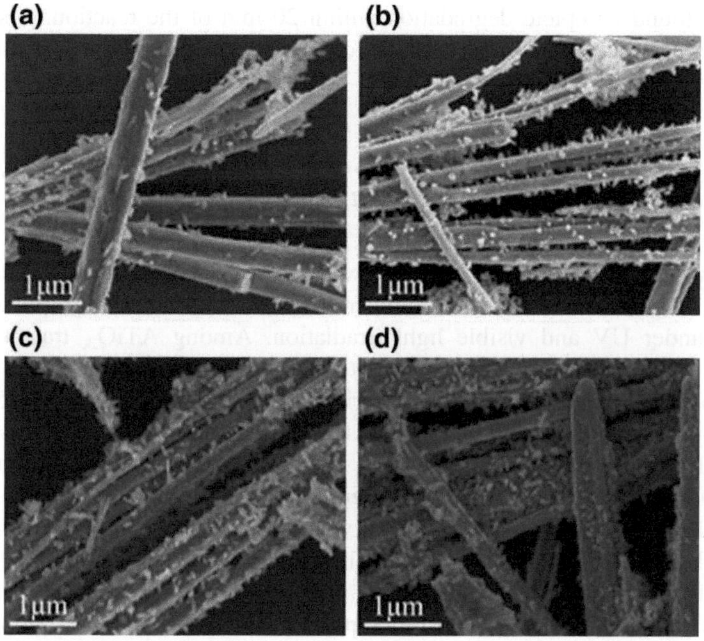

Fig. 10.14 SEM micrographs for $TiO_2/PbTiO_3$ at different wt% of TiO_2. Reproduced from Yu et al. (2015), license no. 4007531506542

This heterostructured composite $TiO_2/PbTiO_3$ shows enhanced photocatalytic degradation of methylene blue compared to $PbTiO_3$ and P25 under visible light irradiation. The enhanced performance is due to its single-crystal component phases, large-scale sharp interfaces with good crystallization between $PbTiO_3$ and TiO_2, which is crucial for photogenerated charge carrier separation and it occurs from $PbTiO_3$ to TiO_2 nanorod. N-doped In_2TiO_5 sensitized by carbon nitride synthesized by heating the precursor material In_2TiO_5 and urea at 400 °C for 2 h (Liu et al. 2011). From the X-ray diffraction study the author reveals that calcination temperature has an important role in the formation of In_2TiO_5. At the calcination temperature of 900 °C there is no peak observed for In_2TiO_5; however, peaks for In_2O_3 and TiO_2 are noticed and this suggests that at this temperature reaction between In_2O_3 and TiO_2 did not occur. But, at higher temperature, i.e., 1000 °C, there occur diffraction peaks for In_2TiO_5. The effect of N with the approach of urea in $InTi_2O_5$ also exhibits the same characteristic peak of $InTi_2O_5$, but the gradual increase in N content decreases the diffraction peak intensity. The increased content of N also helps to shift the light absorption of In_2TiO_5 toward red-end region, which suggests band gap narrowing. Furthermore, the incorporation of carbon nitride to N-In_2TiO_5 also have pivotal effect in the photocatalytic performances, as it increases the separation rate of photoinduced electron–hole pairs. The author studied the photocatalytic performances of the composite toward RhB degradation and they found complete degradation within 20 min of the reaction. Also, during the recycling process they observed this composite has high durable photocatalytic activity and reusable stability.

D Block Mixed Metal Oxide Titanate Nanocomposites

Titanium-based transition metal mixed oxides having perovskite type structures are dragging the interest of many researchers because they show high photocatalytic activity under UV and visible light irradiation. Among $ATiO_3$, transition metal titanates (A = Fe, Co, Ni, Mn, Cu, Zn, etc.) such as $NiTiO_3$, $FeTiO_3$, $MnTiO_3$, etc. have ilmenite structure under atmospheric pressure (Preciado et al. 2015). Out of the titanium-based transition metal oxides, the work on $FeTiO_3$ is extensively done. $FeTiO_3/TiO_2$ heterojunction has been synthesized for the decomposition of 2-propanol in gas phase and 4-chlorophenol in aqueous solution (Gao et al. 2008). The mechanism behind the enhanced activity of the system is explained on the basis of their flat band position (Fig. 10.15). They found that the position of Fe_2O_3 is not suitable for h^+ transfer, but when $FeTiO_3$ is used instead of Fe_2O_3 the hole transfer becomes easier. This results in complete decomposition of 2-propanol and 4-chlorophenol. In the year 2009, a single-crystalline nanodisc $FeTiO_3$ was prepared for the first time by adopting hydrothermal method (Kim et al. 2009). The obtained $FeTiO_3$ hexagonal nanodisc has an average diameter of 400 nm and thickness of 70 nm. They further modified this with TiO_2 to prepare nanodisc $FeTiO_3/TiO_2$ composite. $FeTiO_3/TiO_2$ nanocomposite was able to remove 97% of 2-propanol in 2 h of irradiation. In another work, $FeTiO_3/TiO_2$ nanocomposite was used for photoreduction of CO_2 under both visible and UV light irradiation with a yield of 0.46 μmol/g of CH_3OH (Truong et al. 2012). Authors claimed that the unique band structure, heterojunction effect, and narrow band gap of $FeTiO_3$ were responsible for the high efficiency of CH_3OH production. Using $FeTiO_3/TiO_2$ photocatalyst, reactive dyes can also be degraded in the presence of H_2O_2 (Sivakumar et al. 2013a). $FeTiO_3/TiO_2$ nanocomposite has the ability for the degradation of reactive yellow 84 and reactive black 5 under UV–visible irradiation in the presence of H_2O_2. $FeTiO_3/TiO_2$ nanocomposite has been prepared by taking oxalic acid as an organic linker. They estimated the extent of mineralization from chemical oxygen demand analysis. The result depicts that the materials have good photostability and even in 5th cycle material retains its 97% of activity. By considering oxygen generation as a key step for the photocatalytic overall water splitting, a new hierarchical $FeTiO_3/TiO_2$ hollow sphere by solvothermal process followed by calcination is developed (Han et al. 2015). The formation process of the hierarchical $FeTiO_3/TiO_2$ hollow sphere is shown (Fig. 10.16). Particle diameter of hollow sphere is about 0.5–1 μm. Hollow sphere $FeTiO_3$ retained its structure even after the formation of composite. The transfer mechanism of photogenerated charge carriers is investigated by surface photovoltage, electrochemical characterization, transient-state photovoltage, and fluorescence. According to the results, due to hierarchical porous hollow structure enhancement in the utilization of light, increase in surface active sites, and separation and transfer of photogenerated charge carrier occur easily. A double heterojunction CQD/C-TiO_2/$FeTiO_3$ composite is prepared by loading carbon quantum dot (CQD) onto the prepared C-TiO_2/$FeTiO_3$ for the degradation of methyl orange (MO) (Dadigala et al. 2016). Furthermore, the activity of C-TiO_2/$FeTiO_3$ is increased when a double heterojunction is formed with CQD. CQD during photocatalytic process acts as both electron donor and acceptor; therefore, the photogenerated electron on the surface of C-TiO_2/$FeTiO_3$ can easily be accepted by CQD and it can also donate the electrons generated on its surface to C-TiO_2/$FeTiO_3$. Therefore, the transfer of electrons efficiently suppresses the recombination and thus the enhanced activity is seen in the presence of CQD. The major role of reactive superoxide ($^{\cdot}O_2^-$) and hydroxyl radicals ($^{\cdot}OH$) is demonstrated by quenching effect of scavengers like benzoquinone, isopropyl alcohol, and terephthalic acid. Proposed mechanism for the degradation of MO is demonstrated on the basis of Mulliken electronegativity theory.

Apart from $FeTiO_3$-based nanocomposite, the work on $MnTiO_3$, $ZnTiO_3$, $CoTiO_3$, and $NiTiO_3$ is also reported by some research group. $ZnTiO_3$ is modified with BiOI to form heterojunction-based nanocomposite by precipitation–deposition method for the degradation of Rh 6G (Reddy et al. 2013). The enhanced activity of BiOI/$ZnTiO_3$ is explained on the basis of photosensitization effect of BiOI, which leads to the separation of photogenerated electron–hole pairs. $NiTiO_3$/Ag_3VO_4 nanocomposite (Inceesungvorn et al. 2014) has been synthesized by using a modified precipitation method for degradation of MO. FESEM images of these nanocomposites indicate the presence of a close interface between $NiTiO_3$ and $AgNO_3$. The possible mechanism for the photogenerated electron and hole transfer is explained by using the data obtained from VB-XPS and UV–Vis DRS spectra (Fig. 10.17a) The mechanism is supported by the PL spectra, which shows a decrease in PL intensity after the formation of junction (Fig. 10.17b). Recently, $CoTiO_3$/g-C_3N_4 is prepared using a facile in situ growth method toward hydrogen evolution (Ye et al. 2016). The detailed preparation method is described (Fig. 10.18). As prepared nanocomposite composed of 2D g-C_3N_4 nanosheet and 1D $CoTiO_3$ micro-rod exhibited the enhanced photocatalytic activity that has been described using two possible mechanisms: (i) heterojunction concept and (ii) direct

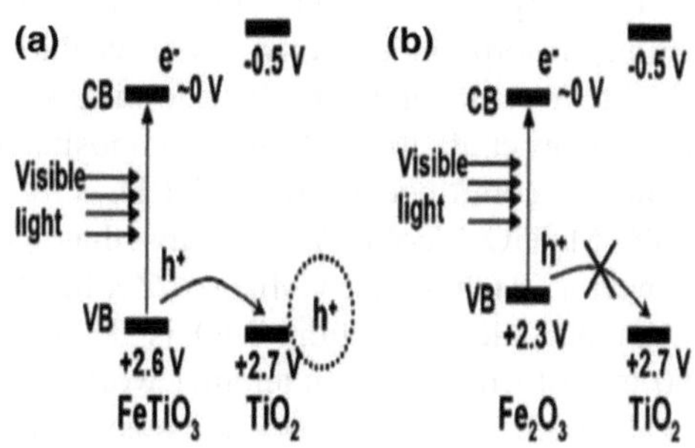

Fig. 10.15 Proposed mechanism for the visible light photocatalytic activity of **a** $FeTiO_3$/TiO_2 and **b** Fe_2O_3/TiO_2. Reproduced from Gao et al. (2008), license no. 4007570728917

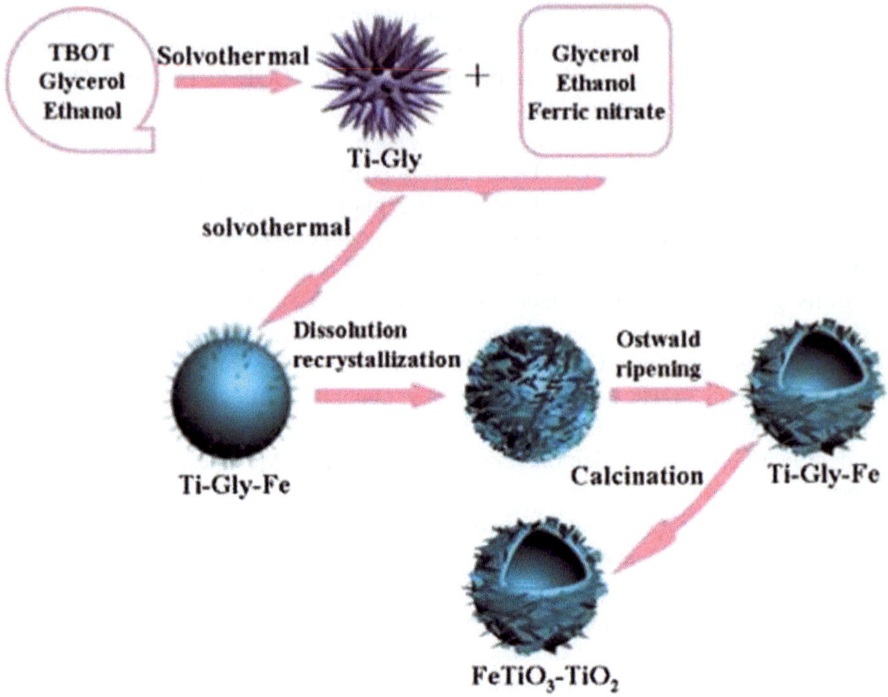

Fig. 10.16 Formation process of the hierarchical $FeTiO_3/TiO_2$ hollow spheres. Reproduced from Han et al. (2015), license no. 4007570968561

Z-scheme type (Fig. 10.19). Authors discarded the occurrence of the mechanism through traditional heterojunction way, as this process supposes to be unfavorable for the formation of active species. This is due to the unsuitable position of CB of $CoTiO_3$ for the reduction of O_2 and H_2O to form $^{\cdot}O_2^-$ and $^{\cdot}OH$, respectively. While by adopting Z-scheme mechanism for hydrogen evolution, there is the transfer of electron from the CB of $CoTiO_3$ to the VB of g-C_3N_4 via solid–solid hetero-interface. During this process, photogenerated electrons recombine with the local holes, thereby accelerating the process of separation of photogenerated electrons and holes in g-C_3N_4. Subsequently, improved activity is seen.

Recently, $MnTiO_3$ has dragged the attention of many researchers owing to its strong absorption capacity under visible light. In a work, a series of $MnTiO_3$/Zeolite-Y nanocomposite by adopting stearic acid gel method has been prepared (Enhessari et al. 2012). In this composite, $MnTiO_3$ exists in a rhombohedral phase, which is confirmed from the XRD. Vibration sample migration (VSM) result of 20% $MnTiO_3$/Zeolite-Y confirms that the nanocomposite has antiferromagnetic behavior. Authors used $MnTiO_3$/Zeolite-Y nanocomposite for the discoloration of MO and calcon solution. MO is decolorized up to 90% and calcon to 63% in 60 min of irradiation. In another work, $MnTiO_3$ is modified with TiO_2 (Sivakumar et al. 2013b) to degrade organic reactive dyes. Authors used the same oxalic acid as

Fig. 10.17 **a** A proposed mechanism toward methyl orange degradation using $NiTiO_3/Ag_3VO_4$ composite, **b** comparison of PL spectra. Reproduced from Inceesungvorn et al. (2014), license no. 4007571326039

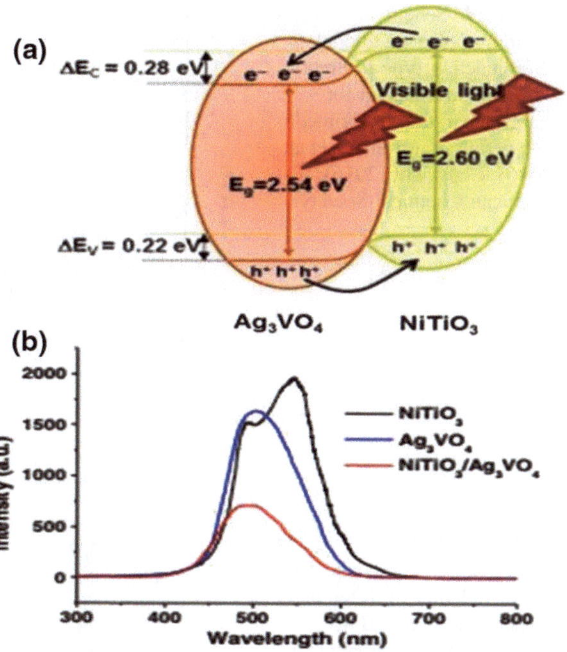

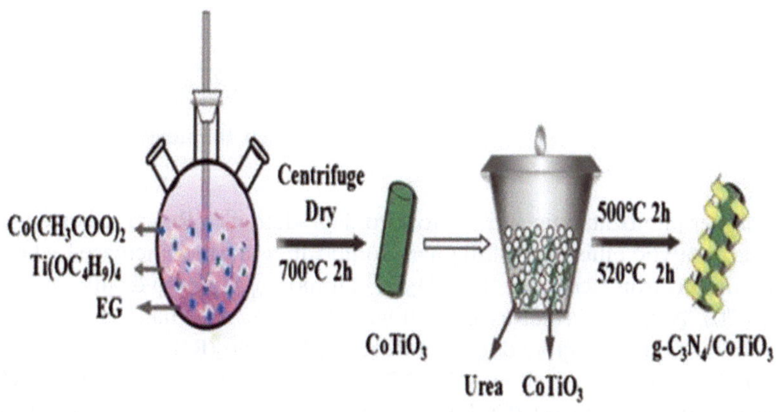

Fig. 10.18 Schematic illustration of procedure for preparing $g\text{-}C_3N_4/CoTiO_3$ photocatalyst. Reprinted with permission from Ye et al. (2016). Copyright (2016) American Chemical Society

a linker, which they used for the synthesis of $FeTiO_3/TiO_2$ in the same year (discussed above). Here different weight ratios of pyrophanite $MnTiO_3$ and TiO_2 are annealed at 300 °C. Among the prepared nanocomposite, 9 wt% $MnTiO_3/TiO_2$

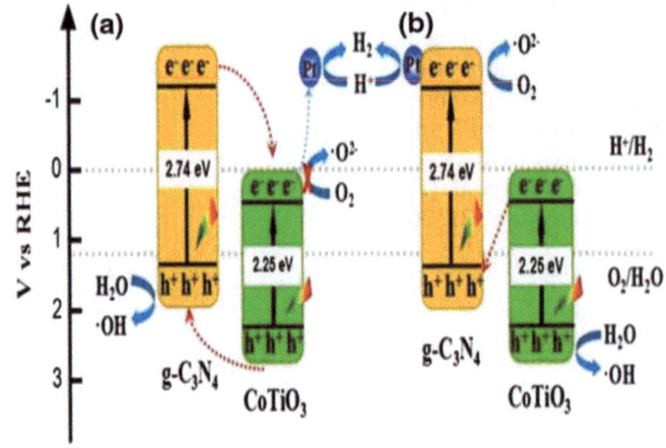

Fig. 10.19 Schematic illustration of **a** traditional heterojunction type **b** direct Z-scheme mechanism for charge separation. Reprinted with permission from Ye et al. (2016). Copyright (2016) American Chemical Society

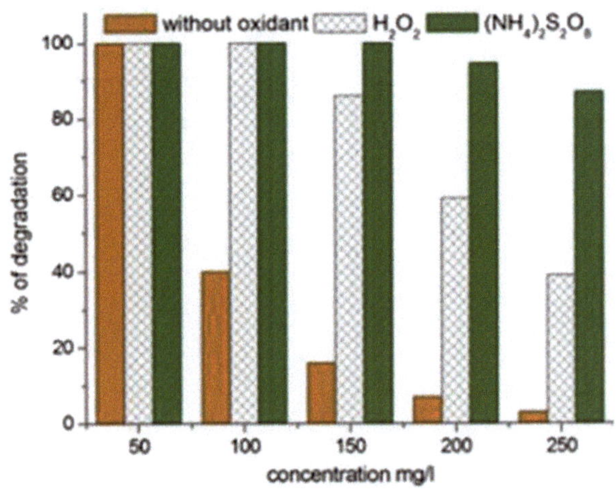

Fig. 10.20 Efficiency of 9 wt % $MnTiO_3/TiO_2$ heterojunction composite for the degradation of different concentrations of RB 4. Reproduced from Sivakumar et al. (2013b), license no. 4010720007771

shows higher photocatalytic activity toward degradation of Reactive Blue 4 (RB 4). The photocatalytic efficiency of the nanocomposite is further increased when small amount of electron acceptor (like H_2O_2, ammonium peroxydisulfate) is added (Fig. 10.20). Authors further used the 9 wt% $MnTiO_3/TiO_2$ nanocomposite for the degradation of other reactive dyes like Reactive Blue 50 (RB 50), Reactive Orange 30 (RO 30), Reactive Yellow 84 (RO 84), and Reactive Red 120 (RR120). Chemical Oxygen Demand (COD) is used for the analysis of the extent of mineralization of four structure reactive dyes (RB 5, RO 30, RY 84, and RR 120). The photocatalytic degradation efficiency of 9 wt% $MnTiO_3/TiO_2$ toward the degradation of various reactive dyes is mentioned above (Fig. 10.21).

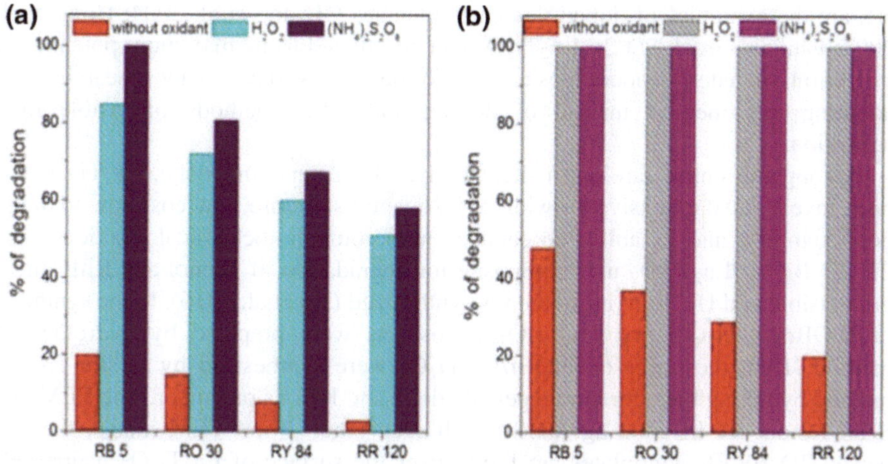

Fig. 10.21 Photocatalytic degradation efficiency of 9 wt% $MnTiO_3/TiO_2$ heterojunction composite on different structured organic reactive dyes in the presence of oxidants after **a** 60 min and **b** 150 min of UV–visible light irradiation. Reproduced from Sivakumar et al. (2013b), license no. 4010720007771

F Block Mixed Metal Oxide Titanate Nanocomposites

F block elements consist of lanthanides and actinides series. They are also called as rare earth elements. Sc and Y are actually d block elements but they are counted as rare earth elements. They are non-lanthanide rare earth elements. Over last 40 years, researches on transition metal oxides especially on TiO_2 have been increased as discussed previously. It is known that modification of TiO_2 with rare earth elements inhibited the phase transformation of TiO_2 (from anatase to rutile) and enhancement in the light-absorbing property is seen when it is modified with rare earth oxides. Rare earth elements like La^{3+}, Ce^{3+}, Ce^{4+}, Nd^{3+}, Sm^{3+}, Pr^{3+}, Eu^{3+}, Gd^{3+}, and Dy^{3+}-doped TiO_2 have been prepared (Stengl et al. 2009) and their photoactivities were studied toward the degradation of Orange II dye. The sample doped with Nd^{3+} shows highest activity when compared to others. The lanthanum (La)-doped $NaTaO_3$ ($NiO/NaTaO_3$:La) shows the maximum apparent quantum efficiency of 56% at $\lambda = 270$ nm for overall water splitting (Kato et al. 2003). They explained the enhanced phototactivity on the ground of small particle size and ordered nanostep structure of $NiO/NaTaO_3$:La.

Nowadays, rare earth titanates, having formula $Ln_2Ti_2O_7$, have been of the topic of interest due to their crystal structure. Moreover, based on their dielectric, ferroelectric, piezoelectric, and ionic conductivity, they have several technological applications (Shcherbakova et al. 1979; Patwe et al. 2015). The enhanced photocatalytic activity of $Ln_2Ti_2O_7$ (Ln = Y, Gd, La) toward water splitting is reported as they contain chains formed by corner-shared octahedral units (TiO_6) of metal cations (Abe et al. 2006). But Cr–Fe-doped $La_2Ti_2O_7$, Co–$La_2Ti_2O_7$, $Sm_2Ti_2S_2O_5$, etc. are the examples of doped rare earth titanates (Hwang et al. 2004; Hong et al. 2007; Ishikawa et al. 2002). $Ln_2Ti_2O_7$ can be utilized to its maximum potential, if rare earth titanates nanocomposites are formed. Nowadays many researches are adopting this method instead of doping and other methods of visible light activation.

Among the entire rare earth titanates based on nanocomposite, $La_2Ti_2O_7$ has been investigated extensively owing to its layered structure, low cost, low toxicity, good stability, and its ability to catalyze numerous photochemical reactions. The 2D-2D BiOBr/$La_2Ti_2O_7$ nanocomposite for degradation of phenol and RhB under both visible and UV light irradiation is synthesized (Ao et al. 2016). In the synthesis of BiOBr/$La_2Ti_2O_7$, first $La_2Ti_2O_7$ nanosheets were prepared by hydrothermal method. Then the series of BiOBr/$La_2Ti_2O_7$ were synthesized by in situ growth method by taking bismuth nitrate pentahydrate and KBr as precursor. The TEM and SEM results are in good agreement with each other. From TEM image, the dispersed 2D BiOBr nanoplates are found over the surface of $La_2Ti_2O_7$ nanosheets (Fig. 10.22a). The presence of heterojunction between BiOBr and $La_2Ti_2O_7$ is confirmed from the HRTEM image (Fig. 10.22b). The presence of lattice fringes with d-spacing value of 0.276 nm and 0.347 nm, respectively, is for (002) plane of $La_2Ti_2O_7$ and (101) plane of BiOBr. The improved activity of BiOBr/$La_2Ti_2O_7$ nanocomposite was illustrated on the ground of the following two factors: first, the photoresponse of BiOBr/$La_2Ti_2O_7$ nanocomposite is increased when compared to bare $La_2Ti_2O_7$, after the decoration of BiOBr nanosheets over $La_2Ti_2O_7$ nanoplates. Second, the formation of a heterojunction between $La_2Ti_2O_7$ and BiOBr is responsible for enhanced activity of this 2D–2D nanocomposite.

A novel visible light active Cu_2ZnSnS_4 (CZTS)/$La_2Ti_2O_7$ nanocomposite was reported (Tian et al. 2014). They first prepared $La_2Ti_2O_7$ by hydrothermal method

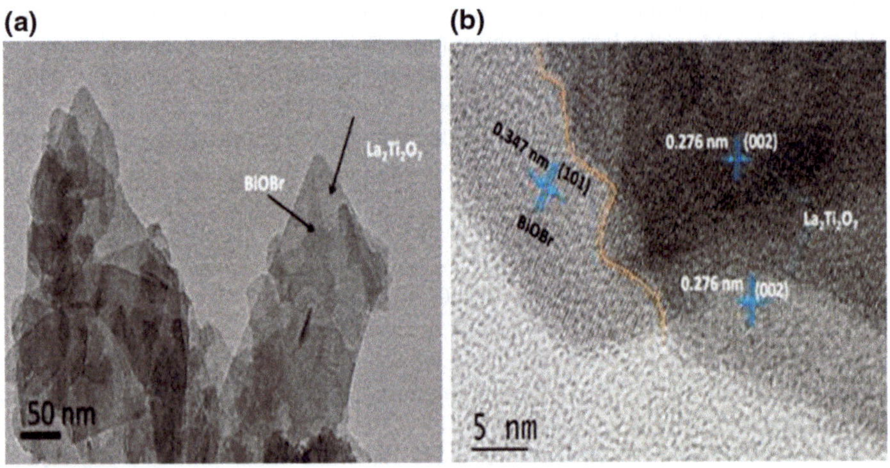

Fig. 10.22 **a** TEM and **b** HRTEM images of BiOBr/$La_2Ti_2O_7$. Reprinted with permission from Ao et al. (2016). Copyright (2016) American Chemical Society

and nanocomposite was prepared by in situ synthesis method. In this method, $La_2Ti_2O_7$ nanosheets were dispersed in ethanol solution, and then appropriate amounts of $CuCl_2$, $ZnCl_2$, $SnCl_2$, and thioacetamide were added to it. Then the mixture was transferred to Teflon-lined stainless steel autoclave and heated at 180 °C for 15 h to get CZTS/$La_2Ti_2O_7$ nanocomposite. The formation of CZTS/ $La_2Ti_2O_7$ nanocomposite is schematically described (Fig. 10.23). From SEM and TEM studies, they found that the irregular spherical particles of about 50 nm were homogeneously dispersed over the surface of $La_2Ti_2O_7$ nanosheet of thickness 10 nm. The band gap of the prepared nanocomposite was estimated by Tauc plot. As the ratio between CZTS and $La_2Ti_2O_7$ increases from 0.2 to 0.5, the band gap decreases from 2.7 eV to 2.5 eV. CZTS is responsible for visible light activation of $La_2Ti_2O_7$. Liu et al. used V_{fb}(NHE) = 2.94-E_g formula (where V_{fb} and E_g represent a flat band potential and band gap, respectively) to calculate the band edge positions of the two semiconductors. As the migration direction of photogenerated charge carrier depends upon the band edge positions of the semiconductors, the detailed mechanism for the degradation of RhB under UV and visible light irradiation is illustrated (Fig. 10.24). Authors concluded that CZTS/$La_2Ti_2O_7$ heterojunction was found to have variable application in the field of environment, energy, and other related issues. CdS/$La_2Ti_2O_7$ nanocomposite photocatalyst was prepared by a simple sonochemical coupled method for the decomposition of MO under UV and visible light irradiation (Wang et al. 2011). Wang et al. adopted the same formula V_{fb} (NHE) = 2.94-E_g as used by Tian et al. for the calculation of band edge positions of the constituent semiconductors. From SEM morphology of $La_2Ti_2O_7$ is found to have plate-like structure. When CdS/$La_2Ti_2O_7$ nanocomposite is formed, the plate structure of $La_2Ti_2O_7$ is retained as it is, which indicates that the ultrasonic processing has no effect on the structure of $La_2Ti_2O_7$. The enhanced photoactivity of nanocomposite is due to the layered structure of CdS/$La_2Ti_2O_7$ and well-matched band potentials of two semiconductors for easy separation of photogenerated charge carriers. The photoactivity toward hydrogen production for 1.5:1 ratio In_2O_3/$La_2Ti_2O_7$ composite enhanced by 29.62 and 6.43 times relative to pure components In_2O_3 and $La_2Ti_2O_7$, respectively (Hu et al. 2014). Enhanced activity of composite is mainly ascribed to the formation of heterojunction between the homogeneously dispersed In_2O_3 nanoparticles on $La_2Ti_2O_7$ nanosheets. They also correlate the magnitude of photocurrent with the trend in hydrogen evolution results of In_2O_3/$La_2Ti_2O_7$ prepared by different ratios. SnS_2/$La_2Ti_2O_7$ is another example of $La_2Ti_2O_7$-based nanocomposite used for the reduction of aqueous Cr(VI) under visible light irradiation (Chen et al. 2015).

The extensive work on rare earth titanate based nanocomposites, e.g., $Gd_2Ti_2O_7$/$GdCrO_3$, $Sm_2Ti_2O_7$/$SmCrO_3$, $Gd_2Ti_2O_7$/In_2O_3, $La_2Ti_2O_7$/CuO, and $La_2Ti_2O_7$/$LaCrO_3$, has been done (Parida et al. 2011; Nashim and Parida 2013; Nashim et al. 2013; Nashim and Parida 2014; Nashim et al. 2014). Various rare earth titanium-based $Ln_2Ti_2O_7$ systems like $Sm_2Ti_2O_7$, $Gd_2Ti_2O_7$, $La_2Ti_2O_7$, etc. are synthesized by solid-state reaction method. The above-mentioned systems are modified with visible active catalysts like In_2O_3, CuO, $LaCrO_3$, $SmCrO_3$, etc. for shifting its absorption edge toward visible region as these are UV active.

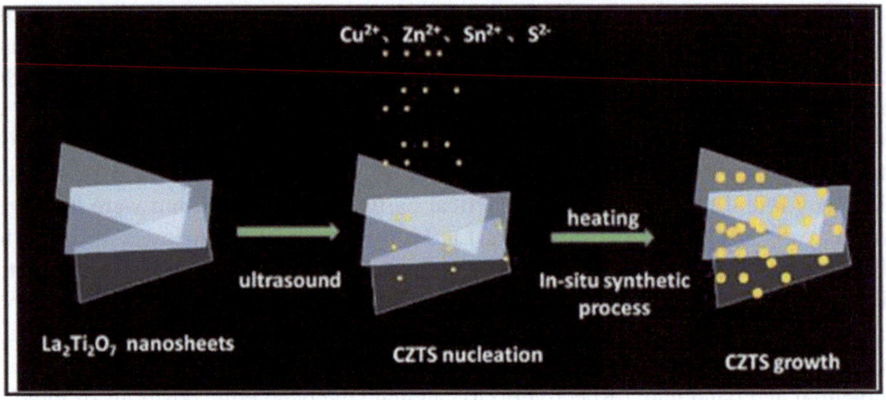

Fig. 10.23 Formation of CZTS/$La_2Ti_2O_7$ nanocomposite. Reproduced from Tian et al. (2014), license no. 4007570255654

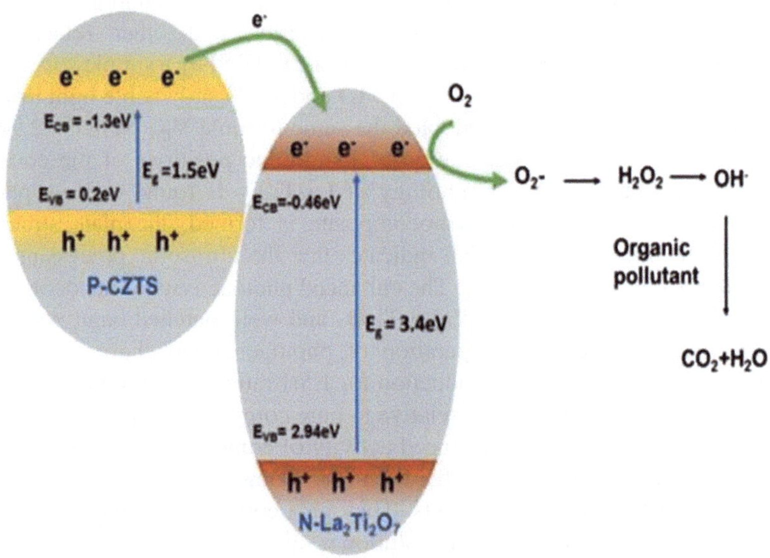

Fig. 10.24 Mechanism for the degradation of RhB over CZTS/$La_2Ti_2O_7$ nanocomposite. Reproduced from Tian et al. (2014), license no. 4007570255654

The prepared materials are characterized by different techniques. The photocatalytic activities of the prepared photocatalysts are tested toward water splitting and photodegradation of organic pollutants. The rare earth titanates are UV active; when a UV active photocatalyst makes heterojunction with the visible active material, its activity is extended toward visible region beside low recombination rate of photogenerated charge carriers. The heterojunction system is attributed to the built-in electric field (p/n junction) in the material; the internal p/n junction in a photocatalyst minimizes an energy wasteful electron–hole recombination by collecting photogenerated charge carriers on different surfaces, thereby enhancing both reduction and oxidation reactions occurring on the photocatalyst surface. For $La_2Ti_2O_7$/CuO system, the position of CuO is not suitable for the reduction of water but after the formation of nanocomposite $La_2Ti_2O_7$/CuO, the system is able to evolve hydrogen under visible light irradiation. During the formation of the heterojunction, alignment of the Fermi level which leads to the bending of the vacuum level makes the water reduction possible. Finally, author concluded that photocatalyst like CuO can be made active toward hydrogen production by coupling with an active photocatalyst.

Factors Affecting the Photoactivity of the Mixed Metal Oxides Nanocomposites

A mixed metal oxide nanocomposite shows superior photoactivity than its constituents as discussed in the above sections. But there are several important factors which influence the photoactivity of nanocomposites like surface area, morphology, band alignments, defect sites, crystallinity, varying percentages of constituents, etc. Generally, in any photocatalytic system, the photoactivity of that system cannot be explained by considering one or more factors. In this section, the factors affecting the photoactivity of the mixed metal oxides nanocomposites will be discussed.

$NaNbO_3$ rods are modified with In_2O_3 nanoparticles via co-precipitation method (Lv et al. 2010). During modification, authors vary the at.% of indium to prepare $NaNbO_3/In_2O_3$ nanocomposite with 12.5 at.% In, 25 at.% In, 37.5 at.% In, 50 at.% In, and 75 at.% In. They tested the activity toward hydrogen evolution under visible light irradiation and water splitting for both hydrogen and oxygen generation under UV light irradiation. The photocatalytic activity of the prepared sample toward hydrogen evolution with their specific surface areas is illustrated in Table 10.1. The photoactivity of $NaNbO_3/In_2O_3$ is increased up to 25 at.% and then it starts decreasing. This may be due to the fact that as the concentration of indium increases, the In_2O_3 particles cover the other In_2O_3 particles already in contact with $NaNbO_3$ rods. In this way, the excessively formed In_2O_3 blocks the light absorption and it becomes difficult for the photogenerated electrons on the outer In_2O_3 to reach $NaNbO_3$. This leads to the decrease in photoactivity. By referring Table 10.1, it can be easily concluded that the specific surface area does not have any obvious relationship with photoactivity. The composites like $Gd_2Ti_2O_7/GdCrO_3$, $Sm_2Ti_2O_7/SmCrO_3$, $Gd_2Ti_2O_7/In_2O_3$, $La_2Ti_2O_7$/CuO, and $La_2Ti_2O_7/LaCrO_3$ (Parida et al. 2011; Nashim and Parida 2013; Nashim et al. 2013; Nashim et al. 2014; Nashim and Parida 2014) follow the same trend that after an optimal concentration of a constituent the photoactivity starts decreasing. Table 10.2 shows the BET surface area and the CH_3OH yield over $FeTiO_3$–TiO_2 photocatalysts (Truong et al. 2012). In this case, the high surface area does not indicate the high photoactivity.

Table 10.1 Specific surface areas and photocatalytic H_2 evolution rates of all prepared samples of $NaNbO_3/In_2O_3$

In (at. %)	Surface area ($m^2\ g^{-1}$)	Photocatalytic H_2 evolution (μ mol $h^{-1}\ g^{-1}$)	
		Full arc	Visible light
0	12.5	12.3	0.3
12.5	17.4	16.9	6.6
25.0	22.9	42.9	16.4
37.5	29.3	44.6	13.2
50.0	32.7	15.6	4.8
75.0	44.0	4.1	1.3
100.0	27.1	4.2	1.7

Table 10.2 BET-specific surface area and CH_3OH yield over $FeTiO_3/TiO_2$ composite

Photocatalyst	BET ($m^2\ g^{-1}$)	CH_3OH yield (μ mol $g^{-1}\ h^{-1}$)	
		UV–Vis irradiation	Visible light irradiation
		$\lambda > 300$ nm	$\lambda > 400$ nm
P25	52.6	0.176	0.045
TiO_2	62.3	0.175	0.141
10% $FeTiO_3/TiO_2$	55.8	0.338	0.319
20% $FeTiO_3/TiO_2$	51.3	0.462	0.432
50% $FeTiO_3/TiO_2$	35.7	0.298	0.352

Recently, researchers are focusing on the synthesis of different types of morphology-based nanocomposites like $La_2Ti_2O_7/In_2O_3$ (nanosheet), $PbTiO_3/TiO_2$ (core–shell particle), $SrTiO_3/TiO_2$ (nanofiber), $FeTiO_3/TiO_2$ (hollow sphere), $FeTiO_3/TiO_2$ (nanodisk) $PbTiO_3/CdS$ (nanorod), $TiO_2/PbTiO_3$ (nanofiber), $CdS/La_2Ti_2O_7$, etc. (Hu et al. 2014; Li et al. 2012; Cao et al. 2011; Han et al. 2015; Kim et al. 2009; Jiang et al. 2015; Yu et al. 2015; Wang et al. 2011). The observations indicate that the nanocomposite with different morphologies shows high surface area than bulk type (high surface to volume ratio), enhancement in the light harvesting properties, cyclic stability and host–guest interaction, and facilitates the channelization of photogenerated charge carriers. These are the factors responsible for the increase in activity of a composite with varying morphology than to that of their corresponding bulk type.

Lattice and energy level matching between the constituents of the nanocomposite is another factor which influences the photoactivity by forming the efficient heterostructure. The experimental finding over $ZnWO_4/BiOI$ (Li et al. 2013b) by considering first-principle calculations shows that there is a matching between the lattice and energy levels between $ZnWO_4$ and BiOI. This directly indicates that there is an efficient heterojunction between $ZnWO_4$ and BiOI. Further the authors concluded that lattice and energy level matching is one of the important factors for the efficient separation of photogenerated charge carriers.

The work on $BiVO_4/InVO_4$, $FeVO_4/Bi_2O_3$, $CaFe_2O_4/MgFe_2O_4$, and graphene–TiO_2 co-modified $Bi_2O_2CO_3$ (Lin et al. 2015; Liu et al. 2016; Borse et al. 2012; Ao et al. 2015) indicates that if a heterojunction system is further modified with nanoparticles, by doping with suitable materials, by loading cocatalyst, etc., the photoactivity of the system will be increased to a greater extend. Authors noticed that when $BiVO_4/InVO_4$ is modified with Ag nanoparticle, the photoactivity of the ternary heterojunction system Ag-$BiVO_4/InVO_4$ toward degradation of 4-chlorophenol shows higher activity than neat $BiVO_4$, $InVO_4$ and $BiVO_4/InVO_4$ composite. Here in this case, surface plasmonic resonance effect of Ag increases the visible light absorption property of the system and acts as a good acceptor. In Au-$FeVO_4/Bi_2O_3$ system, the enhancement in the visible light absorption and efficient separation of photogenerated electron–hole pairs are seen, after loading Au. The photocatalytic degradation of malachite green (Fig. 10.25) by Au-$FeVO_4/Bi_2O_3$ system shows the highest result in comparison to $FeVO_4$, Bi_2O_3, and $FeVO_4/Bi_2O_3$. Table 10.3 depicts the apparent rate constant toward MO degradation under UV light irradiation for graphene–TiO_2 co-modified $Bi_2O_2CO_3$ with their surface areas. The result shows the enhanced activity of ternary systems to that of binary and single systems. The same trend is also observed in Ti-doped $CaFe_2O_4/MgFe_2O_4$ and In_2S_3–Pt–$Na_2Ti_3O_7$ systems (Borse et al. 2012; Liu et al. 2013).

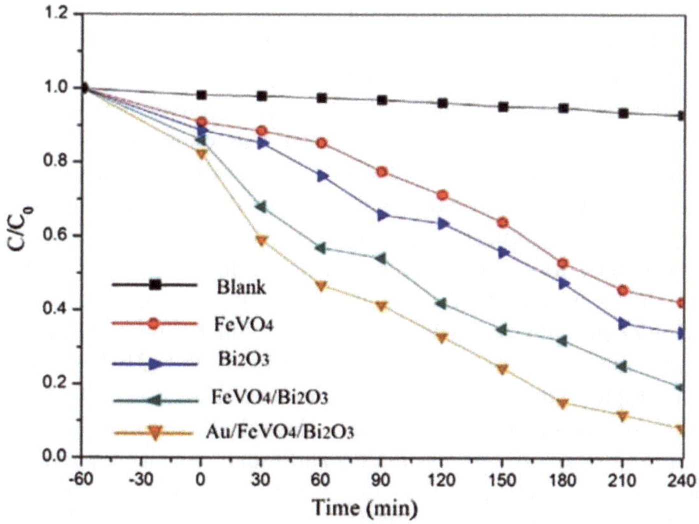

Fig. 10.25 Degradation of MG by different photocatalysts Au/$FeVO_4/Bi_2O_3$ composites. Reproduced from Liu et al. (2016), license no. 4010710501518

Table 10.3 The apparent rate constant and BET surface area for different samples of TiO_2 co-modified $Bi_2O_2CO_3$

Sample	Apparent rate constants (min^{-1})	BET surface area ($m^2\ g^{-1}$)
TiO_2	0.25	40.0491
$Bi_2O_2CO_3$	0.071	0.8363
$GR/Bi_2O_2CO_3$	0.11	12.9639
$GR/Bi_2O_2CO_3/TiO_2$-1	0.47	58.0916
$GR/Bi_2O_2CO_3/TiO_2$-2	0.57	69.0508
$GR/Bi_2O_2CO_3/TiO_2$-3	0.97	73.2312
$GR/Bi_2O_2CO_3/TiO_2$-4	0.88	90.0736

Reproduced from Ao et al. (2015), license no. 4010710309361

Other Mixed Metal Oxides Nanocomposites and Their Photocatalytic Performances

In addition to the “titanium-based mixed metal oxide nanocomposites” a number of other mixed metal oxides nanocomposites are reported and their photocatalytic performances have been evaluated under visible light irradiation. A very short description on recently synthesized mixed metal oxides nanocomposite is summarized below.

Hollow $ZnFe_2O_4$ and $ZnFe_2O_4/TiO_2$ composites have been prepared by co-precipitation and microemulsion method (Xu et al. 2015). From the SEM and TEM microscopic studies, the author confirms that TiO_2 nanoparticles are well coated on $ZnFe_2O_4$. The photocatalytic performance of $ZnFe_2O_4/TiO_2$ composites was studied for MO and rhodamine B degradation under visible light irradiation. The author reveals that degradation of the $ZnFe_2O_4/TiO_2$ composite strongly depends on the content of Ti loading and also has a correlation with reaction time, pH value of the solution and sintering temperature. A novel heterojunction $MgFe_2O_4/ZnO$ has been fabricated and employed for photocatalytic Rhodamine B degradation (Su et al. 2014). $MgFe_2O_4/ZnO$ heterojunction has been synthesized by the following two-step methods, in the first step flower-like nanotube bundles of ZnO were prepared through a simple solution method at low temperature and in second step through chemical co-precipitation method $MgFe_2O_4$ coated on ZnO surface. The photoactivity of the heterojunction composite is studied toward RhB degradation under visible light irradiation and an enhanced activity found for $MgFe_2O_4/ZnO$ heterojunction, which is due to the interconnected heterojunction of $MgFe_2O_4$ and ZnO. A p–n type heterojunction $InVO_4/g\text{-}C_3N_4$ has been fabricated by using a simple hydrothermal synthesis process (Shi et al. 2014). The author evaluated its photocatalytic performances toward Rhodamine B degradation in comparison to pristine $g\text{-}C_3N_4$ and $InVO_4$ and found enhanced activity for $InVO_4/g\text{-}C_3N_4$ heterojunction. The effective charge transportation due to the matched band edge potential at the interface of $InVO4/g\text{-}C_3N_4$ heterojunction (Fig. 10.26) suppresses the recombination of charge carriers, which effectively enhances the photocatalytic performance of $InVO_4/g\text{-}C_3N_4$ heterojunction.

Similar type of work with g-C_3N_4 such as g-C_3N_4/$BiVO_4$ and $CoFe_2O_4$/g-C_3N_4 has been carried out (Ji et al. 2014; Huang et al. 2015) and they evaluated its performances under visible light irradiation toward Rhodamine B and Methylene blue degradation. In both the cases, the composite material shows enhanced photocatalytic performances than to its pristine. The 41.4% mass fraction of $CoFe_2O_4$ in $CoFe_2O_4$/g-C_3N_4 degrades 97.3% in 3 h. In both the studies, the enhanced performance has been explained on the basis of stability of the heterostructure and the energy level matching that significantly reduces the charge recombination as well as efficiently affects the photocatalytic performances.

Magnetically separable Fe_3O_4/SiO_2/Bi_2MoO_6 composite has been prepared via hydrothermal synthesis process (Hou et al. 2015). From the SEM observations, the authors found flower-like three-dimensional (3D) Bi_2MoO_6 microspheres were decorated with Fe_3O_4/SiO_2 magnetic nanoparticles. The Fe_3O_4/SiO_2/Bi_2MoO_6 composite has been evaluated for Rhodamine B degradation under visible light irradiation and the efficiency of degradation reached 96% and almost 100% within 120 min. Moreover, the author investigated photocatalytic performances through 2,4-dichlorophenol degradation and found 70% decomposed after 5 h of visible light irradiation. Not only Fe_3O_4/SiO_2/Bi_2MoO_6 composite showed enhanced photocatalytic performance but also it has excellent stability with a slight decrease in its photocatalytic activity after being used for five cycles.

The visible light active mixed metal oxide, i.e., $BiVO_4$, has been modified with different semiconductor-based materials like CuO, Bi_2WO_6, and $InVO_4$ (Li et al. 2014; Chaiwichian et al. 2014; Guo et al. 2015). The author demonstrates that the morphology of the composite materials strongly depends upon their synthesis process. From the SEM observation of $BiVO_4$/$InVO_4$ composite, the author noticed that with an increase in the content of the $BiVO_4$ in the composite, the morphology

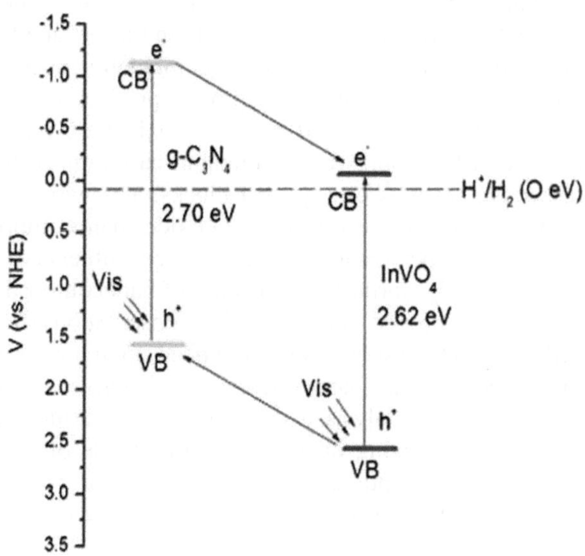

Fig. 10.26 Schematic diagram of the separation and transfer of photogenerated charges in the $InVO_4$/g-C_3N_4 heterojunction under visible light irradiation. Reproduced from Shi et al. (2014), license no. 4014080312801

changes from hierarchical microsphere to leaf-like nanostructure. Also, an interesting observation has been found for $Bi_2WO_6/BiVO_4$ heterojunction, where 0.5 Bi_2WO_6/0.5 $BiVO_4$ composite shows multi-shaped morphology and includes plate, rod, and red blood cell-like shapes. In the case of $CuO/BiVO_4$ composite, the author observes $BiVO_4$ which retains its dumbbell-like shape after the incorporation of CuO. These observed morphologies have tremendous importance toward photocatalytic activity. In all these composites, the monoclinic phase of $BiVO_4$ has been retained which is beneficial for photocatalytic reaction. The photoactivity of all these composites have been evaluated toward dye degradation and the enhanced performance has been explained on the basis of separation of their charge carriers because of the formation of effective heterojunction. Some more examples of mixed metal oxides nanocomposites include Bi_2O_3/Bi-$NaTaO_3$, RGO/N-GZn, RGO/$In_2Ga_2Zn_2O_7$, MgZnO/RGO, RGO/$InVO_4$, etc. (Reddy et al. 2012; Padhi et al. 2015; Martha et al. 2014; Wu et al. 2015; Shen et al. 2013) which shows that the research on non-titanium-based mixed metal oxides nanocomposites is in full swing and even they have potential to emerge as a future photocatalyst.

Summary

In this chapter, photocatalytic activities of various titanium-based mixed metal oxides nanocomposite by categorizing blockwise (s, p, d, f) have been reviewed in detail. Generally, titanium-based mixed metal oxides having ABO_3 formula adopt perovskite structure while a very few have ilmenite structure. On the other hand, rare earth titanates consist of two types of structures depending upon the ionic radii of the A and B cation, i.e., pyrochlore and layered perovskite structure. Modification of titanium-based mixed metal oxides with other semiconductor material to form nanocomposite is found to be an effective method for suppression of photogenerated electron–hole recombination in comparison to the other methods. Photocatalytic activity like hydrogen production, CO_2 reduction, degradation of reactive dyes and other dyes, water oxidation, removal of 2-propanol, etc. have been described using various titanium-based mixed metal oxides nanocomposite in respective sections of s, p, d, f blocks. The various factors which affect the photocatalytic activity of mixed metal oxides nanocomposite like morphology, surface area, varying percentage of one of the constituents of composite, lattice and energy level matching, further modification of nanocomposite, etc. have been discussed in detail with proper examples. The observation showed that the photocatalytic activity of any system does not depend upon a particular issue. The work on other mixed metal oxides nanocomposite is also given in the last section. Apart from the work on photocatalysis using mixed metal oxides nanocomposite, side by many researches are going on in the various fields of fine chemical synthesis, organic conversion, adsorption, sensor, supercapacitor, etc. Therefore, mixed metal oxides nanocomposite can be considered as versatile materials.

References

Abe R, Higashi M, Sayama K, Abe O, Sugihara H (2006) Photocatalytic activity of R_3MO_7 and $R_2Ti_2O_7$ (R) Y, Gd, La; M) Nb, Ta for water splitting into H_2 and O_2. J Phys Chem B 110:2219–2226

Alexe M, Gruverman A, Harnagea C, Zakharov ND, Pignolet A, Hesse D, Scott JF (1999) Switching properties of self-assembled ferroelectric memory cells. Appl Phys Lett 75:1158–1160

Ao Y, Xu L, Wang P, Wang C, Hou J, Qian J, Li Y (2015) Graphene and TiO_2 co-modified flower-like $Bi_2O_2CO_3$: a novel multi-heterojunction photocatalyst with enhanced photocatalytic activity. Appl Surf Sci 355:411–418

Ao Y, Wang K, Wang P, Wang C, Hou J (2016) Synthesis of novel 2D-2D p-n heterojunction $BiOBr/La_2Ti_2O_7$ composite photocatalyst with enhanced photocatalytic performance under both UV and visible light irradiation. Appl Catal B Environ 194:157–168

Armstrong AR, Armstrong G, Canales J, Bruce PG (2004) TiO_2-B nanowires. Angew Chem Int Ed 43:2286–2288

Borse PH, Kim JY, Lee JS, Lim KT, Jeong ED, Bae JS, Yoon JH, Yu SM, Kim HG (2012) Ti-dopant-enhanced photocatalytic activity of a $CaFe_2O_4/MgFe_2O_4$ bulk heterojunction under visible-light irradiation. J Korean Phys Soc 61:73–79

Burbure NV, Salador PA, Rohrer GS (2010) Photochemical reactivity of titania films on $BaTiO_3$ substrates: origin of spatial selectivity. Chem Mater 22:5823–5830

Cao T, Li Y, Wang C, Shao C, Liu Y (2011) A facile in situ hydrothermal method to $SrTiO_3/TiO_2$ nanofiber heterostructures with high photocatalytic activity. Langmuir 27:2946–2952

Chaiwichian S, Inceesungvorn B, Wetchakun K, Phanichphant S, Kangwansupamonkon W, Wetchakun N (2014) Highly efficient visible-light-induced photocatalytic activity of Bi_2WO_6/ $BiVO_4$ heterojunction photocatalysts. Mater Res Bull 54:28–33

Chen X, Shen S, Guo L, Mao SS (2010) Semiconductor-based photocatalytic hydrogen generation. Chem Rev 110:6503–6570

Chen J, Liu S, Zhang L, Chen N (2015) New $SnS_2/La_2Ti_2O_7$ heterojunction photocatalyst with enhanced visible-light activity. Mater Lett 150:44–47

Dadigala R, Gangapuram BR, Bandi R, Dasari A, Gutten V (2016) Synthesis and characterization of C-$TiO_2/FeTiO_3$ and CQD/C-$TiO_2/FeTiO_3$ photocatalysts with enhanced photocatalytic activities under sunlight irradiation. Acta Metall Sin (Engl Lett) 29:17–27

Enhessari M, Kargar-Razi M, Moarefi P, Parviz A (2012) Synthesis, characterization and photocatalytic properties of $MnTiO_3$-Zeolite-Y nanocomposites. J Nanostruct 1:119–125

Fujishima A, Honda K (1972) Electrochemical photolysis of water at a semiconductor electrode. Nature 238:37–38

Gao B, Kim YJ, Chakraborty AK, Lee WI (2008) Efficient decomposition of organic compounds with $FeTiO_3/TiO_2$ heterojunction under visible light irradiation. Appl Catal B-Environ 83:202–207

Gawande MB, Pandey RK, Jayaram RV (2012) Role of mixed metal oxides in catalysis science-versatile applications in organic synthesis. Catal Sci Technol 2:1113–1125

Giocondi JL, Rohrer GS (2001) Spatial separation of photochemical oxidation and reduction reactions on the surface of ferroelectric $BaTiO_3$. J Phys Chem B 105:8275–8277

Guo F, Weilong S, Lin X, Yan X, Guo Y, Che G (2015) Novel $BiVO_4/InVO_4$ heterojunctions: Facile synthesis and efficient visible-light photocatalytic performance for the degradation of rhodamine B. Sep Purif Technol 141:246–255

Han T, Chen Y, Tian G, Wang JQ, Ren Z, Zhou W, Fu H (2015) Hierarchical $FeTiO_3$-TiO_2 hollow spheres for efficient simulated sunlight-driven water oxidation. Nanoscale 7:15924–15934

Han C, Liu J, Yang W, Wu Q, Yang H, Xue X (2016) Photocatalytic activity of $CaTiO_3$ synthesized by solid state, sol-gel and hydrothermal methods. J Sol-Gel Sci Technol. doi:10.1007/s10971-016-4261-3

Hong LI, Gang JC, Zhong LI, Hua ZC (2007) Synthesis and photocatalytic decomposition of water under visible light irradiation of $La_2Ti_{2-x}Co_xO_7$ with pyrochlore structure. Acta Phys Chim Sin 23:761–764

Hou X, Tian Y, Zhang X, Dou S, Pan L, Wang W, Li Y, Zhao J (2015) Preparation and characterization of $Fe_3O_4/SiO_2/Bi_2MoO_6$ composite as magnetically separable photocatalyst. J Alloys Compd 638:214–220

Hu S, Chi B, Pu J, Jian L (2014) Novel heterojunction photocatalysts based on lanthanum titanate nanosheets and indium oxide nanoparticles with enhanced photocatalytic hydrogen production activity. J Mater Chem A 2:19260–19267

Huang S, Xu Y, Xie M, Xu H, He M, Xia J, Huang L, Li H (2015) Synthesis of magnetic $CoFe_2O_4$/g-C_3N_4 composite and its enhancement of photocatalytic ability under visible-light. Colloids Surf A 478:71–80

Hwang DW, Kim HG, Jang SJ, Bae SW, Ji SM, Lee JS (2004) Photocatalytic decomposition of water-methanol solution over metal-doped layered perovskites under visible light irradiation. Catal Today 93:845–850

Inceesungvorn B, Teeranunpong T, Nunkaew J, Suntalelat S, Tantraviwat D (2014) Novel $NiTiO_3$/Ag_3VO_4 composite with enhanced photocatalytic performance under visible light. Catal Commun 54:35–38

Ishikawa A, Takata T, Kondo JN, Hara M, Kobayashi H, Domen K (2002) Oxysulfide $Sm_2Ti_2S_2O_5$ as a stable photocatalyst for water oxidation and reduction under visible light irradiation ($\lambda > 650$ nm). J Am Chem Soc 124:13547–13553

Ji Y, Cao J, Jiang L, Zhang Y, Yi Z (2014) G-$C_3N_4/BiVO_4$ composites with enhanced and stable visible light photocatalytic activity. J Alloys Compd 590:9–14

Jia Y, Shen S, Wang D, Wang X, Shi J, Zhang F, Han H, Li C (2013) Composite Sr2TiO4/SrTiO3 (La, Cr) heterojunction based photocatalyst for hydrogen production under visible light irradiation. J Mater Chem A 1:7905–7912

Jiang S, Ren Z, Li M, Gong S, Yu Y, Pei J, Wei X, Shen G, Han G (2015) Single-crystal heterostructured $PbTiO_3$/CdS nanorods with enhanced visible-light-driven photocatalytic performance. RSC Adv 5:54454–54459

Kanhere P, Chen Z (2014) A review on visible light active perovskite-based photocatalysts. Molecules 19:19995–20022

Kato H, Asakura K, Kudo A (2003) Highly efficient water splitting into H_2 and O_2 over lanthanum-doped $NaTaO_3$ photocatalysts with high crystallinity and surface nanostructure. J Am Chem Soc 125:3082–3089

Kim YJ, Gao B, Han SY, Jung MH, Chakraborty AK, Ko T, Lee C, Lee WI (2009) Heterojunction of $FeTiO_3$ nanodisc and TiO_2 nanoparticle for a novel visible light photocatalyst. J Phys Chem C 113:19179–19184

Kumar S, Tonda S, Baruah A, Kumar B, Shanker V (2014) Synthesis of novel and stable g-C_3N_4/N-doped $SrTiO_3$ hybrid nanocomposites with improved photocurrent and photocatalytic activity under visible light irradiation. Dalton Trans 43:16105–16114

Li L, Zhang Y, Schultz AM, Liu X, Salvador PA, Rohrer GS (2012) Visible light photochemical activity of heterostructured $PbTiO_3$-TiO_2 core-shell particles. Catal Sci Technol 2:1945–1952

Li R, Li Q, Zong L, Wang X, Yang J (2013a) $BaTiO_3/TiO_2$ heterostructure nanotube arrays for improved photoelectrochemical and photocatalytic activity. Electrochim Acta 91:30–35

Li P, Zhao X, Jia C, Sun H, Sun L, Cheng X, Liu L, Fan W (2013b) $ZnWO_4$/BiOI heterostructures with highly efficient visible light photocatalytic activity: the case of interface lattice and energy level match. J Mater Chem A 1:3421–3429

Li J, Cui M, Guo Z, Liu Z, Zhu Z (2014) Synthesis of dumbbell-like CuO-$BiVO_4$ heterogeneous nanostructures with enhanced visible-light photocatalytic activity. Mater Lett 130:36–39

Lim SH, Luo J, Zhong Z, Ji W, Lin J (2005) Room temperature hydrogen uptake by TiO_2 nanotubes. Inorg Chem 44:4124–4126

Lin X, Guo X, Shi W, Zhao L, Yan Y, Wang Q (2015) Ternary heterostructured Ag-$BiVO_4/InVO_4$ composites: synthesis and enhanced visible-light-driven photocatalytic activity. J Alloys Compd 635:256–264

Liu XC, Hong R, Tian C (2009) Tolerance factor and the stability discussion of ABO_3-type ilmenite. J Mater Sci: Mater Electron 20:323–327

Liu H, Waclawik ER, Zheng Z, Yang D, Ke X, Zhu H, Frost RL (2010) TEM Investigation and FBB model explanation to the phase relationships between titanates and titanium dioxides. J Phys Chem C 114:11430–11434

Liu Y, Chen G, Zhou C, Hu Y, Fu D, Liu J, Wang Q (2011) Higher visible photocatalytic activities of nitrogen doped In_2TiO_5 sensitized by carbon nitride. J Hazard Mater 190:75–80

Liu J, Ding T, Li Z, Zhao J, Li S, Liu J (2013) Photocatalytic hydrogen production over In_2S_3-Pt-$Na_2Ti_3O_7$ nanotube films under visible light irradiation. Ceram Int 39:8059–8063

Liu X, Kang Y, Luo D (2016) Synthesis of novel Au/$FeVO_4$/Bi_2O_3 heterojunction for efficient visible light-driven photocatalysis. Mater Lett 185:189–192

Lv J, Kako T, Li Z, Zou Z, Ye J (2010) Synthesis and photocatalytic activities of $NaNbO_3$ rods modified by In_2O_3 nanoparticles. J Phys Chem C 114:6157–6162

Ma RZ, Fukuda K, Sasaki T, Osada M, Bando Y (2005) Structural features of titanate nanotubes/nanobelts revealed by Raman, X-ray absorption fine structure and electron diffraction characterizations. J Phys Chem B 109:6210–6214

Martha S, Padhi DK, Parida K (2014) Reduced graphene oxide/InGaZn mixed oxide nanocomposite photocatalysts for hydrogen production. Chemsuschem 7:585–597

Misono M (2013) Studies in surface science and catalysis, heterogeneous catalysis of mixed oxides-perovskite and heteropoly catalysts, vol 176, pp 25-65

Nashim A, Parida KM (2013) Novel $Sm_2Ti_2O_7$/$SmCrO_3$heterojunction based composite photocatalyst for degradation of Rhodamine 6G dye. Chem Eng J 215:608–615

Nashim A, Parida KM (2014) n-$La_2Ti_2O_7$/p-$LaCrO_3$: a novel heterojunction based composite photocatalyst with enhanced photoactivity towards hydrogen production. J Mater Chem A 2:18405–18412

Nashim A, Martha S, Parida KM (2013) $Gd_2Ti_2O_7$/In_2O_3: efficient visible-light-driven heterojunction-based composite photocatalysts for hydrogen production. ChemCatChem 5:2352–2359

Nashim A, Martha S, Parida KM (2014) Heterojunction conception of n- $La_2Ti_2O_7$/p-CuO in the limelight of photocatalytic formation of hydrogen under visible light. RSC Adv. 4:14633–14643

Padhi DK, Parida K, Singh SK (2015) Facile fabrication of RGO/N-GZ mixed oxide nanocomposite for efficient hydrogen production under visible light. J Phys Chem C 119:6634–6646

Parida KM, Nashim A, Mahanta SK (2011) Visible-light driven $Gd_2Ti_2O_7$/$GdCrO_3$ composite for hydrogen evolution. Dalton Trans 40:12839–12845

Park BH, Kang BS, Bu SD, Noh TW, Lee J, Jo W (1999) Lanthanum-substituted bismuth titanate for use in nonvolatile memories. Nature 401:682–684

Patwe SJ, Katari V, Salke NP, Deshpande SK, Rao R, Gupta MK, Mittal Achary SN, Tyagi AK (2015) Structural and electrical properties of layered perovskite type $Pr_2Ti_2O_7$: experimental and theoretical investigations. J Mater Chem C 3:4570–4584

Pena MA, Fierro JLG (2001) Chemical structures and performance of perovskite oxides. Chem Rev 101:1981–2018

Preciado MAR, Kassiba A, Morales-Acevedoc A, Makowska-Janusik M (2015) Vibrational and electronic peculiarities of $NiTiO_3$ nanostructures inferred from first principle calculations. RSC Adv 5:17396–17404

Razali MH, Noor AFM, Mohamed AR, Sreekantan S (2012) Morphological and structural studies of titanate and titania nanostructured materials obtained after heat treatments of hydrothermally produced layered titanate. J Nanomater 2012: Article ID 962073, 10p

Reddy KH, Martha S, Parida KM (2012) Facile fabrication of Bi_2O_3/Bi-$NaTaO_3$ photocatalysts for hydrogen generation under visible light irradiation. RSC Adv 2:9423–9436

Reddy KH, Martha S, Parida KM (2013) Fabrication of novel p–BiOI/n-$ZnTiO_3$ heterojunction for degradation of rhodamine 6G under visible light irradiation. Inorg Chem 52:6390–6401

Shao Z, Saitzek S, Roussel P, Desfeux R (2012) Stability limit of the layered-perovskite structure in $Ln_2Ti_2O_7$ (Ln = lanthanide) thin films grown on (110)-oriented $SrTiO_3$ substrates by the so-gel route. J Mater Chem 22:24894–24901

Shcherbakova LG, Mamsurova LG, Sukhanova GE (1979) Lanthanide titanates. Russ Chem Rev 48:228–242

Shen J, Li X, Huang W, Li N, Ye M (2013) Synthesis of novel photocatalytic RGO-$InVO_4$ nanocomposites with visible light photoactivity. Mater Res Bull 48:3112–3116

Shi W, Guo F, Chen J, Che G, Lin X (2014) Hydrothermal synthesis of $InVO_4$/Graphitic carbon nitride heterojunctions and excellent visible-light-driven photocatalytic performance for rhodamine B. J Alloys Compd 612:143–148

Sivakumar S, Selvaraj A, Ramasamy AK (2013b) Photocatalytic degradation of organic reactive dyes over $MnTiO_3/TiO_2$ heterojunction composites under UV-visible irradiation. Photochem Photobiol 89:1047–1056

Sivakumar S, Selvaraj A, Ramasamy AK, Balasubramanian A (2013a) Enhanced photocatalytic degradation of reactive dyes over $FeTiO_3/TiO_2$ heterojunction in the presence of H_2O_2. Water Air Soil Pollut 224:1529–1542

Stengl V, Bakardjieva S, Murafa N (2009) Preparation and photocatalytic activity of rare earth doped TiO_2 nanoparticles. Mater Chem Phys 114:217–226

Su NR, Lv P, Li M, Zhang X, Li M, Niu J (2014) Fabrication of $MgFe_2O_4$-ZnO heterojunction photocatalysts for application of organic pollutants. Mater Lett 122:201–204

Sun X, Li Y (2003) Synthesis and characterization of ion-exchangeable titanate nanotubes. Chem Eur J 9:2229–2238

Tian X, Liu J, Wang H, Yan H (2014) $La_2Ti_2O_7$ nanoplates decorated with Cu_2ZnSnS_4 nanoparticles for enhanced visible-light-driven photocatalytic activity. CrystEngComm 16:8517–8522

Truong QD, Liu JY, Chung CC, Ling YC (2012) Photocatalytic reduction of CO_2 on $FeTiO_3/TiO_2$ photocatalyst. Catal Commun 19:85–89

Ueda K, Yanagi H, Noshiro R, Hosono H, Kawazoe H (1998) Vacuum ultraviolet reflectance and electron energy loss spectra of $CaTiO_3$. J Phys: Condens Matter 10:3669–3677

Wang R, Xu D, Liu J, Li KW, Wang H (2011) Preparation and photocatalytic properties of CdS/ $La_2Ti_2O_7$ nanocomposites under visible light. Chem Eng J 168:455–460

Wang R, Zhu Q, Wang W, Fan C, Xu A (2015) $BaTiO_3$-graphene nanocomposites: synthesis and visible light photocatalytic activity. New J Chem 39:4407–4413

Wu HB, Lou XW, Hng HH (2012) Synthesis of uniform layered protonated titanate hierarchical spheres and their transformation to anatase TiO_2 for lithium-ion batteries. Chem-Eur J 18:2094–2099

Wu Y, Wang LQ, Yang X (2015) Photocatalytic properties of Ag-modified MgZnO/RGO composites. Mater Res Innovations 19:318–321

Xian T, Yang H, Huo YS (2014) Enhanced photocatalytic activity of $CaTiO_3$-graphene nanocomposites for dye degradation. Phys Scr 89:115801 (6 pp)

Xian T, Yang H, Di LJ, Dai JF (2015) Enhanced photocatalytic activity of $BaTiO_3$@g-C_3N_4 for the degradation of methyl orange under simulated sunlight irradiation. J Alloys Compd 622:1098–1104

Xian T, Yang H, HuoYS Ma JY, Zhang HM, Su JY, Feng WJ (2016) Fabrication of Ag-decorated $CaTiO_3$ nanoparticles and their enhanced photocatalytic activity for dye degradation. J Nanosci Nanotechnol 16:570–575

Xu Q, Feng J, Li L, Xiao Q, Wang J (2015) Hollow $ZnFe_2O_4/TiO_2$ composites: high-performance and recyclable visible-light photocatalyst. J Alloys Compd 641:110–118

Yang DJ, Zheng ZF, Zhu HY, Liu HW, Gao XP (2008) Titanate nanofibers as intelligent absorbents for the removal of radioactive ions from water. Adv Mater 20:2777–2781

Ye R, Fang HB, Zheng YZ, Li N, Wang Y, Tao X (2016) Fabrication of $CoTiO_3$/g-C_3N_4 hybrid photocatalysts with enhanced H_2 evolution: Z-scheme photocatalytic mechanism insight. ACS Appl Mater Interfaces 8:13879–13889

Yu Y, Ren Z, Li M, Gong S, Yin S, Jiang S, Li X, Wei X, Xu G, Shen G, Han G (2015) Facile synthesis and visible photocatalytic activity of single-crystal $TiO_2/PbTiO_3$ heterostructured nanofiber composites. CrystEngComm 17:1024–1029

Yue X, Zhang J, Yan F, Wang X, Huang F (2014) A situ hydrothermal synthesis of $SrTiO_3/TiO_2$ heterostructure nanosheets with exposed (001) facets for enhancing photocatalytic degradation activity. Appl Surf Sci 319:68–74

Zhang H, Chen G, Li Y, Teng Y (2010) Electronic structure and photocatalytic properties of copper-doped $CaTiO_3$. Int J Hydrogen Energy 3:52713–52716

Zhang Z, Liu G, Mao Y (2013) Improved separation efficiency of photogenerated carriers for $Fe_2O_3/SrTiO_3$ heterojunction semiconductor. Int J Hydrogen Energy 38:9349–9354

Zhang G, Gang L, Wang L, Irvine JTS (2016a) Inorganic perovskite photocatalysts for solar energy utilization. Chem Soc Rev 45:5951–5984

Zhang Q, Huang Y, Xu L, Cao J, Ho W, Lee SC (2016b) Visible-light-active plasmonic Ag-$SrTiO_3$ nanocomposites for the degradation of NO in air with high selectivity. ACS Appl Mater Interfaces 8:4165–4174

Zhao B, Lin L, He D (2013) Phase and morphological transitions of titania/titanate nanostructures from an acid to an alkali hydrothermal environment. J Mater Chem A 1:1659–1668

Zhu W, Han D, Niu L, Wu T, Guan H (2016) Z-scheme Si/$MgTiO_3$ porous heterostructures: noble metal and sacrificial agent free photocatalytic hydrogen evolution. Int J Hydrogen Energy 41:14713–14720

Chapter 11
Novel Applications and Future Perspectives of Nanocomposites

Zsolt Kása, Tamás Gyulavári, Gábor Veréb, Gábor Kovács, Lucian Baia, Zsolt Pap and Klára Hernádi

Abstract As the present chapter of the book is located in the concluding section, it was important to highlight the main applications of composite materials focusing especially on applications, which exploit other peculiarities of the materials besides photocatalysis. This will be done, by introducing those materials and their composites that are most studied, or were found to exhibit interesting behavior. In many of the presented cases, the main structural, morphological, or optical property of the given composite will be discussed to understand its functioning mechanism, and its role in the current scientific approaches. Additionally, this chapter aims to give a perspective regarding the composite-based nanoscience, and points out important research directions for the further developments of composite materials.

Zsolt Kása, Tamás Gyulavári, Gábor Veréb, Gábor Kovács, Lucian Baia—These authors contributed equally.

Z. Kása · Z. Pap (✉)
Institute of Environmental Science and Technology,
University of Szeged, Szeged, Hungary
e-mail: pzsolt@chem.u-szeged.hu

T. Gyulavári · G. Kovács · K. Hernádi (✉)
Department of Applied and Environmental Chemistry,
University of Szeged, Szeged, Hungary
e-mail: hernadi@chem.u-szeged.hu

G. Veréb
Department of Process Engineering, Faculty of Engineering,
University of Szeged, Szeged, Hungary

G. Kovács · L. Baia · Z. Pap
Institute for Interdisciplinary Research on Bio-Nano-Sciences,
Babeş-Bolyai University, Cluj-Napoca, Romania

G. Kovács · L. Baia · Z. Pap
Faculty of Physics, Babeş-Bolyai University, Cluj-Napoca, Romania

M.M. Khan et al. (eds.), *Nanocomposites for Visible Light-induced Photocatalysis*,
Springer Series on Polymer and Composite Materials,
DOI 10.1007/978-3-319-62446-4_11

Keywords Photocatalysis · Application spectrum · Nanocomposites · Semiconductors · Metals · Doping · Mixed oxides

The Most Frequently Applied Photocatalysts and Their Composites. Overview of the Current Hot Topics

It is well known that TiO_2-based photocatalysts have been widely investigated for water treatment applications due to their ability to decompose various kinds of organic pollutants even toxic, chemically stable, and persistent contaminants like phenol (Antoniou and Dionysiou 2007; Herrmann et al. 1999; Kun et al. 2009; Veréb et al. 2012, 2013a, b), organic dyes (Bahnemann et al. 2007; Qamar et al. 2014), pesticides (Bahnemann et al. 2007; Kovács et al. 2016), and pharmaceuticals (Gar Alalm et al. 2016; Lin et al. 2016). However, it is important to mention the main drawback of conventional TiO_2 that it can only be excited efficiently by UV light [≈3–5% of the solar spectrum (Wu et al. 2013a, b)] due to its relatively large band gap [≈3.20 eV for anatase and ≈3.02 eV for rutile (Banerjee et al. 2006)], which limits its practical application possibilities. Hence, it is important to extend titanium dioxide light absorption into the visible-light region (≈45% of the solar spectrum) to enable outdoor usage and to utilize a broader range of the sunlight's spectrum. To overcome this apparent problem, there are many different approaches in the literature to prepare visible-light-active TiO_2-s, like doping with various elements (Pelaez et al. 2012; Veréb et al. 2012, 2013a, b), sensitizing with dyes (Cho et al. 2001; Pelaez et al. 2012; Savinkina et al. 2014), deposition of noble metals (Karácsonyi et al. 2013; Lin et al. 2016; Pelaez et al. 2012), surface modification, or by preparing effective composites. TiO_2 is often combined with other metal oxides to create various composite photocatalysts with superior photoactivity via improved charge separation, light absorption, structural properties, and surface chemistry (Marszewski et al. 2015). For example, if a large bandgap semiconductor is coupled with a small bandgap semiconductor with a more negative conduction band (CB) level, the CB electrons can be injected from the small bandgap semiconductor to the large one achieving wide electron-hole separation (Malato et al. 2009). In addition, charge carrier separation results in the separation of redox reactions as well which prevents detrimental reactions and increases the product yield. With properly chosen incorporated metal oxide the properties of TiO_2 can be fine-tuned (e.g., if the incorporated metal oxide has smaller band gap than TiO_2 or if it is nanostructured, then it improves its light absorption or structural parameters, porosity and surface area). Naturally, composites include TiO_2 in most cases, since it is a versatile photocatalyst, therefore the emphasis in most cases is on the photocatalytic applications. Figure 11.1 demonstrates the contact mode possibilities of the particles in the case of two semiconductors and one noble metal. This question

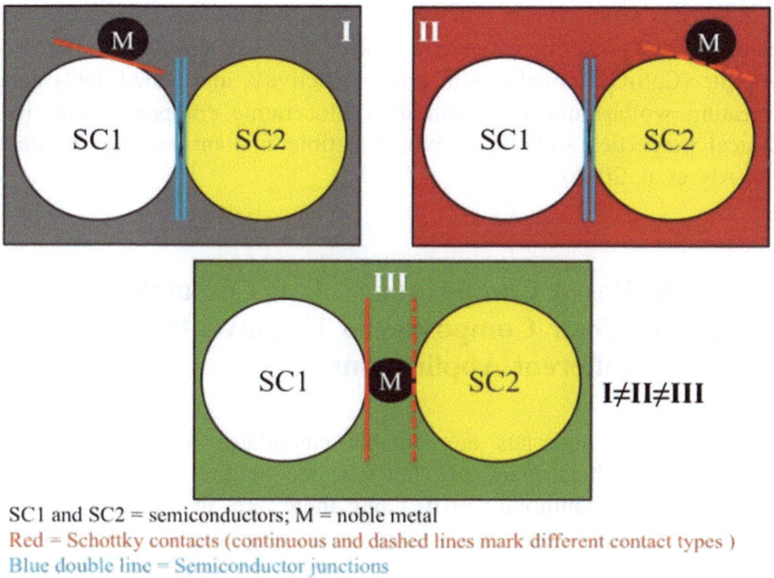

Fig. 11.1 Contact possibilities of two semiconductor crystallites and one noble metal particle (Karácsonyi et al. 2013)

is critical due to simple electronic contact issues. Different contact nature means different charge carrier dynamics, which also leads to different catalytic/photocatalytic activity.

TiO_2 can be loaded containing different amounts of WO_3 in order to obtain better photocatalytic performances and increased stability (Ramos-Delgado et al. 2013). TiO_2-reduced graphene oxide (rGO) composites were proven to be efficient in the disinfection of *Escherichia coli* and *F. colani* (Fernández-Ibáñez et al. 2015), while carbon-modified TiO_2 nanocomposites showed improved H_2 generation rates (Parayil et al. 2012). Composite (Ag, Au, W) titania coatings on glass showed great potential as self-cleaning surfaces due to its resistance against cleaning and reuse (Kafizas et al. 2009). Alalm et al. evaluated the photocatalytic activity of bare TiO_2 and TiO_2 immobilized on activated carbon (TiO_2/AC) for degradation of pharmaceuticals, which was used as a model contaminant in the wastewater, generated by pharmaceutical industries. Results pointed out that complete removal of amoxicillin and ampicillin was achieved by TiO_2/AC after 120 and 180 min of irradiation respectively (Gar Alalm et al. 2016).

Although, TiO_2 is used dominantly as a photocatalyst or in composite photocatalysts, there are numerous other applications. Ram et al. fabricated SnO_2 and/or TiO_2 composite films on glass and interdigitated electrodes to use it as CO sensor (Ram et al. 2005). Cozzi et al. 2014 prepared organically functionalized TiO_2/Nafion composite proton exchange membranes for fuel cell applications. By embedding TiO_2 nanoparticles decorated with propylsulfonic acid groups increased the selectivity and power density significantly. Aly et al. reported the preparation of wollastonite ($CaSiO_3$) powder with good bioactivity, and added TiO_2 nanofibers thus creating wollastonite/TiO_2 nanofiber bioceramic composite with increased mechanical properties suitable as biocompatible implant material to treat bone defects (Aly et al. 2016).

Cobalt Oxide-Based Composites. Cobalt Containing Materials and Their Composites as Unconventional Materials for Different Applications

Cobalt oxide-based materials are suitable candidates for the construction of solid-state sensors, heterogeneous catalysts, electrochromic devices, and solar energy absorbers (Mahmoud 2016), and there are just a few cases where cobalt-based materials were used in photocatalytic applications. Interestingly, metallo-organic materials based on Co, such as triphenylamine functionalized bithiazole-cobalt complex were used in composites with Ag nanoparticles (Co-2TPABTz) and showed high efficiency in hydrogen generation under visible-light illumination with outstanding long-term stability (Huo and Zeng 2016). The photocatalytic efficiency differed from the amount of Ag nanoparticles, which was attributed to the synergistic effects of the unique porous structure of Co-2TPABTz and strong surface plasmon resonance effect of Ag nanoparticles (Huo and Zeng 2016).

Sahu et al. fabricated cobalt-doped neutralized red mud nanocomposite materials (Co/NRM) by impregnation method, and determined its photocatalytic efficiency in case of different Co:NRM ratios with the photodegradation of MB (methylene blue, a model contaminant of wastewater) dye under solar light irradiation (Sahu and Patel 2016). The degradation efficiency was improved with the increasing amount of cobalt, which was ascribed to the accelerated photoinduced electron-hole transfer and separation, decreased recombination rate and band energy.

Madhu et al. fabricated pongam seed shells-derived activated carbon (PSAC) and cobalt oxide nanocomposites and then it was employed as nonenzymatic glucose sensor and supercapacitor (Madhu et al. 2015). Similarly, for nonenzymatic glucose sensing S. Premlatha et al. prepared cobalt-multiwalled carbon nanocomposites with high stability and sensitivity, while Zhang et al. prepared iron-doped cobalt oxide nanocomposite films by electrodeposition and applied them as electrocatalyst for oxygen reduction reactions which can be useful in fuel cells (oxygen reduction reaction plays a significant role in fuel cells because it dominates the overall performance of these energy storage and conversion systems (Zhang et al. 2014).

Nickel Containing Materials. Current and Future Perspectives

Well-dispersed metallic nickel nanoparticles were obtained in graphitized carbon matrix (Ni@C) by the pyrolysis of metal–organic frameworks (Fang et al. 2017). The improved photocatalytic activities through loading Ni@C was attributed to the cooperative work of Ni nanoparticles and graphene layers, which facilitated the separation of photogenerated carriers and suppressed the recombination of the electron-hole pairs. Nickel oxide (NiO) nanoparticles obtained by sol–gel method were efficient for the photocatalytic oxidation of phenol, although ultraviolet laser irradiation (Hayat et al. 2011) was used as a light source. NiO nanoparticles were proven to be efficient in the degradation of fluorescein (a synthetic dye) under sunlight, UV, and CFL (compact fluorescent light) irradiation (Perumal Raj et al. 2016). TiO_2-NiO nanocomposite coatings were obtained by mechanical coating technique and subsequent heat oxidation at different temperatures (Lu et al. 2014). The photocatalytic activity of the composite coatings was evaluated by the degradation of MB solution under UV irradiation.

Nickel oxide materials are known to be suitable for counter electrodes in complementary electrochromic devices and are typically prepared with multiple additives to enhance performance (Lin et al. 2014). Nickel and manganese oxide composites were obtained with a uniformly dispersed microspherical structure (Wang et al. 2016a, b, c) and it was employed as a sensing material for nonenzymatic glucose detection. Wu et al. prepared caterpillar structured $Ni(OH)_2@MnO_2$ core/shell (Fig. 11.2) nanocomposite arrays on nickel foams (CS $Ni(OH)_2@MnO_2$ NFs) (Wu and Liang 2016) which were applied as anode for lithium-ion batteries, with high specific capacity in the initial discharge process and retained a reasonable reversible capacity even after 80 cycles. Zhang et al. (2015a, b) developed high performance/high sensitivity methane gas sensor based on nickel oxide (NiO)/reduced graphene oxide (rGO) nanocomposite film.

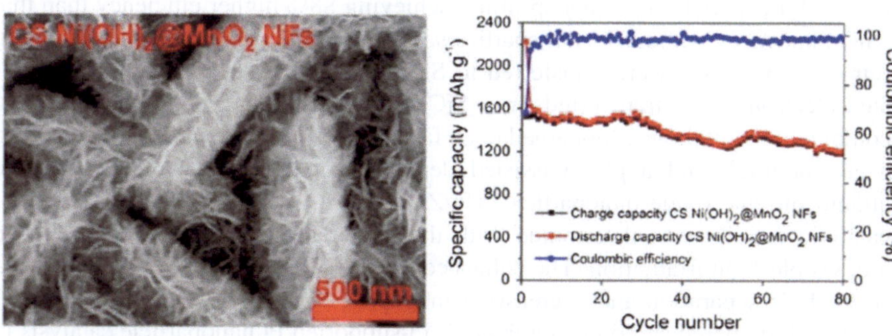

Fig. 11.2 Caterpillar structured $Ni(OH)_2@MnO_2$ core/shell nanocomposite arrays on nickel foam as high performance anode materials for lithium-ion batteries (Zhang et al. 2015a, b)

Palladium in Photocatalytic and Other Composite Materials. Elemental Palladium and Potential New Materials/Composites

It is well known that Pd is usually present in composites alongside with different photocatalysts (predominantly titania) and it is fully documented in the literature. Therefore, the current section is focused on other than titania composites with Pd. Pd deposited-ZrO_2-MWCNTs (multiwalled carbon nanotubes) were applied efficiently in the degradation of acid blue 40 dye in aqueous solution under simulated solar light (Anku et al. 2016). Moreover, bimetallic nanocomposites were obtained (Ag or Pd/-TiO_2/CNT) and efficiently applied in the degradation of MB (Hintsho et al. 2014).

An interesting material consists of palladium oxide (rarely considered in photocatalysis) and nitrogen-doped titanium oxide (PdO/N-TiON) was obtained by sol–gel process, and the photocatalytic activity was determined by the inactivation of *E. coli*, *Pseudomonas aeruginosa*, and *Staphylococcus aureus* under visible-light irradiation (Wu et al. 2009). Disinfection data indicated that PdO/TiON composite photocatalysts had much better photocatalytic activity than either palladium-doped or nitrogen-doped titanium oxide (TiON).

Platinum, the Elite Cocatalyst in Photocatalysis. More Rarely Discussed Composites of Pt

For photocatalytic applications platinum was only used as a dopant (Egerton and Mattinson 2008; Shivalingappa et al. 1997) or deposited as nanoparticles. For instance, among the uncommon cases, platinum nanoparticle-decorated SiC nanowire photocatalyst were used for photocatalytic water splitting under simulated sunlight irradiation (Wang et al. 2014a, b, c). Pt/SiC catalyst showed enhanced photocatalytic activity for water splitting, achieving 88% higher efficiency than that of SiC nanowires without Pt nanoparticles. It was concluded that the surrounding electrons of C atoms were transferred to Si and Pt active sites and the photogenerated electrons were transferred from SiC to Pt quickly and efficiently avoiding recombination with holes. Mohamed et al. fabricated zinc oxide nanoparticles 'via a sol–gel method', and a photo-assisted deposition method was used to prepare platinum on zinc oxide nanoparticles (Pt/ZnO) (Mohamed et al. 2016). The photocatalytic activity was determined with the degradation of malachite green dye under visible-light irradiation. The enhanced activity was attributed to the enhanced electron-hole separation and decreased bandgap.

Lin et al. prepared carbon nanotube (CNT)-supported Pt nanoparticle catalysts in supercritical carbon dioxide using platinum(II) acetylacetonate as metal precursor to be involved as electrocatalysts for low-temperature fuel cells (Lin et al. 2005).

Immobilized glucose oxidase (GOx) in platinum-multiwalled carbon nanotube-alumina-coated silica (Pt-MWCNT-ACS) nanocomposite modified glassy carbon electrode was used as a selective amperometric glucose biosensor (Tsai and Tsai 2009). The prepared electrode was found to be better than the common GOx-Pt-CNT nanobiocomposite modified electrodes, moreover, the prepared biosensor had good anti-interferent ability and long-term storage stability after coating with Nafion.

ZrO_2, the Ceramic Photocatalyst and Its Composites. Other Zr-Based Nanomaterials and Composites

The significance of ZrO_2 lies in its specific optical and electrical properties (Pourbeyram 2016), the potential applications as fuel cells (Pourbeyram 2016), electrodes (Lin et al. 2014), photocatalysts (Carević et al. 2016; Zinatloo-Ajabshir and Salavati-Niasari 2016), or the production of ceramics and dental crowns (Silva et al. 2016). There are a lot of publications in the literature where ZrO_2 was used as photocatalyst or as part of nanocomposite systems as the following two examples show. Zinatloo–Ajabshir et al. prepared nanostructured ZrO_2 by a sonochemical-assisted process and investigated the photocatalytic behavior in the decomposition of eriochrome black T dye which was used as model contaminant of water, and 72% of the dye was degraded after 2 h irradiation of UV light (Zinatloo-Ajabshir and Salavati-Niasari 2016). As composites, palladium-doped-zirconium oxide-multiwalled carbon nanotubes were used for the photocatalytic degradation of acid blue 40 dye under simulated solar light achieving total degradation after 3 h (Anku et al. 2016). Carević et al. synthesized ZnO_2 nanopowders with incorporated Si^{4+} ions by hydrothermal method, which showed significant photocatalytic activity regarding the degradation of triclorophenol, which was explained by the decreased absorption energies caused by the Si ions. An interesting alternative Zr containing photocatalyst emerged recently, namely zirconium–titanium phosphate, which was used with ZnS as efficient visible-light-active material for hydrogen evolution via water splitting (Biswal et al. 2011).

Zirconium oxide and composite materials containing zirconia are widely used in different applications, such as piezoelectric devices, ceramic condensers, thermal barrier coatings, electrode and oxygen sensors, orthopedic implants, etc. (Deshmane and Adewuyi 2012). Pourbeyram obtained graphene oxide-zirconium dioxide nanocomposite (GO-Zr-P), with remarkable adsorption capabilities to remove heavy metals from aqueous solutions (Pourbeyram 2016) (e.g., Cr, Cd, Hg, As, Ni, Cu, and Zn) which in general are not biodegradable unlike organic contaminants, and usually accumulates in living organisms. Lou et al. prepared GO modified hydrated zirconium oxide nanoparticles by hydrothermal coprecipitation reaction with great regeneration abilities to remove As(III) and As(V) ions from drinking water (Luo et al. 2013).

Hafnium—The Non-photocatalytic Material. A Knowledge Void in Photocatalysis

Hafnium-dioxide's applications are based on its intriguing properties such as high melting point, large refractive index, high chemical stability, nontoxicity, and the most important aspect, high dielectric constant that makes it a suitable replacement for silicon oxide (Manikantan et al. 2016; Ramadoss et al. 2012). Most of the hafnium produced is used as control rods for nuclear reactors. It is rarely used in photocatalytic applications, in these cases it is used mainly as dopants and part of complexes, or composite materials.

Manikandan et al. synthesized HfO_2 by coprecipitation method with high absorption capabilities making it suitable to use it as UV blocking layers in contact lens and in optoelectronic devices (Manikandan et al. 2016). HfO_2 nitride binary/ternary nanocomposites can be used as a counter electrode for dye-sensitized solar cells with impressive catalytic activity for I_3^- reduction making it appropriate to replace noble metal Pt in dye-sensitized solar-cell systems (Yun et al. 2013). Reddy et al. used hafnium to prepare doped ceria nanocomposite oxide as a redox additive for three-way catalysts. The hafnium cation in the core structure led to higher activity in CO oxidation via more oxygen vacancies, lattice defects, high oxide ion mobility, and easy reducibility, which is of emphasized importance in three-way converters as they transform CO into nontoxic products in automobiles (Reddy et al. 2007a, b).

Extra Expensive Materials—Rh-Based Photocatalytic Materials

Rhodium is a noble metal and it is one of the rarest and most valuable precious metals. Due to its rarity, utilizing rhodium in different applications is rather expensive. IrO_2-loaded $SrTiO_3$ doped with rhodium and antimony was synthesized, and it was used in water splitting under visible- and simulated sunlight irradiation utilizing its potential to convert light energy into chemical energy (Asai et al. 2014). Li et al. (2015a, b, c) studied the modification of surface active site of Rh in $RhSn_xO_y$ composite oxide cocatalyst by the addition of Sn for photocatalytic hydrogen evolution under visible-light irradiation. Interestingly, the $RhSn_xO_y$ composite oxides' photocatalytic activity was achieved by the optimization of the Rh sites via Sn addition.

Li et al. synthesized mesoporous rhodium oxide/alumina hybrid and used it as high sensitivity and low power consumption methane catalytic combustion microsensor (Li et al. 2012). The importance of the work lies within the development of sensitive and reliable sensors for combustible gases and organic vapors below the lower explosion limit as it has gained considerable attention. Figure 11.3

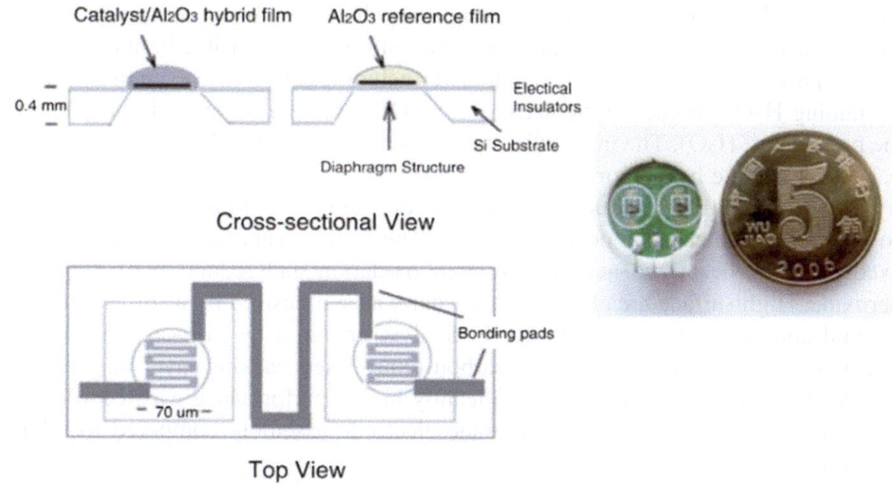

Fig. 11.3 Schematic diagrams and the photograph of the micro-electro-mechanical system for methane catalytic combustion sensor (Li et al. 2012)

shows the assembled MEMS methane catalytic combustion sensor where the Rh_2O_3/Al_2O_3 materials were coated as catalyst on a micro-heater.

Besides sensor applications, also catalytic applications are known. Chandra et al. (2010) prepared rhodium–graphene nanocomposite and examined its catalytic activity by the hydrogenation of benzene. The synthesized composite material reached 100% conversion after 6 h. Hung et al. (2009) fabricated platinum–palladium–rhodium composite oxide catalysts by the coprecipitation of H_2PtCl_6, Pd $(NO_3)_3$ and $Rh(NO_3)_3$ and tested it in catalytic wet oxidation reaction to remove ammonia from aqueous solutions, resulting 99% of the ammonia degraded during wet oxidation over the developed catalyst.

Iridium-Based Nanocomposites. Complexes and Complex Composites for Different Applications

Iridium-based nanomaterials can be used in photocatalytic applications. Fihri et al. (2008) developed photocatalytic systems based on diimine derivatives of ruthenium, cyclometallated iridium, or tricarbonylrhenium as photosensitizers and cobaloxime H_2-evolving catalytic centers, which are among the most efficient molecular systems for hydrogen production. The successful development of a novel photocatalyst was reported, namely the mononuclear iridium(III) terpyridine (tpy) 2-phenylpyridine (ppy) complex $[Ir(tpy)(ppy)Cl]^+$ ([Ir-ppy]), which selectively reduced CO_2 to CO under visible light at 480 nm without additional photosensitizers whereas in case of Re complexes combination with photosensitizers is needed to reduce CO_2

(Sato et al. 2013). Compared to Re complexes the Ir complex had greater photocatalytic activity for CO_2 reduction with the possibility to use visible light irradiation, and the photocatalytic activity is maintained (including selectivity) even in a solution containing H_2O (photocatalytic activity for CO_2 reduction is generally very low in the presence of H_2O). The most efficient photocatalyst possessed a quantum yield of 0.21, which is the best reported value (in 2013).

Tian et al. (2002) introduced iridium powder into sol–gel process to fabricate a sol–gel derived metal composite electrode. The iridium ceramic electrode showed excellent electrocatalytic activity for both oxidation and reduction of hydrogen peroxide. High surface area IrO_2 nanopowder was synthesized by a sulfite complex method and added to a 50% Pt–Ru/C catalyst prepared by the same procedure. A catalytic ink was deposited on a carbon cloth-based backing layer and used as composite anode in a DMFC. A significantly higher performance was recorded for the composite electrode-based membrane electrode assembly compared to a bare one at 60 and 90 °C.

Iron and Iron-Oxide-Based Materials—Large Application Spectrum Composites

The reduction of nitrobenzenes to anilines is a widely used reaction in the industry for the production of dyes, biologically active compounds, pharmaceuticals, and agricultural chemicals (Kumar et al. 2016). The traditional methods (such as catalytic hydrogenation, electrolytic reduction, and metal catalyzed reductions) have some drawbacks, such as the usage of high-pressure reactors, hazardous reagents, high reaction temperature, etc. The photocatalytic reduction of nitrobenzene into aniline can be carried out at mild and ambient temperature conditions. Since most widely investigated TiO_2-based photocatalysts can be effectively activated only by UV photons, the development of suitable, visible-light-active photocatalysts are highly recommended. In 2016, Kumar et al. (2016) synthesized iron (II) bipyridine complex by immobilizing it on rGO, and used successfully in visible-light-assisted photocatalytic reduction of nitrobenzenes at room temperature (Fig. 11.4). The iron complex with its good visible-light absorbance and the rGO with its enhanced charge transportation exhibited synergistic effect to boost the reaction and they were found to be highly stable and easily recoverable.

The World Health Organization (WHO) has classified arsenic as one of the ten chemical species of greatest concern to public health, therefore arsenic water pollution is a worldwide problem. The two main formation of As are arsenite AsO_3^{3-} and arsenate AsO_4^{3-}. Since As(V) is more stable and less toxic than As(III) and the elimination of As(V) is also easier, the first step of common As removal methods is the oxidation of As(III) to As(V) using classical oxidants such as chlorine, ozone,

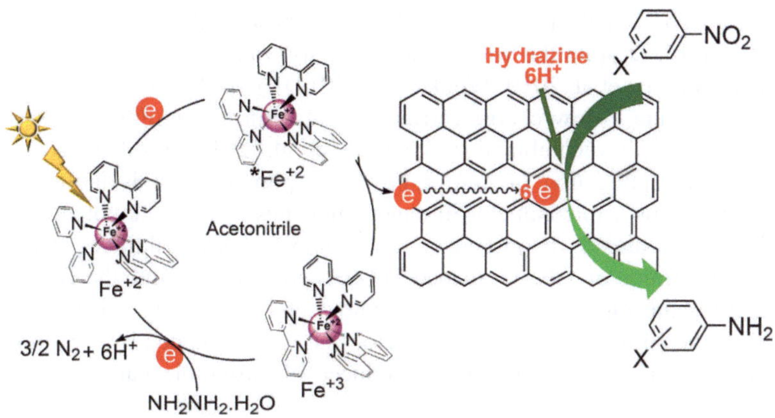

Fig. 11.4 Plausible reaction mechanism of reduction of nitro-compounds using photocatalyst (Kumar et al. 2016)

etc. (López-Muñoz et al. 2016). After the oxidation step, the elimination of arsenate can be carried out by adsorption (using, e.g., activated carbon, alumina, etc.), precipitation, anion exchange, or reverse osmosis. Alternatively, the oxidation of aqueous As(III) to As(V) can be carried out by heterogeneous photocatalysis, process that avoids the addition of toxic chemicals since photocatalysts can be reused.

Zero-valent iron (ZVI) has been efficiently used to remove arsenic compounds (Mohan and Pittman 2007). Important advantages of ZVI include its low cost, simplicity in handling, and the ability for simultaneous removal of As(V) and As (III) without pre-oxidation (López-Muñoz et al. 2016; Sun et al. 2006). Nguyen et al. (2008) and López-Munoz et al. (2016) reported the simultaneous use of TiO_2 and ZVI nanoparticles for the removal of As(III) from aqueous systems. Results evidenced the significant influence of the pH on both the oxidation rate of As(III) and the effectiveness for overall As elimination. Compared to ZVI, the rate of As (III) oxidation was always higher when TiO_2 was present, achieving a complete depletion of arsenite concentration. Fan et al. (2015) prepared nano-iron/oyster shell composite (NI/OS) to explore another efficient treatment technology for arsenic (As) contaminated wastewater. Under the suitable reaction conditions of a pH = 6.8, $T = 20$ °C, and a 1.8 mg L^{-1} initial concentration of As(III) it was almost completely removed from the simulation wastewater.

Iron oxide can exist in several forms such as iron(II) oxide (FeO), unstable amorphous α-FeOOH and γ-FeOOH (with high surface area), and crystalline phases such as hematite (α-Fe_2O_3) maghemite (γ-Fe_2O_3), and magnetite (Fe_3O_4) which exhibit magnetic properties (Can et al. 2012; Cudennec and Lecerf 2005; Wodka et al. 2014).

Hematite (α-Fe_2O_3) is an environment friendly n-type semiconductor (Sun et al. 2016a, b; Wang and Huang 2016), with low cost and having narrow band gap of 2.2 eV (Sun et al. 2016a, b; Wang and Huang 2016; Wodka et al. 2014;

Zhang et al. 2012) and might be one of the cheapest semiconductor material (Liu et al. 2015a, b). Despite the wide light absorption range [30–40% of the incident solar light (Sharma et al. 2010; Wodka et al. 2014)], pure Fe_2O_3 is a relatively poor photocatalyst, due to low charge carrier mobility, short hole diffusion length and rapid electron-hole recombination (Sharma et al. 2010; Wodka et al. 2014). Therefore, the application of pure α-Fe_2O_3 as photocatalyst is limited, but the application of this semiconductor in nanocomposites has a huge potential.

α-Fe_2O_3 was already deposited on ZnO nanoparticles (Sakthivel et al. 2002). The light absorption of α-Fe_2O_3 loaded ZnO system extended into the visible region compared to bare ZnO nanoparticles and photocatalytic activity was also higher in the case of the composite material, behavior that was explained by the authors as a result of the improved separation of produced electron/hole pairs caused by the deposited α-Fe_2O_3 (Fig. 11.5).

Among others (Achouri et al. 2014; Hsu et al. 2015; Liao et al. 2010; Maya-Treviño et al. 2014; Sharma et al. 2010) Xie et al. (2015) also highlighted increased photocatalytic activity of α-Fe_2O_3/ZnO composites under UV-vis light irradiation in comparison with pure ZnO nanoparticles in the case of the degradation of persistent pentachlorophenol. Moreover, authors prepared the composites via a simple and rapid method using only ferric nitrate, zinc nitrate, and sodium hydroxide as reactants. Pt deposited α-Fe_2O_3 nanoring (Fig. 11.6) composites were also obtained and showed more than two times higher photocatalytic activity than pure α-Fe_2O_3 nanorings. The acquired results were explained by the presence of Pt nanoparticles that caused reduced recombination of photoinduced electron/hole pairs.

The photocatalytic activity of generally efficient Evonik Aeroxide P25 TiO_2 can also be increased by creating Fe_2O_3/TiO_2 nanocomposites (containing 1.0 wt% of iron(III) oxide), as it is described by Wodka et al. (2014). However, authors proved that Fe_2O_3 did not change the bandgap energy. Nevertheless, a cycle was noticed of reduction/oxidation of Fe ions. Therefore, the high activity of Fe_2O_3/TiO_2 composite can be explained by the so-called photo-Fenton reaction.

Photocatalytic water treatment processes are generally carried out by the application of semiconductor nanoparticles in suspension form, but this approach is time consuming (filtration and separation of the catalyst particles) and expensive (Gumy et al. 2006; Rao et al. 2004; Veréb et al. 2014; Wang et al. 2010). To solve

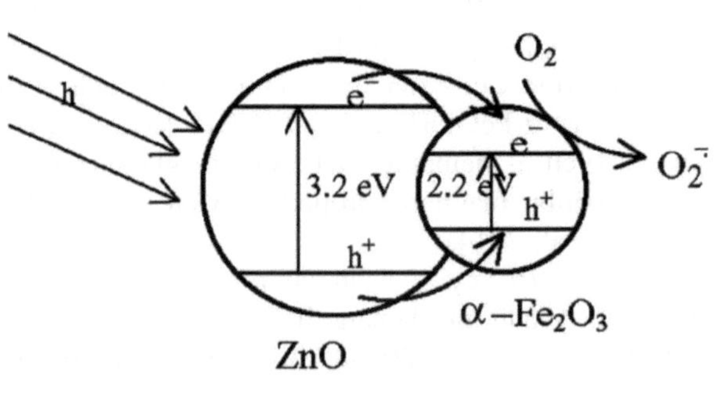

Fig. 11.5 The principle of charge separation in α-Fe_2O_3/ZnO systems in the presence of oxygen (Sakthivel et al. 2002)

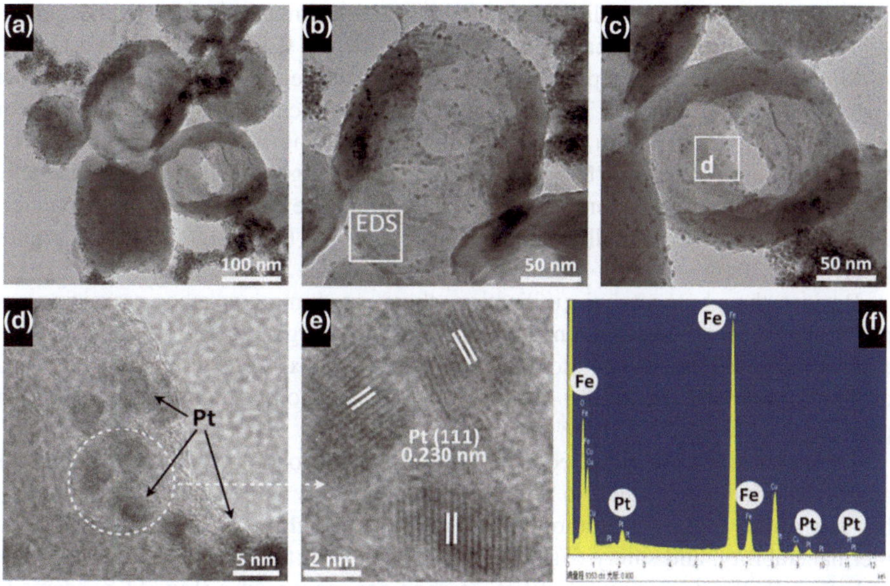

Fig. 11.6 TEM images (**a**, **b**, **c**), HRTEM images (**d**, **e**), and EDS pattern (**f**) of α-Fe_2O_3/Pt nanorings (Liang et al. 2014)

this problem, nanoparticles can be immobilized and photoactive surfaces can be applied in fixed bed flow reactors (Shan et al. 2010; Singh et al. 2013; Veréb et al. 2014); however, the disadvantage of this method is the limitation of mass transfer rate (Wang et al. 2010). Another solution to avoid the filtration is the utilization of magnetic semiconductors (Pang et al. 2016; Wang et al. 2010). In this case, the nanoparticles can be separated easily in magnetic field after the purification (Shylesh et al. 2010). Therefore, magnetic particles with both magnetic and photocatalytic properties have increasing attention in recent years (Pang et al. 2016; Wang et al. 2010). Since crystalline phases of iron oxide such as hematite (α-Fe_2O_3), maghemite (γ-Fe_2O_3), and magnetite (Fe_3O_4) have magnetic and in some cases photocatalytic properties, iron-oxide-based nanocomposites are very promising in magnetic separation combined photocatalytic water treatment technologies. Most of the published and efficient magnetic photocatalysts are nanocomposites (Christoforidis et al. 2016; Cong et al. 2014; Karunakaran et al. 2013; Niu et al. 2010; Preethi and Kanmani 2014; Shylesh et al. 2010; Wang et al. 2010, 2016a, b, c; Zhang et al. 2013) containing two or more components. These composites could offer enhanced photocatalytic activity due to the suppressed recombination rate and promoted transportation rate of photogenerated charge carriers (Cong et al. 2014; Pang et al. 2016), or even higher visible-light activity (Christoforidis et al. 2016; Zhang et al. 2013).

Furthermore, in the publication of Preethi and Kanmani (2014) novel, magnetically separable core-shell nano-photocatalysts, such as CdS/Fe_2O_3, ZnS/Fe_2O_3 and $(CdS + ZnS)/Fe_2O_3$ were prepared and their hydrogen evolution activity under visible light was examined. The developed composites can be potential catalysts for recovery of hydrogen from industrial sulfide containing waste streams. Similarly, Wang et al. (2016a, b, c) developed a highly stable paramagnetic Fe_3O_4@hydrophilic graphene (Fe_3O_4@HG) composite (synthesized by coprecipitating methods) with excellent photo-Fenton activity (Fig. 11.7) used for photocatalysis in 2016.

Besides the photocatalytic utilization, there are numerous other promising applications of iron-containing nanocomposites. Rechargeable Li-ion cells are still extensively used in energy storage; however, the next generation of anode and cathode materials (including iron-containing nanocomposites) ensures further promising advantages of these batteries.

Zhou et al. (2011) developed an α-Fe_2O_3/SnO_2 nanocomposite that was used as anode material and resulted in improved Li-ion battery performance. Pyrite (FeS_2) has lots of advantages as cathode material such as high theoretical capacity, good thermal stability, and safety (Tan et al. 2016). In order to improve the performance of FeS_2 cells the development of FeS_2/carbon nanocomposites (Fig. 11.8) seems to be very promising. Liu et al. (2014) developed FeS_2@porous C-nanooctahedra composites, which exhibit superior rate capability and stable cycling performance. The composite of FeS_2 microspheres@rGO was used as Li-ion cell by Son and coworkers, who showed that it exhibited high capacity and long life performance (Son et al. 2014).

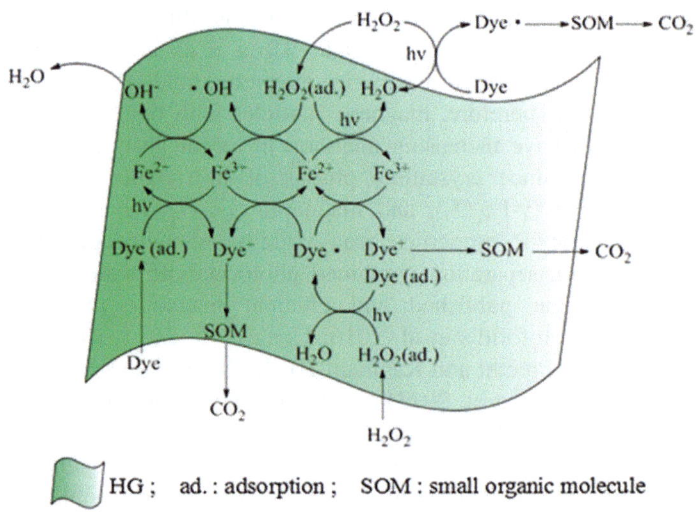

Fig. 11.7 The photo-Fenton catalytic cycle of Fe_3O_4@HG (Wang et al. 2016a, b, c)

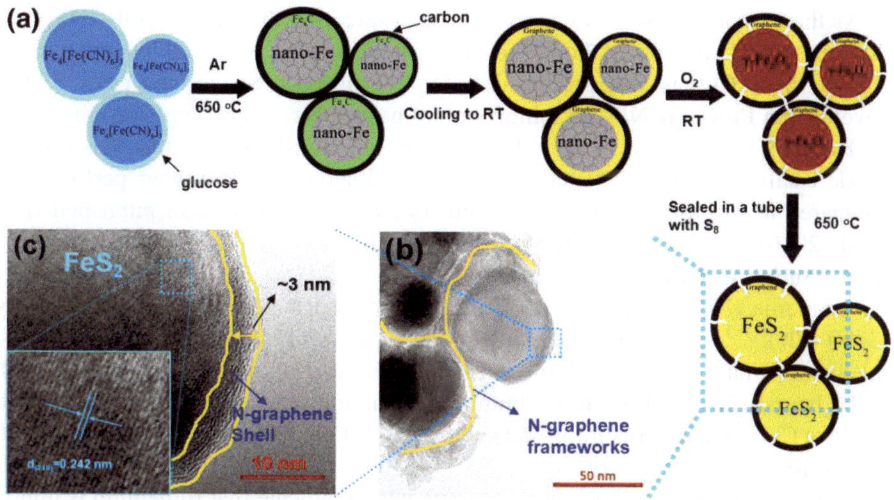

Fig. 11.8 Brief illustration of the fabrication of FeS_2@N_2-graphene particles system (**a**), its TEM image (**b**), and its high resolution TEM image (**c**)

Further promising insights on the utilization of iron-containing nanocomposites are the supercapacitors, as showed by Song et al. (2016a, b). They prepared a nanocomposite comprising Fe_2O_3 anchored on reduced graphene oxide aerogel (rGOA). The Fe_2O_3/rGOA composite exhibited excellent electrochemical performance in negative potential (due to the synergistic effect of Fe_2O_3 particles and rGOA), which makes it a perfectly promising anode material for supercapacitors. The biomedical utilization of magnetic iron-oxide-based nanocomposites are also widely investigated (Akbarzadeh et al. 2012; Boyer et al. 2010; Gupta and Gupta 2005; Sharifi et al. 2012) such as magnetic resonance imaging contrast enhancement (Haw et al. 2010), tissue repair, immunoassay, detoxification of biological fluids, hyperthermia (Laurent et al. 2011; Sharifi et al. 2012), drug delivery, cell separation, etc. Some of iron oxides are generally biocompatible [such as Fe_3O_4 and γ-Fe_2O_3 (Sharifi et al. 2012)] and can be relative easily functionalized (Boyer et al. 2010) and they can be covered by many biocompatible coatings (Boyer et al. 2010; Laurent et al. 2011; Sharifi et al. 2012).

Ruthenium-Based Nanomaterials. A One-Sided Composite Component

Since the utilization of fossil energy sources produces high amount of CO_2 that highly contributes to the global warming, and the amount of available fossil energy sources is limited, the sources of hydrogen-based clean alternative energy are receiving increasing attention.

Methanol can be used as energy source in direct-methanol fuel cells (DMFCs) (Steigerwalt et al. 2001); however, this solution does not produce truly clean energy, since CO_2 is also produced during the process. Steigerwalt et al. (2001) developed a Pt/Ru/GCNFs (graphitic carbon nanofibers) nanocomposite containing Pt–Ru alloy nanoclusters widely dispersed on the GCNF support that was used as anode catalysts in direct-methanol fuel cell and showed 50% higher performance compared to unsupported Pt–Ru colloid. Lee and coworkers also published (Lee et al. 2010) excellent catalytic performance in methanol electrooxidation using RuO_2–SnO_2 nanocomposite electrodes.

Ammonia (NH_3) is a clean (CO_x-free) compound that has high energy density (3000 $Wh \cdot kg^{-1}$) and high hydrogen storage capacity (17.7 wt%) in comparison to methanol (Zhang et al. 2006a, b), and it can be easily stored and delivered (Varisli and Elverisli 2014). Therefore, ammonia is an important alternative material to produce CO_x-free hydrogen for fuel cell uses (Li et al. 2007; Yin 2004; Zhang et al. 2006a, b).

There are several studies in the literature which indicate that ruthenium is highly efficient to produce hydrogen from ammonia (Li et al. 2007; Yin 2004; Zhang et al. 2006a, b); however, its high price and low availability means a limitation in its usage (Varisli and Elverisli 2014). Nonetheless, there are several ruthenium-based nanocomposites in the literature such as Ru/Carbon (Li et al. 2007), Ru/Al_2O_3 (Zheng et al. 2007), Ru/MgO (Zhang et al. 2006a, b), Ru/TiO_2 (Akbayrak et al. 2014), Ru/CNT (Yin 2004), Ru/SiO_2 (Varisli and Elverisli 2014), etc., which were sufficiently efficient for the catalytic generation of hydrogen.

In the recent years, ammonia borane has also increased attention as a hydrogen source (Akbayrak et al. 2014; Fan et al. 2016) because of its higher hydrogen storage capacity [19.6 wt% (Akbayrak et al. 2014)], and its good solubility and stability in water. In 2016 Fan et al. (2016) developed a Ru/nanodiamond nanocomposite that has a large quantity of oxygen containing functional groups (such as hydroxyl and carboxylic groups) on its surface and form a homogeneous suspension in aqueous phase. The above-mentioned nanocomposite exhibited high catalytic activity for ammonia borone hydrolysis, with a turnover frequency number as high as 229 mol $H_2 \cdot (\text{mol Ru min})^{-1}$.

Ru-containing nanocomposites have several other promising application opportunities, too. For example magnetic-mesoporous composite loaded with emissive Ru(II) complex was used for oxygen sensing (Li et al. 2016). Hwang and coworkers demonstrated a simple one-step process for the synthesis and processing of laser-scribed graphene/RuO_2 nanocomposites into electrodes that exhibited ultrahigh energy and power densities (Hwang et al. 2015). The created graphene/RuO_2 film was used directly as a hybrid electrochemical capacitor electrode and demonstrated much-improved cycling stability and rate capability.

Mn-Based Composites and Composite Components in Various Applications

Manganese can be used in economic production of graphene, which is a very promising material for many application fields such as nanoelectronics, hybrid materials, various batteries, and sensors, due to its huge surface area, chemical stability, and unique electronic and mechanical properties (Li et al. 2014). There are several methods to prepare graphene, such as chemical vapor deposition (Arco et al. 2010), cutting carbon nanotubes (Jiao et al. 2009), and thermal and chemical reduction of GO. Since manganese can easily react with other chemicals under acidic conditions it can be a suitable material to reduce GO to RGO (Li et al. 2014). Li et al. (2014) developed an easy and fast process using Mn powder as a reducing agent (which can be carried out in large scale) to produce MnO_2/RGO composites from GO, which can be used to fabricate electrodes for a supercapacitor. These electrodes displayed excellent electrochemical activity and stability. MnO_2/carbon nanotube composites were also used in supercapacitors with high electrochemical performance (Wang et al. 2016a, b, c). A symmetric supercapacitor assembled with the electrodes made from MnO_2/CNTs showed a capacitance decay rate of only 8.7% after 5000 cycles.

Incorporation of manganese into composites can be used to improve the efficiency of silicon solar cells as it was published by Dai Prè et al. (2013), who obtained a stable and narrowly distributed ($d \approx 3$ nm) dispersion of Mn-doped ZnS nanoparticles via chemical precipitation. The synthesized nanocomposites showed photoluminescence in the visible region when excited with UV light, and this effect was used to improve the silicon solar-cell efficiency.

Carbon dioxide emission (caused by intensive utilization of fossil energy sources) progressively increases resulting obvious effects on the global climate. Since the total elimination of fossil energy utilization cannot be solved, the stabilization of CO_2 level in the atmosphere is one of the greatest challenge for nowadays (Al-Dossary et al. 2015; Jiang et al. 2010) and recycling CO_2 to produce renewable hydrocarbons is one of the most interesting catalytic alternative options. A simple conversion route is the Fischer–Tropsch synthesis (FTS), which provides clean synthetic fuels when starting with CO_2-rich feeds (Al-Dossary et al. 2015; Chew et al. 2014). Iron-based catalysts are widely used in FTS, while manganese has been widely used as a promoter of iron catalysts used in FTS (particularly for producing low olefins) (Al-Dossary et al. 2015; Dorner et al. 2010). Al-Dossary et al. (2015) developed mesoporous MnFeO nanocomposites in 2015 that were used as catalysts in CO_2 hydrogenation via Fischer–Tropsch reactions for the production of valuable hydrocarbons. Authors concluded that Mn loading causes strong influence on CO_2 conversion and selectivity to different products. The 0.05 MnFe catalyst displayed the best performance with reduced CO and CH_4 formation and improved selectivity to C_2–C_5 and C_{6+} hydrocarbons.

Another interesting utilization of manganese in composites is *E. coli* sensing (Abdullah et al. 2016) based on polyaniline (PANI) with Ag–Mn nanocomposite thin films prepared by a simple sol–gel method. Authors found that the sample PANI-$Ag_{0.8}$-$Mn_{0.2}$, which contains a higher concentration of Ag than Mn, exhibited the highest sensitivity on *E. coli* detection.

Manganese can also be used in the preparation of magnetic nanoparticles, which are widely used in many application fields as it was previously discussed. For example, Nikolic et al. (2014) developed in 2014 a Magnetite/Mn-ferrite nanocomposite with improved magnetic properties.

Rhenium—Rare Material Composite

Since rhenium has some advantageous properties such as refractoriness, high melting point, mechanical and temperature strength, and chemical inertness, rhenium-containing nanoparticles can be utilized in electronics and electrical engineering (Yurkov et al. 2012), while composites with photocatalytic activity are currently missing.

Yue et al. (2015) created rhenium coating on fabric-based pierced Carbon/Carbon composite substrate using chemical vapor deposition and authors concluded that rhenium coating had good thermal shock resistance. Iridium is also a promising coating on C/C composites for high temperature applications. However, the poor adhesion and thermal stress induced by thermal expansion mismatch between Ir coating and C/C substrate restrict their application. Zhu et al. (2013) prepared rhenium coating on the C/C substrate by chemical vapor deposition (CVD) resulting an interlayer material between the Ir coating and C/C substrate, improving thus the adhesion of Ir coating and relieving the thermal stress.

Zn Containing Materials. Composites with ZnO—The Shape Tailored, UV Active Material, with a Vast Application Spectrum

As it is known ZnO is one of the vastly applied semiconductor in many applications. Therefore, the discussion from the present chapter will not be focused on it. Instead, the attention will be drawn to more rarely discussed ZnO-based composites.

Ag nanoparticles can be applied in order to improve the photocatalytic efficiency of ZnO. In this work, the dispersion and the immobilization of Ag/ZnO nanoparticles were performed in a PAA matrix as shown in Fig. 11.9. However, because of their hydrophobic surface and high surface energy, these nanoparticles, were firstly treated with a silane coupling agent, (3-Glycidyloxypropyl)trimethoxysilane

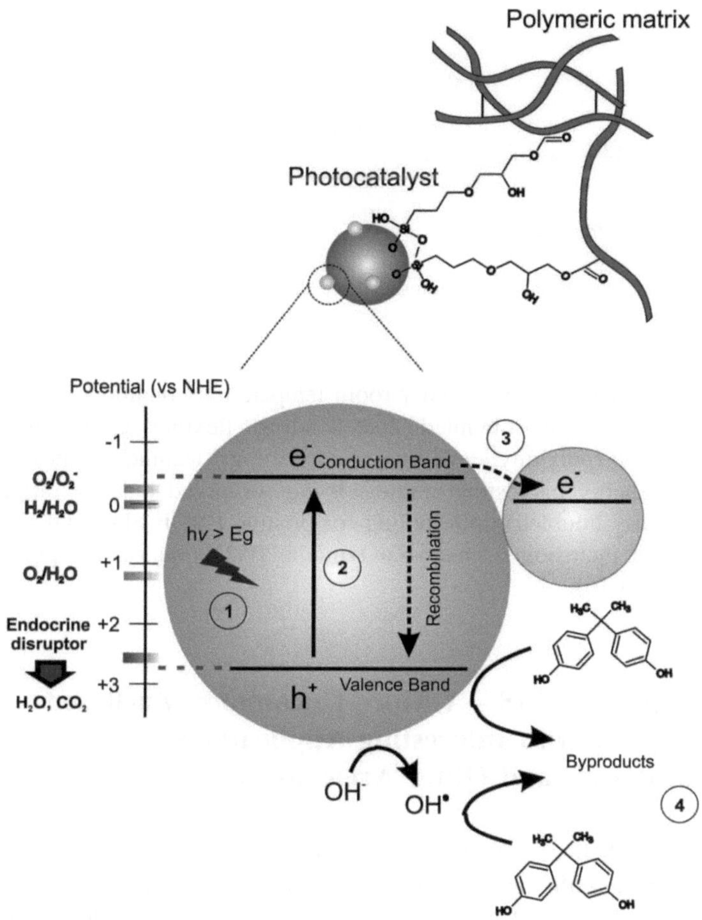

Fig. 11.9 Scheme of the mechanism proposed for the photocatalytic degradation of BPA using Ag/ZnO–PAA composites (Jasso-Salcedo et al. 2016)

(GLYMO) to limit their aggregation and to introduce organic functional groups. An esterification reaction was proposed between the opened ring of the epoxy group of Ag/ZnO-GLYMO and the carboxylic group of PAA (Jasso-Salcedo et al. 2016).

ZnO/CuO nanocomposites were also immobilized on γ-Al_2O_3 by heterogeneous precipitation method. This material was efficient in the removal of MO–methyl orange (Hassanzadeh-Tabrizi et al. 2016). ZnO was likewise immobilized/deposited on GO (Fig. 11.10), forming ZnO/GO nanocomposite, which consisted of flower-like ZnO nanoparticles on GO sheets, and this material also proved to be efficient (Li et al. 2012).

Among the interesting future applications one can be found those in which ZnO is applied as oxygen gas sensors in ZnO/CdS and ZnO/ CdS-EDTA nanostructured thin films by the sol–gel spin-coating method. Moreover, good stability and

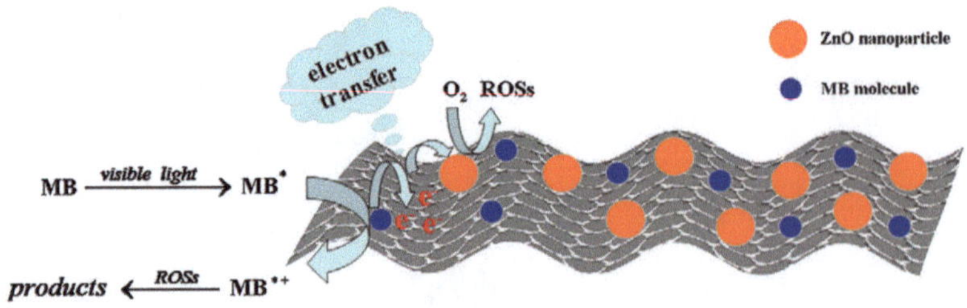

Fig. 11.10 Possible mechanism of photosensitized degradation of dye over ZnO/GO nanocomposite under visible light (Li et al. 2012)

reproducibility were achieved under room temperature conditions (Arunraja et al. 2016). Bhunia et al. obtained free-standing flexible composite films of nano-ZnO/PVDF (polyvinylidene fluoride) by sol–gel technique. These films were subjected to energy harvesting studies. It was observed that the output voltage increased (nearly doubled) upon poling for pristine PVDF films and increased 10 times with the inclusion of nano-ZnO in the PVDF matrix for poled samples (Bhunia et al. 2016).

Composites with ZnS—A More Uncommon Zn-Based Semiconductor with Interesting Applications in Photocatalysis and Other Areas as Well

ZnS/Ni_2P core/shell composites were obtained using a hydrothermal approach. Compared with ZnS microspheres, ZnS/Ni_2P core/shell composites showed enhanced photocatalytic degradation activity for pyronine B under UV irradiation. This may be related to the effective separation of photogenerated electron-hole pairs in ZnS/Ni_2P composites, which can greatly reduce the chance of their recombination. Furthermore, superoxide anions and hydroxyl radicals can be more easily produced through ZnS/Ni_2P composites, which was also beneficial for the degradation of pyronine B (Liu et al. 2016a, b).

Xiong et al. synthetized Bi_2S_3/ZnS nanoplates, which showed a mesoporous structure with high specific surface area of 101.3 $m^2\ g^{-1}$ and exhibited high adsorption capability and photocatalytic activity for MB degradation under UV light irradiation. A tentative mechanism for degradation of MB over Bi_2S_3/ZnS was proposed involving $^{\cdot}$OH radical and photoinduced holes as the active species, which was confirmed by using methanol or ammonium oxalate as scavengers (Xiong et al. 2016). ZnS was deposited on reduced graphene oxide also (ZnS–rGO) via microwave-assisted crystallization. The photocatalytic activity of the prepared ZnS–rGO nanocomposite was examined by the degradation of two model dyes: Methylene Blue and Rhodamine B (RhB). The designed ZnS–rGO nanocomposite

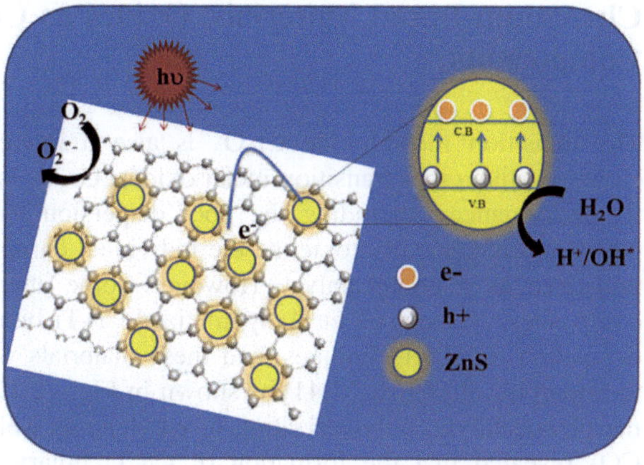

Fig. 11.11 Schematic representation of photocatalysis process of ZnS–rGO nanohybrid (Thangavel et al. 2016)

showed a superior photocatalytic activity with 1.47- and 2.92- fold higher reaction rates for MB and RhB degradation, respectively, than that of the pure ZnS nanoparticles. A plausible mechanism for the enhanced properties of ZnS–rGO nanohybrid was discussed using photoluminescence spectra (Thangavel et al. 2016).

The superior photocatalytic performance of the ZnS–rGO nanohybrid can be explained as follows. The formation of electron-hole pairs at the semiconductor's surface upon light irradiation leads to a series of chain reaction mediated via reactive oxygen species ($^{\cdot}OH$ and O_2^{-}) generation which further results in the photodegradation of dye or pollutants (Fig. 11.11). Hence, the separation of charge carrier recombination can increase the photocatalytic efficiency. The role of rGO in the ZnS–rGO nanohybrid can effectively decrease the recombination rate of photogenerated electron-hole pairs formed at the ZnS surface, due to the metallic nature of rGO (act as electron acceptor).

ZnS can be applied as sensor as shown by Yasushi et al., who fabricated a thick-film NO_2 sensor constructed from Zn–Sn–Sb–O composite materials using a screen-printing method and further studied its characteristics. It was found that the sensor resistance and the NO_2 detection characteristics largely depend on the composition of the gas-sensing materials. The sensor showed a high sensitivity and excellent selectivity for ppm levels of NO_2 gas, while it had no sensitivity toward i-C_4H_{10} and CO gases at all (Yasushi et al. 2003). Nidhi et al. created an enzyme sensor using immobilized ZnS nanocomposites. They used acetylthiocholine chloride (ATCl) as a substrate, while ZnS promoted electron transfer reactions at a lower potential and catalyzed electrochemical oxidation of enzymatically formed thiocholine, thus increasing detection sensitivity.

Chromium-Based Materials. Oxides as Composite Components

The transition metal oxide Cr_2O_3 is also a low-cost supercapacitive material. However, like other transition metal oxides, Cr_2O_3 has some problems that restrict its performances, including the poor dispersion and conductivity. Furthermore, chromium is generally toxic, so the individual chromium oxide is for the moment undesired as a photocatalyst. However, some attempts were made with Cr_2O_3 composites. Kamegawa et al. synthetized Cr–Ti binary oxides by chemical vapor deposition (CVD) techniques, and these materials were successfully anchored on mesoporous silica (MCM-41) as shown in Fig. 11.12. The comparative studies on the photocatalytic polymerization of ethylene as well as the oxidation of CO into CO_2 revealed that the formation of Cr–Ti binary oxide with direct interaction between Ti^{4+} and Cr^{6+} oxide species were effective for the enhancement of catalytic activities in these reactions through the changes in the reactivity of tetrahedral Cr^{6+} oxide species (Kamegawa et al. 2010).

Kim et al. prepared efficient visible-light-active photocatalysts of a porous CrO_x–$Ti_{1.83}O_4$ nanohybrid by the hybridization of exfoliated titanate and narrow-bandgap chromia species. The photocatalytic activity of this nanohybrid was examined by monitoring the decomposition of organic pollutants (acid orange 7 and phenol) under both visible or UV-vis radiation.

Other applications of chromium oxides were also investigated by Song et al. who prepared Cr_2O_3/rGO composites. GO can be reduced by Cr powder, and this special method of introducing rGO leads to a well-dispersed Cr_2O_3 decorated rGO. Meanwhile, the introduction of rGO not only increased the surface area, but also can improve the conductivity of the hybrids. Therefore, the Cr_2O_3/rGO hybrids show more excellent electrochemical performances than the pure Cr_2O_3 (Song et al. 2016a, b). Graphitic carbon containing Cr_2O_3 nanoribbons were also used as support material for Ni–Pd electrocatalyst in direct-methanol fuel cell application. The synthesized Ni–Pd nanoparticles were loaded on the support surface via sonication. Methanol electrooxidation reactions were evaluated through cyclic voltammetry and it was found that these materials were very efficient (Khan et al. 2016).

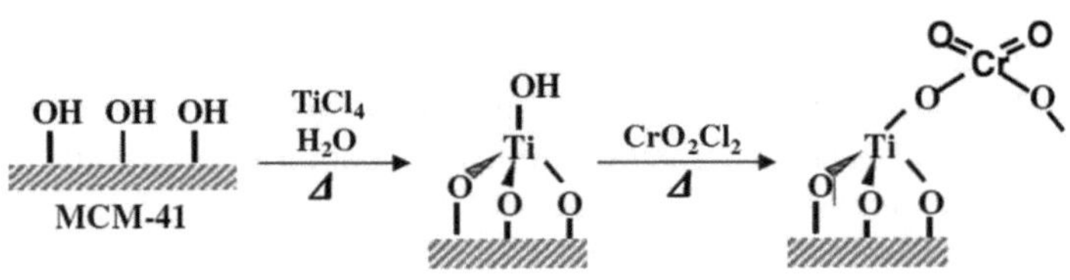

Fig. 11.12 The schematic diagram of the procedures for preparation of Cr–Ti/MCM-41 (Kamegawa et al. 2010)

Molybdenum-Based Nanomaterials. MoO_2—Applicability in Composites, Application Outlook

To date, various MoO_2 nanostructures, including nanowires, nanospheres, nanoparticles, nanorods, and nanocrystals have been successfully synthesized by different synthetic approaches, such as electrospinning, solid reduction reaction, hydrothermal reaction, and solvothermal route (Liu et al. 2004; Luo et al. 2011; Shi et al. 2009). However, previously reported MoO_2 nanostructures were rarely studied to be supercapacitor electrodes and photocatalysts, though some researchers have studied the superior lithium storage for MoO_2 nanosheets and the oxidation of gasoline. Therefore, it is urgent to explore new MoO_2 nanostructures and extend their potential applications in energy storage and environmental issues (Koziej et al. 2011).

Aravind et al. dispersed MoO_2 over graphene layers by a facile and environment friendly route involving co-reduction of metal salt and graphite oxide in the presence of sunlight. These nanocomposites were effective in hydrogen evolution reactions (Aravind et al. 2014). Molybdenum dioxide nanoparticles with the size of 200 nm were obtained and directly functionalized and then used as supercapacitor (SC) electrodes and photocatalysts. They exhibited good cyclic performance with 90% capacity retention after 1000 cycles. The photocatalytic activities were evaluated by the degradation of Methylene blue and Rhodamine B, respectively, and the nanoparticles demonstrated preferred selectivity on the degradation of RhB (70%) in contrast to that of MB (30%) (Zhou et al. 2016).

MoO_2 nanoparticles can be used in other interesting applications as well. Wu et al. obtained outstanding results for MoO_2/CNTs nanocomposites applied as a supercapacitor electrode. Each component in the composite has a crucial influence on the electrochemical properties of the material. MoO_2 nanoparticles can provide higher specific capacitance, while CNTs improve the electrical conductivity and structural stability of the composite (Wu et al. 2016). Yuan et al. developed MoO_2@TiO_2@CNT nanocomposites as presented in Fig. 11.13. The electrochemical test results showed that the unique MoO_2@TiO_2@CNT demonstrated superior performance in specific capacity and cycling stability, compared with that of the TiO_2@CNT material. It is anticipated that the MoO_2@TiO_2@CNT

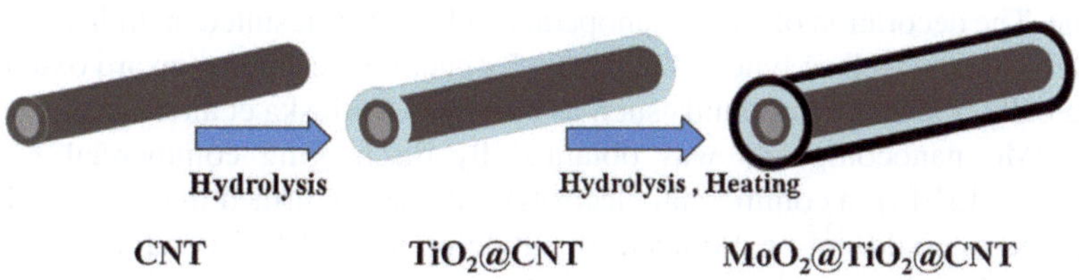

Fig. 11.13 Schematic illustration of fabrication processes of sandwich structured MoO_2@TiO_2@CNT nanocomposites

nanocomposite will have a promising potential for application in stable and high-power lithium-ion batteries (Yuan et al. 2015).

The CNT and highly conductive MoO_2 under/on the TiO_2 layer are served as flexible and strong electronic paths for rapid electron and ion transport. The resulting MoO_2@TiO_2@CNT hybrid structures show improved specific capacity and cycling stability compared with TiO_2@CNT.

MoS_2—Applicability in Composites, Application Outlook

Recently, more and more researchers are paying attention to the layered MoS_2 structures, which shows a similar structure to graphene. Each layer of the MoS_2 nanosheets consists molybdenum atoms sandwiched between two layers of hexagonally close-packed sulfur atoms, that the adjacent atomic sandwiches are held together weakly by van der Waals forces. The photocatalytic and field emission abilities of MoS_2 are not productive enough for large-scale applications in industry because of the relatively narrow band gap (1.8 eV), a rapid recombination rate of photogenerated electrons and holes and lacking of effective emission sites (Hu et al. 2010). However, lots of previous studies reveal the fact that forming composites this barrier can be raised (Bai et al. 2015; Hu et al. 2012).

As a composite photocatalyst component it was applied by Hu et al. who prepared nano-MoS_2/kaolin nanostructures. The obtained nano-MoS_2/kaolin composite showed a good absorption in the visible-light region, which led to high catalytic activity in MO degradation. The composite catalyst may be regenerated by filtration and reactivated by heating in H_2. This composite is one of the promising photocatalytic materials for the removal of organic chemicals from wastewater, such as organic dyes and phenols (Hu et al. 2010). The same group obtained nano-MoS_2/bentonite composite by depositing nano-MoS_2 on the acidified surface of bentonite. The as-prepared composite demonstrated an excellent performance for the removal of the above-mentioned organic dyes (Hu et al. 2012).

Dolinska et al. pointed out that MoS_2 can be applied in other fields as well. They prepared AuNPs/MoS_2 nanopetals on ITO electrode surface. These structures were stable due to the electrostatic interactions between negative surface charge of MoS_2 nanopetals and positively charged functionalities of Au nanoparticles. This approach produces materials with well-developed surface where both components face external volume. The decoration of MoS_2 nanopetals with AuNPs resulted in higher electronic conductivities as well as synergistic effects for catalytic reactions toward oxidation of biologically relevant compounds such as cysteine (Dolinska et al. 2015).

Li_2S/Mo nanocomposite was obtained by discharging commercial MoS_2 to 0.01 V (vs. Li^+/Li) in commercial electrolyte. It was confirmed that the metallic Mo nanoparticles could not be oxidized to sulfide up to 3.00 V. Therefore, Li_2S and S were the sole redox couple in the deeply discharged MoS_2/Li cell, $S + 2Li \leftrightarrow Li_2S$. These results confirmed the feasibility of using Li_2S/metal and S/metal nanocomposites as alternatives to sulfur cathode in batteries (Fang et al. 2012).

W-Based Semiconductors Are a Large Family of Semiconductors for Diverse Composites with High Degree of Applicability. WO_3 Is the Multitasking Semiconductor

WO_3, a semiconductor with a bandgap between 2.4 and 2.8 eV, is a visible-light-responsive catalyst with stability in acidic conditions, which makes it a suitable choice for the photocatalytic degradation of organic pollutants under solar irradiation. It has been reported that the particle morphology, which includes the shape and size of a particular photocatalyst, significantly affects its photocatalytic activity. WO_3 has found useful applications also in semiconductor gas devices, electrochromic devices and, as shown above, photocatalysis (Song et al. 2007). Cui et al. obtained for the first time IR light-driven photocatalysts based on WO_2–NaxWO$_3$ ($x > 0.25$) that exhibited a band structure as presented below (Fig. 11.14), and was directly applicable in sea water (Cui et al. 2015).

Hu et al. (2016) prepared metal/WO_3 composite nanostructures, which involved a procedure of charging electrons into WO_3 and a follow-up procedure of using these electrons to deposit metals on the surface of WO_3. Flower-like CuS and WO_3 nanorods were used to obtain visible-light-active CuS–WO_3 composite photocatalyst. The experimental results demonstrated that the photocatalytic degradation of Methylene Blue over CuS-added WO_3 was much higher than that of either pure CuS or WO_3. The main reason of the high photocatalytic activity is that the CB potential of CuS is more negative than that of WO_3. Thus, the photoinduced electron transfer from CB of CuS to WO_3 and at the same time there is a transfer of the hole from the VB of WO_3 to CuS. Therefore, the electron-hole recombination is reduced and it leads to an increase in the interfacial charge-transfer reactions for the degradation of adsorbed dye molecules (Theerthagiri et al. 2015).

Pt/WO_3 nanocomposites were deposited on the surface of the zeolites. These hybrid materials showed photocatalytic activity for acetaldehyde and trichloroethylene removal under UV and Vis light. Hybridation of the Pt/WO_3 with ordered microporous materials resulted in the promotion of the photocatalytic activity of bare Pt/WO_3. In this system, the zeolites can help the adsorption process

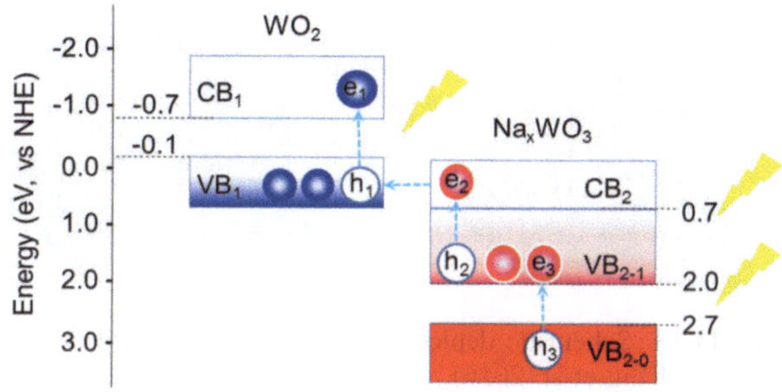

Fig. 11.14 Energy-level diagrams of WO_2–Na_xWO_3 (Cui et al. 2015)

of the model pollutant. Furthermore, the adsorbed species can migrate to the active site where the photocatalytic reaction takes place (Jansson et al. 2016).

Karácsonyi et al. obtained titania/tungsten (VI) oxide/noble metal (gold and platinum) nanocomposites by a selective photodeposition method, which provided composites with gold or platinum deposited either on the TiO_2 or WO_3 surface. The controlled design of the nanoarchitectures by adjusting the Au and Pt particles' localization proved to be critical in many aspects: The light absorption properties were changing if the localization of Au and Pt changed. If the noble metals were deposited onto the TiO_2 surface, the bandgap was not influenced significantly by the WO_3 content, while in the case of depositing on the WO_3 surface a fine decreasing trend of the bandgap energy was observed. The photocatalytic activity of the nanocomposites also changed with the localization of Au and Pt nanoparticles (Karácsonyi et al. 2013).

WO_3-based composites were also important in many other applications, such as sensors. Liu et al. fabricated hybrid sensor of Pt nanoparticles (NPs) functionalized WO_3 nanorods by hydrothermal synthesis of WO_3 nanorods followed by the Pt NPs decoration. Gas-sensing tests demonstrated that the Pt–WO_3 sensor exhibited excellent sensing performance with high response and fast response–recovery speed as well (Liu et al. 2011). Zhang et al. obtained cactus-like SiNWs/WO_3 nanowire composite structures (Fig. 11.15) for NO_2 gas sensors. The optimal SiNWs/WO_3 nanowires composite sensor exhibits good responses to various concentrations of NO_2 gas at room temperature (Zhang et al. 2016).

One conceivable mechanism that could explain this behavior is that the high-density p-n heterointerfaces of the sensor surface provided more active adsorption sites. Since NO_2 is an electron withdrawing oxidizing gas, NO_2

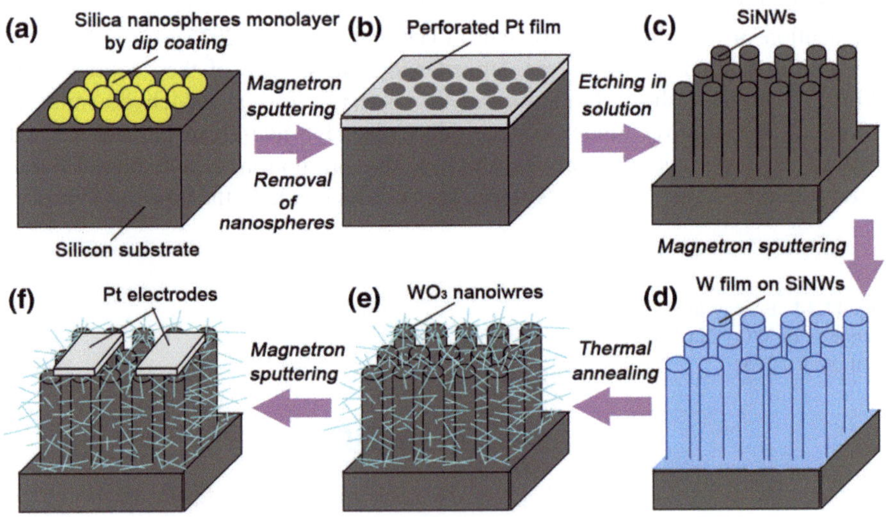

Fig. 11.15 Schematic depiction of the fabrication process of SiNWs/WO_3 nanowires composite sensor (Zhang et al. 2016)

molecules can effectively adsorb and capture the electrons from surface of sensor. It is reported that oxygen species on the WO_3 surface, with which NO_2 molecules could preferably interact, are not active at low temperature. When oxidizing NO_2 gas comes in contact with the sensor surface, NO_2 gas molecules are adsorbed on the surface of sensor. The unpaired electron of the NO_2 molecule reacts with the dangling bonds available on surface of SiNWs, trapping the lone-pair electrons of the dangling bond and forming $(NO_2)^-$, which results in an increase in concentration of holes in the valence band (VB) of SiNWs. The effect can be generally understood as NO_2 withdrawing electrons from silicon and thus causing holes' accumulation.

Bi_2WO_6—A Special Mixed Oxide in Composites

Bi_2WO_6 is one of the simplest Aurivillius oxides, which possess layered structure. Due to nontoxicity, strong oxidizing power, and visible-light responsiveness, Bi_2WO_6 exhibits excellent photocatalytic properties for the decomposition of a great variety of organic pollutants in environmental purification applications (Maczka et al. 2008; Zhang et al. 2010).

Zou et al. have developed a novel electrostatically driven hydrothermal approach to synthesize a heterostructured composite of few layered WS_2-Bi_2WO_6/$Bi_{3.84}W_{0.16}O_{6.24}$ with an enhanced photocatalytic performance and good stability (Zou et al. 2015). Silicon-modified GO/Bi_2WO_6 (Si-GO/BWO) nanoplates were obtained by a facile one-pot hydrothermal method. The photocatalytic experimental results revealed that Si-GO/BWO nanoplates showed high efficiency to degrade high-concentration RhB solution (Xiao et al. 2015). $BiVO_4$/Bi_2WO_6 composite photocatalysts were also obtained by a simple solvothermal synthesis without using any surfactant or template. The morphology formation and photocatalytic activity of the composites were explored as a function of the molar percentage of $BiVO_4$. The results showed that the $BiVO_4$/Bi_2WO_6 composite with a 50% molar percentage of $BiVO_4$ exhibited an excellent photocatalytic degradation of RhB—91% within 60 min (Xue et al. 2015).

In this system, the two components have different function in the photocatalysis. A possible photocatalytic mechanism was proposed for the photodegradation of the used RhB over the composites. The formation of the $BiVO_4$/Bi_2WO_6 heterojunctions can enhance the separation and transport of the photogenerated electron-hole pairs near the heterojunction interfaces. The proposed photodegradation mechanism can be simply summarized as follows (Xue et al. 2015):

$$BiVO_4 + hv \rightarrow BiVO_4(h^+ + e^-)$$

$$BiVO_4(e^-) + Bi_2WO_6 \rightarrow Bi_2WO_6(e^-) + BiVO_4$$

$$Bi_2WO_6(e^-) + O_2 \rightarrow Bi_2WO_6 + {}^{\cdot}O_2^-$$

$$H^+ + {}^{\cdot}O_2^- \rightarrow HOO\cdot$$

$$2\,HOO\cdot \rightarrow 2\cdot OH + O_2$$

$$h^+/{}^{\cdot}OH + RhB \rightarrow CO_2 + H_2O$$

Spherical Bi_2WO_6 nanoarchitectures (2–4 μm) were prepared by a hydrothermal reaction by Wang et al. Ag was deposited on the surface of Bi_2WO_6 via a facile photoreduction process. The results revealed that the monodispersed metallic Ag nanoparticles with average size of 10–15 nm were deposited on the surface of the spherical Bi_2WO_6 nanoarchitectures. The photocatalytic activities of the samples were evaluated by the photodegradation of RhB and thiophene under visible-light irradiation. The results showed that Ag loading greatly improved the photocatalytic activity of Bi_2WO_6. While the Ag-loaded Bi_2WO_6 photocatalyst is being irradiated, CB electrons of Bi_2WO_6 easily flow into the metal Ag through the Schottky barrier because the CB of Bi_2WO_6 is higher than that of the loaded metal Ag. The process of electron transfer is faster than the electron-hole recombination between the valance band and the CB of Bi_2WO_6. Thus, plenty of electrons can be stored in the Ag component (Wang et al. 2012).

Although photocatalytic applications are slightly dominant, some other interesting perspectives can be observed for this material and their composites. Liu et al. obtained three-dimensional flower-like Bi_2WO_6 microspheres by a simple hydrothermal method and subsequently these were used to fabricate a mediator-free biosensor for the detection of H_2O_2. Due to unique morphology of the flower-like microspheres, the biosensor displayed a detection limit as low as 0.18 μM H_2O_2 and a wide linear range from 0.5 to 250 μM for H_2O_2 (Liu et al. 2016a, b). Bi_2WO_6/graphene composites were obtained by sol–gel process. The material (Zhang et al. 2015a, b) exhibited high capacity, excellent cycle performance when used as anode material for batteries. The superior electrochemical performance was due to graphene performing as a mechanical buffer and a conductive network in the composite.

WS_2—Properties in Composite Materials

Tungsten disulfide (WS_2) is an emerging transition metal dichalcogenide (TMD) that has received great attention in various research fields (Huang et al. 2014). The optical properties of exfoliated WS_2 can be modulated by varying its thickness eventually leading to a transition from indirect bandgap in bulk WS_2 to direct bandgap in monolayer WS_2. Monolayer WS_2 is able to emit photoluminescence (PL) and also strongly absorbs in the visible region of the electromagnetic spectrum. However, to our best knowledge, there has been no report thus far

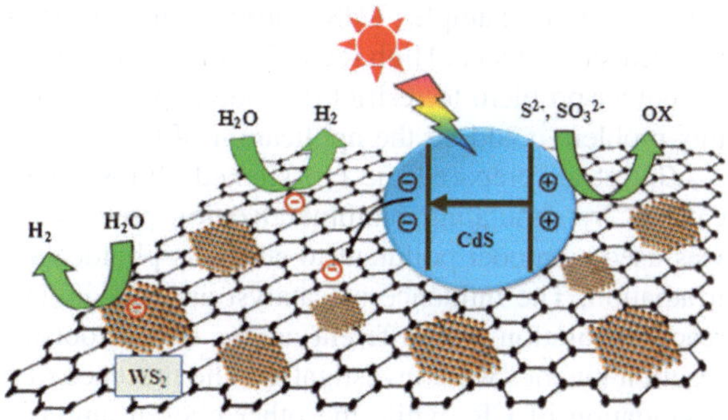

Fig. 11.16 A schematic illustration of the tentative mechanism proposed for the high H_2-production activity of the ternary CWG composite

regarding the chemical transformation of organic compounds photocatalyzed by WS_2 monolayer. A few reports showed the photocatalytic transformation of amines to imines by several types of photosensitizers (Hanbicki et al. 2015; Quan et al. 2016). Although, Cao et al. synthetized nanosized tungsten disulfide (WS_2) sensitized titanium dioxide (TiO_2) by a simple yet facile hydrothermal process. The nanocomposite exhibited a wide and intensive absorption in the visible-light region of 400–700 nm, and may have a potential application as a visible photocatalyst (Cao et al. 2015).

CdS/WS_2/graphene (CWG) nanocomposite photocatalysts were also obtained (Fig. 11.16) with different amount of layered WS_2/graphene hybrid (WG) cocatalyst. The characterization results demonstrated that the hierarchical layered WG offers an excellent supporting matrix for CdS nanorods and makes a strong interaction between WG and CdS nanorods. This study demonstrates for the first time that a noble-metal-free hierarchical layered WG nanosheets hybrid can be used as an effective cocatalyst for photocatalytic water splitting (Xiang et al. 2016).

The composite samples display stronger broad background absorption in the visible-light region with increasing WG content, which can be attributed to the optical absorption of the black WG. In addition, the CWG composites show almost the same absorption edge as that of pure CdS, implying that WS_2 and graphene were not incorporated to the lattice of CdS, and the layered WG was only a substrate for deposition of CdS nanorods.

Cd-Based Nanomaterials. Special Case of CdS and Its Composites

Due to its suitable bandgap (2.4 eV) corresponding with the spectrum of sunlight CdS has been extensively studied. It has been applied as biological sensors, solar cells, field-effect transistors (FET), environment purification, and hydrogen evolution (Arunraja et al. 2016; Reutergådh and Iangphasuk 1997; Xiang et al. 2016). Nevertheless, there are still several issues that limit the utilization of pure CdS particles. For example, CdS particles are prone to aggregation, resulting in a reduced surface area. High recombination rate of photogenerated electron-hole pairs is another problem to restrict its wide application (Ran et al. 2011). A solution for this problem could be the application of CdS in composites.

Zhu et al. prepared the crosslinked chitosan/nano-CdS (CS/n-CdS) composite catalyst by simulating biomineralization process. An azo dye, Congo Red (CR), was used as model pollutant to study its photocatalytic activity under visible-light irradiation. The influence of catalyst amount, initial CR concentrations, pH of the reaction solution and different anions on CR decolorization, and also degradation reaction kinetics were investigated. The presence of NO_3^- accelerated evidently the degradation of CR, while the other chosen anions (Br^-, SO_4^{2-}, and Cl^-) had an inhibitory effect on the decolorization process. Recycling experiments confirmed the relative stability of the catalyst.

In another study, a near infrared photocatalyst ($NaYF_4$:Yb,Tm) and a low energy bandgap semiconductor CdS was combined (Fig. 11.17). Energy transfer from $NaYF_4$:Yb,Tm to CdS was confirmed by the upconversion and fluorescence decay properties. Hydroxyl radicals were generated upon NIR irradiation on the composite material and the degradation of RhB and MB was carried out successfully (Li et al. 2010).

The mechanism for the NIR-driven photocatalysis is shown in Fig. 11.17. The absorption of pump photons populates the $^2F_{5/2}$ level in Yb^{3+}. A Tm^{3+} ion is excited to the $^3H_5$ level by the energy transferred from the excited Yb^{3+}, and then relaxes non-radiatively to the $^3F_4$ level. Energy transfer from another Yb^{3+} ion to the Tm^{3+} causes the formation of a $^3F_2$ level. Subsequently, the Tm^{3+} relaxes to the $^3H_4$ level and then is excited to the $^1G_4$ level by absorbing energy from another excited Yb^{3+}. The sequential energy absorption from two excited Yb^{3+} ions promotes Tm^{3+} to $^1D_2$ and $^3P_2$. Then the excited Tm^{3+} ions fall to lower energy levels. $^1D_2 \rightarrow {}^3F_4$, $^1G_4 \rightarrow {}^3H_6$, $^3P_0 \rightarrow {}^3F_4$, $^1D_2 \rightarrow {}^3H_6$, and $^1G_4 \rightarrow 3F_4$ transitions produce the blue emissions at 450 nm, 470 nm, UV emissions at 350 nm, 361 nm, and red emission at 645 nm, respectively. Here the $^1G_4$ level is the predominant excited state in this material as suggested by the strongest blue emission. For CdS, the energy gap from

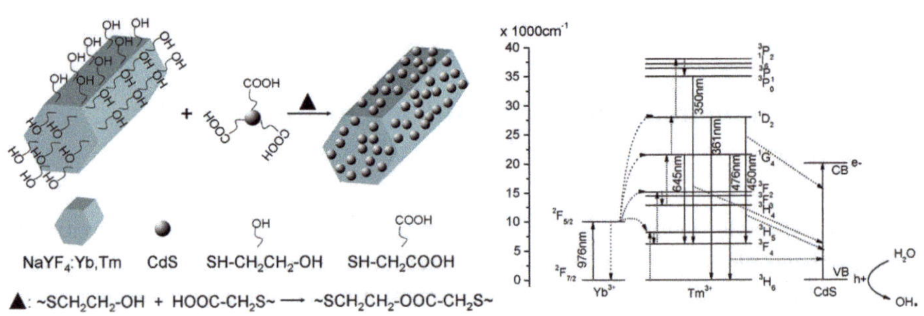

Fig. 11.17 Strategy for the preparation of $NaYF_4$:Yb, Tm/CdS composite and schematic illustration of the energy transfer mechanism (Li et al. 2010)

VB to CB is about 2.5 eV, which is lower than the emitted blue and UV photon energy. As it is well known, semiconductors could be excited by the photons with an energy equivalent to or higher than the band gap energy, which causes the formation of photoelectrons and holes. According to the SEM micrographs, $NaYF_4$:Yb,Tm and CdS are very close to each other. These allow energy transfer from $NaYF_4$:Yb,Tm to CdS. Then the excited CdS triggers photocatalysis (Li et al. 2010).

Gao et al. created Graphene–CdS (G–CdS) composites by a simple solvothermal method. The formed CdS nanospheres were homogeneously scattered on the surface of graphene sheets. Fluorescence quenching effect of the G–CdS composites indicated effective transfer of photoexcited electrons from CdS to graphene, suppressed the recombination of photogenerated electron-hole pairs, so that the enhanced visible-light-induced photodegradation activity for RhB was achieved (Gao et al. 2012).

As alternative applications for oxygen gas sensors based on ZnO/CdS and ZnO/CdS-EDTA nanostructured thin films were developed by sol–gel spin-coating method. ZnO/CdS and ZnO/CdS-EDTA thin films were obtained with optical band gaps of 2.61 and 2.78 eV, respectively. Photoluminescence spectra of the films exhibited blue and red band emissions. The FESEM of the thin films revealed randomly oriented grains in the range of nanometer size, which is very advantageous for gas-sensing applications. The sensor consists of ZnO/CdS-EDTA thin film has a very high variation in sensor response with an increase in oxygen gas concentration (Arunraja et al. 2016).

Sb-Based Materials, Sb_2S_3—A Sparsely Investigated Semiconductor Photocatalyst

In recent years, investigations on binary chalcogenides of group 15 elements of type, A_2B_3 (A = As, Sb, Bi; B = S, Se, Te) have been paid great attention by scientific researchers owning to their distinctive physical and chemical properties (Pawar et al. 1983; Savadogo and Mandal 1992). Some of these metallic sulfides with excellent optical and electrical properties can be potentially applied in photovoltaic and thermoelectric devices in the near future (Arivuoli et al. 1988). Among these binary sulfides, antimony trisulfide (Sb_2S_3) is of particular significance because of its potential technological applications, such as television cameras, solar cells, microwave devices, switching sensors, and thermoelectric and optoelectronic devices (Rajpure and Bhosale 2000; Wu et al. 2013a, b). The bandgap of Sb_2S_3 is reported at around 1.6 eV (Han et al. 2011a, b). As a result of the narrow bandgap, Sb_2S_3 is considered as a promising candidate for solar cells, semiconductor sensors, and photovoltaic devices due to that it can widely absorb the visible and near infrared radiation range of the solar energy. In addition, semiconductor binary sulfide nanomaterials with various morphologies, which can significantly improve specific surface area and possess a suitable bandgap, are quite possible to be applied in photocatalysis, especially in visible-light-driven photodegradation (Wu et al. 2011; Xie et al. 2013). Nevertheless, photocatalysis based on Sb_2S_3 nanostructures have been seldom reported for organic compounds degradation, and in most of the cases, in composite forms.

Tao et al. created graphene-Sb_2S_3 (G-Sb_2S_3) composites via a facile solvothermal method with GO, which played an important role in controlling the size and the distribution of the formed Sb_2S_3 nanoparticles on the graphene sheets with different density. Due to the negative surface charge, smaller Sb_2S_3 particles size, and efficient electrons transfer from Sb_2S_3 to graphene, the composites demonstrated improved photodegradation activity on RhB. Hydroxyl radicals (·OH) derived (Fig. 11.18) from CB electrons of Sb_2S_3 was suggested to be responsible for the photodegradation of RhB (Tao et al. 2013).

A series of carbon-modified antimony sulfide (Sb_2S_3) composites were obtained and tested in the degradation of Methyl orange dye under irradiation of visible light. The higher photocatalytic activity of the carbon-modified Sb_2S_3 photocatalysts was attributed to their higher adsorption capacity and higher separation efficiency of the photogenerated carriers. Visible-light-driven Sb_2S_3/WO_3 photocatalysts were also obtained and tested in the RhB degradation using a LED lamp as visible-light source. Compared with pure WO_3 and Sb_2S_3, the significantly enhanced photocatalytic activities of the Sb_2S_3/WO_3 composite particles were attributed to the decrease of the recombination rate of photoinduced electron-hole pairs due to the coupling of Sb_2S_3 and WO_3 within the composite nanoparticles (He et al. 2013).

Other applications were also targeted by researchers such as the hierarchical Sb_2S_3/C composite bundles that were applied as capacitive materials (Zhou et al. 2013). A high reversible capacity of 1084 mA h•g^{-1} at a current density of 100 mA•g^{-1}, a good cyclability of 960 mA h•g^{-1} at a current density of 100 mA•g^{-1} after 30 cycles, and a superior rate capability of 1019 mA h•g^{-1} at a current density of 4000 mA•g^{-1} was achieved for the as-prepared Sb_2S_3/C composite when evaluated as an alternative electrode material for batteries.

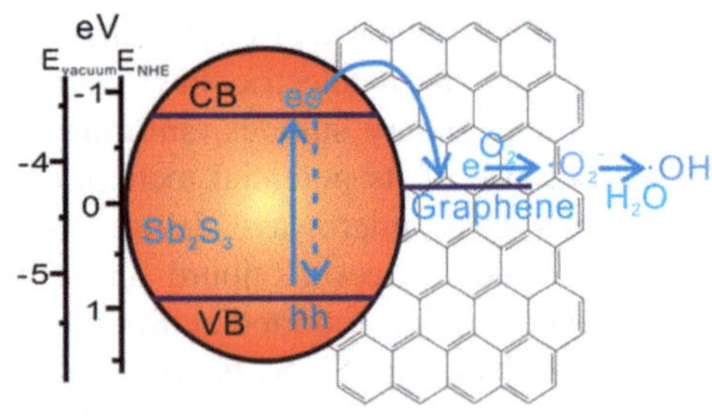

Fig. 11.18 Suggested electron transfer pathway of G–Sb_2S_3 composite (Tao et al. 2013)

Bismuth Containing Composites—Bi_2O_3

Bismuth-based semiconductors are receiving increased attention as photocatalysts that degrade organic pollutants under UV–vis light (Pan et al. 2008; Xiaohong et al. 2007). In particular, bismuth oxide (Bi_2O_3) has a variety of desirable properties, including high band gap (2–3.96 eV), high refractive index, and photoluminescence, and this oxide is implicated in a range of fields such as solid oxide fuel cells, gas sensors, high temperature superconductor materials, functional ceramics, and catalysis (Zhou et al. 2009).

Gou et al. fabricated AgBr@Bi_2O_3 heterojunction photocatalysts by a facile deposition–precipitation method using novel hierarchical α-Bi_2O_3 microrods as substrate. In particular, the photocatalytic activity of AgBr@Bi_2O_3 heterojunction was superior to that of the single visible-light-active components (AgBr, Bi_2O_3) and their mechanical mixture, indicated the presence of a synergic effect between two active components. This material can decolorize about 90% RhB after 60 min of visible-light irradiation. Trapping and photoluminescence experiments show that active h^+, Br^0, and $\cdot O_2^-$ played a major role in RhB degradation while ·OH was confirmed to be insignificant. A possible mechanism of transportation for photon-generated carriers was also proposed as shown in Fig. 11.19 (Gou et al. 2015).

Liu et al. synthetized Bi_2O_3–Ag_2O hybrid photocatalysts by a coprecipitation method. The results showed that Bi_2O_3–Ag_2O hybrid photocatalysts exhibited enhanced photocatalytic performance in the degradation of phenol with a maximum degradation rate of 92% in 60 min under visible-light irradiation, which was ascribed to the increase in light absorption and the reduction in electron–hole pair recombination with the introduction of Ag_2O (Liu et al. 2015a, b).

New visible-light-responsive Bi_2O_3/Co_3O_4 microspheres were obtained, which were assembled from nanosheets with porous structure. According to the

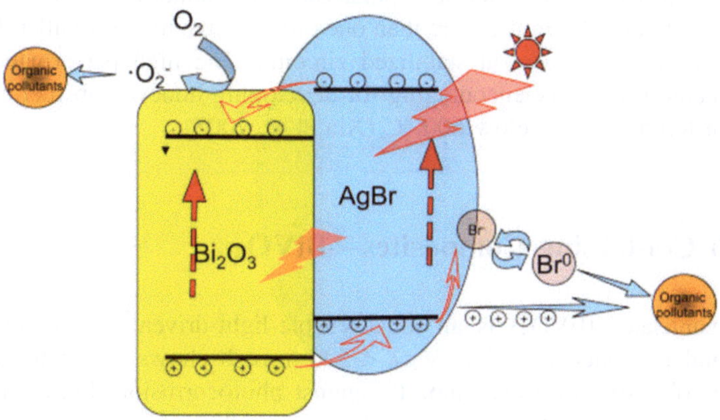

Fig. 11.19 Schematic diagram of photoexcited electron–hole separation process and the possible reaction mechanism over AgBr@Bi_2O_3 photocatalyst (Gou et al. 2015)

photocatalytic results, it could be concluded that the formation of Bi_2O_3/Co_3O_4 composite photocatalyst was more efficient than pristine Bi_2O_3 and Co_3O_4 photocatalyst, respectively, by acquiring better visible-light absorption spectrum. The photocatalysis mechanisms by electron migration from VB to conduction band under visible-light illumination, the dye molecules adsorbed on the photocatalyst can also absorb visible light to produce the excited state. The oxidation potential of the excited state is more negative than the potential of CB of Bi_2O_3/Co_3O_4 particles and an electron is then injected into the CB of Bi_2O_3/Co_3O_4 from the excited state. Therefore, the electrons in the CB of Bi_2O_3/Co_3O_4 can react with absorbed oxygen to produce a superoxide anion radical ($\cdot O_2^-$), which can further decompose the dye. At the same time, the highly oxidative holes in VB not only directly degrade the dye molecules adsorbed on the surface of Bi_2O_3/Co_3O_4, but also are trapped by OH^- to produce hydroxyl radical species ($\cdot OH$) (Hsieh et al. 2013).

CeO_2–Bi_2O_3 composite photocatalysts were also obtained by a two-stage process. They presented the evidence for photocatalytic decomposition of an aqueous solution on these photocatalysts. From the results of the optimization process, it could be concluded that the formation of CeO_2–Bi_2O_3 composite photocatalysts were more efficient than bare Bi_2O_3 photocatalyst (Li and Yan 2009).

Composites with Bi_2O_3 have the potential to be applied in other applications as well. A simple route was employed to prepare nanosized Bi_2O_3 deposited on highly ordered mesoporous carbon. The electrochemical measurements revealed that by loading only 10% Bi_2O_3 on the mesoporous carbon, the specific capacitance of the composite was improved by 62%, with the maximum value reaching 232 $F \cdot g^{-1}$ at a sweep rate of 5 $mV \cdot s^{-1}$. The specific capacitance of Bi_2O_3 was calculated and reached 1305 $F \cdot g^{-1}$ at 1 $mV \cdot s^{-1}$. The cyclic life of composite materials was also measured and the capacitance only declined 21% after 1000 cycles (Yuan et al. 2009).

Silver and yttria stabilized bismuth oxide (YSB) were used as cathodes for low-temperature honeycomb solid oxide fuel cells with stabilized zirconia as electrolytes. At 600 °C, the interfacial polarization resistances of a porous YSB–Ag cathode was about 0.3 $V \cdot cm^2$, more than one order of magnitude smaller than those of other reported cathodes on stabilized zirconia. The high performance of the YSB–Ag cathodes is very encouraging for developing honeycomb fuel cells to be operated at temperatures below 600 °C (Xia et al. 2003).

Bismuth Containing Composites—$BiVO_4$

Bismuth vanadate ($BiVO_4$) is an ideal visible-light-driven semiconductor with narrow band gap energy of 2.4 eV ($\lambda < 520$ nm). It shows sufficient absorption within the solar spectrum and stability against photocorrosion. Furthermore, it is inexpensive, environmentally benign, and can be synthesized using numerous facile methods (Chen et al. 2015). Kudo reported the good photocatalytic activity of

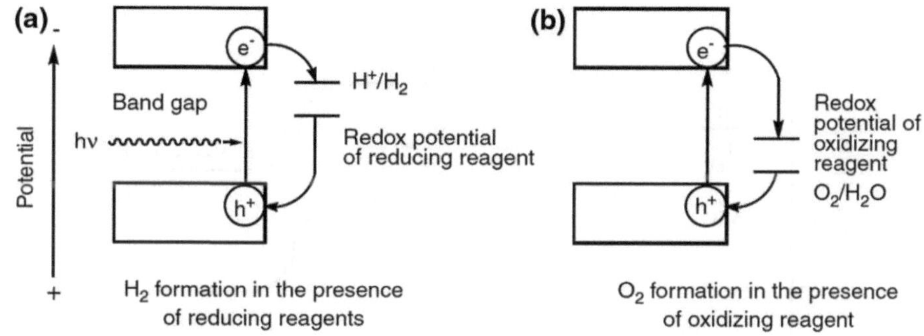

Fig. 11.20 Scheme of photocatalytic H_2 or O_2 evolution in the presence of sacrificial reagents (Kudo 2006)

$BiVO_4$ for O_2 evolution (Fig. 11.20) from an aqueous silver nitrate solution under visible-light irradiation (Kudo 2006).

Though the energy level of the CB may be applicable to the reduction of water, the activity of $BiVO_4$ alone is not sufficiently high due to its difficulty in the separation of photogenerated electron–hole pairs. Therefore, it is necessary to provide a suitable semiconductor or metal in composites with a higher redox potential in order to increase the efficiency of the charge separation of $BiVO_4$ (Pingmuang et al. 2013).

Zhang et al. created Ag/$BiVO_4$ composite photocatalysts by hydrothermal method. The structural studies revealed that all prepared catalysts exhibited the typical pattern for monoclinic scheelite $BiVO_4$ structure. The doped element Ag was confirmed by XPS analysis. It was also found that Ag loading could effectively narrow the bandgap of the catalysts. The photocatalytic activities of the Ag-loaded samples were higher than that of pure $BiVO_4$ and the highest photodegradation efficiency of MO was obtained at 1.0 wt% Ag content (Zhang and Zhang 2010). This can be explained from that the excess Ag may cover the active sites on the $BiVO_4$ surface and thereby reduce the efficiency of charge separation (Zhang and Zhang 2010)

Cu_2O/$BiVO_4$ heterogeneous nanostructures were built by hydrothermal process using polyols. HRTEM investigations indicated that a large number of nano p–n junction heterostructures were formed by the assembly of p-type Cu_2O nanoparticles and n-type $BiVO_4$ (Fig. 11.21). The photocatalytic activity tests demonstrate that these composites exhibited highly efficient visible-light-driven photocatalytic activities as compared to the individual $BiVO_4$ nanocrystals for the degradation of MB and colorless organic phenol under visible-light irradiation (Wang et al. 2013).

Yu et al. prepared Co_3O_4 and CuO/$BiVO_4$ composite photocatalysts by intense ultrasound irradiation method at room temperature. The composite photocatalysts exhibited enhanced photocatalytic activity under visible-light irradiation. As for Co_3O_4/$BiVO_4$, the highest efficiency is observed at 3 wt% content and for CuO/$BiVO_4$, 1 wt% copper content gives the highest photocatalytic activity. The

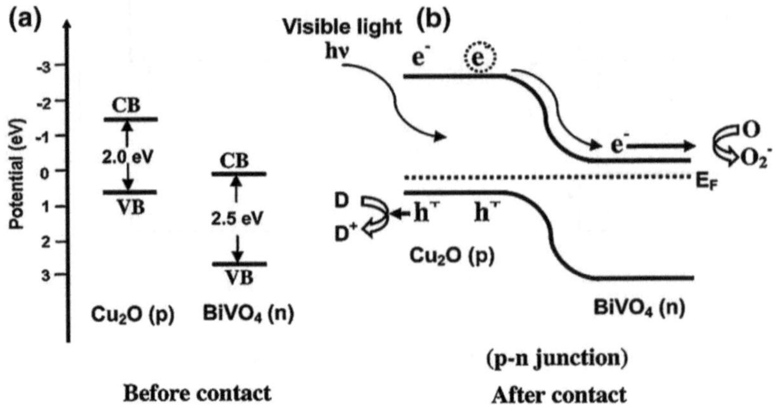

Fig. 11.21 Schematic diagram of charge transfer between p-type Cu_2O and n-type $BiVO_4$: **a** before contact; **b** after formation of the p–n junction (Wang et al. 2013)

enhanced activity could be attributed to the p–n heterojunction semiconductor structure, which effectively restrains the recombination of photogenerated hole–electron pairs (Yu et al. 2011). $BiVO_4$–graphene composite photocatalyst were also obtained with excellent performance by a one-step hydrothermal method without the use of any surfactant. TEM observations indicate that graphene sheets were fully exfoliated and decorated with leaf-like $BiVO_4$ lamellas, due to the fact that the graphene sheets play the role of template to allow two-dimensional planar growth. The photocatalytic activity measurements demonstrate that the $BiVO_4$–graphene photocatalysts show superior photoactivity in degradation of MB, RhB, MO, and active black BL-G under visible-light irradiation (Fu et al. 2011).

Bismuth Containing Composites—BiOX

Xiao et al. created series of BiOI/BiOCl composite photocatalysts with different amounts of BiOI by a one-pot solvothermal method. The as-synthesized BiOI/BiOCl catalysts exhibit 3D hierarchical microsphere morphologies and heterojunction structures (Fig. 11.22), and the EG solvent and urea are crucial for the formation of such structures. The BiOI/BiOCl samples demonstrated strong photoabsorption of UV and visible light, and the samples showed clearly enhanced photocatalytic activities under visible-light irradiation. The highest degradation efficiency was observed over the BiOI/BiOCl composite with 90% BiOI, and the reaction rate constant for this degradation was more than 4- and 20-fold greater than that of pure BiOI and the commercially available Degussa P25, respectively (Xiao et al. 2012).

Many reports on the photocatalytic degradation of BPA over TiO_2 have suggested that BPA degradation mainly proceeds through demethylation and

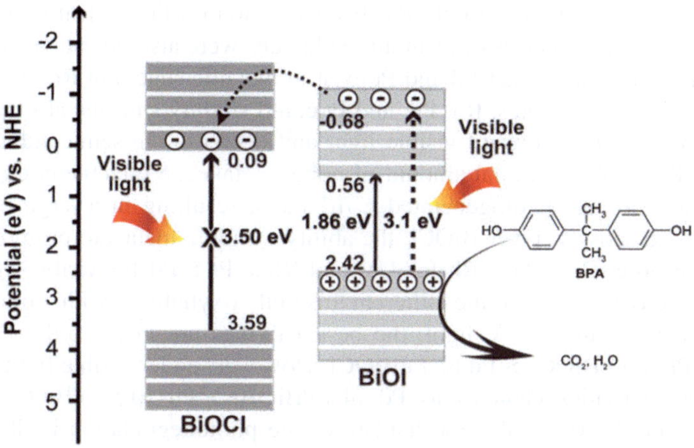

Fig. 11.22 Diagram of the energy band structure and the charge separation of the BiOI/BiOCl heterostructures as well as a possible pathway of photodegradation of BPA under visible-light irradiation (Xiao et al. 2012)

hydroxylation because of the formation of a reactive hydroxyl radical (·OH). However, in this visible-light-induced BiOI/BiOCl system, no hydroxylated compounds were identified. This observation may be ascribed to the lower standard redox potential of Bi^{5+}/Bi^{3+} ($E^0 = 1.59$ V at pH = 0) compared to that of ·OH/OH − ($E^0 = 1.99$ V at pH = 0). Therefore, the holes that are photogenerated on the surface of BiOI/BiOCl could not react with OH^-/H_2O to form ·OH. Therefore, the BPA photodegradation of BiOI/BiOCl composites may be dominated by the direct hole oxidation rather than the oxidation by ·OH.

Ag/BiOI catalysts were obtained with excellent photocatalytic performance by a hydrothermal process combined with a photodeposition method. Due to silver, Ag/BiOI showed substantial improvement in the photocatalytic activity for the degradation of dyes under visible-light irradiation. The loaded Ag nanoparticles on BiOI could form Schottky barriers between their contact regions, which enhanced the surface electron excitation and electron–hole separation, thus promoting the photocatalytic activity. In addition, the surface plasmon resonance of Ag metals on BiOI excited by visible light could also contribute to the enhancement in the photocatalytic activity. It is suggested that Ag/BiOI catalysts are promising visible-light-driven photocatalysts for environmental applications (Liu et al. 2012).

Wang et al. synthetized Ag_2CO_3/BiOBr composite. The as-synthesized Ag_2CO_3/BiOBr composite exhibits more efficient photocatalytic performances for the degradation of organic dyes, compared with BiOBr under visible-light irradiation. The narrower bandgap, which is beneficial in harvesting more light for the Ag_2CO_3/BiOBr composite, and the heterojunction structure between Ag_2CO_3 and BiOBr, which facilitated electron–hole separation, led to the remarkably enhanced photocatalytic activity (Wang et al. 2014a, b, c).

The effects of noble metal (Rh, Pd, Pt) deposition on the optical properties and photocatalytic performances of bismuth oxyhalides were also investigated (Yu et al. 2013). The results show that Pd and Pt exist in metallic state, but Rh exists in both metallic and oxidized states. Rh, Pd, and Pt could slightly shift the absorption edge of bismuth oxyhalides toward visible light and decrease the semiconductor bandgaps. Loadings of an optimal amount of noble metals can effectively suppress the recombination of the photogenerated e^-/h^+ pairs, resulting in a large increase in photocatalytic activity. Over BiOCl, the ability of noble metals to promote activity follows the order Pt > Pd > Rh for UV and Rh > Pt > Pd for visible-light irradiation, respectively. As for the different bismuth oxyhalides with optimal noble metal deposition, under UV light, the best activities are in the order Pd (0.5%)/BiOBr > Pt (1%)/BiOCl > Pd (2%)/BiOCl. However, under visible light, the order of the best activities changes to Pd (4%)/BiOBr > Pd (0.5%)/BiOI > Rh(1%)/BiOCl (Yu et al. 2013). The mechanism of the photodegradation is illustrated in Fig. 11.23.

Shamaila et al. obtained BiOCl powder by a low-temperature hydrolysis method. The control over hydrolysis rate was achieved with in situ-generated ammonia from urea. The WO_3/BiOCl system developed in this work was a new heterojunction-type photocatalyst working efficiently under visible light. The WO_3/BiOCl demonstrates notably high photocatalytic activity over a wide composition range in decomposing RhB in aqueous solution, whereas the individual BiOCl and WO_3 showed a negligible efficiency. Moreover, WO_3/BiOCl induces complete decomposition of RhB as compared to Degussa P25. It is considered for the WO_3/BiOCl composite that BiOCl works as main photocatalyst while the role of WO_3 is a sensitizer absorbing visible light. The electrons from the valence band (VB) of WO_3 are excited to the CB in visible irradiation. Thereby the electrons from VB of BiOCl are transferred to the VB of WO_3 and created the holes in VB of BiOCl. These holes initiate photocatalytic oxidation reactions (Shamaila et al. 2011).

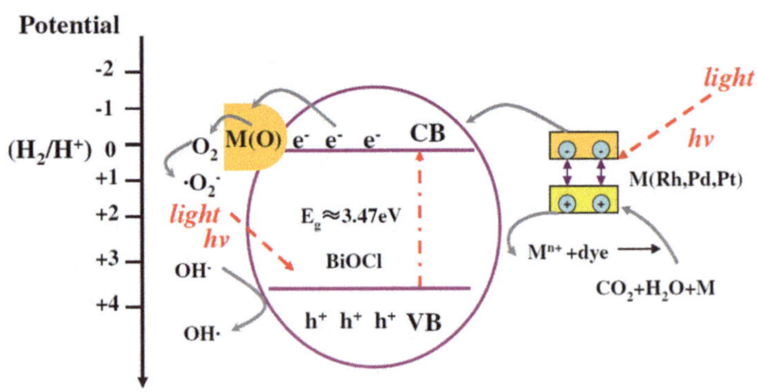

Fig. 11.23 Suggested mechanism of noble metals (Rh, Pd, Pt) enhancing the photocatalytic activity of BiOCl (Yu et al. 2013)

Other applications are also interesting for these materials, including flexible heterojunction photoanode architectures by in situ decorating BiOI nanoflake array films with Bi_2S_3 nanospheres through a simple solution process at a relatively low temperature. It was highlighted and carefully investigated, the photoelectrochemical properties of these BiOI/Bi_2S_3 heterojunction films, for use in solar cells. Owing to the increase of photoabsorption and more efficient charge separation efficiency, while retaining the similar charge transport efficiency, a short-circuit current density of 1.82 mA cm^{-2} and a photo-to-electricity conversion efficiency of 0.36% were achieved by partially decorating the BiOI nanoflake array film with a little amount of Bi_2S_3 nanospheres (BiOI/$Bi_2S_{3-0.5}$) as the photoanode, which is nearly three times higher than the bare BiOI nanoflake array-based cell. Along with more Bi_2S_3 particles formed, the perfect BiOI nanoflake framework gradually broke down and resulted in lower conversion efficiency, which demonstrates that an appropriate transformation degree of BiOI to Bi_2S_3 is critical to improve the photoelectrochemical cell efficiency (Fang et al. 2015).

V, Nb, Ta Containing Materials as Composite Components

Among the transition metal oxides, vanadium pentoxide (V_2O_5) is attracting great attention due to its good property for applications in electronics (e.g., lithium-ion batteries) and Photocatalysis (Wang et al. 2014a, b, c). Various interconnected electronic and structural factors are contributing to the unusual chemistry of vanadium oxide. Vanadium has partially filled d-orbitals, which are responsible for a varied range of its electronic and catalytic properties. The variety of stable oxidation states of V results in easy conversion between oxides of different stoichiometry by oxidation/reduction. Therefore, it is believed, that it can be an important factor for the oxide to be an efficient catalyst in selective oxidation (Asim et al. 2009).

In order to obtain V_2O_5-based nanomaterials, different precursors [e.g., vanadyl sulfate hydrate, ammonium metavanadate (Asim et al. 2009)], and surfactants can be used, such as CTAB (Asim et al. 2009), polyethyleneglycol (PEG) (Reddy et al. 2007a, b), polyvinylpyrrolidone (PVP) (Sakunthala et al. 2011), or other chemicals such as hydrogen peroxide (Avansi et al. 2010), oxalic acid (Wang et al. 2014a, b, c), and $KBrO_3$ (Zhou et al. 2008). The first attempt (for the best knowledge of the authors) to use V_2O_5 as photocatalyst was described by Moshfegh and Ignatiev (Moshfegh and Ignatiev 1990). Their communication focused on the effect of both UV-visible photo-irradiation and thermal heating of a pure V_2O_5 catalyst (with a relatively low surface area) in order to study the heterogeneous catalytic decomposition of isopropanol. They have observed a 2.5-fold enhancement of the catalytic dehydration of isopropanol on vanadia, compared to the thermal heating of the catalyst. This opto-catalytic enhancement of V_2O_5 was attributed to the low bandgap value of the nanomaterial (2.35 eV).

In a more recent study Karunakaran and Karuthapandian investigated the oxidation of diphenylamine to N-phenyl-benzoquinonimine in ethanol under UV and solar sunlight. They have found that the process is successful under natural and artificial sunlight as well. They have found that vanadia mixed with either ZnO or CdO shows an increased efficiency, which may be due to the interparticle charge transfer (Karunakaran and Karuthapandian 2015).

Tungsten-doped vanadium dioxide thin films were used by Liang et al. as smart windows with self-cleaning and energy-saving functions. The vanadium dioxide thin films, obtained by the reaction of vanadyl acetylacetonate, methanol, and tungsten chloride, constituted a polycrystalline VO_2-lattice, showed excellent hydrophilicity with a water contact angle of 12° and a luminous transmittance of ≈80%, having a transition temperature at 32 °C, giving a potential applicability in energy-saving smart windows.

As it was already described above, vanadia itself has some limitations regarding its photocatalytic activity, therefore in most of the studies it is used as "partner" in composites. The most investigated nanocomposites using V_2O_5 are those with $BiVO_4$ (Jiang et al. 2009; Su et al. 2011; Sun et al. 2014), TiO_2 (Sun et al. 2014), and g-C_3N_4 (Jayaraman et al. 2015), but it has to be mentioned, that also relatively new photocatalytic materials are involved in these researches, like $SmVO_4$ (He et al. 2009). From these studies it can be concluded that the composites usually have much better photocatalytic activity than its bulk components, as it is described in detail by Jamaran et al. (2015). They have used these materials for photodecomposition of an organic dye (Direct red 81—DR81), because of its harmful nature. It was observed that vanadia itself was inactive under visible irradiation, while g-C_3N_4 has demonstrated some photocatalytic activity, which was attributed to the moderate bandgap of this material. Furthermore, it was noted, that the composite exhibited significantly higher photocatalytic activity toward DR81, degrading more than 90% of the organic dye in two hours. In the Sm-based composite (He et al. 2009), they have reached an increase from 15 to 60–100% for different V_2O_5–$SmVO_4$ composites on photodegradation of acetone under visible light.

Another, less "famous," but promising class of nanomaterials with photocatalytic applications has started its adventure in Japan, from the early '90s, the niobates, niobium-based materials (Domen et al. 1990a, b; Kudo et al. 1989). Kudo et al. synthetized nickel-loaded niobates ($K_4Nb_6O_{17}$) and found that this type of nanomaterial is a layered compound, which possesses two kind of interlayer spaces (interlayers I and II) alternately and K^+ ions at the interlayer spaces, which can be exchanged for other cations. This photocatalyst can decompose intercalated water into H_2 and O_2 at the interlayer spaces with high efficiency, regarded as a "two-dimensional" photocatalyst (Kudo et al. 1989).

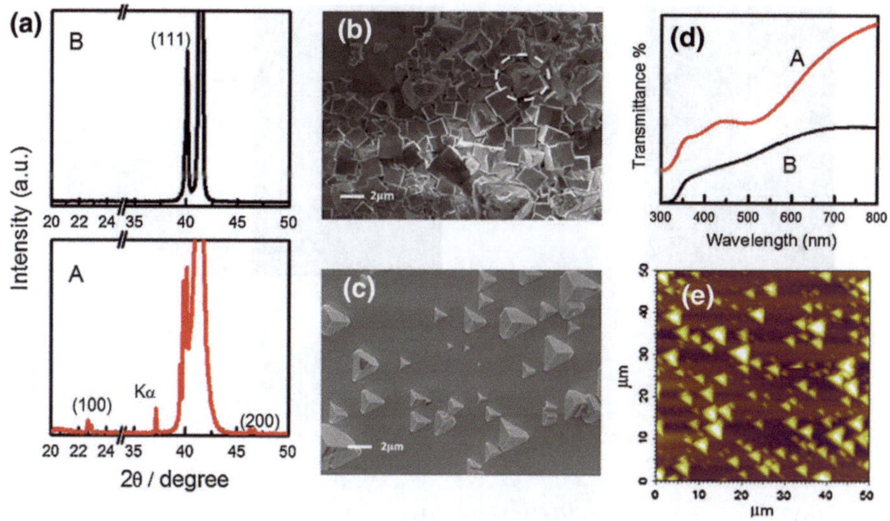

Fig. 11.24 **a** X-ray diffraction patterns, **b**, **c** *top view* SEM images of $NaNbO_3$ samples, **d** UV-vis transmittance spectra, and **e** AFM image of the sample, while *A* and *B* represent two cases, without (*A*) and with (*B*) the presence of $LaAlO_3$ growth substrate (Yu et al. 2016)

A novel Nb-based material was investigated by Yu et al., too, namely the $NaNbO_3$. They have obtained triangular pyramids and cubes on homogeneous and heterogeneous substrates (Fig. 11.24), exhibiting photocatalytic oxidation activities for 2,4-dichlorophene (DCP) and RhB. The ·OH radicals coming from $\cdot O_2^-$ with the intermediate of H_2O_2 was the dominant oxide species in this case (Yu et al. 2016).

A relatively new and less investigated field is the (photo)catalytic properties of niobic acid (HNb_3O_8). This material can be responsive for visible light, even more, if it is doped with nitrogen (from urea, as source) (Li et al. 2008). Because of the favorable properties like the layered structure, the light adsorption and the protonic acidity, the HNb_3O_8 has showed superior activity in comparison to the "legendary" P25 and the nitrogen-doped Nb_2O_5 samples under visible-light irradiation toward RhB.

Another promising family member of niobates is the oxide of Nb, namely the Nb_2O_5 that had proven to have relatively high photocatalytic activity. Kominami et al. prepared this oxide by solvothermal synthesis, using niobium (V) pentabutoxide in toluene at 300 °C. The amorphous powders, having a relatively large surface area, were more efficient than their crystalline analogs for the oxidative decomposition of oxalic acid (Kominami et al. 2001).

The last non-synthetic "member" of the elements from the group V is the tantalum. Its oxide, Ta_2O_5 with a relatively low bandgap (3.0 eV) can be used seldom as photocatalyst, having good performance for degradation of gaseous formaldehyde under UV irradiation, used in the form of nanopowders (TEM micrographs of the samples are presented below on Fig. 11.25), obtained from $TaCl_5$ as precursor (Zhu et al. 2005).

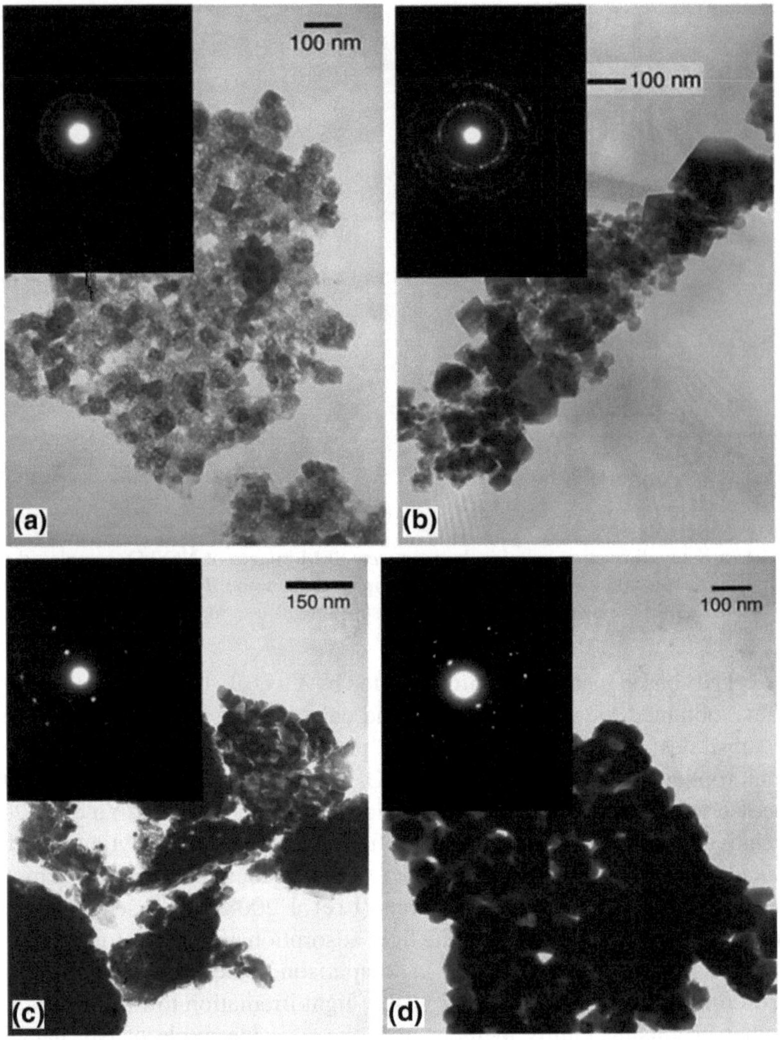

Fig. 11.25 TEM micrographs of Ta_2O_5, calcined at various temperatures (**a** 500, **b** 600, **c** 700, **d** 800 °C) (Zhu et al. 2005)

B, Al, Ga, In, Tl Containing Materials as Composite Components

Group 13 has a great number of nanomaterials and nanocomposites with a large applicability spectrum, from IT industry, pharmaceutical industry to photocatalytic applications.

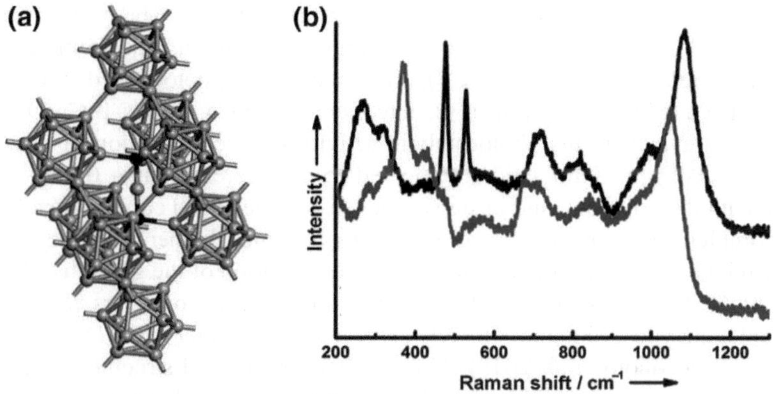

Fig. 11.26 **a** Idealized elementary cell of boron carbides (B-*gray*, C-*black*) and **b** FT-Raman spectra of $B_{13}C_2$ (*gray*) and $B_{4.3}C$ (*black*) (Liu et al. 2013)

The "head" of the group has a relatively low number of photocatalytic materials, if we are not taking into account the B-doped nanomaterials/composites. One of these investigations are described by Liu et al. relating the synthesis and application of some boron carbides as efficient, metal-free, and visible-light-responsive photocatalysts (Liu et al. 2013). They have successfully synthetized two boron carbides (Raman spectrum and elementary cell of the material is presented in Fig. 11.26), which were demonstrated to perform photocatalytic H_2 evolution using photoelectrochemical water reduction under visible light. The main novelty of this approach/material is that it does not need any noble metal as cocatalyst. The $B_{4.3}C$ exhibited better efficiency and stability than $B_{13}C_2$, providing new opportunities for the future development of efficient and stable photocatalysts and solar-cell materials.

Another recently described approach was described by Fan et al. (2014), where a bulk boron-based photocatalytic material was investigated ($K_3B_6O_{10}Br$), demonstrating excellent catalytic activity in UV-induced dechlorination of chlorophenols. The as-obtained efficiencies were two order of magnitudes higher than those obtained for the commercial P25 catalyst, under UV light irradiation, suggesting that these types of nonlinear materials, which have high polarizability, inducing internal electric fields at the space-charge regions, facilitating in this way the charge separation of the electron-hole pairs, can open new windows for designing efficient catalysts.

The next member of the "family" has even lower number of studies involving photocatalytic activities. Just few articles are dealing with Al-containing photocatalysts, mostly as cocatalyst in V_2O_5 and other, more active semiconductor-based materials (Li et al. 2015a, b, c; Teramura et al. 2004). On the other hand, it has to be mentioned, that Al_2O_3 is widely used as supporting or coating material for the "real" catalyst, like CdS (Hirai et al. 2002; Sinha et al. 2001), ZnO, SnO (Miwa et al. 2010), and TiO_2 (Choi et al. 2006; Kim et al. 2009).

Gallium-containing materials are more famous if it is about photocatalytic materials and other composites with an enlarged applicability spectrum, even if its bandgap is relatively large (4.8 eV), therefore being photoexcited only by UV irradiation ($\lambda < 260$ nm). Hou et al. have synthetized a porous β-Ga_2O_3 photocatalyst and evaluated the photocatalytic activity by decomposition of benzene in air, under UV light illumination, compared to commercial titania (P25), mineralizing the model pollutant, and its derivatives (e.g. toluene and ethylbenzene) to CO_2 under ambient conditions, with an order of magnitude higher efficiency than the commercial catalyst, without any noticeable deactivation of the gallium oxide (Hou et al. 2006). On the other hand, Ga_2O_3, containing Ag nanoclusters were used for the photocatalytic reduction of CO_2 with water, forming CO, H_2, and O_2 as products (Yamamoto et al. 2015). Three polymorph crystal structures were investigated by Hou et al. (2007), for the decomposition of benzene, toluene, and ethylbenzene in dry air stream, under UV irradiation. The catalysts showed much higher activity than the reference P25 material. The efficiency of the polymorphs followed the sequence β-Ga_2O_3 > γ-Ga_2O_3 > α-Ga_2O_3. The superior performance of the first semiconductor was attributed to its high crystallinity and geometric structure (Di Paola et al. 2012).

Between indates, MIn_2O_4 (where M = Ca, Sr, Ba) were demonstrated as efficient photocatalysts, which degraded MB under visible-light irradiation and exhibited the following activity-order: $CaIn_2O_4$ > $SrIn_2O_4$ > $BaIn_2O_4$, result that was correlated with the band-structure calculations, a decreasing of the activity being observed with the radius of the M ion (Tang et al. 2004).

In_2O_3 hollow microspheres were synthetized and used as photocatalysts and gas sensors by Benxia et al. (Li et al. 2006). They have prepared first the $In(OH)_3$ hollow microspheres via a novel surfactant-free route, having a size of about 80 nm. The desired In_2O_3 microspheres were obtained from annealing the prepared precursor, as mentioned above. The as-obtained In_2O_3 microspheres (Fig. 11.27) performed better than the In_2O_3 dispersive nanocubes, as gas-sensing material for ethanol and formaldehyde and as efficient photocatalysts for degradation of RhB, possibly due to the larger surface area that provides sufficient space and more active sites for reactions.

Nanosized $InVO_4$ with orthorhombic structure was successfully synthetized by Zhang et al., using a medium–high calcination temperature (600 °C) with an amorphous heteronuclear complex as precursor. The as-obtained photocatalysts were evaluated by photodegradation of formaldehyde, under UV and visible-light irradiation (Zhang et al. 2006a, b). After the evaluation of crystallinity, particle size, morphology, and photocatalytic activity, they have found that the nanoparticles prepared by their synthesis pathway and thermal treatment showed higher photocatalytic efficiency under visible-light irradiation than the sample prepared by traditional solid-state reaction, having the potential to be used in preparation of other complex oxide photocatalyst powders ant thin-film photoelectrodes.

Even if it is less discussed in the literature, thallium-based materials are the "youngest" nanostructures with photocatalytic properties. Tl_2O_3 nanostructures were successfully synthetized by Goudarzi et al. from thallium-acetate in aqueous

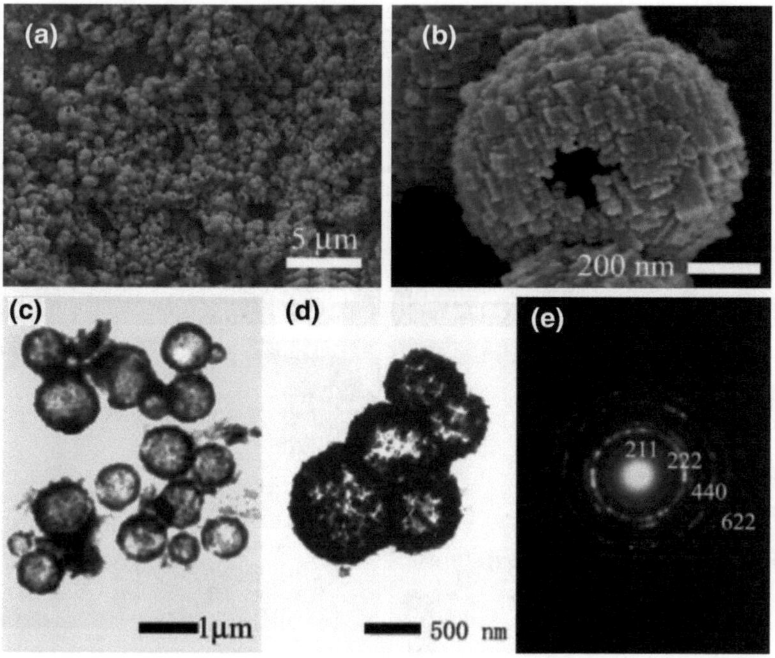

Fig. 11.27 **a**, **b** SEM and **c**, **d** TEM micrographs of In_2O_3 hollow spheres from annealing $In(OH)_3$ hollow spheres. **e** SAED pattern taken from an In_2O_3 hollow microsphere

solution, through two different approaches. The products were analyzed by X-ray diffraction, SEM, and energy dispersive X-ray microanalysis. The as-synthetized nanostructures were used as efficient catalysts in degradation of methyl orange (with 92% degradation within 40 min.) (Goudarzi et al. 2015). $Tl_2S_2O_3$ nanostructures (Fig. 11.28) were fabricated by the same group (Goudarzi and Salavati–Niasari 2016), using a solvothermal method, with water as solvent. These rare-earth compounds had outstanding photocatalytic activities, which were attributed to the increased adsorption of organic substrate and the low recombination rate of the photogenerated electrons and holes.

C, Si, Ge, Sn, Pb-Based Materials and Their Composites

Group 14 can be called the home of one of the highest number of nanostructures involved in photocatalytic applications with a large applicability spectrum.

The first member of the group has, at the moment, one of the most "trendy" photocatalytic nanostructures, the heptazine-based polymer, namely the g-C_3N_4, the so-called "melon" (Lau et al. 2016). The breakthrough of this material can be dated

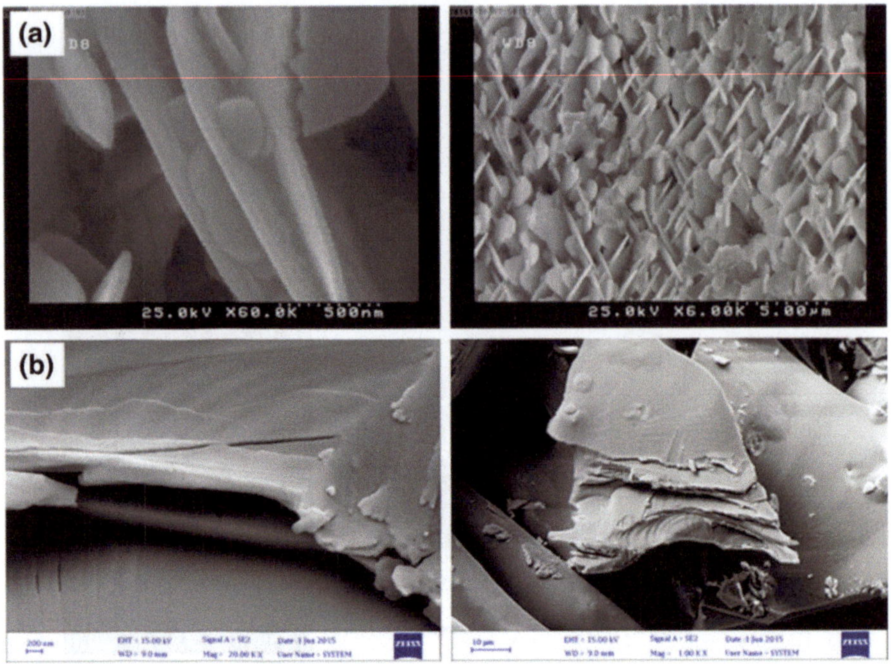

Fig. 11.28 SEM images of $Tl_2S_2O_3$ nanostructures obtained at **a** 160 °C and **b** 180 °C for 10 h (Goudarzi and Salavati-Niasari 2016)

from the early years of the present decade,[1] actually, more than 1500 articles are dealing with this material (Lam et al. 2016; Sun et al. 2016a, b; Ye et al. 2015; Zhao et al. 2015). It can be prepared easily from relatively inexpensive precursors (urea, thiourea, melamine, cyanamide, etc.), possesses a unique two-dimensional structure, excellent chemical stability and tunable electronic structure with a medium bandgap (2.7 eV). Among the several approaches to synthetize C_3N_4-based materials one of the first successful investigations was proposed by Niu et al. (1993), having the evidence for the formation of C_3N_4-based nanostructures by pulsed ablation of graphite, combined with a high-flux, atomic nitrogen source, resulting thin films (Fig. 11.29), where the nitrogen content was controlled systematically by means of the N source, obtaining in this way a thermally robust crystallites.

On the other hand, pure g-C_3N_4 has also some bottlenecks, like the relatively rapid recombination of photogenerated electron-hole pairs, small specific surface area, and low efficiency in using visible light. For the moment, several approaches

[1]According to the search on Scopus (28.08.2016), only less than 40 articles were published about g-C_3N_4-based materials 'til 2010, while the "big boom" has occurred in 2014/2015, when more than 300/500 research articles were published about this structure.

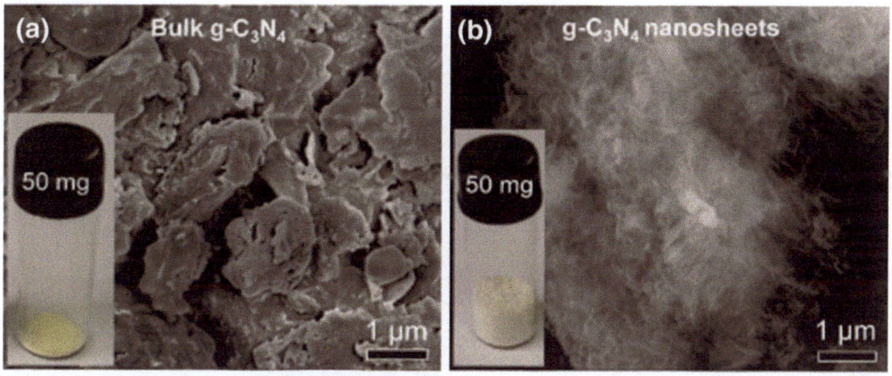

Fig. 11.29 SEM images of bulk g-C_3N_4 and g-C_3N_4 nanosheets (Niu et al. 2012)

have been employed, in order to increase the adsorption of visible light and enhancement of photocatalytic performances, such as formation of surface coupling hybridization using Bi_2WO_6 (Ge et al. 2011), graphene (Xiang et al. 2011), increasing its specific surface area/"building" of mesoporous structure (Su et al. 2010), doping with metal and nonmetal species, e.g., B (Sagara et al. 2016), S (Liang et al. 2016), P (Hu et al. 2014), Ag (Hu et al. 2015), Au (Yin et al. 2015), and Pd (Chang et al. 2013).

According to these, large variability of g-C_3N_4, a large applicability spectrum can be discussed. As a "simple" catalyst can be used as catalyst in oxygen-reduction reactions, water splitting, and can be involved in Friedel-Crafts reactions but also in light emitting devices, photocatodes, and optical sensors (Dong et al. 2014) as well.

The next member of group 14 is no less "famous" than the carbon-based materials. Silicon-based materials, especially SiO_2, even if, so far, no silicate photocatalyst has been reported, it is able to boost the activity of the main catalyst. Sol–gel-based synthesis method can give porous nanostructures with increased specific surface area, allowing an increased efficiency of the adsorption of the model pollutants, improving the activity of the catalytic material.

A new approach to synthetize silicon-based photoactive material was recently by Zhu et al. (2016) and Lou et al. (2014). They have studied silver-silicate-based materials, and they have proved that the unique electronic configuration of Ag^+ ions can take part in the composition and hybridization of the energy band in the Ag-based compounds, which can be beneficial, when the main aim is to adjust the bandgap and the light absorption properties of these materials. An $Ag_6Si_2O_7$ was prepared (Fig. 11.30) slowly adding Na_2SiO_3 into an $AgNO_3$ solution, obtaining a reddish brown powder in this way by the hydrolysis and ion-exchange between silver ions and SiO_3^{2-} ions in the solution, resulting irregular-shaped nanoparticles ($\approx$100 nm). The as-prepared photocatalyst can be successfully used for the photodegradation of MB (decomposing more than 90% in less than 10 min), having an increase by a factor of 5–9 comparing to Ag_2O and by a factor of 9–11 over Ag_3PO_4. Unfortunately, the recyclability of the material decreases, as they were

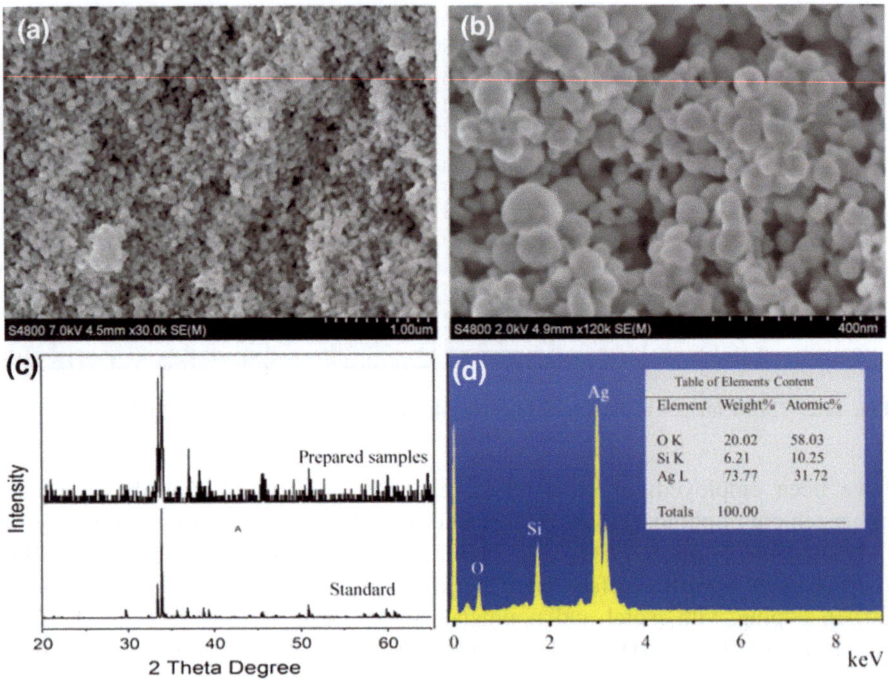

Fig. 11.30 **a**, **b** SEM images, **c** XRD patterns and **d** EDS spectra of $Ag_6Si_2O_7$ samples, prepared by precipitation method Zhou et al. (2016)

able to obtain a degradation of 90% in 70 min during the third "cycle", while this value was obtained in less than 10 min, when the catalyst was used for the first time (Lou et al. 2014).

In the second approach, they have synthetized $Ag_{10}Si_{14}O_{13}$ catalyst through a facile, solid-state reaction, mixing $AgNO_3$ and Na_2SiO_3 in a molar ratio of 3:1, transferring the mixture into deionized water, collecting the resulted precipitation, and treating thermally in order to obtain a higher crystallinity rate. The resulting material had a low bandgap energy (2.0 eV), and the as generated electron–hole pairs had enough ability to oxidize the organic dye to CO_2 and also to oxidize H_2O to generate O_2 (Zhu et al. 2016).

Germanium-based composites are also well investigated in the literature. On the analog of the previously described g-C_3N_4, Maeda et al. have studied germanium nitride (β-Ge_3N_4) as a non-oxide photocatalyst for overall water splitting, by loading RuO_2 nanoparticles to achieve functionality as photocatalyst for stoichiometric decomposition of water under UV irradiation ($\lambda > 200$ nm) (Maeda et al. 2007). The (β-Ge_3N_4) was prepared by heating GeO_2 powder under NH_3 flow, for 10 h and then grinding into powder. During the water splitting, the β-Ge_3N_4 absorbs photon energy greater than the bandgap, generating electron/hole pairs. The electrons are injected into RuO_2 nanoparticles to reduce adsorbed H^+ into H_2, whereas holes can oxidize H_2O into O_2 on the surface of germanium nitride. Although, this type of construction for this nanocomposite is promising, more effective electron transfer from β-Ge_3N_4 to RuO_2 nanoparticles will be necessary, to obtain better photocatalytic efficiencies.

Another investigated photocatalytic material with promising prospects is the Zn_2GeO_4. As the Ge_3N_4-based materials, after coupling with RuO_2 it is photoactive for water splitting to generate hydrogen fuel (Sato et al. 2004). But, Zn_2GeO_4 by itself can also exhibit as an environmental photocatalyst for mineralization of volatile aromatic hydrocarbons (benzene, toluene, and ethylbenzene) in gas phase. Therefore, Huang et al. have synthetized Zn_2GeO_4 nanorods (using CTAB, GeO_2, and Zn-$(CH_3COO)_2 \cdot 2H_2O$) for the decomposition of MO, salicylic acid, and 4-chlorophenol (Huang et al. 2008). They have concluded that the better photocatalytic efficiency of nanorods (Fig. 11.31) in comparison with bulk Zn_2GeO_4 can be attributed to its special geometry, electronic structure, and large specific surface area (32 $m^2\ g^{-1}$ vs. <1 $m^2\ g^{-1}$), which favored the formation of reactive hydroxyl radicals. They have also observed that the as-obtained material is stable/recyclable, as it has maintained its activity even after six successive cycles.

Abundantly available and environmentally nontoxic **tin**, Sn_3O_4 can efficiently be used to catalyze the targeted water splitting in aqueous solution under irradiation of visible light. It belongs to a series of layered tin oxides (SnO and Sn_2O_3) of which crystal structure consists of stacking of alternating atomic layers of tin and oxygen (Seko et al. 2008). Manikandan et al. have prepared Sn_3O_4 nanocrystals by hydrothermal synthesis using sodium citrate as ligand (Manikandan et al. 2014). SEM/TEM investigations have shown that the material consisted of thin, highly crystalline flexes (500 × 500 × 10 nm), which were surrounded by {110} facets, having an orange color and absorbed more efficiently the visible light, than the

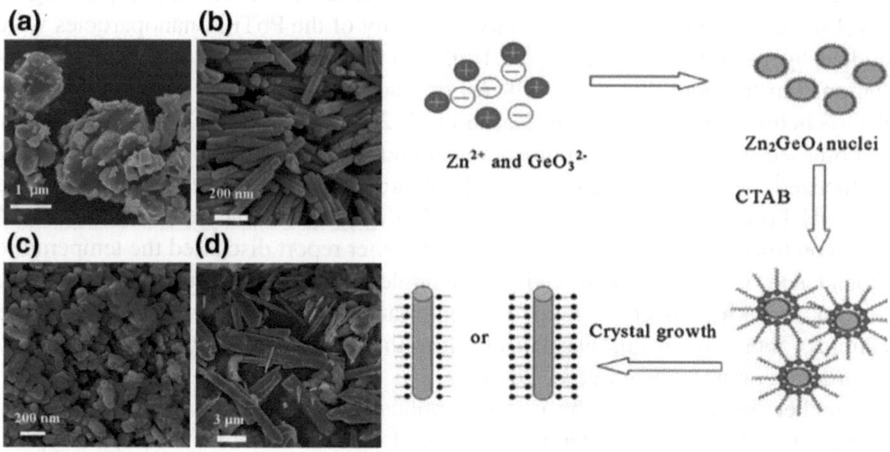

Fig. 11.31 SEM micrographs of **a** bulk Zn_2GeO_4 and **b**, **c**, **d** Zn_2GeO_4 nanorods prepared using different reaction parameters and the growth scheme of Zn_2GeO_4 nanostructures by hydrothermal process with CTAB as template (Huang et al. 2008)

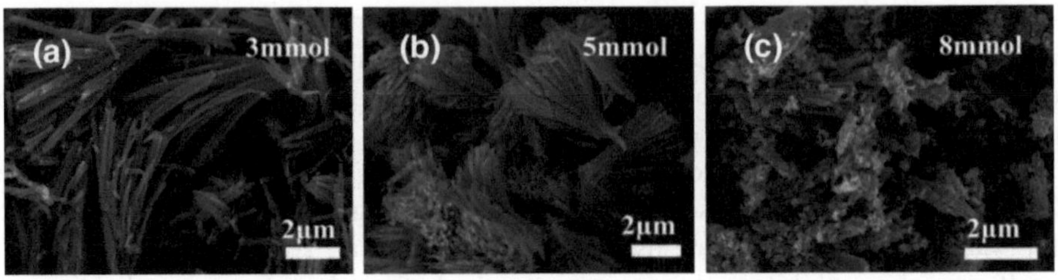

Fig. 11.32 Growth control of the SnO_2 morphology with different dosages of Sn salt added into reaction solution (Han et al. 2011a, b)

control SnO_2. H_2 evolution tests have been performed in aqueous solution, under visible-light irradiation, and it was demonstrated that Sn_3O_4 can promote water splitting with a significant efficiency, whereas neither SnO nor SnO_2 is active in these circumstances.

SnO_2 itself is an important n-type semiconductor, possessing interesting optical, photocatalytic, and optoelectronic properties, being widely studied for electronic and optoelectronic devices. Hence, Han et al. have successfully synthetized porous SnO_2 nanowire bundles (Fig. 11.32) for photocatalysts and Li-ion battery applications (Han et al. 2011a, b). The resulting nanowire bundles exhibited remarkable photocatalytic effect for the degradation of RhB under UV illumination. Electrochemical measurements have shown that these bundles, when are used as anode material, can demonstrate a high de-lithiation capacity and good cycle stability, being useful in Li-ion batteries.

Even if Pb-based materials are a bit "stepchild" in materials science due to its health and environmental issues, few articles can be found which are dealing with lead-based nanomaterials/composites. $PbTiO_3$ nanoparticles were produced by Sobhani–Nasab et al., using lead-acetate, trimeric acid, and $Ti(OC_4H_9)_4$ through a novel sol–gel method: The photocatalytic activity of the $PbTiO_3$ nanoparticles were studied by performing bleaching of RhB and MB solutions under UV light irradiation. Furthermore, the solar cell, using lead-titanate showed promising photoelectrochemical results (Sobhani-Nasab et al. 2015).

Pb-based perovskites for photochemical and solar cells-based applications were synthetized by Li et al. (2015a, b, c), obtaining a flat and uniform perovskite ($CH_3NH_3PbI_3$) layer on TiO_2 film, exhibiting a high efficiency (11.2% under AM 1.5G conditions) in a photovoltaic device. Another report discussed the temperature dependence of thermal conductivity of a single and polycrystalline organometallic perovskite, taking under the loupe the importance of the resonant scattering of the samples, obtaining in this way an ultralow thermal conductivity (0.5 W/(mK) at room temperature) (Pisoni et al. 2014).

Another interesting synthesis and application of lead-based materials was described by Fan et al.the literature (Fan et al. 2010), where $Pb_3Nb_4O_{13}$, a typical multi-metal oxide, which has high nucleation density during the crystallization process was synthetized in its mesoporous form, having a large specific surface area (51–95 $m^2 \cdot g^{-1}$), and exhibiting high activity in photodegradation of 2-propanol.

O, S,

Se in Photocatalytic Materials and Other Composites

The present section of this book chapter has to be started with an explanation: the first, and undoubtedly, the most important member of the group has so many variabilities in nanostructures and nanocomposites with different applications, that the largeness of the present work is way too small to be able to discuss in an adequate manner, therefore, discussion over oxygen-based nanostructures will be omitted from the present work.

Sulfur-based nanomaterials are attracting more and more attention in the last decade. For example, more than 30 kinds of sulfide materials can act as efficient photocatalysts and even more, as efficient catalysts for hydrogen production via water splitting, almost all of the sulfide catalysts consisting of metal cations with d^{10} configuration (Cu, Ag, Zn, Cd, Ga, In, Ge, and Sn) (Zhang and Guo 2013).

Selenium is the only chemical element, which exhibits photocatalytic activity just by "itself," without the necessity to be used in the form of compound. Recently, two groups have dealt with the synthesis of selenium nanoparticles. Zhang et al. (Lv et al. 2015) have prepared one-dimensional single crystalline trigonal selenium nanorods, by dispersing the prepared amorphous α-Se spheres in ethanol. This structure can be excited under visible light, as its band gap energy is very low (1.56 eV). The Se nanorods were found for the first time to display noticeable photocatalytic activity under visible light toward photodegradation of MO. Chiou et al. have also prepared Se nanorods (Chiou and Hsu 2011), via chemical reduction approach at room temperature with noticeable activity toward MB degradation in dark environment after subjected to a short period of irradiation, attributed to the "memory effect" related to pre-irradiation treatment. Compared to the P25/TiO_2 powder, the as-synthetized Se nanorods exhibited superior performance under UV illumination, demonstrating their potential as active photocatalyst in redox reactions. Furthermore, the recycling tests revealed that the Se nanoparticles could be promisingly used in long-term course of photocatalysis.

Conclusions and Final Remarks

This chapter filed several composite materials for several application areas. During this documentation process the following main research threads can be formulated:

- Some of the composite materials are extremely intensively investigated for photocatalytic purposes, while other types of applications are secondary to this.
- In the case of the composites where other applications are in focus, the photocatalytic exploitation is still needed to be carried out.

These two points show that the knowledge transfer between photocatalysis and the other research areas dealing with composites is rather weak. Of course, there are exceptions, such as the case of WO_3 where the electron acceptor capacity is exploited both in photocatalysis and in gas sensors and smart windows as well.

Furthermore, it is clear that the future of photocatalytic materials lies in the composites, as nearly all the possible compounds' individual photocatalytic properties are exploited. An efficient way for doing it, is by the careful examination of the nanomaterials' other application areas. By carrying out this screening, important properties could be imported from other applications. This includes, specific adsorption capacities, special electron conduction mechanisms, innovative band engineering, and specific redox couples.

To achieve the above-mentioned knowledge fusion a large amount of work is needed as composite materials require several optimization processes, such as: choosing the right components, selecting the buildup methodology, tuning the individual component properties (crystal phase composition, crystal size, morphology, surface functionalization properties, etc.) and finally choosing the right composite supports which can facilitate their applicability.

If the above-mentioned large amount of work can be done than it sure that composite photocatalysts will be a real and widespread tool in water and air decontamination at large scale, replacing currently available technologies.

References

Abdullah H, Naim NM, Hamid AA, Umar AA (2016) Characterization and fabrication of nanocomposite thin films of PANI embedded with Ag-Mn alloy for *E. coli* sensor. Mater Today: Proc 3:538–544

Achouri F, Corbel S, Aboulaich A, Balan L, Ghrabi A, Ben Said M, Schneider R (2014) Aqueous synthesis and enhanced photocatalytic activity of ZnO/Fe2O3 heterostructures. J Phys Chem Solids 75:1081–1087

Akbarzadeh A, Samiei M, Davaran S (2012) Magnetic nanoparticles: preparation, physical properties, and applications in biomedicine. Nanoscale Res Lett 7:1–13

Akbayrak S, Tanyıldızı S, Morkan İ, Özkar S (2014) Ruthenium(0) nanoparticles supported on nanotitania as highly active and reusable catalyst in hydrogen generation from the hydrolysis of ammonia borane. Int J Hydrogen Energy 39:9628–9637

Al-Dossary M, Ismail AA, Fierro JLG, Bouzid H, Al-Sayari SA (2015) Effect of Mn loading onto MnFeO nanocomposites for the CO2 hydrogenation reaction. Appl Catal B 165:651–660

Aly IHM, Abed Alrahim Mohammed L, Al-Meer S, Elsaid K, NaM Barakat (2016) Preparation and characterization of wollastonite/titanium oxide nanofiber bioceramic composite as a future implant material. Ceram Int 42:11525–11534

Anku WW, Oppong SO-B, Shukla SK, Agorku ES, Govender PP (2016) Palladium-doped–ZrO_2–multiwalled carbon nanotubes nanocomposite: an advanced photocatalyst for water treatment. Appl Phys A 122

Antoniou MG, Dionysiou DD (2007) Application of immobilized titanium dioxide photocatalysts for the degradation of creatinine and phenol, model organic contaminants found in NASA's spacecrafts wastewater streams. Catal Today 124:215–223

Aravind SSJ, Costa M, Pereira V, Mugweru A, Ramanujachary K, Vaden TD (2014) Molybdenum/graphene—based catalyst for hydrogen evolution reaction synthesized by a rapid photothermal method. Int J Hydrogen Energy 39:11528–11536

Arco LGD, Zhang Y, Schlenker CW, Ryu K, Thompson ME, Zhou C (2010) Continuous, highly flexible, and transparent graphene films by chemical vapor deposition for organic photovoltaics. ACS Nano 4:2865–2873

Arivuoli D, Gnanam FD, Ramasamy P (1988) Growth and microhardness studies of chalcogneides of arsenic, antimony and bismuth. J Mater Sci Lett 7:711–713

Arunraja L, Thirumoorthy P, Karthik A, Rajendran V, Edwinpaul L (2016) EDTA-decorated nanostructured ZnO/CdS thin films for oxygen gas sensing applications. J Electron Mater 45:4100–4107

Asai R, Nemoto H, Jia Q, Saito K, Iwase A, Kudo A (2014) A visible light responsive rhodium and antimony-codoped SrTiO3 powdered photocatalyst loaded with an IrO2 cocatalyst for solar water splitting. Chem Commun (Camb) 50:2543–2546

Asim N, Radiman S, Yarmo MA, Banaye Golriz MS (2009) Vanadium pentoxide: Synthesis and characterization of nanorod and nanoparticle V2O5 using CTAB micelle solution. Microporous Mesoporous Mater 120:397–401

Avansi W Jr, Ribeiro C, Leite ER, Mastelaro VR (2010) Growth kinetics of vanadium pentoxide nanostructures under hydrothermal conditions. J Cryst Growth 312:3555–3559

Bahnemann W, Muneer M, Haque MM (2007) Titanium dioxide-mediated photocatalysed degradation of few selected organic pollutants in aqueous suspensions. Catal Today 124:133–148

Bai S, Wang L, Chen X, Du J, Xiong Y (2015) Chemically exfoliated metallic MoS2 nanosheets: A promising supporting co-catalyst for enhancing the photocatalytic performance of TiO2 nanocrystals. Nano Research 8:175–183

Banerjee S, Gopal J, Muraleedharan P, Tyagi AK, Raj B (2006) Physics and chemistry of photocatalytic titanium dioxide: Visualization of bactericidal activity using atomic force microscopy. Res Commun 90:1378–1383

Bhunia R, Das S, Dalui S, Hussain S, Paul R, Bhar R, Pal AK (2016) Flexible nano-ZnO/polyvinylidene difluoride piezoelectric composite films as energy harvester. Appl Phys A 122:1–13

Biswal N, Das DP, Martha S, Parida KM (2011) Efficient hydrogen production by composite photocatalyst CdS–ZnS/Zirconium–titanium phosphate (ZTP) under visible light illumination. Int J Hydrogen Energy 36:13452–13460

Boyer C, Whittaker MR, Bulmus V, Liu J, Davis TP (2010) The design and utility of polymer-stabilized iron-oxide nanoparticles for nanomedicine applications. NPG Asia Mater 2:23–30

Can MM, Coşkun M, Fırat T (2012) A comparative study of nanosized iron oxide particles; magnetite (Fe_3O_4), maghemite (γ-Fe_2O_3) and hematite (α-Fe_2O_3), using ferromagnetic resonance. J Alloy Compd 542:241–247

Cao S, Liu T, Hussain S, Zeng W, Peng X, Pan F (2015) Hydrothermal synthesis, characterization and optical absorption property of nanoscale WS2/TiO2 composites. Physica E 68:171–175

Carević MV, Abazović ND, Novaković TB, Pavlović VB, Čomor MI (2016) Zirconium dioxide nanopowders with incorporated Si4 + ions as efficient photocatalyst for degradation of trichlorophenol using simulated solar light. Appl Catal B 195:112–120

Chandra S, Bag S, Bhar R, Pramanik P (2010) Sonochemical synthesis and application of rhodium–graphene nanocomposite. J Nanopart Res 13:2769–2777

Chang C, Fu Y, Hu M, Wang C, Shan G, Zhu L (2013) Photodegradation of bisphenol A by highly stable palladium-doped mesoporous graphite carbon nitride (Pd/mpg-C3N4) under simulated solar light irradiation. Appl Catal B 142–143:553–560

Chen L, Wang J, Meng D, Xing Y, Tian X, Yu X, Xu K, Wu X (2015) Effects of citric acid and urea on the structural and morphological characteristics of BiVO4 synthesized by the sol–gel combustion method. J Sol-Gel Sci Technol 76:562–571

Chew LM, Ruland H, Schulte HJ, Xia W, Muhler M (2014) CO_2 hydrogenation to hydrocarbons over iron nanoparticles supported on oxygen-functionalized carbon nanotubes. J Chem Sci 126:481–486

Chiou YD, Hsu YJ (2011) Room-temperature synthesis of single-crystalline Se nanorods with remarkable photocatalytic properties. Appl Catal B 105:211–219

Cho Y, Choi W, Lee C-H, Hyeon T, Lee H-I (2001) Visible light-induced degradation of carbon tetrachloride on dye-sensitized TiO_2. Environ Sci Technol 35:966–970

Choi H, Stathatos E, Dionysiou DD (2006) Sol-gel preparation of mesoporous photocatalytic TiO_2 films and TiO_2/Al_2O_3 composite membranes for environmental applications. Appl Catal B 63:60–67

Christoforidis KC, Montini T, Bontempi E, Zafeiratos S, Jaén JJD, Fornasiero P (2016) Synthesis and photocatalytic application of visible-light active β-Fe_2O_3/g-C_3N_4 hybrid nanocomposites. Appl Catal B 187:171–180

Cong Y, Chen M, Xu T, Zhang Y, Wang Q (2014) Tantalum and aluminum co-doped iron oxide as a robust photocatalyst for water oxidation. Appl Catal B 147:733–740

Cozzi D, De Bonis C, D'epifanio A, Mecheri B, Tavares AC, Licoccia S (2014) Organically functionalized titanium oxide/Nafion composite proton exchange membranes for fuel cells applications. J Power Sour 248:1127–1132

Cudennec Y, Lecerf A (2005) Topotactic transformations of goethite and lepidocrocite into hematite and maghemite. Solid State Sci 7:520–529

Cui G, Wang W, Ma M, Xie J, Shi X, Deng N, Xin J, Tang B (2015) IR-driven photocatalytic water splitting with WO_2–NaxWO_3 hybrid conductor material. Nano Lett 15:7199–7203

Dai Prè M, Morrow I, Martin DJ, Mos M, Del Negro A, Padovani S, Martucci A (2013) Preparation and characterization of down shifting ZnS:Mn/PMMA nanocomposites for improving photovoltaic silicon solar cell efficiency. Mater Chem Phys 139:531–536

Deshmane VG, Adewuyi YG (2012) Synthesis of thermally stable, high surface area, nanocrystalline mesoporous tetragonal zirconium dioxide (ZrO2): Effects of different process parameters. Microporous Mesoporous Mater 148:88–100

Di Paola A, García-López E, Marcì G, Palmisano L (2012) A survey of photocatalytic materials for environmental remediation. J Hazard Mater 211–212:3–29

Dolinska J, Chidambaram A, Taleat Z, Adamkiewicz W, Lisowski W, Palys B, Holdynski M, Andryszewski T, Sashuk V, Rassaei L, Opallo M (2015) Decoration of MoS_2 nanopetal stacks with positively charged gold nanoparticles for synergistic electrocatalytic oxidation of biologically relevant compounds. Electrochim Acta 182:659–667

Domen K, Kudo A, Tanaka A, Onishi T (1990a) Overall photodecomposition of water on a layered niobiate catalyst. Catal Today 8:77–84

Domen K, Yoshimura J, Sekine T, Tanaka A, Onishi T (1990b) A novel series of photocatalysts with an ion-exchangeable layered structure of niobate. Catal Lett 4:339–343

Dong G, Zhang Y, Pan Q, Qiu J (2014) A fantastic graphitic carbon nitride (g-C3N4) material: Electronic structure, photocatalytic and photoelectronic properties. J Photochem Photobiol, C 20:33–50

Dorner RW, Hardy DR, Williams FW, Willauer HD (2010) K and Mn doped iron-based CO_2 hydrogenation catalysts: detection of KAlH4 as part of the catalyst's active phase. Appl Catal A 373:112–121

Egerton TA, Mattinson JA (2008) The influence of platinum on UV and 'visible' photocatalysis by rutile and Degussa P25. J Photochem Photobiol, A 194:283–289

Fan X, Gao J, Wang Y, Li Z, Zou Z (2010) Effect of crystal growth on mesoporous Pb_3Nb_4O 13 formation, and their photocatalytic activity under visible-light irradiation. J Mater Chem 20:2865–2869

Fan X, Zang L, Zhang M, Qiu H, Wang Z, Yin J, Jia H, Pan S, Wang C (2014) A bulk boron-based photocatalyst for efficient dechlorination: K 3B6O10Br. Chem Mater 26:3169–3174

Fan L, Zhang S, Zhang X, Zhou H, Lu Z, Wang S (2015) Removal of arsenic from simulation wastewater using nano-iron/oyster shell composites. J Environ Manage 156:109–114

Fan G, Liu Q, Tang D, Li X, Bi J, Gao D (2016) Nanodiamond supported Ru nanoparticles as an effective catalyst for hydrogen evolution from hydrolysis of ammonia borane. Int J Hydrogen Energy 41:1542–1549

Fang X, Guo X, Mao Y, Hua C, Shen L, Hu Y, Wang Z, Wu F, Chen L (2012) Mechanism of lithium storage in MoS_2 and the feasibility of using Li_2S/Mo nanocomposites as cathode materials for lithium–sulfur batteries. Chem—An Asian J 7:1013–1017

Fang M, Jia H, He W, Lei Y, Zhang L, Zheng Z (2015) Construction of flexible photoelectrochemical solar cells based on ordered nanostructural BiOI/Bi2S3 heterojunction films. Phys Chem Chem Phys 17:13531–13538

Fang LJ, Wang XL, Li YH, Liu PF, Wang YL, Zeng HD, Yang HG (2017) Nickel nanoparticles coated with graphene layers as efficient co-catalyst for photocatalytic hydrogen evolution. Appl Catal B 200:578–584

Fernández-Ibáñez P, Polo-López MI, Malato S, Wadhwa S, Hamilton JWJ, Dunlop PSM, D'sa R, Magee E, O'shea K, Dionysiou DD, Byrne JA (2015) Solar photocatalytic disinfection of water using titanium dioxide graphene composites. Chem Eng J 261:36–44

Fihri A, Artero V, Pereira A, Fontecave M (2008) Efficient H2-producing photocatalytic systems based on cyclometalated iridium—and tricarbonylrhenium-diimine photosensitizers and cobaloxime catalysts. Dalton Trans 5567–5569

Fu Y, Sun X, Wang X (2011) BiVO4–graphene catalyst and its high photocatalytic performance under visible light irradiation. Mater Chem Phys 131:325–330

Gao Z, Liu N, Wu D, Tao W, Xu F, Jiang K (2012) Graphene–CdS composite, synthesis and enhanced photocatalytic activity. Appl Surf Sci 258:2473–2478

Gar Alalm M, Tawfik A, Ookawara S (2016) Enhancement of photocatalytic activity of TiO_2 by immobilization on activated carbon for degradation of pharmaceuticals. J Environ Chem Eng 4:1929–1937

Ge L, Han C, Liu J (2011) Novel visible light-induced g-C3N4/Bi2WO6 composite photocatalysts for efficient degradation of MO. Appl Catal B 108–109:100–107

Gou W, Wu P, Jiang D, Ma X (2015) Synthesis of AgBr@Bi2O3 composite with enhanced photocatalytic performance under visible light. J Alloy Compd 646:437–445

Goudarzi M, Salavati-Niasari M (2016) Controllable synthesis of new Tl2S2O3 nanostructures via hydrothermal process; Characterization and investigation photocatalytic activity for degradation of some anionic dyes. J Mol Liq 219:851–857

Goudarzi M, Salavati-Niasari M, Hosseinpour-Mashkani SM, Mir N (2015) Controlled synthesis of Tl_2O_3 nanostructures via microwave route by a novel pH adjuster and investigation of its photocatalytic activity. J Mater Sci: Mater Electron 26:5326–5334

Gumy D, Rincon A, Hajdu R, Pulgarin C (2006) Solar photocatalysis for detoxification and disinfection of water: different types of suspended and fixed TiO_2 catalysts study. Sol Energy 80:1376–1381

Gupta AK, Gupta M (2005) Synthesis and surface engineering of iron oxide nanoparticles for biomedical applications. Biomaterials 26:3995–4021

Han Q, Sun S, Sun D, Zhu J, Wang X (2011a) Room-temperature synthesis from molecular precursors and photocatalytic activities of ultralong Sb2S3 nanowires. RSC Advances 1:1364–1369

Han Y, Wu X, Ma Y, Gong L, Qu F, Fan H (2011b) Porous SnO_2 nanowire bundles for photocatalyst and Li ion battery applications. CrystEngComm 13:3506–3510

Hanbicki AT, Currie M, Kioseoglou G, Friedman AL, Jonker BT (2015) Measurement of high exciton binding energy in the monolayer transition-metal dichalcogenides WS2 and WSe2. Solid State Commun 203:16–20

Hassanzadeh-Tabrizi SA, Motlagh MM, Salahshour S (2016) Synthesis of ZnO/CuO nanocomposite immobilized on γ-Al_2O_3 and application for removal of MO. Appl Surf Sci 384:237–243

Haw CY, Mohamed F, Chia CH, Radiman S, Zakaria S, Huang NM, Lim HN (2010) Hydrothermal synthesis of magnetite nanoparticles as MRI contrast agents. Ceram Int 36:1417–1422

Hayat K, Gondal MA, Khaled MM, Ahmed S (2011) Effect of operational key parameters on photocatalytic degradation of phenol using nano nickel oxide synthesized by sol–gel method. J Mol Catal A: Chem 336:64–71

He Y, Sheng T, Chen J, Fu R, Hu S, Wu X (2009) Photodegradation of organics over a new composite catalyst $V_2O_5/SmVO_4$. Catal Commun 10:1354–1357

He G-H, Liang C-J, Ou Y-D, Liu D-N, Fang Y-P, Xu Y-H (2013) Preparation of novel Sb_2O_3/WO_3 photocatalysts and their activities under visible light irradiation. Mater Res Bull 48:2244–2249

Herrmann J-M, Matos J, Disdier J, Guillard C, Laine J, Malato S, Blanco J (1999) Solar photocatalytic degradation of 4-chlorophenol using the synergistic effect between titania and activated carbon in aqueous suspension. Catal Today 54:255–265

Hintsho N, Petrik L, Nechaev A, Titinchi S, Ndungu P (2014) Photo-catalytic activity of titanium dioxide carbon nanotube nano-composites modified with silver and palladium nanoparticles. Appl Catal B 156–157:273–283

Hirai T, Bando Y, Komasawa I (2002) Immobilization of CdS nanoparticles formed in reverse micelles onto alumina particles and their photocatalytic properties. J Phys Chem B 106:8967–8970

Hou Y, Wang X, Wu L, Ding Z, Fu X (2006) Efficient decomposition of benzene over a β-Ga2O 3 photocatalyst under ambient conditions. Environ Sci Technol 40:5799–5803

Hou Y, Wu L, Wang X, Ding Z, Li Z, Fu X (2007) Photocatalytic performance of α-, β-, and γ-Ga2O3 for the destruction of volatile aromatic pollutants in air. J Catal 250:12–18

Hsieh S-H, Lee G-J, Chen C-Y, Chen J-H, Ma S-H, Horng T-L, Chen K-H, Wu JJ (2013) Hydrothermal synthesis of mesoporous Bi_2O_3/Co_3O_4 microsphere and photocatalytic degradation of orange II Dyes by visible light. Top Catal 56:623–629

Hsu Y-K, Chen Y-C, Lin Y-G (2015) Novel ZnO/Fe_2O_3 core–shell nanowires for photoelectrochemical water splitting. ACS Appl Mater Interfaces 7:14157–14162

Hu KH, Liu Z, Huang F, Hu XG, Han CL (2010) Synthesis and photocatalytic properties of nano-MoS_2/kaolin composite. Chem Eng J 162:836–843

Hu K-H, Zhao D-F, Liu J-S (2012) Synthesis of nano-MoS2/bentonite composite and its application for removal of organic dye. Trans Nonferrous Metals Soc China 22:2484–2490

Hu S, Ma L, You J, Li F, Fan Z, Wang F, Liu D, Gui J (2014) A simple and efficient method to prepare a phosphorus modified g-C 3N4 visible light photocatalyst. RSC Adv 4:21657–21663

Hu SW, Yang LW, Tian Y, Wei XL, Ding JW, Zhong JX, Chu PK (2015) Simultaneous nanostructure and heterojunction engineering of graphitic carbon nitride via in situ Ag doping for enhanced photoelectrochemical activity. Appl Catal B 163:611–622

Hu D, Diao P, Xu D, Wu Q (2016) Gold/WO_3 nanocomposite photoanodes for plasmonic solar water splitting. Nano Res 9:1735–1751

Huang J, Ding K, Hou Y, Wang X, Fu X (2008) Synthesis and photocatalytic activity of Zn_2GeO_4 nanorods for the degradation of organic pollutants in water. Chemsuschem 1:1011–1019

Huang K-J, Liu Y-J, Wang H-B, Gan T, Liu Y-M, Wang L-L (2014) Signal amplification for electrochemical DNA biosensor based on two-dimensional graphene analogue tungsten sulfide–graphene composites and gold nanoparticles. Sens Actuators B: Chem 191:828–836

Hung CM (2009) Catalytic wet oxidation of ammonia solution: activity of the nanoscale platinum-palladium-rhodium composite oxide catalyst. J Hazard Mater 163:180–186

Huo J, Zeng H (2016) Silver nanoparticles-sensitized cobalt complex for highly-efficient photocatalytic activity. Appl Catal B 199:342–349

Hwang JY, El-Kady MF, Wang Y, Wang L, Shao Y, Marsh K, Ko JM, Kaner RB (2015) Direct preparation and processing of graphene/RuO_2 nanocomposite electrodes for high-performance capacitive energy storage. Nano Energy 18:57–70

Jansson I, Yoshiiri K, Hori H, García-García FJ, Rojas S, Sánchez B, Ohtani B, Suárez S (2016) Visible light responsive Zeolite/WO_3–Pt hybrid photocatalysts for degradation of pollutants in air. Appl Catal A 521:208–219

Jasso-Salcedo AB, Meimaroglou D, Hoppe S, Pla F, Escobar-Barrios VA (2016) Surface modification and immobilization in poly(acrylic acid) of Ag/ZnO for photocatalytic degradation of endocrine-disrupting compounds. J Appl Polym Sci 133:n/a–n/a

Jayaraman T, Arumugam Raja S, Priya A, Jagannathan M, Ashokkumar M (2015) Synthesis of a visible-light active V_2O_5-g-C_3N_4 heterojunction as an efficient photocatalytic and photoelectrochemical material. New J Chem 39:1367–1374

Jiang H, Nagai M, Kobayashi K (2009) Enhanced photocatalytic activity for degradation of MB over V_2O_5/$BiVO_4$ composite. J Alloy Compd 479:821–827
Jiang Z, Xiao T, Kuznetsov VL, Edwards PP (2010) Turning carbon dioxide into fuel. Philos Trans R Soc A: Math Phys Eng Sci 368:3343–3364
Jiao L, Zhang L, Wang X, Diankov G, Dai H (2009) Narrow graphene nanoribbons from carbon nanotubes. Nature 458:877–880
Kafizas A, Kellici S, Darr JA, Parkin IP (2009) Titanium dioxide and composite metal/metal oxide titania thin films on glass: a comparative study of photocatalytic activity. J Photochem Photobiol, A 204:183–190
Kamegawa T, Shudo T, Yamashita H (2010) Preparation of Cr–Ti binary oxide anchored mesoporous silica by CVD method and their photocatalytic activities. Top Catal 53:555–559
Karácsonyi É, Baia L, Dombi A, Danciu V, Mogyorósi K, Pop LC, Kovács G, Coşoveanu V, Vulpoi A, Simon S, Pap Z (2013) The photocatalytic activity of TiO_2/WO_3/noble metal (Au or Pt) nanoarchitectures obtained by selective photodeposition. Catal Today 208:19–27
Karunakaran C, Karuthapandian S (2015) V_2O_5-photocatalyzed oxidation of diphenylamine. In: Materials Science Forum, pp 81–90
Karunakaran C, Sakthiraadha S, Gomathisankar P (2013) Photocatalytic and bactericidal activities of hydrothermally and sonochemically prepared Fe_2O_3–SnO_2 nanoparticles. Mater Sci Semicond Process 16:818–824
Khan IA, Ullah S, Nasim F, Choucair M, Nadeem MA, Iqbal A, Badshah A, Nadeem MA (2016) Cr_2O_3–carbon composite as a new support material for efficient methanol electrooxidation. Mater Res Bull 77:221–227
Kim W, Tachikawa T, Majima T, Choi W (2009) Photocatalysis of dye-sensitized TiO_2 nanoparticles with thin overcoat of Al_2O_3: enhanced activity for H2 production and dechlorination of CCl_4. J Phys Chem C 113:10603–10609
Kominami H, Oki K, Kohno M, Onoue S-I, Kera Y, Ohtani B (2001) Novel solvothermal synthesis of niobium(v) oxide powders and their photocatalytic activity in aqueous suspensions. J Mater Chem 11:604–609
Kovács K, Farkas J, Vereb G, Arany E, Simon G, Schrantz K, Dombi A, Hernadi K, Alapi T (2016) Comparison of various advanced oxidation processes for the degradation of phenylurea herbicides. J Environ Sci Health B 51:205–214
Koziej D, Rossell MD, Ludi B, Hintennach A, Novák P, Grunwaldt J-D, Niederberger M (2011) Interplay between size and crystal structure of molybdenum dioxide nanoparticles—synthesis, growth mechanism, and electrochemical performance. Small 7:377–387
Kudo A (2006) Development of photocatalyst materials for water splitting. Int J Hydrogen Energy 31:197–202
Kudo A, Sayama K, Tanaka A, Asakura K, Domen K, Maruya K, Onishi T (1989) Nickel-loaded K4Nb6O17 photocatalyst in the decomposition of H_2O into H_2 and O_2: Structure and reaction mechanism. J Catal 120:337–352
Kumar A, Kumar P, Paul S, Jain SL (2016) Visible light assisted reduction of nitrobenzenes using Fe(bpy)3 + 2/rGO nanocomposite as photocatalyst. Appl Surf Sci 386:103–114
Kun R, Tarján S, Oszkó A, Seemann T, Zöllmer V, Busse M, Dékány I (2009) Preparation and characterization of mesoporous N-doped and sulfuric acid treated anatase TiO_2 catalysts and their photocatalytic activity under UV and Vis illumination. J Solid State Chem 182:3076–3084
Lam SM, Sin JC, Mohamed AR (2016) A review on photocatalytic application of g-C_3N_4/semiconductor (CNS) nanocomposites towards the erasure of dyeing wastewater. Mater Sci Semicond Process 47:62–84
Lau VW-H, Moudrakovski I, Botari T, Weinberger S, Mesch MB, Duppel V, Senker J, Blum V, Lotsch BV (2016) Rational design of carbon nitride photocatalysts by identification of cyanamide defects as catalytically relevant sites. Nat Commun 7
Laurent S, Dutz S, Häfeli UO, Mahmoudi M (2011) Magnetic fluid hyperthermia: Focus on superparamagnetic iron oxide nanoparticles. Adv Coll Interface Sci 166:8–23

Lee J-M, Han S-B, Lee Y-W, Song Y-J, Kim J-Y, Park K-W (2010) RuO_2–SnO_2 nanocomposite electrodes for methanol electrooxidation. J Alloy Compd 506:57–62
Li L, Yan B (2009) CeO_2–Bi_2O_3 nanocomposite: two step synthesis, microstructure and photocatalytic activity. J Non-Cryst Solids 355:776–779
Li B, Xie Y, Jing M, Rong G, Tang Y, Zhang G (2006) In_2O_3 hollow microspheres: synthesis from designed $In(OH)_3$ precursors and applications in gas sensors and photocatalysis. Langmuir 22:9380–9385
Li L, Zhu ZH, Yan ZF, Lu GQ, Rintoul L (2007) Catalytic ammonia decomposition over Ru/carbon catalysts: the importance of the structure of carbon support. Appl Catal A 320:166–172
Li X, Kikugawa N, Ye J (2008) Nitrogen-doped lamellar niobic acid with visible light-responsive photocatalytic activity. Adv Mater 20:3816–3819
Li C, Wang F, Zhu J, Yu JC (2010) NaYF4:Yb, Tm/CdS composite as a novel near-infrared-driven photocatalyst. Appl Catal B 100:433–439
Li B, Liu T, Wang Y, Wang Z (2012) ZnO/graphene-oxide nanocomposite with remarkably enhanced visible-light-driven photocatalytic performance. J Colloid Interface Sci 377:114–121
Li X, Xu X, Xia F, Bu L, Qiu H, Chen M, Zhang L, Gao J (2014) Electrochemically active MnO_2/RGO nanocomposites using Mn powder as the reducing agent of GO and the MnO_2 precursor. Electrochim Acta 130:305–313
Li FT, Zhao Y, Wang Q, Wang XJ, Hao YJ, Liu RH, Zhao D (2015a) Enhanced visible-light photocatalytic activity of active Al_2O_3/g-C_3N_4 heterojunctions synthesized via surface hydroxyl modification. J Hazard Mater 283:371–381
Li L, Niu S, Qu Y, Zhang Q, Li H, Li Y, Zhao W, Shi J (2012) J Mater Chem 22:9263
Li SG, Jiang KJ, Su MJ, Cui XP, Huang JH, Zhang QQ, Zhou XQ, Yang LM, Song YL (2015b) Inkjet printing of $CH_3NH_3PbI_3$ on a mesoscopic TiO_2 film for highly efficient perovskite solar cells. J Mater Chem A 3:9092–9097
Li Z, Kong C, Lu G (2015c) Rhodium tin composite oxides co-catalyst for high efficient photocatalytic hydrogen evolution. Int J Hydrogen Energy 40:9061–9068
Li J, Yang C, Wu Y, Wang B, Sun W, Shao T (2016) On a magnetic-mesoporous composite loaded with emissive Ru(II) complex for oxygen sensing application: Construction, characterization and emission response to oxygen molecules. Inorg Chim Acta 442:111–118
Liang H, Jiang X, Chen W, Wang S, Xu B, Wang Z (2014) α-Fe_2O_3/Pt hybrid nanorings and their enhanced photocatalytic activities. Ceram Int 40:5653–5658
Liang Q, Zhang M, Liu C, Xu S, Li Z (2016) Sulfur-doped graphitic carbon nitride decorated with zinc phthalocyanines towards highly stable and efficient photocatalysis. Appl Catal A 519:107–115
Liao Y, Li H, Liu Y, Zou Z, Zeng D, Xie C (2010) Characterization of photoelectric properties and composition effect of TiO_2/ZnO/Fe_2O_3 composite by combinatorial methodology. J Comb Chem 12:883–889
Lin Y, Cui X, Yen C, Wai CM (2005) Platinum/Carbon nanotube nanocomposite synthesized in supercritical fluid as electrocatalysts for low-temperature fuel cells. J Phys Chem B 109:14410–14415
Lin F, Montano M, Tian C, Ji Y, Nordlund D, Weng T-C, Moore RG, Gillaspie DT, Jones KM, Dillon AC, Richards RM, Engtrakul C (2014) Electrochromic performance of nanocomposite nickel oxide counter electrodes containing lithium and zirconium. Sol Energy Mater Sol Cells 126:206–212
Lin W, Zheng H, Zhang P, Xu T (2016) Pt deposited TiO_2 films with exposed 001 facets for photocatalytic degradation of a pharmaceutical pollutant. Appl Catal A 521:75–82
Liu J, Zhang Z, Pan C, Zhao Y, Su X, Zhou Y, Yu D (2004) Enhanced field emission properties of MoO_2 nanorods with controllable shape and orientation. Mater Lett 58:3812–3815
Liu X, Zhang J, Yang T, Guo X, Wu S, Wang S (2011) Synthesis of Pt nanoparticles functionalized WO_3 nanorods and their gas sensing properties. Sens Actuators B: Chem 156:918–923
Liu H, Cao W, Su Y, Wang Y, Wang X (2012) Synthesis, characterization and photocatalytic performance of novel visible-light-induced Ag/BiOI. Appl Catal B 111–112:271–279

Liu J, Wen S, Hou Y, Zuo F, Beran GJO, Feng P (2013) Boron carbides as efficient, metal-free, visible-light-responsive photocatalysts. Angewandte Chemie—Intl Edn 52:3241–3245

Liu J, Wen Y, Wang Y, Van Aken PA, Maier J, Yu Y (2014) Carbon-encapsulated pyrite as stable and earth-abundant high energy cathode material for rechargeable lithium batteries. Adv Mater 26:6025–6030

Liu X, Chen K, Shim J-J, Huang J (2015a) Facile synthesis of porous Fe_2O_3 nanorods and their photocatalytic properties. J Saudi Chem Soc 19:479–484

Liu X, Liu J, Chu H, Li J, Yu W, Zhu G, Niu L, Sun Z, Pan L, Sun CQ (2015b) Enhanced photocatalytic activity of Bi_2O_3–Ag_2O hybrid photocatalysts. Appl Surf Sci 347:269–274

Liu H, Guo K, Duan C, Chen X, Zhu Z (2016a) A novel biosensor based on the direct electrochemistry of horseradish peroxidase immobilized in the three-dimensional flower-like Bi2WO6 microspheres. Mater Sci Eng, C 64:243–248

Liu S, Ma L, Zhang H, Ma C (2016b) ZnS/Ni_2P core/shell composites: Simple hydrothermal synthesis, characterization and its photocatalytic degradation of pyronine B. Mater Res Bull 77:271–278

López-Muñoz MJ, Arencibia A, Segura Y, Raez JM (2016) Removal of As(III) from aqueous solutions through simultaneous photocatalytic oxidation and adsorption by TiO_2 and zero-valent iron. Catal Today

Lou Z, Huang B, Wang Z, Ma X, Zhang R, Zhang X, Qin X, Dai Y, Whangbo M-H (2014) $Ag_6Si_2O_7$: a Silicate Photocatalyst for the Visible Region. Chem Mater 26:3873–3875

Lu Y, Hao L, Matsuzaka K, Yoshida H, Asanuma H, Chen J, Pan F (2014) Titanium dioxide–nickel oxide composite coatings: preparation by mechanical coating/thermal oxidation and photocatalytic activity. Mater Sci Semicond Process 24:138–145

Luo W, Hu X, Sun Y, Huang Y (2011) Electrospinning of carbon-coated MoO_2 nanofibers with enhanced lithium-storage properties. Phys Chem Chem Phys 13:16735–16740

Luo X, Wang C, Wang L, Deng F, Luo S, Tu X, Au C (2013) Nanocomposites of graphene oxide-hydrated zirconium oxide for simultaneous removal of As(III) and As(V) from water. Chem Eng J 220:98–106

Lv Y, Chen J, Zheng RK, Song J, Zhang T, Li X, Shi X, Chen L (2015) Photo-induced enhancement of the power factor of Cu2S thermoelectric films. Sci Rep 5:16291

Maczka M, Hanuza J, Paraguassu W, Gomes Souza Filho A, Tarso Cavalcante Freire P, Mendes Filho J (2008) Phonons in ferroelectric Bi[sub 2]WO[sub 6]: Raman and infrared spectra and lattice dynamics. Appl Phys Lett 92:112911

Madhu R, Veeramani V, Chen SM, Manikandan A, Lo AY, Chueh YL (2015) Honeycomb-like porous carbon-cobalt oxide nanocomposite for high-performance enzymeless glucose sensor and supercapacitor applications. ACS Appl Mater Interfaces 7:15812–15820

Maeda K, Saito N, Daling L, Inoue Y, Domen K (2007) Photocatalytic properties of RuO_2-Loaded β-Ge 3N4 for overall water splitting. J Phys Chem C 111:4749–4755

Mahmoud KH (2016) Synthesis and spectroscopic investigation of cobalt oxide nanoparticles. Polym Compos 37:1881–1885

Malato S, Fernández-Ibáñez P, Maldonado MI, Blanco J, Gernjak W (2009) Decontamination and disinfection of water by solar photocatalysis: Recent overview and trends. Catal Today 147:1–59

Manikandan M, Tanabe T, Li P, Ueda S, Ramesh GV, Kodiyath R, Wang J, Hara T, Dakshanamoorthy A, Ishihara S, Ariga K, Ye J, Umezawa N, Abe H (2014) Photocatalytic water splitting under visible light by mixed-valence Sn 3O4. ACS Appl Mater Interfaces 6:3790–3793

Manikantan J, Ramalingam HB, Shekar BC, Murugan B, Kumar RR, Santhoshi JS (2016) Wide band gap of Strontium doped Hafnium oxide nanoparticles for opto-electronic device applications—synthesis and characterisation. Mater Lett

Marszewski M, Cao S, Yu J, Jaroniec M (2015) Semiconductor-based photocatalytic CO_2conversion. Mater Horiz 2:261–278

Maya-Treviño ML, Guzmán-Mar JL, Hinojosa-Reyes L, Ramos-Delgado NA, Maldonado MI, Hernández-Ramírez A (2014) Activity of the ZnO–Fe_2O_3 catalyst on the degradation of Dicamba and 2,4-D herbicides using simulated solar light. Ceram Int 40:8701–8708

Miwa T, Kaneco S, Katsumata H, Suzuki T, Ohta K, Chand Verma S, Sugihara K (2010) Photocatalytic hydrogen production from aqueous methanol solution with CuO/Al_2O_3/TiO_2 nanocomposite. Int J Hydrogen Energy 35:6554–6560

Mohamed RM, Mckinney D, Kadi MW, Mkhalid IA, Sigmund W (2016) Platinum/zinc oxide nanoparticles: enhanced photocatalysts degrade malachite green dye under visible light conditions. Ceram Int 42:9375–9381

Mohan D, Pittman CU (2007) Arsenic removal from water/wastewater using adsorbents—a critical review. J Hazard Mater 142:1–53

Moshfegh AZ, Ignatiev A (1990) Photo-enhanced catalytic decomposition of isopropanol on V_2O_5. Catal Lett 4:113–122

Nguyen T, Vigneswaran S, Ngo H, Kandasamy J, Choi H (2008) Arsenic removal by photo-catalysis hybrid system. Sep Purif Technol 61:44–50

Nikolic AS, Boskovic M, Spasojevic V, Jancar B, Antic B (2014) Magnetite/Mn-ferrite nanocomposite with improved magnetic properties. Mater Lett 120:86–89

Niu C, Lu YZ, Lieber CM (1993) Experimental realization of the covalent solid carbon nitride. Science 261:334–337

Niu M, Huang F, Cui L, Huang P, Yu Y, Wang Y (2010) Hydrothermal Synthesis, Structural Characteristics, and Enhanced Photocatalysis of SnO_2/Fe_2O_3 Semiconductor Nanoheterostructures. ACS Nano 4:681–688

Niu P, Zhang L, Liu G, Cheng HM (2012) Graphene-like carbon nitride nanosheets for improved photocatalytic activities. Adv Func Mater 22:4763–4770

Pan C, Li X, Wang F, Wang L (2008) Synthesis of bismuth oxide nanoparticles by the polyacrylamide gel route. Ceram Int 34:439–441

Pang YL, Lim S, Ong HC, Chong WT (2016) Research progress on iron oxide-based magnetic materials: synthesis techniques and photocatalytic applications. Ceram Int 42:9–34

Parayil SK, Kibombo HS, Wu C-M, Peng R, Baltrusaitis J, Koodali RT (2012) Enhanced photocatalytic water splitting activity of carbon-modified TiO_2 composite materials synthesized by a green synthetic approach. Int J Hydrogen Energy 37:8257–8267

Pawar SH, Bhosale PN, Uplane MD, Tamhankar S (1983) Growth of Bi_2S_3 film using a solution-gas interface technique. Thin Solid Films 110:165–170

Pelaez M, Nolan NT, Pillai SC, Seery MK, Falaras P, Kontos AG, Dunlop PSM, Hamilton JWJ, Byrne JA, O'shea K, Entezari MH, Dionysiou DD (2012) A review on the visible light active titanium dioxide photocatalysts for environmental applications. Appl Catal B: Environ 125:331–349

Perumal Raj K, Thangaraj V, Uthirakumar AP (2016) Enhanced photocatalytic behaviour of synthesized nickel oxide nanoparticles on Fluorescein under different irradiations. Optik—Intl J Light Electron Opt 127:2631–2634

Pingmuang K, Wetchakun N, Kangwansupamonkon W, Ounnunkad K, Inceesungvorn B, Phanichphant S (2013) Photocatalytic mineralization of organic acids over visible-light-driven Au/$BiVO_4$ photocatalyst. Int J Photoenergy 2013:1–7

Pisoni A, Jaćimović J, Barišić OS, Spina M, Gaál R, Forró L, Horváth E (2014) Ultra-low thermal conductivity in organic-inorganic hybrid perovskite $CH_3NH_3PbI_3$. J Phys Chem Lett 5:2488–2492

Pourbeyram S (2016) Effective removal of heavy metals from aqueous solutions by graphene oxide-zirconium phosphate (GO–Zr–P) nanocomposite. Ind Eng Chem Res 55:5608–5617

Preethi V, Kanmani S (2014) Photocatalytic hydrogen production using Fe_2O_3-based core shell nano particles with ZnS and CdS. Int J Hydrogen Energy 39:1613–1622

Qamar M, Merzougui B, Anjum D, Hakeem AS, Yamani ZH, Bahnemann D (2014) Synthesis and photocatalytic activity of mesoporous nanocrystalline Fe-doped titanium dioxide. Catal Today 230:158–165

Quan X, Hu M, Gao X, Fu Y, Weng L, Wang D, Jiang D, Sun J (2016) Friction and wear performance of dual lubrication systems combining WS_2–MoS_2 composite film and low volatility oils under vacuum condition. Tribol Int 99:57–66

Rajpure KY, Bhosale CH (2000) Sb_2S_3 semiconductor-septum rechargeable storage cell. Mater Chem Phys 64:70–74

Ram MK, Yavuz O, Lahsangah V, Aldissi M (2005) CO gas sensing from ultrathin nano-composite conducting polymer film. Sensor Actuat B-Chem 106:750–757

Ramadoss A, Krishnamoorthy K, Kim SJ (2012) Facile synthesis of hafnium oxide nanoparticles via precipitation method. Mater Lett 75:215–217

Ramos-Delgado NA, Gracia-Pinilla MA, Maya-Trevino L, Hinojosa-Reyes L, Guzman-Mar JL, Hernandez-Ramirez A (2013) Solar photocatalytic activity of TiO_2 modified with WO_3 on the degradation of an organophosphorus pesticide. J Hazard Mater 263(Pt 1):36–44

Ran J, Yu J, Jaroniec M (2011) Ni(OH)2 modified CdS nanorods for highly efficient visible-light-driven photocatalytic H_2 generation. Green Chem 13:2708–2713

Rao KVS, Subrahmanyam M, Boule P (2004) Immobilized TiO_2 photocatalyst during long-term use: decrease of its activity. Appl Catal B 49:239–249

Reddy BM, Bharali P, Saikia P, Khan A, Loridant S, Muhler M, Grünert W (2007a) Hafnium doped ceria nanocomposite oxide as a novel redox additive for three-way catalysts. J Phys Chem C 111:1878–1881

Reddy CVS, Wei J, Quan-Yao Z, Zhi-Rong D, Wen C, Mho S-I, Kalluru RR (2007b) Cathodic performance of (V_2O_5 + PEG) nanobelts for Li ion rechargeable battery. J Power Sources 166:244–249

Reutergådh LB, Iangphasuk M (1997) Photocatalytic decolourization of reactive azo dye: a comparison between TiO_2 and us photocatalysis. Chemosphere 35:585–596

Sagara N, Kamimura S, Tsubota T, Ohno T (2016) Photoelectrochemical CO_2 reduction by a p-type boron-doped g-C_3N_4 electrode under visible light. Appl Catal B 192:193–198

Sahu MK, Patel RK (2016) Novel visible-light-driven cobalt loaded neutralized red mud (Co/NRM) composite with photocatalytic activity toward MB dye degradation. J Ind Eng Chem 40:72–82

Sakthivel S, Geissen SU, Bahnemann DW, Murugesan V, Vogelpohl A (2002) Enhancement of photocatalytic activity by semiconductor heterojunctions: α-Fe_2O_3, WO_3 and CdS deposited on ZnO. J Photochem Photobiol, A 148:283–293

Sakunthala A, Reddy MV, Selvasekarapandian S, Chowdari BVR, Selvin PC (2011) Energy storage studies of bare and doped vanadium pentoxide, (V1.95M0.05)O5, M = Nb, Ta, for lithium ion batteries. Energy Environ Sci 4:1712–1725

Sato J, Kobayashi H, Ikarashi K, Saito N, Nishiyama H, Inoue Y (2004) Photocatalytic activity for water decomposition of RuO_2-dispersed Zn_2GeO_4 with d10 configuration. J Phys Chem B 108:4369–4375

Sato S, Morikawa T, Kajino T, Ishitani O (2013) A highly efficient mononuclear iridium complex photocatalyst for CO_2 reduction under visible light. Angew Chem Int Ed Engl 52:988–992

Savadogo O, Mandal KC (1992) Studies on new chemically deposited photoconducting antimony trisulphide thin films. Sol Energy Mater Sol Cells 26:117–136

Savinkina E, Obolenskaya L, Kuzmicheva G (2014) Efficiency of sensitizing nano-titania with organic dyes and peroxo complexes. Appl Nanosci 5:125–133

Seko A, Togo A, Oba F, Tanaka I (2008) Structure and stability of a homologous series of tin oxides. Phys Rev Lett 100:045702

Shamaila S, Sajjad AKL, Chen F, Zhang J (2011) WO_3/BiOCl, a novel heterojunction as visible light photocatalyst. J Colloid Interface Sci 356:465–472

Shan AY, Ghazi TIM, Rashid SA (2010) Immobilisation of titanium dioxide onto supporting materials in heterogeneous photocatalysis: a review. Appl Catal A 389:1–8

Sharifi I, Shokrollahi H, Amiri S (2012) Ferrite-based magnetic nanofluids used in hyperthermia applications. J Magn Magn Mater 324:903–915

Sharma P, Kumar P, Deva D, Shrivastav R, Dass S, Satsangi VR (2010) Nanostructured Zn-Fe_2O_3 thin film modified by Fe-TiO_2 for photoelectrochemical generation of hydrogen. Int J Hydrogen Energy 35:10883–10889

Shi Y, Guo B, Corr SA, Shi Q, Hu Y-S, Heier KR, Chen L, Seshadri R, Stucky GD (2009) Ordered mesoporous metallic MoO_2 materials with highly reversible lithium storage capacity. Nano Lett 9:4215–4220

Shivalingappa L, Sheng J, Fukami T (1997) Photocatalytic effect in platinum doped titanium dioxide films. Vacuum 48:413–416

Shylesh S, Schünemann V, Thiel WR (2010) Magnetically separable nanocatalysts: bridges between homogeneous and heterogeneous catalysis. Angew Chem Int Ed 49:3428–3459

Silva GAF, Da Luz EC, Dos Reis Goyatá F, Da Silva Concilio LR, Neves ACC, Vitti RP, Cunha LG (2016) Influence of surface treatments on topography and bond strength of densely-sintered zirconium-oxide ceramic. Ceramics International 42:8136–8139

Singh S, Mahalingam H, Singh PK (2013) Polymer-supported titanium dioxide photocatalysts for environmental remediation: a review. Appl Catal A 462–463:178–195

Sinha ASK, Sahu N, Arora MK, Upadhyay SN (2001) Preparation of egg-shell type Al_2O_3-supported CdS photocatalysts for reduction of H_2O to H_2. Catal Today 69:297–305

Sobhani-Nasab A, Rangraz-Jeddy M, Avanes A, Salavati-Niasari M (2015) Novel sol–gel method for synthesis of PbTiO3 and its light harvesting applications. J Mater Sci: Mater Electron 26:9552–9560

Son S-B, Yersak TA, Piper DM, Kim SC, Kang CS, Cho JS, Suh S-S, Kim Y-U, Oh KH, Lee S-H (2014) A stabilized PAN-FeS_2 cathode with an EC/DEC liquid electrolyte. Adv Energy Mater 4:1300961

Song XC, Zheng YF, Yang E, Wang Y (2007) Large-scale hydrothermal synthesis of WO_3 nanowires in the presence of K_2SO_4. Mater Lett 61:3904–3908

Song C, Gui Y, Xing X, Zhang W (2016a) Well-dispersed chromium oxide decorated reduced graphene oxide hybrids and application in energy storage. Mater Chem Phys 173:460–466

Song Z, Liu W, Wei W, Quan C, Sun N, Zhou Q, Liu G, Wen X (2016b) Preparation and electrochemical properties of Fe_2O_3/reduced graphene oxide aerogel (Fe_2O_3/rGOA) composites for supercapacitors. J Alloy Compd 685:355–363

Steigerwalt ES, Deluga GA, Cliffel DE, Lukehart CM (2001) A Pt-Ru/graphitic carbon nanofiber nanocomposite exhibiting high relative performance as a direct-methanol fuel cell anode catalyst. J Phys Chem B 105:8097–8101

Su F, Mathew SC, Lipner G, Fu X, Antonietti M, Blechert S, Wang X (2010) Mpg-C_3N_4-catalyzed selective oxidation of alcohols using O_2 and visible light. J Am Chem Soc 132:16299–16301

Su J, Zou XX, Li GD, Wei X, Yan C, Wang YN, Zhao J, Zhou LJ, Chen JS (2011) Macroporous V_2O_5-BiVO4 composites: effect of heterojunction on the behavior of photogenerated charges. J Phys Chem C 115:8064–8071

Sun H, Wang L, Zhang R, Sui J, Xu G (2006) Treatment of groundwater polluted by arsenic compounds by zero valent iron. J Hazard Mater 129:297–303

Sun J, Li X, Zhao Q, Ke J, Zhang D (2014) Novel V_2O_5/BiVO4/TiO_2 nanocomposites with high visible-light-induced photocatalytic activity for the degradation of toluene. J Phys Chem C 118:10113–10121

Sun T-W, Zhu Y-J, Qi C, Ding G-J, Chen F, Wu J (2016a) α-Fe_2O_3 nanosheet-assembled hierarchical hollow mesoporous microspheres: microwave-assisted solvothermal synthesis and application in photocatalysis. J Colloid Interface Sci 463:107–117

Sun YP, Ha W, Chen J, Qi HY, Shi YP (2016) Advances and applications of graphitic carbon nitride as sorbent in analytical chemistry for sample pretreatment: a review. TrAC—Trends Anal Chem

Tan R, Yang J, Hu J, Wang K, Zhao Y, Pan F (2016) Core–shell nano-FeS_2@N-doped graphene as an advanced cathode material for rechargeable Li-ion batteries. Chem Commun 52:986–989

Tang J, Zou Z, Katagiri M, Kako T, Ye J (2004) Photocatalytic degradation of MB on MIn_2O_4 (M = alkali earth metal) under visible light: effects of crystal and electronic structure on the photocatalytic activity. Catal Today 93–95:885–889

Tao W, Chang J, Wu D, Gao Z, Duan X, Xu F, Jiang K (2013) Solvothermal synthesis of graphene-Sb_2S_3 composite and the degradation activity under visible light. Mater Res Bull 48:538–543

Teramura K, Tanaka T, Hosokawa T, Ohuchi T, Kani M, Funabiki T (2004) Selective photo-oxidation of various hydrocarbons in the liquid phase over V_2O_5/Al_2O_3. Catal Today 96:205–209

Thangavel S, Krishnamoorthy K, Kim S-J, Venugopal G (2016) Designing ZnS decorated reduced graphene-oxide nanohybrid via microwave route and their application in photocatalysis. J Alloy Compd 683:456–462

Theerthagiri J, Senthil RA, Malathi A, Selvi A, Madhavan J, Ashokkumar M (2015) Synthesis and characterization of a CuS-WO_3 composite photocatalyst for enhanced visible light photocatalytic activity. RSC Adv 5:52718–52725

Tian F (2002) Sol–gel derived iridium composite glucose biosensor. Sens Actuators B: Chem 86:266–270

Tsai M-C, Tsai Y-C (2009) Adsorption of glucose oxidase at platinum-multiwalled carbon nanotube-alumina-coated silica nanocomposite for amperometric glucose biosensor. Sens Actuators B: Chem 141:592–598

Varisli D, Elverisli EE (2014) Synthesizing hydrogen from ammonia over Ru incorporated SiO_2 type nanocomposite catalysts. Int J Hydrogen Energy 39:10399–10408

Veréb G, Ambrus Z, Pap Z, Kmetykó Á, Dombi A, Danciu V, Cheesman A, Mogyorósi K (2012) Comparative study on UV and visible light sensitive bare and doped titanium dioxide photocatalysts for the decomposition of environmental pollutants in water. Appl Catal A 417–418:26–36

Veréb G, Manczinger L, Bozsó G, Sienkiewicz A, Forró L, Mogyorósi K, Hernádi K, Dombi A (2013a) Comparison of the photocatalytic efficiencies of bare and doped rutile and anatase TiO_2 photocatalysts under visible light for phenol degradation and E. coli inactivation. Appl Catal B 129:566–574

Veréb G, Manczinger L, Oszkó A, Sienkiewicz A, Forró L, Mogyorósi K, Dombi A, Hernádi K (2013b) Highly efficient bacteria inactivation and phenol degradation by visible light irradiated iodine doped TiO_2. Appl Catal B 129:194–201

Veréb G, Ambrus Z, Pap Z, Mogyorósi K, Dombi A, Hernádi K (2014) Immobilization of crystallized photocatalysts on ceramic paper by titanium(IV) ethoxide and photocatalytic decomposition of phenol. React Kinet Mech Catal 113:293–303

Wang C, Huang Z (2016) Controlled synthesis of α-Fe_2O_3 nanostructures for efficient photocatalysis. Mater Lett 164:194–197

Wang C, Ao Y, Wang P, Hou J, Qian J (2010) A facile method for the preparation of titania-coated magnetic porous silica and its photocatalytic activity under UV or visible light. Colloids Surf, A 360:184–189

Wang D, Xue G, Zhen Y, Fu F, Li D (2012) Monodispersed Ag nanoparticles loaded on the surface of spherical Bi2WO6 nanoarchitectures with enhanced photocatalytic activities. J Mater Chem 22:4751

Wang W, Huang X, Wu S, Zhou Y, Wang L, Shi H, Liang Y, Zou B (2013) Preparation of p–n junction $Cu_2O/BiVO_4$ heterogeneous nanostructures with enhanced visible-light photocatalytic activity. Appl Catal B 134–135:293–301

Wang J, Dong C, Jiang BB, Wu KL, Sun J, Li XZ, Zhang WJ, Zhang B, Wei XW (2014a) Preparation of visible light-driven Ag_2CO_3/BiOBr composite photocatalysts with universal degradation abilities. Mater Lett 131:108–111

Wang J, Yu W, Xu S, Dai S, Wang J, Wang C, Zeng W, Cao P (2014b) A study on the precursor of vanadium pentoxide by the hydrothermal method. Ceram Int 40:317–321

Wang M, Chen J, Liao X, Liu Z, Zhang J, Gao L, Li Y (2014c) Highly efficient photocatalytic hydrogen production of platinum nanoparticle-decorated SiC nanowires under simulated sunlight irradiation. Int J Hydrogen Energy 39:14581–14587

Wang D, Cai D, Wang C, Liu B, Wang L, Liu Y, Li H, Wang Y, Li Q, Wang T (2016a) Muti-component nanocomposite of nickel and manganese oxides with enhanced stability and catalytic performance for non-enzymatic glucose sensors. Nanotechnology 27:255501

Wang J-W, Chen Y, Chen B-Z (2016b) Synthesis and control of high-performance MnO_2/carbon nanotubes nanocomposites for supercapacitors. J Alloy Compd 688:184–197

Wang P, Wang L, Sun Q, Qiu S, Liu Y, Zhang X, Liu X, Zheng L (2016c) Preparation and performance of Fe_3O_4@hydrophilic graphene composites with excellent Photo-Fenton activity for photocatalysis. Mater Lett 183:61–64

Wodka D, Socha RP, Bielańska E, Elżbieciak-Wodka M, Nowak P, Warszyński P (2014) Photocatalytic activity of titanium dioxide modified by Fe_2O_3 nanoparticles. Appl Surf Sci 319:173–180

Wu T, Liang K (2016) Caterpillar structured $Ni(OH)_2$@MnO_2core/shell nanocomposite arrays on nickel foam as high performance anode materials for lithium ion batteries. RSC Adv. 6:15541–15548

Wu P, Xie R, Imlay JA, Shang JK (2009) Visible-light-induced photocatalytic inactivation of bacteria by composite photocatalysts of palladium oxide and nitrogen-doped titanium oxide. Appl Catal B 88:576–581

Wu P, Dai Y, Sun T, Ye Y, Meng H, Fang X, Yu B, Dai L (2011) Impurity-dependent photoresponse properties in single CdSe nanobelt photodetectors. ACS Appl Mater Interfaces 3:1859–1864

Wu D, Wang L, Song X, Tan Y (2013a) Enhancing the visible-light-induced photocatalytic activity of the self-cleaning TiO_2-coated cotton by loading Ag/AgCl nanoparticles. Thin Solid Films 540:36–40

Wu L, Xu H, Han Q, Wang X (2013b) Large-scale synthesis of double cauliflower-like Sb2S3 microcrystallines by hydrothermal method. J Alloy Compd 572:56–61

Wu J-Z, Li X-Y, Zhu Y-R, Yi T-F, Zhang J-H, Xie Y (2016) Facile synthesis of MoO_2/CNTs composites for high-performance supercapacitor electrodes. Ceram Int 42:9250–9256

Xia C, Zhang Y, Liu M (2003) Composite cathode based on yttria stabilized bismuth oxide for low-temperature solid oxide fuel cells. Appl Phys Lett 82:901–903

Xiang Q, Yu J, Jaroniec M (2011) Preparation and enhanced visible-light photocatalytic H_2-production activity of graphene/C_3N_4 composites. J Phys Chem C 115:7355–7363

Xiang Q, Cheng F, Lang D (2016) Hierarchical layered WS2/graphene-modified CdS nanorods for efficient photocatalytic hydrogen evolution. Chemsuschem 9:996–1002

Xiao X, Hao R, Liang M, Zuo X, Nan J, Li L, Zhang W (2012) One-pot solvothermal synthesis of three-dimensional (3D) BiOI/BiOCl composites with enhanced visible-light photocatalytic activities for the degradation of bisphenol-A. J Hazard Mater 233–234:122–130

Xiao Y, Chen CS, Cao SY, Qian GP, Nie XB, Yu WW (2015) Enhanced sunlight-driven photocatalytic activity of graphene oxide/Bi2WO6 nanoplates by silicon modification. Ceram Int 41:10087–10094

Xiaohong W, Wei Q, Weidong H (2007) Thin bismuth oxide films prepared through the sol–gel method as photocatalyst. J Mol Catal A: Chem 261:167–171

Xie H, Que W, He Z, Zhong P, Liao Y, Wang G (2013) Preparation and photocatalytic activities of Sb_2S_3/TiO_2 nanotube coaxial heterogeneous structure arrays via an ion exchange adsorption method. J Alloy Compd 550:314–319

Xie J, Zhou Z, Lian Y, Hao Y, Li P, Wei Y (2015) Synthesis of α-Fe_2O_3/ZnO composites for photocatalytic degradation of pentachlorophenol under UV–vis light irradiation. Ceram Int 41:2622–2625

Xiong D-N, Huang G-F, Zhou B-X, Yan Q, Pan A-L, Huang W-Q (2016) Facile ion-exchange synthesis of mesoporous Bi_2S_3/ZnS nanoplate with high adsorption capability and photocatalytic activity. J Colloid Interface Sci 464:103–109

Xue S, Wei Z, Hou X, Xie W, Li S, Shang X, He D (2015) Enhanced visible-light photocatalytic activities and mechanism insight of $BiVO_4$/Bi_2WO_6 composites with virus-like structures. Appl Surf Sci 355:1107–1115

Yamamoto M, Yoshida T, Yamamoto N, Nomoto T, Yamamoto Y, Yagi S, Yoshida H (2015) Photocatalytic reduction of CO_2 with water promoted by Ag clusters in Ag/Ga_2 O_3 photocatalysts. J Mater Chem A 3:16810–16816

Yasushi Y, Katsuji Y, Yumi M, Masami O (2003) A Thick-Film NO_2 sensor fabricated using Zn–Sn–Sb–O composite material. Jpn J Appl Phys 42:7594

Ye S, Wang R, Wu MZ, Yuan YP (2015) A review on g-C_3N_4 for photocatalytic water splitting and CO_2 reduction. Appl Surf Sci

Yin S (2004) Nano Ru/CNTs: a highly active and stable catalyst for the generation of COx-free hydrogen in ammonia decomposition. Appl Catal B 48:237–241

Yin H, Zhou Y, Li B, Li X, Yang Z, Ai S, Zhang X (2015) Photoelectrochemical immunosensor for microRNA detection based on gold nanoparticles-functionalized g-C_3N_4 and anti-DNA: RNA antibody. Sens Actuators, B: Chem

Yu C, Yang K, Yu JC, Cao F, Li X, Zhou X (2011) Fast fabrication of Co_3O_4 and CuO/$BiVO_4$ composite photocatalysts with high crystallinity and enhanced photocatalytic activity via ultrasound irradiation. J Alloy Compd 509:4547–4552

Yu C, Cao F, Li G, Wei R, Yu JC, Jin R, Fan Q, Wang C (2013) Novel noble metal (Rh, Pd, Pt)/ BiOX(Cl, Br, I) composite photocatalysts with enhanced photocatalytic performance in dye degradation. Sep Purif Technol 120:110–122

Yu Q, Zhang F, Li G, Zhang W (2016) Preparation and photocatalytic activity of triangular pyramid $NaNbO_3$. Appl Catal B 199:166–169

Yuan D, Zeng J, Kristian N, Wang Y, Wang X (2009) Bi_2O_3 deposited on highly ordered mesoporous carbon for supercapacitors. Electrochem Commun 11:313–317

Yuan D, Yang W, Ni J, Gao L (2015) Sandwich structured MoO_2@TiO_2@CNT nanocomposites with high-rate performance for lithium ion batteries. Electrochim Acta 163:57–63

Yue G, Hong-Yu X, Zeng-Rong J, Zhi-Xiang X (2015) Mechanical properties and thermal shock resistance of rhenium coating in iridium/rhenium/carbon-carbon composites. Procedia Eng 99:1407–1414

Yun S, Zhou H, Wang L, Zhang H, Ma T (2013) Economical hafnium oxygen nitride binary/ternary nanocomposite counter electrode catalysts for high-efficiency dye-sensitized solar cells. J Mater Chem A 1:1341–1348

Yurkov GY, Kozinkin AV, Koksharov YA, Fionov AS, Taratanov NA, Vlasenko VG, Pirog IV, Shishilov ON, Popkov OV (2012) Synthesis and properties of rhenium–polyethylene nanocomposite. Compos B Eng 43:3192–3197

Zhang K, Guo L (2013) Metal sulphide semiconductors for photocatalytic hydrogen production. Catal Sci Technol 3:1672–1690

Zhang A, Zhang J (2010) Synthesis and characterization of Ag/$BiVO_4$ composite photocatalyst. Appl Surf Sci 256:3224–3227

Zhang J, Xu H, Ge Q, Li W (2006a) Highly efficient Ru/MgO catalysts for NH3 decomposition: synthesis, characterization and promoter effect. Catal Commun 7:148–152

Zhang L, Fu H, Zhang C, Zhu Y (2006b) Synthesis, characterization, and photocatalytic properties of $InVO_4$ nanoparticles. J Solid State Chem 179:804–811

Zhang Z, Wang W, Shang M, Yin W (2010) Low-temperature combustion synthesis of Bi_2WO_6 nanoparticles as a visible-light-driven photocatalyst. J Hazard Mater 177:1013–1018

Zhang G-Y, Feng Y, Xu Y-Y, Gao D-Z, Sun Y-Q (2012) Controlled synthesis of mesoporous α-Fe_2O_3 nanorods and visible light photocatalytic property. Mater Res Bull 47:625–630

Zhang Y, Zhang Y, Tan J (2013) Novel magnetically separable AgCl/iron oxide composites with enhanced photocatalytic activity driven by visible light. J Alloy Compd 574:383–390

Zhang J, Wang X, Qin D, Xue Z, Lu X (2014) Fabrication of iron-doped cobalt oxide nanocomposite films by electrodeposition and application as electrocatalyst for oxygen reduction reaction. Appl Surf Sci 320:73–82

Zhang D, Chang H, Li P, Liu R (2015a) Characterization of nickel oxide decorated-reduced graphene oxide nanocomposite and its sensing properties toward methane gas detection. J Mater Sci: Mater Electron 27:3723–3730

Zhang L, Bai Q, Jin K, Wang L, Zhang Y, Yanhua S (2015b) Synthesis and electrochemical performance of Bi_2WO_6/graphene composite as anode material for lithium-ion batteries. Mater Lett 141:88–91

Zhang W, Hu M, Liu X, Wei Y, Li N, Qin Y (2016) Synthesis of the cactus-like silicon nanowires/tungsten oxide nanowires composite for room-temperature NO_2 gas sensor. J Alloy Compd 679:391–399

Zhao Z, Sun Y, Dong F (2015) Graphitic carbon nitride based nanocomposites: a review. Nanoscale 7:15–37

Zheng W, Zhang J, Xu H, Li W (2007) NH_3 decomposition kinetics on supported Ru clusters: morphology and particle size effect. Catal Lett 119:311–318

Zhou F, Zhao X, Yuan C, Li L (2008) Vanadium pentoxide nanowires: hydrothermal synthesis, formation mechanism, and phase control parameters. Cryst Growth Des 8:723–727

Zhou L, Wang W, Xu H, Sun S, Shang M (2009) Bi_2O_3 hierarchical nanostructures: controllable synthesis, growth mechanism, and their application in photocatalysis. Chem—Eur J 15:1776–1782

Zhou W, Cheng C, Liu J, Tay YY, Jiang J, Jia X, Zhang J, Gong H, Hng HH, Yu T, Fan HJ (2011) Epitaxial growth of branched α-Fe_2O_3/SnO_2 nano-heterostructures with improved lithium-ion battery performance. Adv Func Mater 21:2439–2445

Zhou X, Bai L, Yan J, He S, Lei Z (2013) Solvothermal synthesis of Sb_2S_3/C composite nanorods with excellent Li-storage performance. Electrochim Acta 108:17–21

Zhou E, Wang C, Zhao Q, Li Z, Shao M, Deng X, Liu X, Xu X (2016) Facile synthesis of MoO_2 nanoparticles as high performance supercapacitor electrodes and photocatalysts. Ceram Int 42:2198–2203

Zhu Y, Yu F, Man Y, Tian Q, He Y, Wu N (2005) Preparation and performances of nanosized Ta_2O_5 powder photocatalyst. J Solid State Chem 178:224–229

Zhu LA, Bai S, Zhang H, Ye Y, Gao W (2013) Rhenium used as an interlayer between carbon–carbon composites and iridium coating: Adhesion and wettability. Surf Coat Technol 235:68–74

Zhu X, Wang P, Huang B, Ma X, Qin X, Zhang X, Dai Y (2016) Synthesis of novel visible light response $Ag_{10}Si_4O_{13}$ photocatalyst. Appl Catal B 199:315–322

Zinatloo-Ajabshir S, Salavati-Niasari M (2016) Facile route to synthesize zirconium dioxide (ZrO_2) nanostructures: structural, optical and photocatalytic studies. J Mol Liq 216:545–551

Zou J-P, Ma J, Luo J-M, Yu J, He J, Meng Y, Luo Z, Bao S-K, Liu H-L, Luo S-L, Luo X-B, Chen T-C, Suib SL (2015) Fabrication of novel heterostructured few layered WS2-Bi2WO6/Bi3.84W0.16O6.24 composites with enhanced photocatalytic performance. Appl Catal B 179:220–228

Index

C

Carbon-based nanocomposites, 9, 11, 132, 205
Carbon nanotubes, 3, 44, 62, 74, 109, 204, 205, 232, 236, 339, 349, 268
Catalysis, 3, 10, 25, 70, 84, 89, 94, 95, 111, 130, 160, 205, 232, 365
Charge recombination, 44, 45, 207, 236, 255, 325

D

Doped metal oxides, 34, 54, 107, 109, 113, 119, 153, 296–298, 308, 324, 326

E

Electron–hole pairs, 6, 8, 42, 43, 46, 47, 51, 56, 57, 61, 62, 87, 99, 108, 120, 151, 163, 179, 186, 212, 216, 217, 222, 234, 239, 241, 252, 256, 279, 282, 285, 289, 292, 304, 311, 313, 323, 337, 344, 352, 353, 362, 367, 375, 378, 380
Electrons, 3, 4, 6–9, 24, 27, 28, 35, 36, 41–46, 48, 50, 51, 53, 58, 59, 62, 71–75, 82–99, 110, 116, 119–123, 125, 130, 131, 134, 138, 145, 150, 151, 176, 178, 182, 185, 187, 204, 206, 207, 213, 220, 222, 227, 228, 235, 238, 240, 253–257
Environmental remediation, 70, 227, 296

F

Fullerenes, 203–205, 255
Future perspectives of nanocomposites, 335, 338, 340, 344, 345, 347–349, 351, 355, 358, 383

G

Graphene, 1, 3, 11, 45–48, 58, 74, 123, 131, 132, 141, 204–212, 214, 216–218, 220, 221, 223, 231, 232, 255, 263, 264, 267, 307, 341, 347, 349, 355, 360, 363
Graphene oxide, 45, 47, 58, 89, 141, 203, 205, 209, 210, 214, 217, 221, 223, 236
Graphitic nanocomposites, 4, 133, 269, 204, 210, 225, 226, 254, 348

H

Holes, 9, 24, 27, 28, 35, 36, 42–46, 58, 62, 71, 75, 83–85, 87–89, 96, 99, 110, 116, 120, 125, 130, 131, 138, 145, 176, 182, 206, 220, 225, 228, 240, 254, 257, 258, 273, 282, 289, 307, 338, 352, 356, 359, 363, 369, 371, 381

I

Importance in photocatalysis, 41, 49

M

Mechanism of photocatalysis, 20, 26, 41, 130
Metal doping, 7, 34, 109, 297
Mixed metal oxides, 113, 115, 119, 296, 297, 300, 321, 324, 326
Mixed metal oxides nanocomposites, 109, 126, 321, 326

N

Nanocomposites, 1, 5, 9, 42, 45, 48, 56, 62, 109–111, 116, 118, 121, 131, 132, 137, 146, 149, 162, 163, 189, 203, 205, 209, 237, 251, 252, 264, 278, 297, 313, 319, 321, 326, 336, 346, 353
Nanocomposites of g-C_3N_4, 256, 263, 274, 275, 278, 282, 285, 287
Nanomaterials, 1–3, 5, 12, 29, 75, 81, 119, 203, 204, 242, 341, 363, 371, 372, 374, 382, 383
Nanoparticles, 3, 10, 44, 46, 51, 55, 57, 59, 71, 78, 86, 98, 111, 118, 130, 133, 140, 143, 146, 149, 152, 155, 161, 184, 194, 205, 213, 221, 230, 232, 233, 235, 236, 239,

M.M. Khan et al. (eds.), *Nanocomposites for Visible Light-induced Photocatalysis*, Springer Series on Polymer and Composite Materials,
DOI 10.1007/978-3-319-62446-4

256, 282, 285, 301, 309, 319, 324, 336, 337, 343, 350, 355, 360, 376, 382
Nanoporous materials, 130, 132, 163
Nanoporous Nanocomposite materials for Photocatalysis, 10, 132, 166
Novel applications of nanocomposites, 346, 355, 383

O
Oxidation, 4, 23, 25, 28, 42, 49–51, 53, 55, 56, 61, 70, 71, 78, 85, 93, 100, 120, 153, 177, 178, 183, 184, 191, 206, 229, 231, 237, 257, 264, 321, 326, 337, 340–343, 353, 354, 356, 369, 371

P
Photocatalysis, 3, 9, 11, 20, 24, 25, 32, 34, 42, 43, 53, 59, 70, 74, 83, 97, 109, 130, 132, 138, 157, 175, 176, 179, 180, 182, 184, 187, 190, 191, 195, 210, 216, 229, 240, 257, 278, 279, 296, 298, 300, 308, 326, 343, 359, 364, 371, 383
Photocatalysts, 4, 11, 12, 29, 44, 47, 57, 61, 70, 75, 81, 90, 109, 111, 123, 126, 131, 146–148, 157, 159, 161, 166, 180, 194, 204, 217, 225, 227, 229, 234, 240, 241, 252, 259, 285, 290, 296, 320, 334, 339, 343, 354, 359, 364–366, 368, 376, 383
Plasmonic photocatalyst, 51, 71, 74, 75, 79, 81, 90, 99, 303
Pollutants, 11, 20, 21, 25, 32, 34, 46, 81, 115, 116, 120, 131, 176, 180, 195, 216, 226, 238, 240, 241, 266, 334, 353, 354, 357, 359, 379
Pollution, 19, 20, 22, 70, 81, 112, 130, 180, 227, 252, 257
Polymeric materials, 182, 252, 253, 255, 256, 261, 285
Polymeric nanocomposites, 11, 252, 255, 282, 292

R
Reduction, 4, 8, 27, 28, 36, 42, 44–46, 57–59, 71, 75, 81, 91, 93, 100, 110, 120, 130, 141, 149, 159, 160, 178, 182, 204, 206, 212, 221, 229, 252, 257, 258, 262–264, 266, 267, 282, 292, 308, 319, 321, 336, 342, 349, 355, 365, 371, 376, 379, 383
Role of Metal Nanoparticles, 79, 83, 85, 94

S
Selective transformation
Semiconductors, 5–7, 33–35, 43, 46, 71, 74, 79, 89, 109, 110, 122, 130, 144, 163, 180, 181, 206, 227, 240, 297, 319, 334, 365
Surface plasmon activity on nanocomposites
Surface plasmon resonance, 52, 54, 71, 72, 122, 336, 369

T
Titanium, 7, 136, 138, 141, 142, 146, 152, 153, 155, 158, 233, 297, 300, 308, 312, 319, 326, 338, 361
Titanium-based mixed metal oxide nanocomposites, 297, 324

V
Visible light, 3, 4, 6, 9, 12, 21, 34, 35, 44, 46, 48, 51, 53–55, 57, 62, 70, 71, 81, 83, 85, 86, 88, 90, 92, 94, 96, 98, 100, 108, 109, 111, 116, 117, 120, 122–124, 126, 131, 133, 143, 150, 162, 180–182, 185, 193, 194, 204, 210, 213, 216, 218, 219, 221, 226–229, 231, 233, 236–240, 256–258, 267, 272, 276, 279, 281, 285, 292, 302, 306, 307, 310, 317, 319, 323, 325, 334, 338, 340, 342, 345, 354, 357, 360, 363, 365, 367, 369, 373, 376, 382
Visible light-induced catalysis, 9
Visible-light induced photocatalyst, 6

W
Water remediation, 109, 122, 126

MIX
Papier aus verantwortungsvollen Quellen
Paper from responsible sources
FSC® C105338

If you have any concerns about our products,
you can contact us on
ProductSafety@springernature.com

In case Publisher is established outside the EU,
the EU authorized representative is:
Springer Nature Customer Service Center GmbH
Europaplatz 3, 69115 Heidelberg, Germany

Printed by Libri Plureos GmbH
in Hamburg, Germany